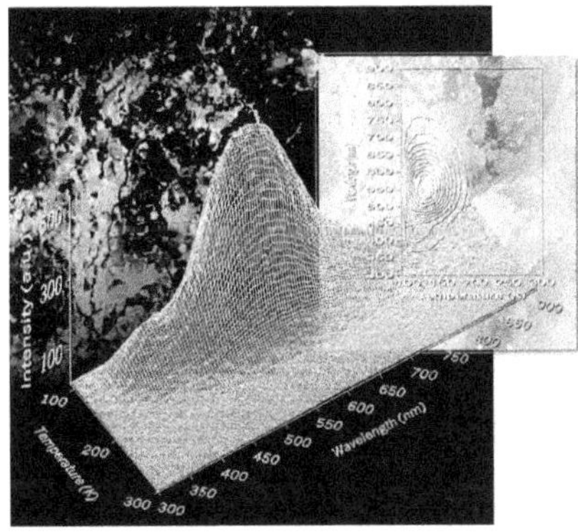

HANDBOOK OF THERMOLUMINESCENCE

Second Edition

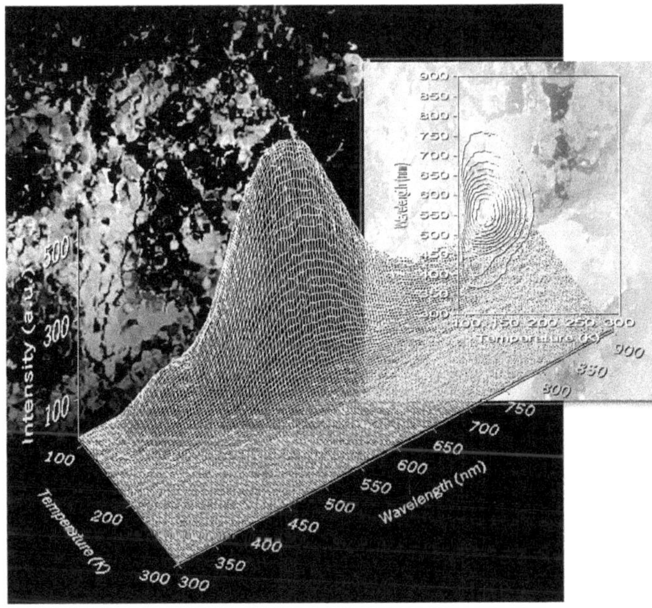

HANDBOOK OF THERMOLUMINESCENCE

Second Edition

Claudio Furetta

Touro University Rome, Italy
Division of Touro College New York

World Scientific

NEW JERSEY • LONDON • SINGAPORE • BEIJING • SHANGHAI • HONG KONG • TAIPEI • CHENNAI

Published by

World Scientific Publishing Co. Pte. Ltd.
5 Toh Tuck Link, Singapore 596224
USA office: 27 Warren Street, Suite 401-402, Hackensack, NJ 07601
UK office: 57 Shelton Street, Covent Garden, London WC2H 9HE

British Library Cataloguing-in-Publication Data
A catalogue record for this book is available from the British Library.

HANDBOOK OF THERMOLUMINESCENCE (2nd Edition)

ISBN-13 978-981-283-891-9
ISBN-10 981-283-891-0

I am deeply grateful to my wife Maria Clotilde for her constant and loving support of my scientific work: This book is dedicated to her.

PREFACE

This book on thermoluminescence (TL) is born from the idea to provide to experts, teachers, students and technicians practical support for research, study, routine work and terminology.

The term "handbook" of the title does not mean that this book is a "summa" of thermoluminescence. Actually, the goal is to be dynamic, fluid and of easy consultation on several subjects.

This book collects a certain number of subjects, mainly referring to the thermoluminescence models, to the methods for determining the kinetic parameters, to the procedures to follow for characterizing a thermoluminescent dosimetric system and to the definition of terms commonly used in TL literature. Furthermore, the analytical treatments of the various TL models are fully developed.

Subjects concerning solid state physics as well as TL dating are not considered because they are widely treated in many fundamental books which can be easily found on the market.

In general, the subjects considered here are dispersed in specialized journals which are not always available to everybody.

The arguments are given in alphabetic order to make the research easy. The words in bold indicate the item to which the reader is referred to for more information.

CONTENTS

CHAPTER D

CHAPTER E

CHAPTER F

CHAPTER G

CHAPTER H

CHAPTER I

CHAPTER K

CHAPTER L

CHAPTER M

CHAPTER N

CHAPTER Q

CHAPTER R

CHAPTER S

CHAPTER T

CHAPTER V

A

Accuracy (definition)

Errors of measurement are of two types, random and systematic. For a given set of measurement conditions a source of random error is variable in both magnitude and sign, whereas a source of systematic error has a constant relative magnitude and is always of the same sign.

The accuracy is affected by both systematic and random uncertainties.

Accuracy is related to the closeness of a measurement, within certain limits, with the true value of the quantity under measurement. For instance, the accuracy of dose determination by TLD is given by the difference between the measured value of the dose (TL reading) and the true dose given to the dosemeter.

A method of combining systematic and random uncertainties has been suggested in a BCS document: both systematic and random errors are combined by quadratic addition but the result for systematic errors is multiplied by 1.13. This factor is necessary to ensure a minimum confidence level of 95%.

Activation energy (definition and properties)

It is the energy, E, expressed in eV, assigned to a **metastable state** or level within the forbidden band gap between the **conduction band** (CB) and the **valence band** (VB) of a crystal. This energy is also called trap depth. The metastable level can be an electron trap, near to the CB, or a hole trap, near the VB, or a **luminescence centre**, more or less in the middle of the band gap. The metastable levels are originated by **defects** of the crystal structure. A crystal can contain several kinds of traps and luminescence centres. If E is such that E > several kT, where k is the Boltzmann's constant, then the trapped charge can remain in the trap for a long period. For an electron trap, E is measured, in eV, from the trap level to the bottom of the CB. For a hole trap, it is measured from the trap to the top of the VB.

Fig. 1 shows the simplest band structure of an isolant containing defects acting as traps or luminescence centres.

Bombarding the solid with an ionizing radiation, this produces free charges which can be trapped at the metastable states. Supposing the solid previously excited is heated, a quantity of energy is supplied in the form of thermal energy and the trapped charges can be released from the traps. The rate of such thermally stimulated process is usually expressed by the **Arrhenius** equation which leads to the concept of the activation energy, E, which can be seen as an energy barrier which must be overcome to reach equilibrium.

Considering the maximum condition using the first order kinetics:

$$\frac{\beta \cdot E}{k \cdot T_M^2} = s \exp\left(-\frac{E}{kT_M}\right)$$

it is easily observed that T_M increases as E increases. In fact, for $E \succ\succ kT_M$, T_M increase almost linearly with E. This behaviour agree to the Randall-Wilkins model where, for deeper traps, more energy and, in turn, a higher temperature, is required to detrap the electrons [1-4].

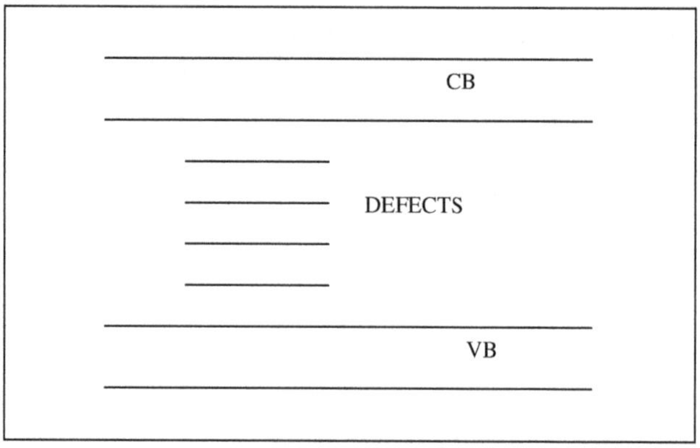

Fig. 1. A simple band structure of an isolant with defect levels in the band gap.

References

1. Bräunlich P. in Thermally Stimulated Relaxion in Solids, P. Bräunlich editor, Spring-Verlag, Berlin (1979)
2. Chen R. and McKeever S.W.S., Theory of Thermoluminescence and Related Phenomena, World Scientific (1997)
3. Chen R. and Kirsh Y., Analysis of Thermally Stimulated Processes, Pergamon Press (1981)
4. McKeever S.W.S., Thermoluminescence of Solids, Cambridge University Press (1985)

Activator

There are several luminescent materials but not all are efficient enough for practical purposes. To enhance the luminescence efficiency of the material is necessary to add an element, called activator (i.e., Dy in $CaSO_4$), to the host crystal. The activator acts then as **luminescent centre**.

Adirovitch model

Adirovitch, in 1956, used a set of three differential equations to explain the decay of phosphorescence in the general case. The same model has been used by Haering-Adams (1960) and Halperin-Braner (1960) to describe the flow charge between localized energy levels and delocalized bands during trap emptying The energy level diagram is shown in Fig. 1.

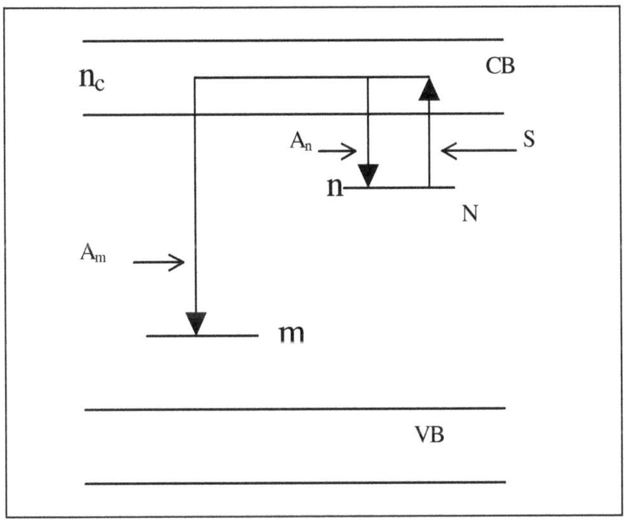

Fig. 1. Energy level diagram concerning the phosphorescence decay according to Adirovitch. The meaning of the symbols is given in the text.

With the assumption that retrapping of electrons occurs into trapping states of the same kind as those from which they had been released, the intensity of emission, I , is given by

$$I = -\frac{dm}{dt} = A_m m n_c \tag{1}$$

where

m is the concentration of recombination centers (holes in centers), (cm^{-3});

n_c is the concentration of free electrons in the conduction band, (cm^{-3});

A_m is the recombination probability (cm^3 sec^{-1}).

This equation states that the recombination rate is proportional to the number of free electrons, n_c, and the number of active recombination centers, m.

A second equation deals with the population variation of electrons in traps, n (cm^{-3}), and it takes into account of the excitation of electrons into the conduction band as well as the possible retrapping. Then we have:

$$\frac{dn}{dt} = -sn\exp\left(-\frac{E}{kT}\right) + n_c(N-n)A_n \tag{2}$$

where A_n ($cm^3\ s^{-1}$) is the retrapping probability and N (cm^{-3}) is the total concentration of traps. A_m and A_n are assumed to be independent of temperature.

The third equation relates to the charge neutrality. It can be expressed as

$$\frac{dn_c}{dt} = \frac{dm}{dt} - \frac{dn}{dt} \tag{3}$$

or better, using Eq. (1) and (2), as

$$\frac{dn_c}{dt} = sn\exp\left(-\frac{E}{kT}\right) - n_c mA_m - n_c(N-n)A_n \tag{4}$$

Eq. (4) states that the rate of change of n_c is given by the rate of release of electrons from N, minus the rate of recombination in m and retrapping in N.

While Adirovitch used the previous equations to explain the decay of phosphorescence, Halperin and Braner were the formers to apply the same equations to the case of thermoluminescence, that is to say when the light emission is measured during the heating of the sample, when one trapping state and one kind of recombination center are involved.

Two basic assumptions have been made for solving the previous set of equations:

$$n_c << n \tag{5}$$

and

$$\left|\frac{dn_c}{dt}\right| << \left|\frac{dn}{dt}\right| \tag{6}$$

The condition (6) means that the concentration of carriers in the conduction band does not change; that is to say

$$dn_c = 0 \tag{7}$$

In this case Eq. (4) gets

$$n_c = \frac{sn\exp\left(-\frac{E}{kT}\right)}{mA_m + (N-n)A_n} \tag{8}$$

and then the intensity is given by

$$I = -\frac{dm}{dt} = \frac{sn\exp\left(-\frac{E}{kT}\right)}{mA_m + (N-n)A_n} \cdot mA_m \tag{9}$$

Introducing the retrapping-recombination cross-section ratio

$$\sigma = \frac{A_n}{A_m} \tag{10}$$

equation (9) becomes

$$I = -\frac{dm}{dt} = ns\exp\left(-\frac{E}{kT}\right)\left[1 - \frac{\sigma(N-n)}{\sigma(N-n)+m}\right] \tag{11}$$

Equation (11) gets the general one-trap equation (GOT) for the TL intensity. The term preceding the square brackets is the number of electrons thermally released to the conduction band per unit time. The term in square brackets is the fraction of conduction band electrons undergoing recombination. From this equation it is possible to obtain the first and second order kinetics equations.

Indeed, the first order kinetics is the case when recombination dominates and this means that

$$mA_m >> (N-n)A_n \tag{12}$$

or

$$\sigma = 0 \tag{13}$$

The equation of intensity becomes then

$$I = -\frac{dm}{dt} = ns\exp\left(-\frac{E}{kT}\right) \tag{14}$$

The assumption (7) gives

$$\frac{dm}{dt} = \frac{dn}{dt} \qquad \text{or} \qquad m = n + const$$

and so Eq. (14) becomes

$$I = -\frac{dn}{dt} = ns\exp\left(-\frac{E}{kT}\right) \tag{15}$$

that is the same of equation of the **first order kinetics**.

The second order expression can be derived from Eq. (11) using two assumptions which both include the restrictive assertion $m = n$.

Remembering the **Garlick and Gibson's retrapping assumption**, the first condition can be written as

$$mA_m << (N-n)A_n$$

and then the intensity is given by

$$I = -\frac{dm}{dt} = \frac{mA_m ns\exp\left(-\frac{E}{kT}\right)}{(N-n)A_n} \tag{16}$$

Secondly, assuming that the trap is far from saturation, which means $N \gg n$, we obtain

$$I = -\frac{dm}{dt} = \frac{mA_m ns\exp\left(-\frac{E}{kT}\right)}{NA_n} \tag{17}$$

Using the condition $m = n$ the last equation becomes

$$I = -\frac{dn}{dt} = \frac{n^2 sA_m \exp\left(-\frac{E}{kT}\right)}{NA_n} \tag{18}$$

which, with $s' = sA_m/NA_n$, is the **Garlick and Gibson equation**.

Assuming now equal recombination and retrapping probabilities $A_m = A_n$, as suggested by Wrzesinska, one obtains the same equation of Garlick and Gibson with $s' = s/N$:

$$I = -\frac{dn}{dt} = n^2 s' \exp\left(-\frac{E}{kT}\right) \tag{19}$$

Reference

Adirovitch E.I.A., J. Phys. Rad. **17** (1956) 705

Afterglow

Afterglow is the term used to indicate the luminescence emitted from a TL phosphor immediately after irradiation. If this effect is thermally dependent, according to the equation

$$\tau = s^{-1} \exp\left(\frac{E}{kT}\right)$$

it is more properly termed phosphorescence.

The emission spectrum of the afterglow is the same as that of thermoluminescence: this fact indicates that the same luminescence centers are involved. Zimmerman found a correlation between the anomalous (athermal) fading and the afterglow [1-3].

References

1. Zimmermann D.N., Abstract Symp. Archaeometry and Archaeological Prospection, Philadelfia (1977)
2. Zimmermann D.N., PACT **3** (1979) 257
3. Visocekas R., Leva T., Marti C., Lefaucheux F. and Robert M.C., Phys. Stat. Sol. (a) **35** (1976) 315

Aluminium oxide (Al_2O_3)

Chromium substituting for some of the aluminum atoms in Al_2O_3 changes sapphire into ruby, which exhibits TL properties studied since the sixties [1-5].

Investigations on the TL of ruby, whose effective atomic number is 10.2, are performed by using synthetic crystals of Al_2O_3 containing various known concentrations of Cr_20_3 (typically 0.01 to 0.2 wt%).

TL glow curve of ruby consists of a main glow peak at 347°C (shifting toward lower temperatures for high exposures) and a less intense peak at 132°C (in the same region as the peak reported for sapphire). High chromium concentrations cause a relative increase in the lower temperature portion of the glow curve.

References

1. Gabrysh A.F., Eyring H., Le Febre V. and Evans M.D., J. Appl. Phys. **33** (1962) 3389
2. Maruyama T., Matsuda Y. and Kon H., J. Phys. Soc. Japan **18-II** (1963) 315
3. Buckman W.G., Philbrick C.R. and Underwood N., U.S. Atomic Energy Commission Rep. CONF-680920 (1968)
4. Hashizume T., Kato Y., Nakajima T., Yamaguchi H. and Fujimoto K., Health Phys. **23** (1972) 855
5. Watson J.E., Health Phys. **31** (1976) 47

Annealing (definition)

Annealing is the thermal treatment needs to erase any irradiation memory from the dosimetric material.

Some thermoluminescent material required a complex annealing procedure. LiF:Mg,Ti is one of them. It requires a high temperature anneal, followed by a low temperature anneal. Generally speaking the high temperature anneal is required to clear the **dosimetric traps** of residual signal which may cause unwanted backgrounds during subsequent use of the dosimeters. The low temperature anneal is required to stabilize and aggregate low temperature traps in order to enhance the sensitivity of the main dosimetry traps and to reduce losses of radiation-induced signal due to thermal or optical fading during use. The combination of these two anneals is termed standard anneal.

For lithium fluoride the standard annealing consists in a high temperature anneal at 400°C during 1 hour followed by a low temperature thermal treatment for 20 hours at 80°C. In some laboratories, annealing at 100°C for 2 hours has been used instead of the longer anneal at 80°C.

The TL properties exhibited by a phosphor strongly depend upon the kind of thermal annealing experienced by it prior to the irradiation. It is also true, in general, that more defects are produced ay higher temperatures of annealing.

The number of defects depends also on the cooling rate employed to cool the phosphor to the ambient temperature.

Once the best annealing procedure has been determined, i.e. the highest TL response with the lowest standard deviation, the same procedure must be always followed for reproducible results in TL applications [1,2]

References

1. Driscoll C.M.H., National Radiological Protection Board, Tech. Mem. 5(82)
2. Busuoli G. in Applied Thermoluminescence Dosimetry, ISPRA Courses, Edited by M. Oberhofer and A. Scharmann, Adam Hilger publisher (1981)

Annealing (general considerations)

Before using a thermoluminescent material for dosimetric purposes, it has to be prepared. To prepare a TL material means to erase from it all the information due to any previous irradiation, i.e., to restore in it the initial conditions of the crystal as they were before irradiation. The preparation has also the purpose of to stabilising the trap structure.

In order to prepare a thermoluminescent material for use, it is needed to perform a thermal treatment, usually called **annealing** [1,2], carried out in oven or/and furnace, which consists of heating up the TL samples to a predetermined temperature, keeping them at that temperature for a predetermined period of time and then cooling down the samples to room temperature. It has to be stressed that the thermal history of the thermoluminescent dosimeters is crucial for the performance of any TLD system.

There is a large number of thermoluminescent materials, however the annealing procedures are quite similar. Just few materials, like LiF:Mg,Ti, need a complex annealing procedure.

The thermal treatments normally adopted for the TLDs can be divided into three classes:

- **initialisation treatment**: this treatment is used for new (fresh or virgin) TL samples or for dosimeters which have not been used for long time. The aim of this thermal treatment is to stabilise the trap levels, so that during subsequent uses the intrinsic background and the sensitivity are both reproducible. The time and temperature of the initialisation annealing are, in general, the same as those of the standard annealing.

- **erasing treatment** or **standard annealing** (also called **pre-irradiation annealing** or **post-readout annealing**): this treatment is used to erase any previous residual irradiation effect which is supposed to remain stored in the crystal after the readout. It is carried out before using the TLDs in new measurements. The general aim of this thermal treatment is to bring back the traps - recombination centers structure to the former one obtained after the initialisation procedure. It may consist of one or two thermal treatments (in latter case, at two different temperatures).

- **post-irradiation** or **pre-readout annealing**: this kind of thermal treatment is used to erase the low-temperature peaks, if they are found in the glow-curve structure. Such low-temperature peaks are normally subjected to a quick thermal decay (fading) and possibly must not be included in the readout to avoid any errors in the dose determination.

In all cases, value and reproducibility of the cooling rate after the annealing are of great importance for the performance of a TLD system. In general, the TL sensitivity is increased using a rapid cool down. It seems that the sensitivity reaches the maximum value when a cooling rate of 50-100°C/s is used. To obtain this, the TLDs must be taken out of the oven after the pre-set time of annealing is over and placed directly on a cold metal block. The procedure must be reproducible and unchanged during the whole use of the dosimeters.

It must be noted that the thermal procedures listed above can be carried out in the reader itself. This is important for TL elements embedded in plastic cards as the dosimeters used for large personnel dosimetry services. In fact, the plastic cards are not able to tolerate high temperatures and the in-reader annealing is shortened to a few seconds. However, its efficiency is very low when high dose values are involved. The in-reader annealing procedure should be used only if the dose received by the dosimeter is lower than 10 to 20 mGy. Driscoll suggests in this case a further annealing in oven during 20 hours at 80°C for cards holding LiF:Mg,Ti; at this temperature the plastic holder does not suffer any deformation. Any way, excluding cards, for bare TL solid chips or TL materials in powder form, the annealing must be performed in an oven.

References

1. Busuoli G. in Applied Thermoluminescence Dosimetry, ISPRA Courses, Edited by M. Oberhofer and A. Scharmann, Adam Hilger publisher (1981)
2. Drisoll C.M.H., Barthe J.R., Oberhofer M., Busuoli G. and Hickman C., Rad. Prot. Dos. **14**(1) (1986) 17

Annealing procedures

When a new TL material is going to be used for the first time, it is necessary to perform at first an annealing study which has three main goals:

- to find the good combination of annealing temperature and time to erase any effect of previous irradiation,
- to produce the lowest intrinsic background and the highest sensitivity,
- to obtain the highest reproducibility for both TL and background signals.

The suggested procedures are the following:

1st procedure

- irradiate 10 TLDs samples to a test dose in the range of the field applications,
- anneal the irradiated samples at a given temperature (e.g., 300°C) for a given period of time (e.g., 30 minutes),
- read the samples,
- repeat steps 1), 2) and 3) increasing the annealing temperature of 50°C each time up to the maximum value at which the residual TL (background) will remain constant as the temperature increases,

- plot the data as shown in Fig. 1. As it can be observed, after a threshold temperature value, i.e., T_c , the residual TL signal remains constant,
- repeat now the procedure, keeping constant the temperature at the value T_c and varying the annealing time by steps of 30 minutes and plot the results. The plot should be similar to the previous one,
- choose now the best combination of temperature and time.
- carry out a reproducibility test to verify the goodness of the annealing, in the sense that background must be unchanged during the test.

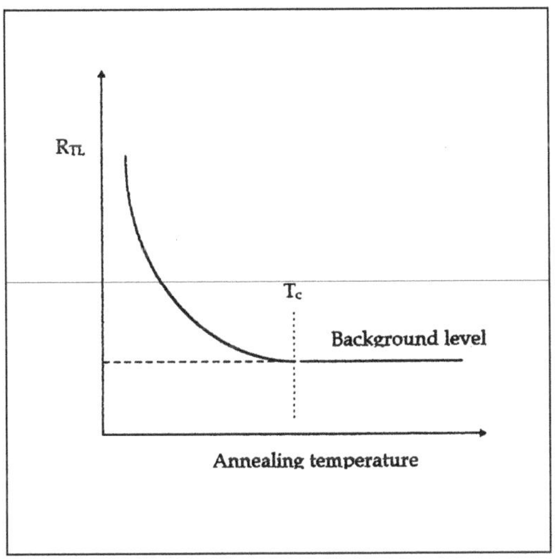

Fig. 1. Decrease of TL response, after irradiation, as a function of the annealing procedure.

2nd procedure

This procedure has been suggested by G. Scarpa [1] who used it for sintered Beryllium Oxide. With this procedure both informations concerning annealing and reproducibility are obtained at once. The procedure consists in changing the temperature, step by step, at a constant annealing time. After annealing at a given temperature, the samples are irradiated and then readout. For each temperature 10 samples are used, cycled 10 times. So that each experimental point in Fig. 2 is based on 100 measurements. From the figure it can be seen that the best reproducibility, i.e., the lowest standard deviation in %, is achieved at around 600°C, whereas the absolute value of the TL output is practically constant between 500 and 700°C. The same procedure can be now carried out for a constant temperature and changing the annealing time. Finally, as before, the best combination of time and temperature should give the optimum annealing procedure.

To be sure that the annealing procedure is useful at any level of dose, it is suggested to repeat the procedure at different doses, according to the specific use of the material.

Figs. 3 and 4 gave other examples of this procedure [2]. Each experimental point correspond to the average over ten samples. The annealing time at each temperature was 1 hr.

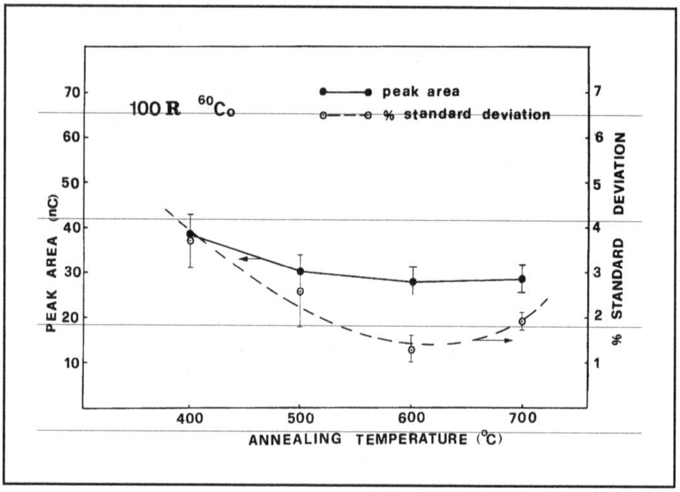

Fig. 2. TL emission (peak area) and corresponding S.D.% Vs annealing temperatures.

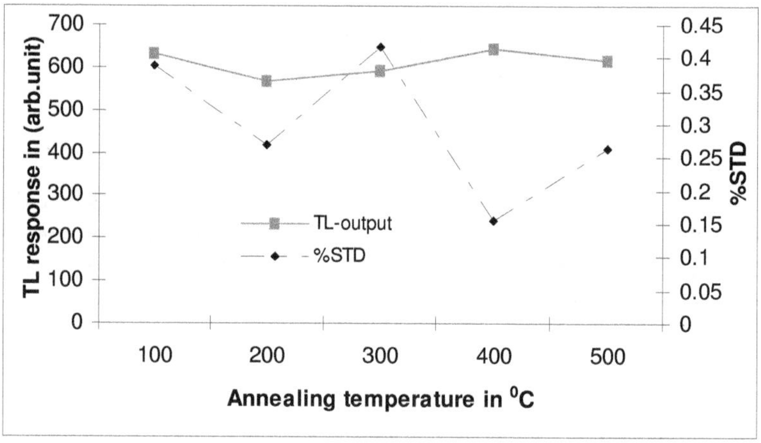

Fig. 3. Behaviour of the TL response and the corresponding standard deviation as a function of the annealing temperature (Ge-doped optical fibre).

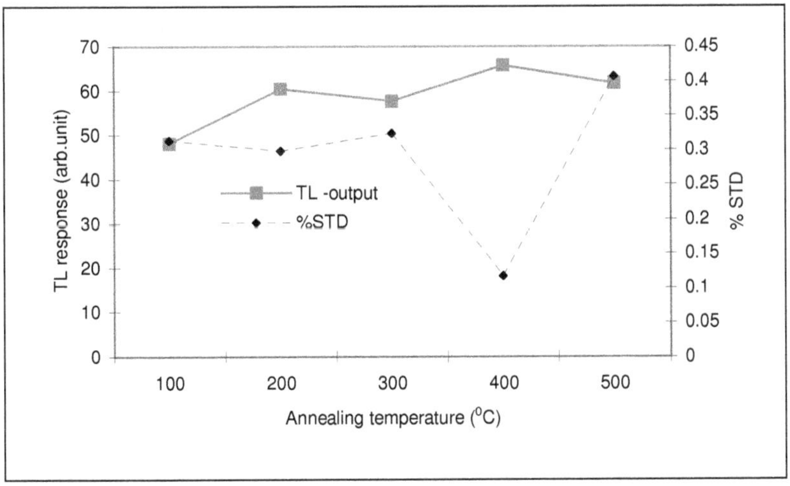

Fig. 4. Behaviour of the TL response and the corresponding standard deviation as a function of the annealing temperature (Eu-doped optical fibre).

The following Tables 1a, 1b and 1c list the annealing and the post-annealing procedures used for most of the thermoluminescent materials.

material	annealing procedure	
	in oven	in reader
LiF:Mg,Ti (TLD100,600,700)	1 h at 400°C + 2 h at 100°C [4] or 1 h at 400°C + 20 h at 80°C [4] fast anneal: 15 min at 400°C +10 min at 100°C [5]	30 sec at 300-400°C (+ 20 h at 80°C in oven) [3]
LiF:Mg,Ti in PTFE (polytetrafluoroethylene)	1 h at 300°C + 20 h at 80°C [6]	30 sec at 300°C (+ 20 h at 80°C in oven)
LiF:Mg,Ti,Na (LiF-PTL)	30 min at 500°C + fast cooling [7]	
LiF:Mg,Cu,P (GR-200A)	10 min at 240°C [8-11] or 15 min at 240°C [12]	30 sec at 240°C
CaF_2:Dy (TLD-200)	1 h at 600°C or 30 min at 450°C or h at 400°C or 1 h at 400°C or 1 h at 400°C + 3 h at 100°C [13,14]	30 sec at 400°C
CaF_2:Tm (TLD-300)	1½ - 2 h at 400°C or 30 min at 300°C [15]	
CaF_2:Mn (TLD-400)	30-60 min at 450-500°C [16]	
$CaSO_4$:Dy (TLD-900)	½ - 1 h at 400°C	
$CaSO_4$:Tm	30 min-1 h at 400°C (PTFE: 2 h at 300°C)	
BeO (Thermal ox 995)	15 min at 400 or 600°C [17,18]	30 sec at 400°C
$Li_2B_4O_7$:Mn (TLD-800)	15 min - 1 h at 300°C	
$Li_2B_4O_7$:Mn,Si	30 min at 400°C [31]	
$Li_2B_4O_7$:Cu	30 min at 300°C [31]	
$Li_2B_4O_7$:Cu,Ag	15 min - 1 h at 300°C	
$Li_2B_4O_7$:Cu,In	30 min at 300°C [31]	

Table 1a. Annealing treatments [3].

material	annealing procedure	
	in oven	in reader
α-Al_2O_3:C	1 h at 400°C + 16 h at 80°C	
Al_2O_3:Cr	15 min at 350°C	
Mg_2SiO_4:Tb	2 - 3 h at 500°C 1 h at 300°C	
MgB_4O_7:Dy/Tm	1 h at 500-600°C [19,20]	
MgB_4O_7:Dy,Na	30 min at 700°C + 30 min at 800°C or 2 h at 550°C [21,22] 1 hr at 400°C [32]	
CVD Diamond	½ h at 300°C [23]	
$KMgF_3$ (various dopants)	1 hr at 400°C [24-28]	
semiconductor-doped Vycor glass	several seconds at 400°C	
RbCl:OH^- RbCl:OH^-	30 min at 600°C [33]	

Table 1b. Annealing treatments [3].

material	pre-readout treatment (post-irradiation anneal)	
	in oven	in reader
LiF:Mg,Ti (TLD-100,600,700)	10 min at 100°C	20 sec at 160°C
LiF:Mg,Ti in PTFE	10 min at 100°C	10-20 sec at 160°C
LiF:Mg,Na (LiF-PTL)		10 sec at 130°C
LiF:Mg,Cu,P (GR-200A)	10 min at 130°C [29]	20-30 sec at 160°C [29]
CaF_2:Dy (TLD-200)	10 min at 110°C or 10 min at 115°C	16 sec at 160°C
CaF_2:Tm (TLD-300)	30 min at 90°C or 10 min at 115°C	16 sec at 160°C
$CaSO_4$:Dy (TLD-900)	20 - 30 min at 100°C or 5 min at 140°C	16 - 32 sec at 120°C
$CaSO_4$:Tm	20 - 30 min at 100°C	16 - 32 sec at 120°C
BeO (Thermalox 995)		1 min at 140°C
$Li_2B_4O_7$:Mn (TLD-800)	10 min at 100°C	
$Li_2B_4O_7$:Mn,Si		20 sec at 160°C
$Li_2B_4O_7$:Cu,Ag		20 sec at 160°C
Al_2O_3:Cr	15 min at 150°C	
MgB_4O_7:Dy/Tm		few sec at 160°C [30]
$KMgF_3$ (various dopants)	30 - 60 min at 50°C [24-28]	

Table 1c. Post-irradiation treatments.

References

1. Benincasa G., Ceravolo L. and Scarpa G., CNEN RT/PROT(74)1
2. Youssef Abdulla, private communication
3. Driscoll C.M.H., Barthe J.R., Oberhofer M., Busuoli G. and Hickman C. Rad. Prot. Dos. **14**(1) (1986) 17

4. Scarpa G. in "Corso sulla termoluminescenza applicata alla dosimetria" University of Rome "La Sapienza", Italy, 15-17 February 1994
5. Scarpa G. in "IV incontro di aggiornamento e di studio sulla dosimetria a termoluminescenza" ENEA, Centro Ricerche Energia Ambiente, S.Teresa (La Spezia), Italy, 18-19 June 1984
6. Horowitz Y.S., "Thermoluminescence and thermoluminescent dosimetry" Vol. I, CRC Press, 1984
7. Portal G., Francois H., Carpenter S., Dajlevic R. Proc. 2nd Int. Conf. Lum. Dos., Gatlinburg USAEC Rep. Conf. 680920, 1968
8. Wang S., Cheng G., Wu F., Li Y., Zha Z., Zhu J., Rad. Prot. Dos. **14**, 223, 1986
9. Driscoll C.M.H., McWhan A.F., O'Hogan J.B., Dodson J., Mundy S.J., Todd C.D.T., Rad. Prot. Dos. **17**, 367, 1986
10. Horowitz Y.S., Horowitz A., Rad. Prot. Dos. 33, 279, 1990
11. Zha Z., Wang S., Wu F., Chen G., Li Y., Zhu J., Rad. Prot. Dos. **17**, 415, 1986
12. Scarpa G. private communication 1991
13. Binder W., Cameron R.J., Health Phys. **17**, 613, 1969
14. Portal G., in Applied Thermoluminescence Dosimetry, ed. M. Oberhofer and A. Sharmann, Adam & Hilger, Bristol, 1981
15. Furetta C., Lee Y.K., Rad. Prot. Dos., **5**, 57, 1983
16. Ginther R.J., Kirk R.D., J. Electrochem. Soc., **104**, 365, 1957
17. Tochilin E., Goldstein, N., Miller W.G., Health Phys. **16**, 1, 1969
18. Busuoli G., Lembo L., Nanni R., Sermenghi I. Rad. Prot. Dos. **6**, 317, 1984
19. Barbina V., Contento G., Furetta C., Molisan C., Padovani R., Rad. Eff. Lett. **67**, 55, 1981
20. Barbina V., Contento G., Furetta C., Padovani R., Prokic M., Proc Third Int. Symp. Soc. Radiol. Prot. (Inverness) 1982
21. Driscoll C.M.H., Mundy S.J., Elliot J.M., Rad. Prot. Dos. **1**, 135, 1981
22. Furetta C., Weng P.S., Hsu P.C., Tsai L.J and Vismara L., Int. Conf. Rad. Dos. & Safety, Taipei, Taiwan, 1997
23. Borchi E., Furetta C., Kitis G., Leroy C. and Sussmann R.S., Rad. Prot. Dos. **65**(1-4), 291, 1996
24. Furetta C., Bacci C., Rispoli B., Sanipoli C. and Scacco A., Rad. Prot. Dos. **33** 107, 1990
25. Bacci C., Fioravanti S., Furetta C., Missori M., Ramogida G, Rossetti R, Sanipoli C. and Scacco A., Rad. Prot. Dos. **47**, 1993, 277
26. Furetta C., Ramogida G., Scacco A, Martini M. and Paravisi S., J. Phys. Chem. Solids **55**, 1994, 1337
27. Furetta C., Santopietro F., Sanipoli C. and Kitis G., Appl. Rad. Isot. 55, 2001 533
28. Furetta C., Sanipoli C. and Kitis G., J. Phys D: Appl. Phys. **34**, 2001, 857
29. Scarpa G., Moscati M., Soriani A. in "Proc. XXVII Cong. Naz. AIRP, Ferrara, Italy, 16-18 Sept., 1991
30. Driscoll C.M.H., Mundy S.J. and Elliot J.M., Rad. Prot. Dos. **1** (1981) 135
31. Kitis G, Furetta C. Prokic M. and Prokic V., J. Phys. D: Appl. Phys. (2000) 1252
32. Furetta C., Prokic M., Salamon R. and Kitis G., Appl. Rad. Isot. **52** (2000) 243
33. Furetta C., Laudadio M.T., Sanipoli C., Scacco A., Gomez Ros J.M. and Correcher V., J. Phys. Chem. Solids **60** (1999) 957

Anomalous fading

The expected mean lifetime, τ, of a charge in a trap having a depth E is given by the following equation, according to a first order kinetics:

$$\tau = s^{-1} \exp\left(\frac{E}{kT}\right)$$

where s is the frequency factor and T is the storage temperature.

For many materials it is often found that the drainage of traps is not accounted for by the previous equation: i.e., the charges are released by the trap at a rate which is much faster than those expected from the equation and the phenomenon is only weakly dependent on the temperature. This kind of fading is known as anomalous fading and it is explained by **tunnelling** of carriers from the trap to the recombination centre [1,2].

The anomalous fading is observed in natural minerals, as well as in TL materials as ZnS:Cu, ZnS:Co, CaF_2:Mn, KCl:Tl, etc.

The characteristic of the anomalous fading is an initial rapid decay followed by a decrease of the decay rate over long storage periods.

The experimental way for detecting a suspected anomalous fading is to perform a long-term fading experiment in order to accumulate a measurable signal loss and to compare the experimental amount of fading to the one calculated taking into account the quantities E, s and the storage temperature.

References

1. McKeever S.W.S., Thermoluminescence of Solids, Cambridge University Press (1985)
2. Chen R. and Kirsh Y., Analysis of Thermally Stimulated Processes, Pergamon Press (1981)

Anomalous thermal fading

This effect has been encountered in dating of meteorites. When very long periods of storage/irradiation are involved, a thermal fading of the TL corresponding to very deep traps becomes significant. This means that the thermal detrapping of these traps takes place at the same time of their filling due to the irradiation. In case of very long periods it can be possible to reach an equilibrium condition between the filling rate and the detrapping rate although a saturation level has not reached.

Area measurement methods (generality)

The area methods are based on the measurements of the integral of the glow-peak; it can be applied when a well isolated and clean peak is available.

Assuming a first order kinetics, the Randall-Wilkins relation in the time domain gives

$$\int_{t_0}^{t} Idt = n = \frac{I}{s\exp(-\frac{E}{kT})} \tag{1}$$

In the temperature domain the same relation leads to

$$\frac{I}{\int_{T}^{T_\infty} IdT} = \frac{s}{\beta}\exp(-\frac{E}{kT}) \tag{2}$$

$$\ln[\frac{I}{\int_{T}^{T_\infty} IdT}] = \ln\frac{s}{\beta} - \frac{E}{kT} \tag{3}$$

where the integral is calculated as shown in Fig. 1.

Again, the ln term is a linear function of $1/T$ and can be plotted on a semilog paper to obtain the slope $-E/k$ and the intercept $\ln(s/\beta)$.

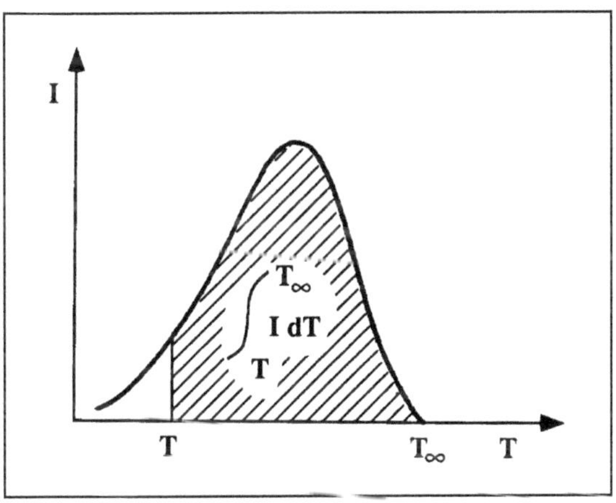

Fig. 1. How to calculate the integral comparing in Eq. (3).

Area measurement method (Maxia et al.)

Maxia et al. [1] have suggested a rather complicated area method for the evaluation of both activation energy and frequency factor. It essentially stems from the fact that the filled traps density, at any temperature T, is proportional to the remaining area of the glow-peak. The main assumption is that the various peaks in a complex glow-curve result from the escape of electrons from a single trap and their recombination into various recombination centres. The mathematical treatment is based on the equation proposed by **Antonov-Romanovskii** [2]:

$$\frac{dm_i}{dt} = -s\exp\left(-\frac{E}{kT}\right)\frac{B_i n m_i}{A(N-n)+B_i m_i} \tag{1}$$

where, in particular

m_i is the concentration of luminescent centers,
A is the probability factor for retrapping,
B_i is the probability for recombination,
N is the electron trap concentration,
n is the trapped electron concentration.

Using a constant heating rate one can write

$$\frac{dm_i}{dT} = \alpha I(T) \tag{2}$$

where α being a constant depending on the light collection efficiency and the used units.

From Eq. (2) one has

$$m_i(T) = \alpha\int_T^{T_f} I(T')dT' = \alpha S(T) \tag{3}$$

and

$$m_{i0} = \alpha\int_{T_0}^{T_f} I(T')dT' = \alpha S_0 \tag{4}$$

in which

T_0 = temperature at beginning of the glow-peak,
T_f = temperature at the end of the glow-peak,
T = temperature between T_0 and T_f,
S_0 = total area from T_0 to T_f,
S = area from T and T_f.

Introducing now the areas σ and Σ defined as

$$n_0 = \alpha\sigma \tag{5}$$

$$N = \alpha\Sigma \tag{6}$$

from a single glow-peak one obtains

$$n_0 = m_{i0} \tag{7}$$

and then

$$\sigma = S_0 \tag{8}$$

where n_0 and m_{i0} represent initial values.

By straight-forward calculations, using Eq. (1) to Eq. (6), one gets

$$\ln\left[I(T)\frac{\cos\theta + S(T)\sin\theta}{S^2(T)+S(T)\Delta} \right] = -\frac{E}{kT} + \mu \tag{9}$$

where

$$\tan\theta = \frac{B-A}{A(\Sigma-\sigma+S_0)} \tag{10}$$

$$\Delta = \sigma - S_0 \tag{11}$$

$$\mu = -\frac{1}{2}\ln\left\{ \left(\frac{\beta}{Bs}\right)^2 \left[A^2(\Sigma-\sigma+S_0)^2 + (B-A)^2 \right] \right\} \tag{12}$$

The plot of Eq. (9) will be rectilinear if an appropriate value of θ is chosen. The angle θ depends on unknown parameters, as one can see from Eq. (10). To determine θ one can put

$$y = \ln\left[I(T)\frac{\cos\theta + S(T)\sin\theta}{S^2(T)+S(T)\Delta} \right] \tag{13}$$

$$x = \frac{1}{T} \tag{14}$$

The searched value for θ is the one corresponding to a linear behaviour of y versus x. Such a behaviour can be carried out considering the covariance, cov(x,y), and the variances, v(x) and v(y), for x and y respectively. In this way the correlation coefficient can be numerically computed as a function of θ. Then, using the angular coefficient of the straight line described by Eq. (9), the activation energy can be found as

$$E = -k\frac{\text{cov}(x,y)}{v(x)} = -k\frac{\text{cov}(x,y)}{v(y)} \tag{15}$$

The Authors claim that this method is also applicable when overlapped peaks are present.

References

1. Maxia V., Onnis S. and Rucci A., J. Lum. **3** (1971) 378
2. Antonov-Romanovskii V.V., Isvest. Akad. Nauk. SSSR Fiz. **10** (1946) 477

Area measurement methods (May and Partridge: general order)

May and Partridge proposed the **area method** in the case of a general order b is considered. In this case the equation is

$$\ln(\frac{I}{n^b}) = \ln(s) - \frac{E}{kT} \tag{1}$$

which is graphically processed in the usual way. By visual examination of such plots, the value of b which gives the best straight line can be ascertained.

Reference

May C.E. and Partridge J.A., J. Chem. Phys. **40** (1964) 1401

Area measurement methods (Muntoni et al.: general order)

Muntoni and co-workers suggested a method base on the glow prak area and on the fact that the filled traps density, at any temperature T, is proportional to the remaining area of the glow-peak $S(T)$. They used the general order kinetics equation in the form

$$I = -a\frac{dm}{dt} = aAm^b \exp(-\frac{E}{kT}) \tag{1}$$

where m is the concentration of recombination centres, A is a proportionality factor, b represents the kinetic order and a is a constant. The integral area, $S(T)$ in the interval from T to T_f in the glow-curve, is given by

$$S(T) = \int_T^{T_f} I dT = -a\beta \int_m^0 dm = a\beta m \tag{2}$$

from which, considering the order b,

$$m^b = \left[\frac{S(T)}{a\beta}\right]^b \tag{3}$$

Equation (1), using Eq. (3), yields to

$$\ln\frac{I}{[S(T)]^b}=c-\frac{E}{kT} \tag{4}$$

A plot of the first term against $1/T$ gives a straight line when the best value for b is chosen.

Reference

Muntoni C., Rucci A. and Serpi A., Ric. Sci. **38** (1968) 762

Area measurement method (Moharil: general order)

This method [1] uses the peak area and it is based on the **Antonov-Romanovskii equation** [2]:

$$I=-\frac{dn}{dt}=s\frac{Bnm}{Bn+A(N-n)}\exp(-\frac{E}{kT}) \tag{1}$$

Considering that:
n_0 is proportional to the area under the total peak (= A_0); n is proportional to the remaining area, A_T, under the glow peak, from a given temperature T to the end of the peak. If $n_0 = N$, saturation case, A_0 is proportional to N: in this case the area is denoted A_S instead of A_0

Eq. (1) transforms in, with the conditions $n = m$, $n = n_0$

$$I=\frac{BA_T^2}{(B-A)A_T+AA_S}s\exp(-\frac{E}{kT}) \tag{2}$$

Considering the intensity at the maximum, I_M, and the half maximum intensities, I_1 and I_2, from Eq. (2) can be derived the following two expressions:

$$\begin{aligned}\ln 2&=-\frac{E}{k}\left(\frac{1}{T_M}-\frac{1}{T_2}\right)+\ln\left(\frac{A_{T_M}^2}{A_{T_2}^2}\cdot\frac{A_{T_2}+(A_S-A_{T_2})\frac{A}{B}}{A_{T_M}+\left(A_S-A_{T_M}\right)\frac{A}{B}}\right)\\ \ln 2&=+\frac{E}{k}\left(\frac{1}{T_1}-\frac{1}{T_M}\right)-\ln\left(\frac{A_{T_1}^2}{A_{T_M}^2}\cdot\frac{A_{T_M}+(A_S-A_{T_M})\frac{A}{B}}{A_{T_1}+\left(A_S-A_{T_1}\right)\frac{A}{B}}\right)\end{aligned} \tag{3}$$

For different values of the ratio A/B, E can be calculated. The correct value of A/B is the one which gives the same value of E from both equations.

The same procedure can be carried out in the case of non-saturation doses [3].

References

1. Moharil S.V., Phys. Stat. Sol. (a) **73** (1982) 509
2. Atonov-Romanoski V.V., Bull. Acad. Sci. USSR Phys. Res. **15** (1951) 673
3. Moharil S.V. and Kathurian S.P., J. Phys. D: Appl. Phys. **16** (1983) 2017

Area measurement method (Moharil: general order, $s = s(T)$)

Moharil suggested a new method to obtain the trapping parameters which does not require a priori knowledge of the order of kinetics. Furthermore, he assumes a temperature dependence of the frequency factor.

Starting from the general order equation, he modifies it as following, taking into account the temperature dependence of the frequency factor:

$$I = -\frac{dn}{dt} = s_0 T^{\alpha} n^{b} \exp(-\frac{E}{kT}) \tag{1}$$

and assumes that the glow-curve consists of a single glow-peak corresponding to traps of only one kind.

If it is assumed that at the end of the glow-curve all the traps are emptied, the number of traps populated at the beginning of the glow-curve, n_0, is proportional to the area under the glow-curve. As a consequence, the number of traps emptied up to the temperature T is given by the area under the glow-curve up to T. Hence, the number of populated traps at temperature T is proportional to the area A_T which is equal to the total area under the glow-curve less the area under the curve between the initial point and T. So, Eq. (1) becomes

$$I = B s_0 T^{\alpha} A_T^{b} \exp(-\frac{E}{kT}) \tag{2}$$

where B is a constant.

Let us say T_M be the temperature at which the TL intensity is maximum and T_1 and T_2 be the temperatures at which the intensity falls to half of its maximum both on the low and high temperature sides of T_M, respectively. Fig. 1 shows the different parameters. Equation (2) gives then

$$I_M = Bs_0 T_M^\alpha A_{T_M}^b \exp\left(-\frac{E}{kT_M}\right)$$

$$I_1 = \frac{I_M}{2} Bs_0 T_1^\alpha A_{T_1}^b \exp\left(-\frac{E}{kT_1}\right) \tag{3}$$

$$I_2 = \frac{I_M}{2} Bs_0 T_2^\alpha A_{T_2}^b \exp\left(-\frac{E}{kT_2}\right)$$

From the previous expressions one obtains

$$\ln 2 = -\alpha \ln\left(\frac{T_2}{T_M}\right) + b \ln\left(\frac{A_{T_M}}{A_{T_2}}\right) - \frac{E}{k}\left(\frac{1}{T_M} - \frac{1}{T_2}\right)$$

$$\ln 2 = -\alpha \ln\left(\frac{T_M}{T_1}\right) + b \ln\left(\frac{A_{T_1}}{A_{T_M}}\right) - \frac{E}{k}\left(\frac{1}{T_1} - \frac{1}{T_M}\right) \tag{4}$$

The terms containing α can be neglected because $\ln(T_2/T_M)$ and $\ln(T_M/T_1)$ are of the order of 10^{-2} and α lies between -2 and +2. Using Eq. (4) E and b can then be obtained. The value of α is obtained by Eq. (2):

$$\ln I = \ln B + \alpha \ln T + b \ln A_T - \frac{E}{kT} \tag{5}$$

Because at $T=T_M$, $d(\ln I)/dT=0$, one gets

$$\alpha = -\left(bT_M \frac{d}{dT}(\ln A_T)_{T=T_M}\right) - \frac{E}{kT_M} \tag{6}$$

where $d(\ln A_T)/dT$ is know experimentally and will be negative. The sign of α depends upon whether E/kT_M is larger or smaller than $\left[bT_M d(\ln A_T)dT\right]_{T=T_M}$.

After having determined the values of α, b and E, one can now obtain the frequency factor starting from the **general order equation including the temperature dependence of the frequency factor**:

$$-\frac{dn}{dt} = -n^b s_0 T^\alpha \exp\left(-\frac{E}{kT}\right) \tag{7}$$

which can be written as, using a linear heating rate $\beta = \frac{dT}{dt}$

$$-\frac{dn}{n^b} = \frac{s_0}{\beta} T^\alpha \exp(-\frac{E}{kT}) dT \tag{8}$$

Integrating this equation between 0 and T and using the condition $n = n_0$ at $T = 0$:

$$\int_{n_0}^{n} \frac{dn}{n^b} = -\frac{s_0}{\beta} \int_0^T T^\alpha \exp\left(-\frac{E}{kT'}\right) dT'$$

$$n^{1-b} = n_0^{1-b} - \frac{s_0(1-b)}{\beta} \int_0^T T^\alpha \exp\left(-\frac{e}{kT'}\right) dT'$$

and the expression of n is obtained:

$$n = n_0 \left[1 + \frac{s_0 n_0^{b-1}(b-1)}{\beta} \int_0^T T^\alpha \exp\left(-\frac{E}{kT'}\right) dT' \right]^{\frac{1}{1-b}} \tag{9}$$

so that the expression of the intensity I is now

$$I = n_0^b s_0 T^\alpha \exp(-\frac{E}{kT}) \left[1 + \frac{(b-1)s_0 n_0^{b-1}}{\beta} \int_0^T T^\alpha \exp(-\frac{E}{kT'}) dT' \right]^{-\frac{b}{b-1}}$$

where $s_0 n_0^{b-1} T^\alpha$ is similar to the frequency factor s of the first-order equation.

Using the substitution

$$RT^\alpha = s \qquad \text{with} \qquad R = s_0 n_0^{b-1}$$

the previous equation for the intensity becomes

$$I = n_0 R\, T^\alpha \exp(-\frac{E}{kT}) \left[1 + \frac{(b-1)R}{\beta} \int_0^T T^\alpha \exp(-\frac{E}{kT'}) dT' \right]^{-\frac{b}{b-1}} \tag{10}$$

Because $dI/dT = 0$ at $T = T_M$, Eq. (10) gives

$$R=\frac{\alpha+\dfrac{E}{kT_M}}{\left(\dfrac{b}{\beta}\right)T_M^{\alpha+1}\exp(-\dfrac{E}{kT_M})-\left(\alpha+\dfrac{E}{kT_M}\right)\left[\dfrac{b-1}{\beta}\right]\displaystyle\int_0^T T^{\alpha}\exp(-\dfrac{E}{kT'})dT'} \tag{11}$$

where the integral is evaluated by graphical method or using the Newton-Raphson method. The frequency factor at any temperature is then calculated by $s=RT^{a}$.

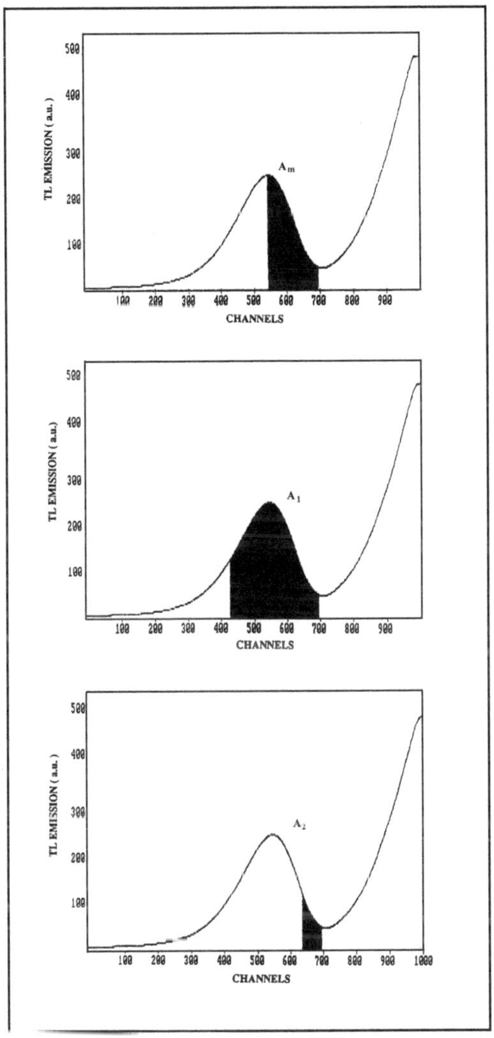

Fig. 1. In this figure the channel number is proportional to the temperature.

Reference

Moharil S.V., Phys. Stat. Sol. (a) **66** (1981) 767

Area measurement method (Rasheedy: general order)

M.S. Rasheedy developed a method for obtaining the trap parameters of a complex TL glow-curve including several peaks [1]. His method is based on a development of **Moharil's method**. The trap parameters are obtained starting from the higher temperature glow peak. The procedure begins by determining the order of kinetics, b, of the higher temperature peak of the glow curve.

It is assumed, at first, that the glow curve consists in a single glow-peak, corresponding to only one kind of traps. Furthermore, it is also assumed that all the traps are emptied at the end of the glow-curve. As usual, the concentration of the trapped charges, at the beginning of the glow-curve, n_0, is proportional to the area A under the glow-curve; then, the concentration of trapped charges at any temperature T_i, during the read out run, is proportional to the area A_i under the glow-curve between T_i and the final temperature, T_f, at which the TL light falls to zero. Fig. 1 gives a sketch of an isolated peak with indication of the different parameters.

Taking into account the equation for a general order kinetics written as following [2]:

$$I = -\frac{dn}{dt} = \frac{n^b}{N^{b-1}} s\exp(-\frac{E}{kT}) \tag{1}$$

and considering the maximum intensity of the peak, I_M, the previous equation becomes

$$I_m = \frac{A_m^b}{N^{b-1}} s\exp\left(-\frac{E}{kT_m}\right) \tag{2}$$

According to the quantities shown in Fig. 1, the following equations can also be written

$$\begin{aligned} I_2 &= \frac{I_M}{2} = \frac{A_2^b}{N^{b-1}} s\exp(-\frac{E}{kT_2}) \\ I_4 &= \frac{I_M}{4} = \frac{A_4^b}{N^{b-1}} s\exp(-\frac{E}{kT_4}) \end{aligned} \tag{3}$$

where A_2 and A_4 are the areas under the glow peak from T_2 to T_f and from T_4 to T_f respectively.

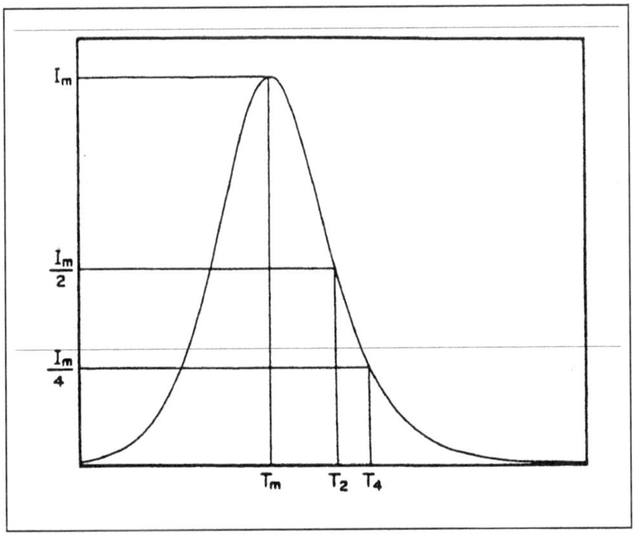

Fig. 1. An isolated glow-peak with the parameters of interest.

Making now the logarithm of Eqs. (2) and (3) one gets

$$E = \frac{[\ln 2 - b\ln(\frac{A_M}{A_2})]kT_M T_2}{T_M - T_2} \tag{4}$$

$$E = \frac{[\ln 2 - b\ln(\frac{A_M}{A_4})]kT_M T_4}{T_M - T_4} \tag{5}$$

Eliminating E from the two previous equations, it is easy to find an expression which gives the kinetics order b:

$$b = \frac{T_2(T_M - T_4)\ln 2 - T_4(T_M - T_2)\ln 4}{T_2(T_M - T_4)\ln(\frac{A_M}{A_2}) - T_4(T_M - T_2)\ln(\frac{A_M}{A_4})} \tag{6}$$

The previous equations can be arranged to determine E and b using any portion of the descending part of the glow-peak. Let us indicate I_x the TL intensity at a temperature T_x of the descending part of the peak and T'_2 and T'_4 the temperatures at which the TL intensities are equal to $I_x/2$ and $I_x/4$ respectively; the new equations are then

$$E = \frac{[\ln 2 - b\ln(\frac{A_x}{A_2})]kT_xT_2^{'}}{T_x - T_2^{'}}$$

$$E = \frac{[\ln 2 - b\ln(\frac{A_x}{A_4})]kT_xT_2^{'}}{T_x - T_4^{'}} \tag{7}$$

$$b = \frac{T_2^{'}(T_x - T_4^{'})\ln 2 - T_4^{'}(T_x - T_2^{'})\ln 4}{T_2^{'}(T_x - T_4^{'})\ln(\frac{A_M}{A_2}) - T_4^{'}(T_x - T_2^{'})\ln(\frac{A_x}{A_4})} \tag{8}$$

The same method is then applied to glow-curves having peaks more than one. In particular, the Author applies his method to BeO which presents a glow-curve with two well resolved peaks. The first step of the method consists in determining the trap parameters of the higher temperature peak. The value of b of the peak is evaluated at different intensities of the descending part of the peak starting from T_M. The pre-exponential factor

$$s'' = s(\frac{n_0}{N})^{b-1} \tag{9}$$

and the relative value n_0 are estimated by the equation

$$S'' = \frac{\beta E \exp(\frac{E}{kT_M})}{bkT_M^2 - E(b-1)\Phi \exp(\frac{E}{kT_M})} \tag{10}$$

and where

$$\Phi = \int_{T_0}^{T_M} \exp(-\frac{E}{kT'})dT'$$

Eq. (10) is obtained by equating the derivative of the following equation to zero

$$I(T) = n_0 s'' \exp\left(-\frac{E}{kT}\right)\left[1 + \frac{s''(b-1)}{\beta}\int_{T_0}^{T} \exp\left(-\frac{E}{kT'}\right)dT'\right]^{\frac{1}{1-b}} \tag{11}$$

Now, substituting s'' in Eq. (11) one obtains

$$I = \frac{n_0 s'' \exp(-\frac{E}{kT_M})}{\left\{1+[\frac{(b-1)s''}{\beta}\int_{T_0}^{T}\exp(-\frac{E}{kT'})dT']\right\}^{\frac{b}{b-1}}} \tag{12}$$

Furthermore, the relative value of n_0 can be found using the maximum intensity I_M. In this case the procedure is the usual one, which means to make the logarithm of Eq. (11), then its derivative respect to the temperature T and finally to equate to zero the derivative at $T = T_M$. In this way the maximum of the intensity, I_M, is given by the following expression:

$$I_M = \frac{n_0 s'' \exp(-\frac{E}{kT_M})}{\left[\frac{kT_M^2 bs''}{\beta E}\exp(-\frac{E}{kT_M})\right]^{\frac{b}{b-1}}} \tag{13}$$

from which the value of n_0 is obtained

$$n_0 = \frac{I_M \exp(\frac{E}{kT_M})}{s''}\left[\frac{bkT_M^2 s''}{\beta E \exp(\frac{E}{kT_M})}\right]^{\frac{b}{b-1}} \tag{14}$$

Reference

1. Rasheedy M.S., J. Phys. D: Appl. Phys. **29** (1996) 1340
2. Rasheedy M.S., J. Phys.: Condens. Matter. **5** (1993) 633

Arrhenius equation

The Arrhenius equation gives the mean time, τ, that an electron spends in a trap at a given temperature T. It is

$$\tau = s^{-1}\exp\left(\frac{E}{kT}\right) \tag{1}$$

where s is the frequency factor (in the case of thermoluminescence the frequency factor is also called attempt-to-escape frequency), E is the energy difference between the bottom of the CB and the trap position in the band gap, also called trap depth or activation energy, k is the Boltzmann's constant.

Eq. (1) can be rewritten as

$$p = \tau^{-1} \tag{2}$$

which gives the probability p, per unit of time, of the release of an electron from the trap.

According to Eq. (1), if the trap depth is such that at the temperature of irradiation, let us say T_i, E results to be much larger than kT_i, electrons produced by irradiation and then trapped will remain in the trap for a long period of time, even after the removal of the irradiation. The Arrhenius equation introduces the concept of an **activation energy**, E, seen as an energy barrier which must be overcome in order to reach equilibrium.

Reference

Bube R.H., Photoconductivity of Solids, Wiley & Sons, N.Y. (1960)

Assessment of random uncertainties in precision of TL measurements (general)

The reproducibility of TL measurements depends on the dose level. The following Fig. 1 shows how the standard deviation, in percentage, behaves as a function of the dose.

From the figure it can be observed that the relative standard deviation in percentage decreases very fast as the dose increase. As the dose increases, the relative standard deviation assumes a minimum constant value.

This behaviour is justified by the competition of two components:

- the intrinsic variability of the TL system, given by the standard deviation of the zero dose readings (background),
- the variation of the TL system at high doses, expressed in terms of standard deviation.

Burkhardt and Piesh [1] and Zarand and Polgar [2,3] used a mathematical formalism to describe the effect of the two components so far introduced. They proposed the following expression

$$\sigma_D = \sqrt{\sigma_{BKG}^2 + \sigma_{r,D}^2 D^2} \tag{1}$$

where

σ_D = standard deviation of the evaluated dose D,

σ_{BKG} = standard deviation of the zero-dose readings, expressed in unit of dose,

$\sigma_{r,D}$ = relative standard deviation of the readings obtained at the dose D, relatively high.

Eq. (1) can be re-written as

$$\frac{\sigma_D}{D} = \sqrt{\frac{\sigma_{BKG}^2}{D^2} + \sigma_{r,D}^2} \tag{2}$$

From Eq. (2) it can be observed that:

- the ratio σ_{BKG}/D becomes almost zero for doses quite large respect to σ_{BKG} and then $\sigma_D/D \approx \sigma_{r,D}$, which takes into account of the minimum and constant value observed,

- for very little doses, the term $\sigma_{r,D}$ becomes negligible and the Eq. (2) assumes the form

$$y = \frac{a}{x} \tag{3}$$

- expression (3), on a log-log scale, is a straight line having a unity negative slope (in the region 1 – 10 μGy of the Fig. 2).

- furthermore, modifying Eq. (2) as in the following

$$\frac{\sigma_D}{D} = \sqrt{\left(\frac{\sigma_{BKG}}{\dfrac{\sigma_{BKG}}{\dfrac{D}{\sigma_{BKG}}}}\right)^2 + \sigma_{r,D}^2}$$

or better

$$\frac{\sigma_D}{D} = \sqrt{\frac{1}{R^2} + B^2} \tag{4}$$

From Eq. (4) results that σ_D/D is independent by the standard deviation, expressed in terms of dose, of the zero-dose readings, but only depends by R and $\sigma_{r,D}$, as it is shown in Fig. 3

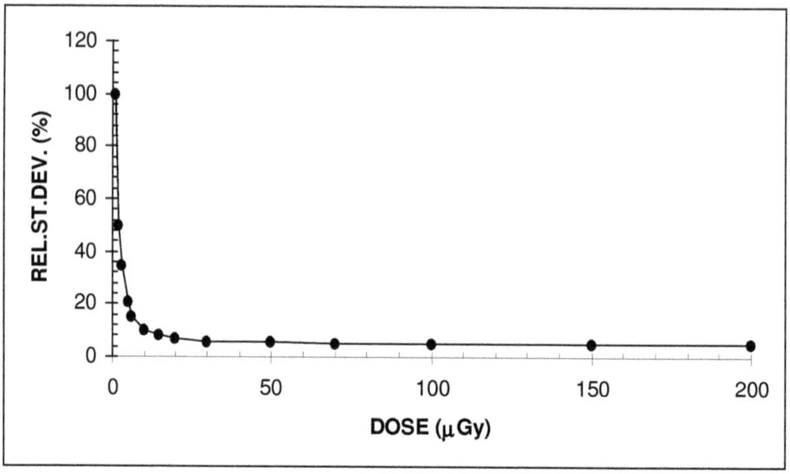

Fig. 1. Behaviour of the Rel. Stand. Dev. of the TL readings as a function of dose.

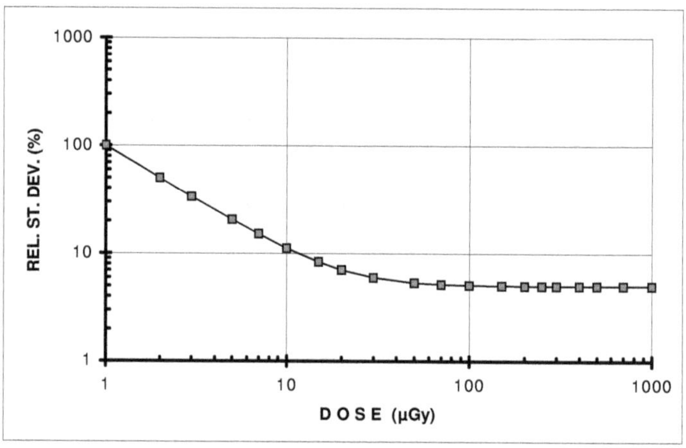

Fig. 2. Same plot of Fig. 1 but in log-log scale.

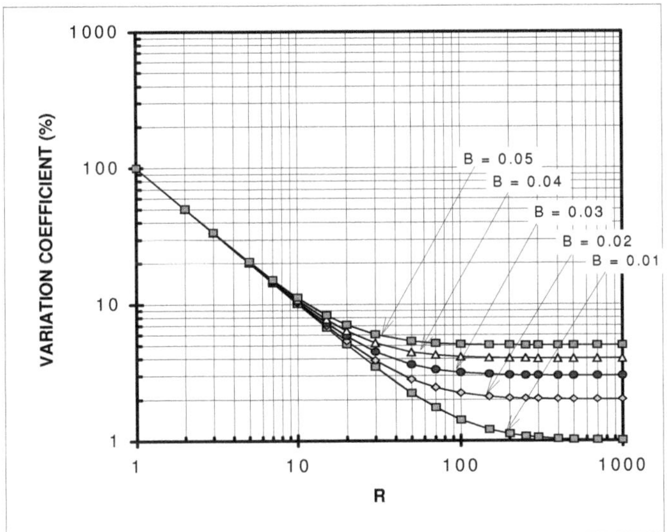

Fig. 3. Behaviour of $\frac{\sigma_D}{D}(\%)$ as a function of $R = \frac{D}{\sigma_{BKG}}$ for given values of $B = \sigma_{r,D}$.

References

1. Burkhardt B. and Piesh E., Nucl. Instr. Meth. **175** (1980) 159
2. Zarand P. and Polgar I., Nucl. Instr. Meth. **205** (1983) 525
3. Zarand P. and Polgar I., Nucl. Instr. Meth. **222** (1984) 567

Atomic number (calculation)

For some practical dosimetric applications, as the wide range of radiological dosimetry, two properties of the TL dosemeters are advantageous for precise measurements. These are high sensitivity and tissue equivalence. High sensitivity thermoluminescent phosphors (i.e. CaF_2 and $CaSO_4$) have high effective atomic numbers, Z_{eff}, so that at photon energies below about 100 keV, the response to a given absorbed dose of radiation becomes significantly greater than that at higher energies. In this region the photoelectric effect is predominant and the cross section per atom depends upon approximately Z^4 for high atomic number materials and on $Z^{4.8}$ for low Z materials. Since each atom contains Z electrons, the coefficient per electron depends upon Z^3 and $Z^{3.8}$ for high and low Z materials respectively.

It is important to know a priori the effective atomic number of a thermoluminescent material, Z, for getting an approximated idea of the expected TL response at different energies. The behaviour of different materials to X and gamma rays depends on the atomic number of the constituents and not on the chemical composition of these constituents.

$$\overline{Z} = \sqrt[x]{a_1 Z_1^x + a_2 Z_2^x + \ldots} \tag{1}$$

$$a_i = \frac{n_i(Z_i)}{\sum_i n_i(Z_i)} \tag{2}$$

$$n_i = N_A \cdot Z_i \tag{3}$$

where a_1, a_2, ... are the fractional contents of electrons belonging to elements Z_1, Z_2 , ... respectively, n_i is the number of electrons, in one mole, belonging to each element Z_i and N_A is the Avogadro's number. The value of x is 2.94.

A numerical example concerning LiF is given below:
1 mole of compounds contains $6.022 \cdot 10^{23}$ atoms so that 1 mole of LiF has $6.022 \cdot 10^{23}$ atoms of Li and $6.022 \cdot 10^{23}$ atoms of F. Now, the number of electrons belonging to each element in 1 mole of compound is given by the atomic number of the element multiplied by the number of atoms:

for Li: $3 \cdot 6.022 \cdot 10^{23} = 1.81 \cdot 10^{24}$ electrons,
for F: $9 \cdot 6.022 \cdot 10^{23} = 5.41 \cdot 10^{24}$ electrons.

The total amount of electrons in LiF is then $7.23 \cdot 10^{24}$.
The partial contents, a_i , are respectively

$$a_{Li} = \frac{1.81 \cdot 10^{24}}{7.23 \cdot 10^{24}} = 0.25$$

$$a_F = \frac{5.41 \cdot 10^{24}}{7.23 \cdot 10^{24}} = 0.75$$

Then

$$Z_{Li}^{2.94} = 3^{2.94} = 25.28$$

$$Z_F^{2.94} = 9^{2.94} = 638.96$$

from which

$$a_{Li} \cdot Z_{Li}^{2.94} = 6.32$$

$$a_F \cdot Z_F^{2.94} = 479.22$$

and finally

$$\bar{Z} \approx 8.2$$

Alternatively, the number of electrons per gram can be acalculated as follows

$$n_i = \left(\frac{N_A \cdot Z_i}{A_{w,i}} \right) \cdot W_i \qquad (4)$$

where N_A is the Avogadro's number, $A_{w,i}$ is the atomic number, W_i is the fractional weight and Z_i is the atomic number of the i-th element in the compound.

The following table shows the atomic number of the main TL materials.

phosphors	effective atomic number
LiF:Mg,Ti LiF:Mg,Ti,Ma LiF:Mg,Cu,P	8.14
$Li_2B_4O_7$:Mn $Li_2B_4O_7$:Cu $Li_2B_4O_7$:Cu,Ag	7.4
MgB_4O_7:Tm MgB_4O_7:Tb	8.4
$CaSO_4$:Dy $CaSO_4$:Tm $CaSO_4$:Mn	15.3
CaF_2:Dy CaF_2:Mn CaF_2 (nat) CaF_2:Tm	16.3
BeO	7.13
Al_2O_3	10.2
ZrO_2	35.6
$KMgF_3$ (various activators)	13.4
CVD diamond	6
$Ca_5F(PO_4)_3$	14
MgF_2	10
Mg_2SiO_4	11
$Na_4P_20_7$	11

Reference

Mayneord W.V., The significance of the Roentgen. Acta Int. Union Against Cancer **2** (1937) 271

B

Basic equation of radiation dosimetry by thermoluminescence

A certain amount of the ionizing radiation energy absorbed by an insulating medium, i.e., a thermoluminescent material, provokes the excitation of electrons from the valence band (VB) to the conduction band (CB) of the material. The free electrons in the CB may be trapped at a site of crystalline imperfection (i.e., impurity atom, lattice vacancy, dislocation). The trapped electrons have a certain probability per unit of time, p, to be released back into the CB which depends on the temperature (T) and on the activation energy (E). This probability is given by the **Arrhenius equation** rewritten as

$$p = s\exp\left(-\frac{E}{kT}\right) \tag{1}$$

where s is a constant for each kind of insulator, called frequency factor, in (s^{-1}), E is the activation energy, called trap depth, in (eV), given as a difference between the trap level and the bottom of the CB, k is the Boltzmann's constant ($0.862\text{x}10^{-4}$ eV/K), T is the temperature in (K).

By heating of the sample, the filled traps can be evacuated by thermal stimulation of the trapped electrons which rise to the CB. From here the free electrons have a certain probability to recombine with a hole at some sites, called luminescent or recombination centres. The recombination event results in the emission of visible light. This emission of light is called TL glow curve which is formed, in general, by some peaks. Each peak reflects a trap type having a defined activation energy. The wavelenght spectra of the emitted light gives information about the recombination centres.

Let us define N as the concentration of empty traps in the material. During irradiation at a dose rate dD/dt the filled traps are

$$N_f = N - n \tag{2}$$

where n is the concentration of the remaining empty traps. So the rate of decrease of n can be written as

$$\frac{dn}{dt} = A \cdot n\frac{dD}{dt} \tag{3}$$

where A is a constant of the material, called radiation suscetibility.

Making the assumption that no trapped electrons are thermally released during the irradiation (i.e., the filled traps are deep enough to resist to a thermal drainage), Eq. (2) can be integrated as it follows, with the initial condition that at $t=0$, $n=N$

$$\int_N^n \frac{dn}{n} = \int_0^t -A\frac{dD}{dt}dt$$

from which

$$n = N\exp(-A\cdot D) \tag{4}$$

where D is the total irradiation dose received by the material during the irradiation time t.

It is now possible to define the constant A considering that if $D_{1/2}$ is the radiation dose needed to fill half of the empty traps, from Eq. (3) we obatin

$$A = \frac{0.693}{D_{1/2}}$$

The filled traps at the end of the irradiation is given by

$$N_f = N\left[1-\exp(-A\cdot D)\right]$$

The heating phase of the irradiated sample, for obtained thermoluminescence, can be expressed as following

$$-\frac{dN_f}{dt} = p\cdot N_f = N_f\cdot s\cdot\exp\left(-\frac{E}{kT}\right)$$

and the intensity of thermoluminescence, $I(D,T)$, is then given by

$$I(D,T) = -C\frac{dN_f}{dt} = C\cdot s\cdot N\left[1-\exp(-A\cdot D)\right]\exp\left(-\frac{E}{kT}\right) \tag{5}$$

If $A\cdot D \prec 1$, for small values of D, $1-\exp(-A\cdot D)$ can be approximated to AD and then Eq. (5) becomes

$$I(D,T) = C\cdot s\cdot N\cdot A\cdot D\cdot\exp\left(-\frac{E}{kT}\right) \tag{6}$$

from which it is easily observed that the TL intensity at a given temperature, i.e., the glow peak temperature, is proportional to the received dose D.

Batch of TLDs

A batch of TLDs is defined as the whole number of dosimeters of the same kind of material and **activator(s)**, as obtained from the manufacturer, having the same thermal and irradiation history and, possibly, produced at the same time (this last requirement is not imperative). Before using a new batch of TLDs, it has to be submitted to an **initialization procedure**.

Braunlich-Scharman model

A more satisfactory physical interpretation of the TL kinetics can be base on a more complex description of the TL centers in the forbidden gap. Braunlich and Scharmann (1966), wrote a set of differential equations describing the traffic of the charge carriers, during the thermal excitation, making reference to the energy level scheme proposed by Schon. This scheme contains one electron trap, one hole trap and retrapping transitions of the freed carriers back into their respective traps. The following Fig. 1 shows the band model used to describe the traffic of the carriers.

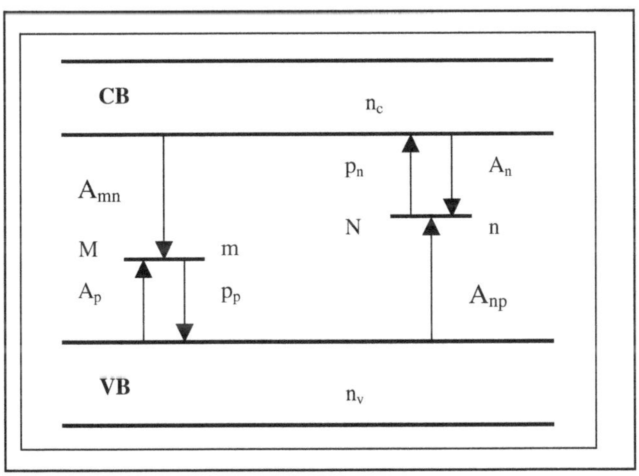

Fig. 1. The energy level scheme proposed by Schon.

Explanation of the symbols:

- n_c = concentration of electrons in CB,
- n_v = concentration of holes in VB,
- n = concentration of trapped electrons,

- N = concentration of electron traps,
- m = concentration of trapped holes,
- M = concentration of recombination centers (hole traps),
- A_n = retrapping probability for electrons in N,
- A_{mm} = recombination probability for electrons in M,
- A_p = retrapping probability for holes in M,
- A_{np} = recombination probability for holes in N,
- $p_n = s_n \exp\left(-\frac{E_n}{kT}\right)$

is the thermal excitation probability for electrons from N to CB,

- $p_p = s_p \exp\left(-\frac{E_p}{kT}\right)$

is the thermal excitation probability for holes from M to VB,

- E_n = electron trap activation energy,
- Ep = hole trap activation energy.

The set of the differential equations is:

$$\frac{dn_c}{dt} = np_n - n_c A_n (N-n) - n_c m A_{mn} \tag{1a}$$

$$\frac{dn_v}{dt} = mp_p - n_v A_p (M-n) - n_v n A_{np} \tag{1b}$$

$$\frac{dn}{dt} = -np_n + n_c A_n (N-n) - n_v n A_{np} \tag{1c}$$

$$\frac{dm}{dt} = -mp_p + n_v A_p (M-m) - n_c m A_{mn} \tag{1d}$$

Considering that, in the most general case, both recombination transitions are radiative, the total TL intensity is given by

$$I = -\frac{dn_c}{dt} - \frac{dn}{dt} = n_c m A_{mn} + n_v n A_{np} \tag{2}$$

Writing the previous equation for the intensity, it has been considered that the transitions of conduction electrons into traps and of holes from the valence band into recombination centers (hole traps) are non radiative.

Two parameters have now to be defined:

$$R_n = \frac{A_n}{A_{mn}} \tag{3a}$$

$$R_m = \frac{A_p}{A_{np}} \tag{3b}$$

which express the ratio of the retrapping probabilities compared to recombination for both electrons and holes.

The neutrality condition is given by

$$n_c + n = n_v + m \tag{4}$$

and furthermore, with the assumptions that

$$n_c \prec\prec n, \qquad n_v \prec\prec m \tag{5}$$

the following relation is also valid:

$$n \approx m \tag{6}$$

Four cases can be now analyzed:

$$\begin{array}{lll} a) & R_n \approx 0, & R_m \approx 0 \\ b) & R_n \succ\succ 1, & R_m \succ\succ 1 \\ c) & R_n \approx 0, & R_m \succ\succ 1 \\ d) & R_n \succ\succ 1, & R_m \approx 0 \end{array}$$

Case a) concerns a situation where recombination prevails over trapping, in case b) retrapping prevails over recombination and the two others cases are intermediate.

The quasi-equilibrium assumption is valid for both electrons and holes:

$$\frac{dn_c}{dt} - \frac{dn_v}{dt} \approx 0 \tag{7}$$

Case (a)

The retrapping rate for both electrons and holes is very small. Then the retrapping terms can be neglected. Furthermore, taking into account the quasi-equilibrium condition the previous Eqs. (1a,b,c,d) become

$$\frac{dn_c}{dt} = np_n - n_c m A_{mn} \tag{8a}$$

$$\frac{dn_v}{dt} = mp_p - n_v n A_{np} \tag{8b}$$

$$\frac{dn}{dt} = -np_n - n_v n A_{np} \tag{8c}$$

$$\frac{dm}{dt} = -mp_p - n_c m A_{mn} \tag{8d}$$

Because n ≈ m, from Eqs. (8a) and (8b) we obtain, taking into account relation (7)

$$n_c \approx \frac{p_n}{A_{mn}} \tag{9}$$

$$n_v \approx \frac{p_p}{A_{np}} \tag{10}$$

Eq. (8c) then reduces to

$$\frac{dn}{dt} = -n\left(p_n + p_p\right) \tag{11}$$

Considering a constant rate $\beta = dT/dt$, Eq. (11) becomes

$$\frac{dn}{n} = -\frac{p_n + p_p}{\beta} dT \tag{12}$$

Integration of Eq. (12) yields

$$n = n_0 \exp\left\{-\frac{1}{\beta}\int_{T_0}^{T}\left[p_n + p_p\right]dT'\right\} \tag{13}$$

Going back to Eq. (2), it can be rewritten, using Eqs. (9) and (10), as:

$$I = p_n m + p_p n \tag{14}$$

and using the relation (6) $n \approx m$

$$I = n(p_n + p_p) \tag{15}$$

which can be rewritten, using Eq. (13)

$$I = n_0 \left[p_n + p_p \right] \exp\left\{ -\frac{1}{\beta} \int_{T_0}^{T} \left[p_n + p_p \right] dT' \right\} \tag{16}$$

which is similar to the **Randall-Wilkins first order equation**. Neglecting the transitions to the valence band, i.e. $p_p = 0$, the Randall-Wilkins equation is obtained.

Case (b)

The retrapping of charge carriers prevails over the recombination transitions.

Eqs. (1a,b,c,d) become now

$$\frac{dn_c}{dt} = np_n - n_c A_n (N - n) \tag{17a}$$

$$\frac{dn_v}{dt} = mp_p - n_v A_p (M - n) \tag{17b}$$

$$\frac{dn}{dt} = -np_n + n_c A_n (N - n) - n_v n A_{np} \tag{17c}$$

$$\frac{dm}{dt} = -mp_p + n_v A_p (M - m) - n_c m A_{mn} \tag{17d}$$

Using now the quasi-equilibrium condition, i.e.

$$\frac{dn_c}{dt} = \frac{dn_v}{dt} \approx 0 \tag{18}$$

and the neutrality condition in the form

$$\frac{dn}{dt} \approx \frac{dm}{dt} \tag{19}$$

from Eqs. (1a,c) we get

$$\frac{dn_c}{dt}+\frac{dn}{dt}\approx\frac{dn}{dt}=-A_{mn}n_c m-A_{np}n_v n \tag{20}$$

which becomes, using $n\approx m$,

$$\frac{dn}{dt}=-n(n_c A_{mn}+n_v A_{np}) \tag{21}$$

Eq. (17a) becomes

$$np_n-n_c A_n(N-n)\approx 0 \tag{22}$$

Because $n << N$ (far from saturation), Eq. (22) gives

$$n_c\approx\frac{np_n}{A_n N} \tag{23}$$

Similarly, considering $m << M$, we obtain for n_v

$$n_v\approx\frac{mp_p}{A_p M} \tag{24}$$

Substituting expressions (23) and (24) in Eq. (21), we obtain, using $n\approx m$:

$$\frac{dn}{dt}\approx-n^2\left(\frac{p_n A_{mn}}{A_n N}+\frac{p_p A_{np}}{A_p M}\right) \tag{25}$$

Using as before a linear heating rate, we get by integration

$$n=\frac{1}{\left\{\frac{1}{n_0}+\frac{1}{\beta}\int_{T_0}^{T}\left[\left(\frac{A_{mn}p_n}{A_n N}\right)+\left(\frac{A_{np}p_p}{A_p M}\right)\right]dT'\right\}} \tag{26}$$

In conclusion, the TL intensity is given by

$$I=-\frac{dn}{dt}=-\frac{1}{\left\{\frac{1}{n_0}+\frac{1}{\beta}\int_{T_0}^{T}\left[\left(\frac{A_{mn}p_n}{A_n N}\right)+\left(\frac{A_{np}p_p}{A_p M}\right)\right]dT'\right\}^2}\left(\frac{p_n A_{mn}}{A_n N}+\frac{p_p A_{np}}{A_p M}\right) \tag{27}$$

This equation is similar to the second order equation given by Garlick and Gibson. It becomes identical to it by neglecting the probability for transitions into the valence band, i.e. supposing $p_p = 0$

Case (c)

The new equations are now:

$$\frac{dn_c}{dt} = np_n - n_c m A_{mn} \tag{28a}$$

$$\frac{dn_v}{dt} = mp_p - n_v A_p (M - m) \tag{28b}$$

$$\frac{dn}{dt} = -np_n + n_c A_n (N - n) - n_v n A_{np} \tag{28c}$$

$$\frac{dn}{dt} = -mp_p + n_v A_p (M - m) - n_c m A_{mn} \tag{28d}$$

From Eqs. (28a) and (28b) we get

$$n_c \approx \frac{p_n}{A_{mn}}, \quad n_v = \frac{mp_p}{A_p (M - m)} \tag{29}$$

From (29) and (28d) we obtain

$$\frac{dm}{dt} = -mp_n \tag{30}$$

and then

$$m \approx m_0 \exp\left(-\frac{1}{\beta} \int_{T_0}^{T} p_n dT' \right) \tag{31}$$

The thermoluminescence intensity is

$$I = n_c m A_{mn} + n_v n A_{np} \tag{32}$$

which transforms in, using (29), $n \approx m$ and $M >> m$

$$I = mp_n + m^2 \frac{p_p A_{np}}{A_p M} \tag{33}$$

and then, the explicit form for I is the following

$$I = m_0 p_n \exp\left(-\frac{1}{\beta}\int_{T_0}^{T} p_n dT'\right) + \frac{m_0^2 p_p A_{np}}{A_p M}\exp\left(-\frac{2}{\beta}\int_{T_0}^{T} p_n dT'\right) \tag{34}$$

which is again the Randall-Wilkins equation for $p_p = 0$.

Case (d)

Eqs. (1a,b,c,d) reduce to

$$\frac{dn_c}{dt} = np_n - n_c A_n (N - n) \tag{35a}$$

$$\frac{dn_v}{dt} = mp_p - n_v n A_{np} \tag{35b}$$

$$\frac{dn}{dt} = -np_n + n_c A_n (N - n) - n_v n A_{np} \tag{35c}$$

$$\frac{dm}{dt} = -mp_p + n_v A_p (M - m) - n_c m A_{mn} \tag{35d}$$

Assuming the quasi-equilibrium condition, i.e.

$$\frac{dn_c}{dt} \approx 0, \qquad \frac{dn_v}{dt} \approx 0 \tag{36}$$

and $N >> n$, n_v and n_c very small, i.e. $m \approx n$, we get from (35a)

$$n_c \approx \frac{np_n}{A_n N} \tag{37}$$

and from (35b)

$$n_v \approx \frac{p_p}{A_{np}} \tag{38}$$

Then

$$\frac{dn}{dt} = -np_n + n_c A_n (N - n) - n_v n A_{np} = -\frac{dn_c}{dt} - \frac{p_p n A_{np}}{A_{np}} = -np_p \tag{39}$$

from which, by integration

$$n = n_0 \exp\left(-\frac{1}{\beta} \int_{T_0}^{T} p_p dT' \right) \tag{40}$$

The TL emission is then given by

$$I = n_c m A_{mn} + n_v n A_{np} \tag{41}$$

which transforms, using approximations (37), (38) and $n \approx m$, in the following expression

$$I = \frac{n^2 p_n A_{mn}}{A_n N} + np_p \tag{42}$$

Using Eq. (40), we get the final expression for the intensity:

$$I = \frac{n_0^2 A_{mn} p_n}{NA_n} \exp\left(-\frac{2}{\beta} \int_{T_0}^{T} p_p dT' \right) + n_0 p_p \exp\left(-\frac{1}{\beta} \int_{T_0}^{T} p_p dT' \right) \tag{43}$$

This equation becomes again the **Randall-Wilkins equation** of the first order, ignoring the thermal release of trapped electrons, i.e. $p_n = 0$.

Reference

Bräunlich P. and Scharmann A., Phys. Stat. Sol. **18** (1966) 307

C

Calcium fluoride (CaF_2)

CaF_2, activated by various dopants, is a TL phophor widely used in many dosimetric applications. It is used as natural CaF_2 or with different activators as Mn, Dy and Tm [1-10].

Preparation of CaF_2:Mn is carried out using the precipitation technique from a solution of $CaCl_2$ and $MnCl_2$ in NH_4F. The precipitate is dried and heated in oven with inert atmosphere at 1200°C, then it is powdered and graded. The final material can be pressed and sintered.

Its atomic number is 16.57. Its sensitivity at the 30 keV of photon energy is 15 times greater than the sensitivity at the ^{60}Co energy.

The linearity of CaF_2 natural is observed up to 50 Gy. CaF_2:Mn, produced by Harshaw under the name TLD-400, gives a linear response up to 2 KGy. CaF_2:Dy has been commercialized by Harshaw under the name TLD-200; it presents a complicate glow curve consisting in six peaks. The TL response is linear up to 1 KGy. CaF_2:Tm, known as TLD-300, shows three resolved peaks, high stability and selective peak sensitivity to the radiation quality.

References

1. Schayes R. and Brooke C., Rev. MBLE **6** (1963) 24
2. Ginther R.J., CONF 650637 (1965)
3. Binder W., Disterhoft S. and Cameron J.R., Proc. 2nd Int. Conf. Lumin. Dos., Gatlinburg (USA), 1968
4. Furetta C. and Lee Y.K., Rad. Prot. Dos. **5(1)** (1983) 57
5. Furetta C., Lee Y.K. and Tuyn J.W.N., Int. J. Appl. Rad. Isot. **36(11)** (1985) 896
6. Furetta C. and Tuyn J.W.N., Rad. Prot. Dos. **11(4)** (1985) 893
7. Furetta C. and Lee K.Y., Rad. Prot. Dos. **11(2)** (1985) 101
8. Furetta C. and Tuyn J.W.N., Int. J. Appl. Rad. Isot. **36(12)** (1985) 1000
9. Furetta C. and Tuyn J.W.N., Int. J. Appl. Rad. Isot. **17(5)** (1986) 458
10. Azorin-Nieto J., Furetta C. and Gutierrez A., J. Phys. D: Appl. Phys. **22** (1989) 458

Calibration factor F_c (definition)

The so called calibration factor, F_c, allows to translate the TL emission from a given phophor to the dose received by the phosphor itself. This factor includes both reader and dosemeter properties.

Many experiments carried out in the field of thermoluminescent dosimetry have well demonstrated that a reduction of uncertainties in the dose determination can be attained using a calibration factor of the dosimetric system.

At first we can introduce an individual calibration factor, F_{ci}, defined for a given quality of the calibration beam. Therefore, an unknown dose D is given by the following relation

$$D = F_{ci} \cdot M_{i,net} \tag{1}$$

where
D is the unknown absorbed dose and,
$M_{i,net}$ is the TL signal, corrected by background, of the ith dosimeter.

The experimental determination of the calibration factor can be carried out in principle in two different ways, according to the methodologies. The first method consists in the determination of a single value of the calibration factor, delivering to TLD a calibration dose D_c which is chosen in the linear region of the TL response of the material used. The second approach consists in determining a calibration curve, obtained with three or more points of dose, always in the linear region.

Calibration factor F_c (procedures)

1st procedure

Let us show now the first procedure consisting in the determination of only one calibration factor. In this case it is necessary to introduce a group of reference dosemeters ($m \geq 10$), belonging to the same batch of the field dosemeters. As stated before, it is very important that the reference and the field dosemeters have the same thermal and irradiation history.

The reference dosemeters have to be prepared and then irradiated with a calibration dose (for every dosemeter the intrinsic background is known), D_c, chosen in the linear range of the TL response. From the calibration factor definition

$$D = F_{ci} \cdot M_{i,net} \tag{1}$$

we obtain the following expression:

$$F_{c,r} = \frac{D_c}{\frac{1}{m}\sum_{r=1}^{m} M_{r,net} \cdot S_r} \tag{2}$$

where S_r is the intrinsic sensitivity factor.

Using Eq. (2), an unknown dose D will be given by

$$D = M_{i,net} \cdot S_i \cdot F_{c,r} \tag{3}$$

Comparing Eqs. (1) and (3) one can observe that

$$F_{c,i} = S_i \cdot F_{c,r} \tag{4}$$

The previous relation means that the individual calibration factor $F_{c,i}$ is depending on two different quantities: the first one is the relative intrinsic sensitivity S_i, which is quite stable during time and then it has to be checked no more than two times per year; the second quantity is the calibration factor $F_{c,r}$, obtained using the reference dosemeters, whose response can vary tremendously from a reading cycle to another because the delay between the moment of $F_{c,r}$ determination and the period of field TLDs measurements, which means that any instability in the reader electronic, for instance due to environmental variations and/or different periods of switch-off/switch-on of the reader, is not taken into account. It has been proved that the $F_{c,r}$ factor can vary a lot along a period of a few months and provoke large errors in the dose determination. Then it is recommended to check the calibration factor before any reading session.

In the case of radiotherapy measurements where the accuracy in the dose determination must be within 2 or 3%, it is imperative to determine the $F_{c,r}$ factor just before a cycle of TLDs readings. In this case, the reference dosemeters are irradiated to the proper value of calibration dose and read togheter with the field dosemeters to avoid any effect of the TLD system instability.

2nd procedure

The second procedure consists in getting a calibration curve at each reading session. The calibration curve is obtained using three or more points of dose.

The procedure is the following.

- choose three different values of dose in the linear range, possibly in a logarithmic scale, noted here as $D_{c,1}$, $D_{c,2}$ and $D_{c,3}$.
- prepare a group of reference dosemeters, at least 5 for each level of dose, and irradiate them.
- read all the dosemeters and correct the readings for background and relative intrinsic correction factor.
- the 5 corrected readings corresponding to the dose $D_{c,1}$ are then averaged. - call these averaged values as

$$\overline{M}_{c,1},\quad \overline{M}_{c,2},\quad \overline{M}_{c,3} \tag{5}$$

- for each value and each dose one obtains

$$F_{c,1} = \frac{D_{c,1}}{\overline{M}_{c,1}},\quad F_{c,2} = \frac{D_{c,2}}{\overline{M}_{c,2}},\quad F_{c,3} = \frac{D_{c,3}}{\overline{M}_{c,3}} \tag{6}$$

with the condition

$$F_{c,1} \cong F_{c,2} \cong F_{c,3} \tag{7}$$

The previous suggested procedures for the determination of the calibration factor must be, in principle, repeated at each reading session. In this way the possible variations in the efficiency of the TLD reader are neglected. However, the stability of the system has to be checked periodically for detecting any possible variation due to environmental conditions and/or related to the reader itself.

Calibration procedure for a batch of TLDs

First procedure

In this procedure, sensitivity differences between samples of the batch are not considered.

- All TLDs are submitted to the appropriate annealing procedure to standardise their sensitivity and background levels
- Read out all the annealed samples to evaluate the mean reading of the background, $\overline{B}$
- Irradiate the TLDs to a calibration dose D_C very much greater than the expected standard deviation of the mean equivalent dose of the background
- Readout all the irradiate TLDs to evaluate the mean reading of the batch, $\overline{M}_C$
- The quantity $\overline{M}_C - \overline{B}$ is the net reading corresponding to the calibration dose D_C
- Some of the TLDs are now annealed again and then irradiated with an unknown dose D_X and then readout. Let the mean reading be $\overline{M}_X$
- The unknown dose D_X is then given by

$$D_X = \frac{\overline{M}_X - \overline{B}}{\overline{M}_C - \overline{B}} \cdot D_C$$

Second procedure

In this procedure differences in sensitivity are taken into account to improve the accuracy of the dosimetric system.

- Provide a special container for the identification of the dosemeter
- All the dosemeters are submitted to the appropriate annealing procedure
- Readout the dosemeter to evaluate the background of each sample, i.e. B_{0i}, with $i = 1, 2, ..., n$ (individual numbers of each dosemeter)

- Irradiate the samples to a given calibration dose D_C and then readout to evaluate the corresponding readings, $M_{C,i}$
- Calculate the net reading for each sample

$$M_{C,i(net)} = M_{C,i} - B_{0i}$$

- Calculate the average of the net readings, $\overline{M}_C$
- Calculate the individual correction factors as

$$S_i = \frac{\overline{M}_C}{M_{C,i(net)}}$$

  This factor is associated to the proper dosemeter multiplying it to the net reading to correct the dosemeter response at any given dose. S_i should remain constant during the use of the dosemeters; any way it should be checked periodically to take into account any possible variations.
- Some of the TLDs are now annealed again, irradiated to an unknown dose D_X and then readout:

$$M_{X,i(corr)} = \left(M_{X,i} - M_{0,i}\right) \cdot S_i$$

- Considered that the given dose is in the linear region of the thermoluminescent batch response, the unknown dose D_X is given by

$$D_x = \frac{M_{X,i(corr)}}{\overline{M}_C} \cdot D_C$$

Competition

Various traps (**competitors**) may be in competition among them for trapping the free carriers produced during irradiation or heating. The process of competion has been used to explain the enhancement of the TL sensitivity and then the phenomenon of supralinearity [1-4].

Fig. 1 shows the competition during irradiation of the TL sample and Fig. 2 shows the mechanism of competition during heating.

During heating (readout), the electrons released from N_1 could be retrapped in N_2 or recombine in M. At higher dose levels, N_2 could saturate and then the released electrons can be involved in the recombination process.

Both models have been also used, among other models, to explain the supralinearity phenomenon.

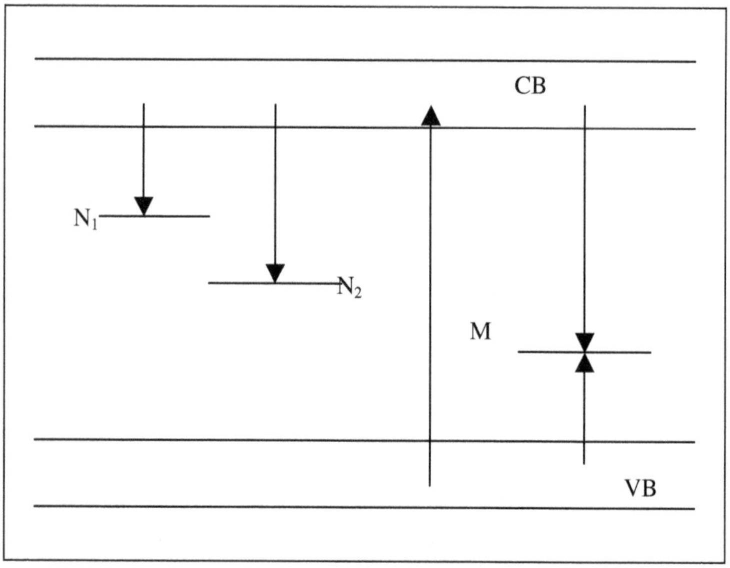

Fig. 1. Competition during irradiation. N_1 = active trap (TL signal), N_2 = competing trap having a trapping probability larger than that of N_1, M = recombination center.

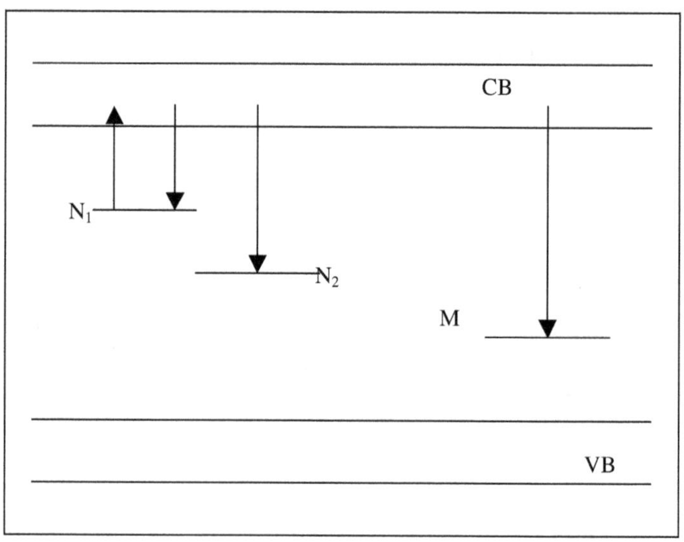

Fig. 2. Competition during heating.

References

1. Suntharalingam N. and Cameron J.R., Report COO-1105-130, USAEC (1967)
2. Aitken M.J., Thompson J. and Fleming S.J. in 2nd Conf. Lumin. Dosim., Gattlinburg, Tennessee (1968)
3. Kristianpoller N., Chen R. and Israeli M., J. Phys. D: Appl. Phys. **7** (1974) 1063
4. Chen R., Yang X.H. and McKeever S.W.S., J. Phys. D: Appl. Phys. **21** (1988) 1452

Competitors

The term competitors indicate traps which are in **competition** over free carriers during irradiation or heating the thermoluminescent samples

Computerised glow curve deconvolution (CGCD): Kitis' expressions

The computerised glow curve deconvolution (CGCD) analysis has been widely applied since 1980 to resolve a complex thermoluminescent glow curve into individual peak components. Once each component is determined, the trapping parameters, activation energy and frequency factor, can be evaluated.

The main problem is that the basic TL kinetics equations, i.e. the Randall-Wikins equation for the first-order kinetics, and the Garlick-Gibson equation for the second-order, give the glow peak TL intensity, *I*, as a function of various parameters:

$$I = I\left(n_0, E, s, T\right) \tag{1}$$

where

$n_0 =$ initial concentration of trapped electrons (cm^{-3})

$E =$ activation energy (eV)

$s =$ frequency factor (s^{-1})

$T =$ absolute temperature (K)

The values of n_0 and s are not known.

Some approximated functions have been proposed for resolving a composite glow curve into its components: i.e. Podgorsak-Moran-Cameron approximation [1], Gaussian peak shape, Asymmetric Gaussian functions and others reviewed by Horowitz and Yossian [2].

From an historical point of view, the PMC approximation was the first. Although it was found that the approximation of PMC function is rather poor, it is the only one which transforms Eq. (1) into the following

$$I = I(I_M, E, T_M, T) \tag{2}$$

where I_M and T_M are the TL intensity and temperature at the glow peak maximum.

The advantage of Eq. (2) is evident: infact it has two only free parameters, namely I_M and T_M , which are directly obtained from the experimental glow curve.

Kitis [3,4] has proposed new functions for describing a glow peak which, keeping the advantage of the PMC equation, have the same accuracy of the basic TL kinetic equations.

First-order expression

The TL intensity of a single glow peak following a first-order process is given by the equation

$$I(T) = sn_0 \exp\left(-\frac{E}{kT}\right)\exp\left[-\frac{s}{\beta}\int_{T_0}^{T}\exp\left(-\frac{E}{kT'}\right)dT'\right] \tag{3}$$

The integral comparing in Eq. (3) cannot be solved in an analytical form, but using successive integration by parts, in a second-order approximation (**integral approximation**) it becomes

$$\int_{T_0}^{T}\exp\left(-\frac{E}{kT'}\right)dT' = \frac{kT^2}{E}\left(1-\frac{2kT}{E}\right)\exp\left(-\frac{E}{kT}\right) \tag{4}$$

Hence, Eq. (3) becomes

$$I(T) = sn_0 \exp\left(-\frac{E}{kT}\right)\exp\left[-\frac{skT^2}{\beta E}\left(1-\frac{2kT}{E}\right)\exp\left(-\frac{E}{kT}\right)\right] \tag{5}$$

from which the **condition at the maximum** is given as

$$\frac{\beta E}{kT_M^2} = s\exp\left(-\frac{E}{kT_M}\right) \tag{6}$$

or

$$s = \frac{\beta E}{kT_M^2}\exp\left(\frac{E}{kT_M}\right) \tag{7}$$

- Inserting Eq. (6) into Eq. (5) one obtains

$$I_M = \frac{n_0 \beta E}{kT_M^2}\exp\left[-\left(1-\frac{2kT_M}{E}\right)\right]$$

or better

$$I_M = \frac{n_0 \beta E}{kT_M^2} \exp\left[-(1-\Delta_M)\right] \tag{8}$$

where $\Delta_M = \frac{2kT_M}{E}$.

Eq. (8) can be rewritten as

$$\frac{n_0 \beta E}{kT_M^2} = I_M \exp(1-\Delta_M) \tag{9}$$

- Inserting Eq. (7) in Eq. (5), after a little algebra, one obtains

$$I(T) = \frac{n_0 \beta E}{kT_M^2} \exp\left[\frac{E}{kT}\left(\frac{T-T_M}{T_M}\right)\right] \times$$
$$\times \exp\left\{-\frac{T^2}{T_M^2}(1-\Delta)\exp\left[\frac{E}{kT}\left(\frac{T-T_M}{T_M}\right)\right]\right\} \tag{10}$$

with $\Delta = \frac{2kT}{E}$

- Eq. (9) can be now inserted into Eq. (10) for getting the final expression of the form $I(I_M, E, T_M, T)$:

$$I(T) = I_M \exp\left[1 + \frac{E}{kT}\cdot\frac{T-T_M}{T_M} - \frac{T^2}{T_M^2}\cdot(1-\Delta_M)\exp\left(\frac{E}{kT}\cdot\frac{T-T_M}{T_M}\right) - \Delta_M\right] \tag{11}$$

<u>*Second order expression*</u>

The second-order kinetic equation is

$$I(T) = sn_0 \exp\left(-\frac{E}{kT}\right)\left[1 + \frac{s}{\beta}\int_{T_0}^{T} \exp\left(-\frac{E}{kT'}\right) dT'\right]^{-2} \tag{12}$$

Inserting the integral approximation given by (4) in Eq. (12), one gets

$$I(T) = sn_0 \exp\left(-\frac{E}{kT}\right)\left[\frac{skT^2}{\beta E}(1-\Delta)\exp\left(-\frac{E}{kT}\right) + 1\right]^{-2} \tag{13}$$

from which the condition at the maximum is given by

$$s = \frac{\beta E}{kT_M^2} \cdot \frac{1}{1+\Delta_M} \exp\left(\frac{E}{kT_M}\right) \tag{14}$$

or in another form

$$s \exp\left(-\frac{E}{kT_M}\right) = \frac{\beta E}{kT_M^2} \cdot \frac{1}{1+\Delta_M} \tag{15}$$

Furthermore, Eq. (13) can be rewritten for the peak at the maximum:

$$I_M = sn_0 \exp\left(-\frac{E}{kT_M}\right)\left[\frac{skT_M^2}{\beta E}(1-\Delta_M)\exp\left(-\frac{E}{kT_M}\right)+1\right]^{-2} \tag{16}$$

- The insertion of Eq. (14) into Eq. (13) gives the following expression for $I(T)$:

$$I(T) = \frac{n_0 \beta E}{kT_M^2} \cdot \frac{1}{1+\Delta_M} \exp\left[\frac{E}{kT} \cdot \left(\frac{T-T_M}{T_M}\right)\right] \times$$
$$\times \left\{\frac{T^2}{T_M^2} \cdot \frac{1-\Delta}{1+\Delta_M} \exp\left[\frac{E}{kT} \cdot \left(\frac{T-T_M}{T_M}\right)\right]+1\right\}^{-2} \tag{17}$$

- Inserting Eq. (15) into Eq. (16) we get a more semplified expression for I_M :

$$I_M = \frac{n_0 \beta E}{kT_M^2} \cdot \frac{1}{1+\Delta_M} \cdot \left(\frac{2}{1+\Delta_M}\right)^2$$

which can be rewritten as

$$\frac{n_0 \beta E}{kT_M^2} \cdot \frac{1}{1+\Delta_M} = I_M \left(\frac{2}{1+\Delta_M}\right)^2 \tag{18}$$

- Eq. (18) is now inserting into Eq. (17) for getting the final expression for the TL intensity:

$$I(T)=4I_M\exp\left(\frac{E}{kT}\cdot\frac{T-T_M}{T_M}\right)\times$$
$$\times\left[\frac{T^2}{T_M^2}(1-\Delta)\exp\left(\frac{E}{kT}\cdot\frac{T-T_M}{T_M}\right)+1+\Delta_M\right]^2 \tag{19}$$

General order

The equation of the TL intensity for a glow peak following a general-order process is:

$$I(T)=sn_0\exp\left(-\frac{E}{kT}\right)\left[1+\frac{s(b-1)}{\beta}\int_{T_o}^{T}\exp\left(-\frac{E}{kT'}\right)dT'\right]^{-\frac{b}{b-1}} \tag{20}$$

It transforms in the following equation using the approximation (4):

$$I(T)=sn_0\exp\left(-\frac{E}{kT}\right)\left[\frac{(b-1)\,skT^2}{\beta E}\cdot(1-\Delta)\exp\left(-\frac{E}{kT}\right)+1\right]^{-\frac{b}{b-1}} \tag{21}$$

The intensity at the peak maximum is then given by

$$I_M=sn_0\exp\left(-\frac{E}{kT}\right)\left[\frac{(b-1)\,skT_M^2}{\beta E}(1-\Delta_M)\exp\left(-\frac{E}{kT_M}\right)+1\right]^{-\frac{b}{b-1}} \tag{22}$$

The maximum condition, obtained from (21), is

$$\frac{\beta E}{kT_M^2}=Z_M s\exp\left(-\frac{E}{kT_M}\right) \tag{23}$$

with

$$Z_M=1+(b-1)\Delta_M \tag{24}$$

Eq. (23) can be rewritten in two different ways:

$$s=\frac{\beta E}{kT_M^2}\cdot\frac{1}{Z_M}\exp\left(\frac{E}{kT_M}\right) \tag{25}$$

or

$$s\exp\left(-\frac{E}{kT_M}\right)=\frac{\beta E}{kT_M^2 Z_M} \tag{26}$$

- Inserting Eq. (25) in Eq. (21) we obtain the following expression for the intensity:

$$I(T)=\frac{n_0\beta E}{kT_M^2 Z_M}\exp\left(\frac{E}{kT}\cdot\frac{T-T_M}{T_M}\right)\times$$
$$\times\left[\frac{b-1}{Z_M}\cdot\frac{T^2}{T_M^2}(1-\Delta)\exp\left(\frac{E}{kT}\cdot\frac{T-T_M}{T_M}\right)+1\right]^{-\frac{b}{b-1}} \tag{27}$$

- Inserting Eq. (27) in Eq. (22) we get, after arrangement, the expression for the intensity at the maximum:

$$I_M=\frac{n_o\beta E}{kT_M^2 Z_M}\cdot\left(\frac{b}{Z_M}\right)^{-\frac{b}{b-1}} \tag{28}$$

from which

$$\frac{n_0\beta E}{kT_M^2 Z_M}=I_M\left(\frac{b}{Z_M}\right)^{-\frac{b}{b-1}} \tag{29}$$

- Insertion of Eq. (29) into Eq. (27) gives the final equation for the TL intensity:

$$I(T)=I_M\,(b)^{\frac{b}{b-1}}\exp\left(\frac{E}{kT}\cdot\frac{T-T_m}{T_M}\right)\times$$
$$\times\left[(b-1)(1-\Delta)\frac{T^2}{T_M^2}\exp\left(\frac{E}{kT}\cdot\frac{T-T_M}{T_M}\right)+Z_M\right]^{-\frac{b}{b-1}} \tag{30}$$

Eqs. (11), (19) and (30) are equations in the form $I(I_M,E,T_M,T)$ which has only two free parameters, I_M and T_M, directly obtained from the experimental peak.

A further develop allows to transform Eqs. (11), (19) and (30) from the $I(I_M,E,T_M,T)$ space into the $I(I_M,\omega,T_M,T)$ space, where $\omega=\frac{\Delta E}{2kT}$.

With the assumption $n_0 = 1$, Eq. (8), first-order, and Eq. (28), general-order, become

$$I_M = \frac{\beta E}{kT_M^2} \frac{1}{e} \exp(\Delta_M) \tag{31}$$

and

$$I_M = \frac{\beta E}{kT_M^2 Z_M} \left[\frac{b}{1+(b-1)\Delta_M} \right]^{-\frac{b}{b-1}} \tag{32}$$

The two quantities

$$c_1 = \frac{1}{e} \exp(\Delta_M) \tag{33}$$

and

$$c_b = \frac{1}{Z_M} \left[\frac{b}{1+(b-1)\Delta_M} \right]^{-\frac{b}{b-1}} \tag{34}$$

vary extremely slowly in a large range of both E, from 0.5 to 2.5 eV, and s, from 10^5 to 10^{25} s^{-1}, so that they can be considered as constants. In turn, Eqs. (31) and (32) assume the following general form

$$I_M = (c_1, c_b) \frac{\beta E}{kT_M^2} \tag{35}$$

Eq. (35) can be solved respect to the activation energy, giving

$$E = \frac{kT_M^2 I_M}{\beta(c_1, c_b)} \tag{36}$$

In the given range of E and s values, Kitis found that the quantity I_M/β can be expressed as

$$\frac{I_M}{\beta} - \frac{c_d}{\omega} \tag{37}$$

where c_d is practically constant.

So, Eq. (36) can be transformed in

$$E = c_f \frac{kT_M^2}{\omega} \tag{38}$$

where

$$c_f = \frac{c_d}{(c_1, c_b)}$$

It must be noted that Eq. (38) is equivalent to the Chen's peak shape formula based on the FWHM. c_f assumes in this case a mean value of 2.4.

Eqs. (11) and (30) can be transformed using Eq. (38) as it follows, using the substitutions:

$$\Delta = \frac{2kT}{E} = \frac{2T\omega}{c_f T_M^2}$$

$$\Delta_M = \frac{2\omega}{c_f T_M}$$

$$\frac{E}{kT} \cdot \frac{T - T_M}{T_M} = \frac{c_f T_M (T - T_M)}{T\omega} = u_f$$

So, Eq. (11) transforms in

$$I(T) = I_M \exp\left\{1 + u_f - \frac{T^2}{T_M^2}\left(1 - \frac{2T\omega}{c_f T_M^2}\right)\exp(u_f) - \frac{2\omega}{c_f T_M}\right\} \tag{39}$$

and Eq. (30) transforms in:

$$I(T) =$$

$$= I_M (b)^{\frac{b}{b-1}} \exp(u_f)\left[1 + (b-1)\left(1 - \frac{2T\omega}{c_f T_M^2}\right)\frac{T^2}{T_M^2}\exp(u_f) + (b-1)\frac{2\omega}{c_f T_M}\right]^{-\frac{b}{b-1}} \tag{40}$$

Kitis investigated the variations of c_b and c_f as a function of $\ln(s)$ and reported that, in case of first-order kinetics, pairs of E and s can be accepted if c_b, or c_f, are within the following limits

$$0.38 < c_b < 0.4$$

$$2.3 < c_f < 2.44$$

taking into account that the minimum value of c_f corresponds to the maximum for c_b.

References

1. Podgorsak E.B., Moran P.R. and Cameron J.R., Proc. 3rd Inter. Conf. On Luminescence Dosimetry, Riso, 11-14 October, 1971
2. Horowitz Y.S. and Yossian D., Rad. Prot. Dos. **60** (1995) (special issue)
3. Kitis G., Gomez-Ros J.M. and Tuyn J.W.N., J Phys. D: Appl. Phys. **31** (1998) 2636
4. Kitis G., J. Radionalyt. Nucl. Chem., **247**(3) (2001) 697

Condition at the maximum (first order)

An important relationship is obtained by the **first order equation**

$$I(T) = n_0 s \exp\left(-\frac{E}{kT}\right) \exp\left[-\frac{s}{\beta}\int_{T_0}^{T} \exp\left(-\frac{E}{kT'}\right) dT'\right] \tag{1}$$

by setting

$$\frac{dI}{dT} = 0 \quad \text{at} \quad T = t_M$$

For practical purposes, the logarithm derivative is considered:

$$\frac{d(\ln I)}{dT} = \frac{1}{T}\frac{dI}{dT}$$

From Eq. (1) we obtain

$$\ln(I) = \ln(n_0 s) - \frac{E}{kT} - \frac{s}{\beta}\int_{T_0}^{T} \exp\left(-\frac{E}{kT'}\right) dT'$$

then

$$\left[\frac{d(\ln I)}{dT}\right]_{T-T_m} = \frac{E}{kT_M^2} - \frac{s}{\beta}\exp\left(-\frac{E}{kT_M}\right) = 0$$

which yields to the expression

$$\frac{\beta E}{kT_M^2} = s\exp\left(-\frac{E}{kT_M}\right) \tag{2}$$

From Eq. (2), the frequency factor is easily determined

$$s = \frac{\beta \cdot E}{k \cdot T_M^2}\exp\left(\frac{E}{kT_M}\right) \qquad [\mathrm{sec}^{-1}] \tag{3}$$

Condition at the maximum (first order): remarks

From the equation at the maximum

$$\frac{\beta E}{kT_M^2} = s\exp\left(-\frac{E}{kT_M}\right) \tag{1}$$

we can obtain some interesting remarks:

- for a constant heating rate T_M shifts toward higher temperatures as E increases or s decreases;
- for a given trap (E and s are constant values) T_M shifts to higher temperatures as heating rate increases;
- T_M is independent of n_o.

Condition at the maximum (general order)

The condition of maximum emission for b-order kinetics can be looked from the general order equation:

$$I(T) = sn_0\exp\left(-\frac{E}{kT}\right)\left[1+\frac{s(b-1)}{\beta}\int_{T_0}^{T}\exp\left(-\frac{E}{kT'}\right)dT'\right]^{-\frac{b}{b-1}} \tag{1}$$

where $s = s'' n_0^{b-1}$ expressed in sec^{-1}.

The logarithm of $I(T)$ is:

$$\ln\left[I(T)\right] = \ln(sn_0) - \frac{E}{kT} - \frac{b}{b-1}\ln\left[1+\frac{s(b-1)}{\beta}\int_{T_0}^{T}\exp\left(-\frac{E}{kT'}\right)dT'\right]$$

then

$$\left.\frac{d(\ln I)}{dT}\right|_{T=T_M} =$$

$$= \frac{E}{kT_M^2} - \frac{b}{b-1}\left[1+\frac{s(b-1)}{\beta}\int_{T_0}^{T_M}\exp\left(-\frac{E}{kT'}\right)dT'\right]^{-1}\frac{s(b-1)}{\beta}\exp\left(-\frac{E}{kT_M}\right)=0$$

from which we obtain

$$\frac{kT_M^2 bs}{\beta E}\exp\left(-\frac{E}{kT_M}\right)=1+\frac{s(b-1)}{\beta}\int_{T_0}^{T_M}\exp\left(-\frac{E}{kT'}\right)dT' \tag{2}$$

From the last equation it is possible to obtain the expression for the pre-exponential factor. Rearranging Eq. (2), we obtain:

$$s = \left[\frac{kT_M^2 b\exp\left(-\frac{E}{kT_M}\right)}{\beta E} - \frac{(b-1)\int_{T_0}^{T_M}\exp\left(-\frac{E}{kT'}\right)dT'}{\beta}\right]^{-1}$$

Using the **integral approximation**, we get

$$s = \left[\frac{kT_M^2\exp\left(-\frac{E}{kT_M}\right)}{\beta E}\left(1+\frac{2kT_M(b-1)}{E}\right)\right]^{-1} \tag{3}$$

expressed in sec^{-1}.

Considering s'':

$$s'' = \frac{1}{n_0^{b-1}}\left[\frac{kT_M^2\exp\left(-\frac{E}{kT_M}\right)}{\beta E}\left(1+\frac{2kT_M(b-1)}{E}\right)\right]^{-1} \tag{4}$$

or

$$s'' = \frac{\beta E \exp\left(\frac{E}{kT_M}\right)}{n_o^{b-1} k T_M^2}\left[1+\frac{2kT_M(b-1)}{E}\right]^{-1} \tag{5}$$

which is expressed in $cm^{3(b-1)}$ sec^{-1}.

Condition at the maximum (second order)

The condition at the maximum is obtained by differentiating the second order equation for the intensity

$$I(T) = -\frac{dn}{dt} = n^2 s' \exp\left(-\frac{E}{kT}\right) = \frac{n_0^2 s' \exp\left(-\frac{E}{kT}\right)}{\left[1+\frac{s' n_0}{\beta}\int_{T_0}^{T} \exp\left(-\frac{E}{kT'}\right) dT'\right]^2} \tag{1}$$

by setting

$$\frac{dI}{dT} = 0 \qquad \text{at} \qquad t = T_M$$

As usual, the logarithm derivative is considered:

$$\ln(I) = \ln(n_0^2 s') - \frac{E}{kT} - 2\ln\left[1+\left(\frac{s' n_0}{\beta}\right)\int_{T_0}^{T} \exp\left(-\frac{E}{kT'}\right) dT'\right]$$

$$\left[d\left(\frac{\ln I}{dT}\right)\right]_{T=T_M} = \frac{E}{kT_M^2} - 2\frac{\frac{s' n_0}{\beta}\exp\left(-\frac{E}{kT_M}\right)}{1+\frac{s' n_0}{\beta}\int_{T_0}^{T_M} \exp\left(-\frac{E}{kT'}\right) dT'} = 0$$

Then

$$\frac{E}{kT_M^2} = 2\frac{\frac{s' n_0}{\beta}\exp\left(-\frac{E}{kT_M}\right)}{1+\frac{s' n_0}{\beta}\int_{T_0}^{T_M} \exp\left(-\frac{E}{kT'}\right) dT'}$$

and rearranging

$$\frac{\beta E}{2kT_M^2}\left[1+\frac{s'n_0}{\beta}\int_{T_0}^{T_M}\exp\left(-\frac{E}{kT'}\right)dT'\right]=s'n_0\exp\left(-\frac{E}{kT_M}\right) \tag{2}$$

From this expression, the pre-exponential factor can be determined:

$$s'=\frac{1}{n_0}\left[\frac{2kT_M^2\exp\left(-\frac{E}{kT_M}\right)}{\beta\cdot E}-\frac{\int_{T_0}^{T_M}\exp\left(-\frac{E}{kT'}\right)dT'}{\beta}\right]^{-1}$$

using the integral approximation, the previous expression becomes

$$s'=\frac{\beta E\exp\left(\frac{E}{kT_M}\right)}{kn_0T_M^2}\left[1+\frac{2kT_M}{E}\right]^{-1} \tag{3}$$

which becomes, introducing $s=s'n_0$

$$s=\frac{\beta E\exp\left(\frac{E}{kT_M}\right)}{kT_M^2}\left[1+\frac{2kT_M}{E}\right]^{-1} \tag{4}$$

expressed in sec $^{-1}$.

Condition at the maximum when $s'=s'(T)$ (second-order kinetics)

To obtain the maximum condition we consider the *I=I(T)* equation:

$$I(T)=\frac{n_0^2s_0'T^\alpha\exp(-\frac{E}{kT})}{\left[1+\frac{s_0'n_0}{\beta}\int_{T_0}^{T}T^\alpha\exp\left(-\frac{E}{kT}\right)dT\right]^2} \tag{1}$$

and its logarithm:

$$\ln(I)=\ln(n_0^2 s_0')+\alpha\ln T-\frac{E}{kT}-2\ln\left[1+\frac{s_0' n_0}{\beta}\int_{T_0}^{T}T^{\alpha}\exp\left(-\frac{E}{kT}\right)dT\right] \tag{2}$$

The derivative equal to zero yields

$$\frac{\alpha}{T_M}+\frac{E}{kT_M^2}-2\left[1+\frac{s_0' n_0}{\beta}\int_{T_M}^{T}T^{\alpha}\exp\left(-\frac{E}{kT}\right)dT\right]^{-1}\left[\left(\frac{s_0' n_0}{\beta}\right)T_M^{\alpha}\exp\left(-\frac{E}{kT_M}\right)\right]=0 \tag{3}$$

Using the integral approximation in the case of s' temperature dependent, we get

$$\frac{\alpha}{T_M}+\frac{E}{kT_M^2}-2\left\{1+\frac{s_0' n_0}{\beta}\frac{kT_M^{\alpha+2}}{E}\left[1-(\alpha+2)\frac{kT_M}{E}\right]\exp\left(-\frac{E}{kT_M}\right)\right\}^{-1}\times$$
$$\frac{s_0' n_0}{\beta}T_M^{\alpha}\exp\left(-\frac{E}{kT_M}\right)=0$$

and rearranging, using $\Delta_M=2kT_M/E$:

$$\left[\frac{\alpha}{T_M}+\frac{E}{kT_M^2}\right]\cdot\left[1+\frac{s_0' n_0}{\beta}\cdot\frac{kT_M^{\alpha+2}}{E}\left[1-\left(1+\frac{\alpha}{2}\right)\Delta_M\right]\exp\left(-\frac{E}{kT_M}\right)\right]=$$
$$=\frac{2s_0' n_0}{\beta}\cdot T_M^{\alpha}\exp\left(-\frac{E}{kT_M}\right)$$

from which the pre-exponential factor can be derived:

$$s_0'=\frac{\beta E\left(1+\frac{\alpha}{2}\Delta_M\right)}{n_0 kT_M^{\alpha+2}}\exp\left(\frac{E}{kT_M}\right) \tag{4}$$

Condition at the maximum when $s''=s''(T)$ (general-order kinetics)

To obtain the maximum condition, the logarithm derivative of Eq. (2) given in **General-order kinetics when $s''=s''(T)$** will be carried out as below:

$$\ln(I) =$$

$$= \ln(s'_0 n_0) + \alpha \ln T - \frac{E}{kT} - \frac{b}{b-1} \ln\left[1 + \frac{s'_0(b-1)}{\beta} \int_{T_0}^{T} T^{\alpha} \exp\left(-\frac{E}{kT}\right) dT\right]$$

Considering

$$\frac{d(\ln I)}{dT} \Big|_{T=T_M} = 0$$

and using $\Delta_M = 2kT_M/E$, one obtains the maximum condition

$$\frac{s'_0 b T_M^{\alpha+2} k \exp\left(-\frac{E}{kT_M}\right)}{\beta E\left(1 + \frac{\alpha}{2}\Delta_M\right)} = \left[1 + \frac{s'_0(b-1)}{\beta} \int_{T_0}^{T_M} T^{\alpha} \exp\left(-\frac{E}{kT}\right) dT\right] \qquad (1)$$

Using now the **integral approximation when $s'' = s''(T)$**, we obtain

$$\frac{s''_0 b T_M^{\alpha+2} k \exp\left(-\frac{E}{kT_M^2}\right)}{\beta E\left(1 + \frac{\alpha}{2}\Delta_M\right)} = 1 + \frac{s''_0(b-1)}{\beta} \cdot \frac{kT_M^{\alpha+2}}{E} \times$$

$$\times \left[1 + (\alpha+2)\frac{kT_M}{E}\right] \exp\left(-\frac{E}{kT_M}\right)$$

which can be rearranged for determining the pre-exponential factor:

$$s''_0 = \frac{\beta E}{kT_M^{\alpha+2}} \left[\frac{1 - \Delta_M(1+\alpha)(1-b)}{1 + \frac{\alpha}{2}\Delta_M}\right]^{-1} \exp\left(\frac{E}{kT_M}\right) \qquad (2)$$

It must be noted that when $\alpha \to 0$ Eq. (2) becomes the no-temperature dependent expression for the pre-exponential factor.

Condition at the maximum when $s = s(T)$ (first-order kinetics)

The condition of maximum TL emission is obtained by the logarithm of the equation

$$I(T) = n_0 s_0 T^\alpha \exp(-\frac{E}{kT}) \exp\left[-\frac{s_0}{\beta} \int_{T_0}^{T} T^\alpha \exp(-\frac{E}{kT}) dT \right] \quad (1)$$

i.e.

$$\ln I = \ln(n_0 s_0) + \alpha \ln T - \frac{E}{kT} + \left[-\frac{s_0}{\beta} \int_{T_0}^{T} T^\alpha \exp(-\frac{E}{kT'}) dT' \right]$$

and its derivative, $\dfrac{d\ln I}{dT}$ equal to zero. Then we have:

$$\frac{\alpha}{T_M} + \frac{E}{kT_M^2} - \frac{s_0 T_M^\alpha}{\beta} \exp\left(-\frac{E}{kT_M} \right) = 0$$

$$\frac{s_0}{\beta} T_M^\alpha \exp(-\frac{E}{kT}) = \frac{\alpha}{T_M} + \frac{E}{kT_M^2}$$

and the final expression is then

$$\frac{\beta}{T_M^{\alpha+2}} = \frac{\exp(-\frac{E}{kT_M})}{\left(\frac{\alpha T_M}{s_0} + \frac{E}{ks_0} \right)} \quad (2)$$

Rearranging Eq. (2) and using $\Delta_M = 2kT_M/E$ one obtains

$$\frac{\beta}{T_M^{\alpha+2}} = \exp(-\frac{E}{kT_M}) \left[\frac{ks_0}{E\left(1 + \frac{\alpha}{2}\Delta_M \right)} \right] \quad (3)$$

From Eq. (3), the frequency factor is obtained

$$s_0 = \frac{\beta E}{kT_M^{\alpha+2}} \left(1 + \frac{\alpha}{2} \Delta_M \right) \exp\left(\frac{E}{kT_M} \right) \quad (4)$$

Considerations on the heating rate

Because the great importance of the heating rate (H.R.) in any kind of thermoluminescent measurements, it is better to report here the most relevant observations on this experimental parameter.

Kelly and co-workers [1] discussed about the validity of the TL kinetic theories when high heating rates are involved: they found that heating rates up to 10^5 °C/s do not invalidate the Shockley-Read statistics on which kinetic theories are based. Gorbics et al. [2] reported studies on thermal quenching of TL by varying the H.R. between 0.07 and about 11°C/s. They found the following results:

- the maximum glow-peak temperature, T_M, is shifted to higher temperatures as the H.R. increases.
- the TL intensity, measured by both integrating and peak height methods, decreases as the H.R. increases.

Other works, not specifically dedicated on the effect of H.R. on the TL intensity, report experimentally results not always in agreement among them.

The H.R. effect on TL glow-peaks has been largely discussed by G. Kitis [3] who considers the H.R. as a dynamic parameter rather than a simple experimental setup variable. His study has been carried out on single, well separated glow peaks, considering the following experimental characteristics: i.e., T_M, full width at half maximum (FWHM), peak intensity and peak integral. The first thing to be considered it is a possible delay between the temperature monitored by the thermocouple, fixed on the heating planchet, and the sample. Furthermore, the possibility of temperature gradients within the measured sample must be considered too. To avoid, totally or partially, these effects, special care has to be taken: i.e., the use of powder instead of solid samples diminishes greatly the gradient effects within the sample as well as between the heating planchet and the sample.

To ensure a good thermal contact between the heating strip and the powder sample the following rules have to be taken into account:

- dimensions of powder grains in the range of 80 - 140 mm
- use no more than 4 mg in weight of powder
- fix the powder on the heating element with silicon oil.

However, a certain gradient between sample and heating strip is emerging when high heating rates are used. Fig. 1 shows that temperature gradients emerge for heating rates greater than 50°C/s.

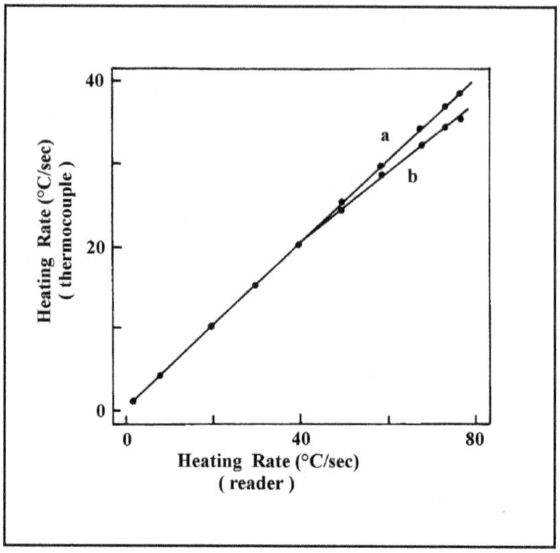

Fig. 1. The temperature gradient between heating try and sample. (a) heating rate on the try, (b) heating rate on the sample [3].

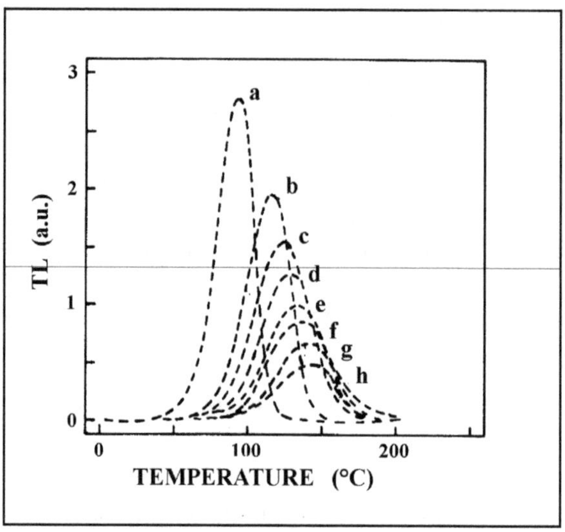

Fig. 2. Change of the peak shape and shift in the peak position as a function of the heating rate. From (a) to (h) = 2, 8, 20, 30, 40, 50, 57, 71°C/s [3].

The TL reader used was a TL analyzer type 711 of the Littlemore Company with a planchet of nicochrom of thickness 0.8 mm. The experimental results of Fig. 1 have been obtained by measuring directly the H.R. on the planchet and on the sample separately with Cr-Al thermocouples fixed on them. The main results of this investigation are concerning the influence of the H.R. on the TL glow-peak and are summarized by the following figures. From Fig. 2 one can observe the behavior of the shape of the experimental glow-curves for the 110°C glow-peak of quartz, obtained using various heating rates between 2°C/s and 70°C/s. As the H.R. increases, the peak height decreases and the peak temperature shifts towards high values of temperature. The shift of T_M is better seen in Fig. 3, where are reported the data concerning the Victoreen a-Al_2O_3:C which has a well isolated main glow-peak [4]. The dashed lines are the theoretical values calculated using the trapping parameters E and s determined with the lowest possible H.R. The solid lines are obtained as the best fit (Minuit program) of the experimental results. The experimental results follows exactly an equation of the form

$$T_M = \alpha \cdot \beta^{\gamma} \tag{1}$$

where β is the heating rate and α and γ are constants. α stands for the T_M value obtained with the lower heating rate. The same equation can fit the value of T_1 and T_2 which are the low and high half maximum temperatures respectively. The theoretical behavior is obtained using the general order equation for the heating rate:

$$\beta = T_M^2 \left(\frac{sk}{E}\right)\left[1+(b-1)\frac{2kT_M}{E}\right]\exp\left(-\frac{E}{kT_M}\right) \tag{2}$$

As above reported, the theoretical behavior has been obtained using the trapping parameters as calculated using the lowest heating rate: i.e., E = 1.339 eV, s = $1.13 \cdot 10^{14}$ s^{-1}, b = 1.45. The experimental values have been fitted according to Eq. (1) where α = 443.7 for H.R. = 1°C/s and γ = 0.025. The plots in Fig. 3 give a measure of the discrepancy between the experimental behavior and the one expected from the kinetic model according to Eq. (2). Figure 4 shows the behavior of FWHM for the peak in a-Al_2O_3:C as a function of the heating rate.

Also in this case the experimental points can be fitted by an equation similar to the Eq. (1):

$$FWHM = \alpha \cdot \beta^{\gamma} \tag{3}$$

where α = 36.5 and γ = 0.165.

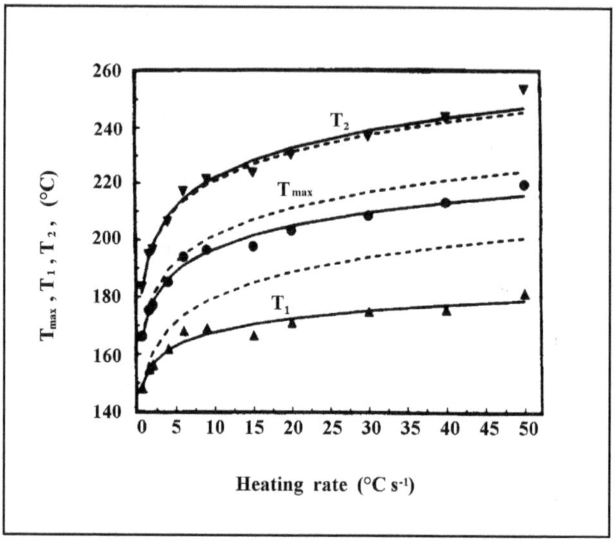

Fig. 3. Behavior of T_1, T_2 and T_M as a function of the heating rate. The dashed lines show the theoretical behavior and the solid lines the experimental one.

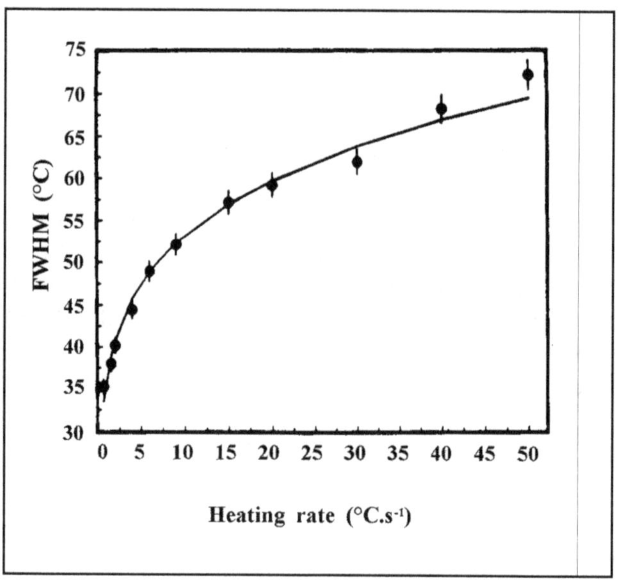

Fig. 4. Behavior of FWHM as a function of the heating rate [4].

More others important data are also reported in the same paper [4]. One of these is concerning the evolution of the integral and the peak height as a function of the H.R. Figure 5 shows the TL response of Al_2O_3 normalized to the response at the lower H.R. (0.6°C/s) as a function of H.R. for both integral (•) and peak height (*).

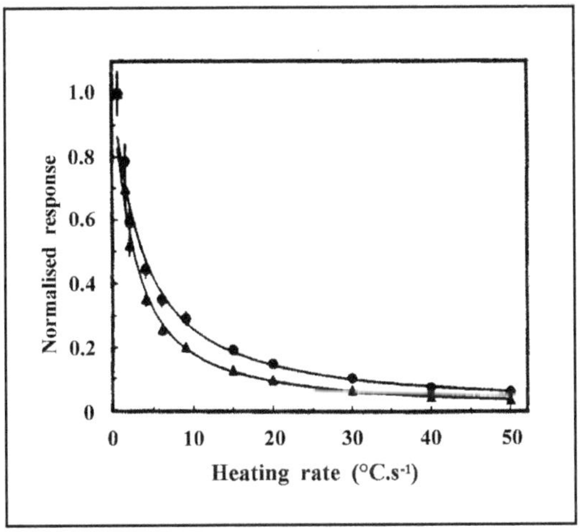

Fig. 5. TL response of Al_2O_3 as a function of H.R. The response has been normalized to the one obtained with the lowest H.R. [4].

The experimental points have been fitted by the equation

$$n = \frac{1}{1 + \alpha \cdot \beta^{\gamma}} \tag{4}$$

where n is the TL emission (integral or peak height) normalized to that at the lower H.R., β is the heating rate, α and γ are constants ($\alpha = 0.366$ and γ is equal to 1 in the case of integral and equal to 1.103 for the peak height). As it can be observed from the Fig. 5, there is a drastic reduction of TL as the heating rate increases. From a kinetic point of view, the peak integral is expected to remain constant as the heating rate increases. On the other hand, the peak height is expected to decrease as the heating rate increases, because the FWHM increases, so that the integral is constant.

The experimental evidence of the reduction of the TL as a function of heating rate is a general phenomenon and it has been observed in many different materials [5-8]. This reduction has been attributed to thermal quenching effect, whose efficiency increases as the temperature increases [2]: since the glow peak shifts to higher temperatures it suffers from thermal quenching. The results indicate that thermal quenching can be a very good explanation of the TL reduction with the heating rate. In fact, the luminescence efficiency of a phosphor, η, is given by

$$\eta = \frac{P_r}{P_r + P_{nr}} \tag{5}$$

and where P_r is the probability of luminescence transitions, temperature independent, and P_{nr} is the probability of non-radiative transitions, which is temperature dependent.

According to [4], Eq. (5) can be rewritten as

$$n = \frac{1}{1 + c\exp(-\frac{\Delta E}{kT})} \tag{6}$$

having replaced the efficiency η with the obtained TL emission, n, where c is a constant and the Boltzman factor exp(-ΔE/kT) replaces P_{nr} owing its dependence from temperature. Using then Eq. (6), the final expression for the luminescence efficiency, related to the maximum temperature T_M, is now

$$n = \frac{1}{1 + c\exp\left(-\frac{\Delta E}{k\alpha\beta^{\gamma}}\right)} \tag{7}$$

Using the values for a and g above reported, Eq. (7) gives an excellent fit of the TL response vs heating rate. The very good fit of the exponential data obtained using Eq. (7) allows to attribute the TL response reduction with H.R. to thermal quenching effect.

References

1. Kelly P., Braunlich P., Abtani A., Jones S.C. and deMurcia M., Rad. Prot. Dos. **6** (1984) 25
2. Gorbics S.G., Nash A.E. and Attix F.H., Proc. 2nd Int. Conf on Lum. Dos., Gatlinburg, TN, USA, 587 (1968)
3. Kitis G., Spiropulu M., Papadopoulos J. and Charalambous S., Nucl. Instr. Meth. **B73** (1993) 367
4. Kitis G., Papadoupoulos J., Charalambous S. and Tuyn J.W.N., Rad. Prot. Dos. **55(3)** (1994) 183
5. Kathuria S.P. and Sunta C.M., J. Phys. D: Appl. Phys. **15** (1982) 497
6. Kathuria S.P. and Moharil S.V., J. Phys. D: Appl. Phys. **16** (1983) 1331
7. Vana N. and Ritzinger G., Rad. Prot. Dos. **6** (1984) 29
8. Gartia R.K., Singh S.J. and Mazumdar P.S., Phys. Stat. Sol. (a) **106** (1988) 291

Considerations on the methods for determining *E*

A critical survey on the methods for determinin E, points out at first how each one of them is applicable as regards one or more of the physical considered parameters. A graphical approach is often made possible by the analytical features the glow-curve can show locally or on its whole. Because of its particular mathematical shape, the unitary order kinetics case is commonly apart from the other ones; the general aim of the analytical techniques is anyhow to extend the domains of application as long as possible.

The ways the glow-curve is taken into account are quite various: its analysis may be local or general; it may regard the peak alone or the whole line; finally the curve or the area it subtends may be, for each case, considered. The temperatures of most interest are however the peak ones and, eventually, those where the curve inflects. The tangents are then pointed out by Ilich [1] as auxiliary plots which might usefully be applied to achieve, as described above, a knowledge of the involved energy.

More in detail, it is possible to group these methods in main sections:

a) Methods based upon maximum temperatures,
b) Methods based upon low temperatures side analysis,
c) Variable heating rates methods,
d) Area measurements methods,
e) Isothermal decay method,
f) Inflection points method,
g) Peak shape geometrical methods.

It is evident, therefore, how anyone of the analytical features of the glow-curve can give, if suitably manipulated, useful information on the quality of the phenomena which the thermoluminescent emission is an overall effect of.

The simplest procedure is that searching for a linear relationship between glow temperature and activation energy. This has lead Randall-Wilkins [2] and Urbach [3] to their formulas; it is on the other side to be noted how the corresponding solutions are approximated; this is due to the fact that they have been computed starting from, as previously said, already known values of s, which the expressions are independent on. For istance, the expression of Urbach ($E = T_M/500$) is a very rough guide and then it is of limited accuracy. As reported in [4], the use of the Urbach's expression is equivalent to the assumption $E/kT_M = 23.2$ and gives energy values which may be wrong by up to a factor of two.

As regards section b), the initial rise method makes use of the existence, in the glow-curve, of a temperature range where, while the integral exponential factor remains practically unitary, the Boltzmann probability factor increases with T and therefore rules the curve shape. A semilog plot of I vs $1/T$, acting as linearizing transformation, gives an E evaluation which doesn't depends on s. It is worthwhile to remark that, when the method is extended to non unitary kinetics order configurations, and, therefore a knowledge of n is required, it is possible to associate this last one with the glow-curve area, thus introducing an integrated parameter; finally, when the order b is unknown, the only way to proceed is to adjust it and, by a repeated procedure, to determine the value giving the best linear fit; thus a related statistical analysis is, for the present situation, required, and, eventually, the application of convenient tests regarding the goodness of fit may constitute an useful numerical tool.

The tangent method is related to the initial rise technique, as far as it starts from the same equation; more emphasis is however attributed to the role of the tangent, the plot of which is important in computing the expression for E.

An eventual limitation of the initial rise method is given by the risk to underestimate the actual E value. This might be caused by non-radiative events which could lead to a computation of an apparent energy, differing from the real one by an amount W connected to the characteristic non-radiative contribution depth. Wintle [5], analyzing the E values obtained by different methods, found discrepancies among them in the sense that the activation energies obtained with initial rise method were always much less than the E values obtained with other methods. Indeed, the initial rise method does not take into account the luminescence efficiency expressed by

$$\eta = \frac{P_r}{P_r + P_{nr}} \tag{1}$$

where P_r is the probability of radiative emission and it is independent of temperature, and P_{nr} is the probability of non-radiative transition, which is temperature dependent and rises with increasing temperature. The resulting decrease of efficiency with temperature rise is called thermal quenching. Wintle suggested that a better expression for the initial rise part is

$$I = sn\exp\left(-\frac{E-W}{kT}\right) \tag{2}$$

Then the E_{ir} value derived from initial rise measurements will be smaller than E by an amount W.

The thermal quenching is experimentally demonstrated by the luminescence emission during irradiation at different temperatures. The W values obtained are the same as the discrepancy observed using different methods.

Other methods make use of the dependence of the glow-peak shape on the heating rate. When increasing it, a shift toward higher temperatures is observed, together with an increase in the peak height. The former effect is mathematically expressible through the glow-peak numerical condition, giving, as a solution, the value for E. This computation can be carried out apart from an s preliminary knowledge, by writing down the equations for two different heating rates and replacing in them the experimental data. By combining the two expressions, s can be dropped, and therefore an independent estimate of E is possible. The frequency factor may be found, after E, by substitution in either expression. It is to be remarked that E, as computed by means of the double heating rate technique doesn't depend on the existence of the non radiative contribution W, described for the initial rise method.. Therefore, by this latter procedure it is feasible to estimate the apparent E; by the double heating rate method, on the other side, a "true" value for E may be found out; therefore a suitable combined use of both systems may give useful information on the W amount.

By generalizing the present method, after Hoogenstraaten [6] and Chen-Winer [7], it is possible to make use of several heating rates; by manipulating, in such cases, the general equations ruling the various kinetics, it is feasible to obtain quite simple shaped plots, respectively for unitary and non unitary configurations. It is to be observed how this technique marks out a graphical approach to the numerical solutions. Its domain of application includes whatever order kinetics cases, within the theoretical limitation above seen.

Moreover, the heating rate itself may be time dependent, although, if constant, the plotting procedure is made quite simpler. Even configuration with an unknown b may be analyzed in this way: in such cases, only an approach by attempts is feasible, and the best statistical value for b is consequently reckoned on the basis of statistical tests.

Finally, it is to be noted how the double heating rate method itself can be extended to non unitary order cases. The choice of the heating rate value is arbitrary, though tied to the practical limits.

The area measurement methods are independent on the glow-curve shape, and only the surface subtended by it, between two given temperatures, is required. An analytical survey on this procedure starts again from first order kinetics, and passes then to include the possible variants and extensions. In the $b = 1$ case a graphical study appears simple and feasible. The analytical remarks regard the use of a linearising logarithmic function, which leads to a parallel E and s evaluation. As in other methods, an expansion to more general configurations is of particular physical interest, and is attainable by referring back to the $I(T)$ expression for the general-order kinetics, where the overall effect of the involved phenomena is considered and synthesized in terms of a first order differential equation. From the May-Partridge area method applied to the case of general-order kinetics [8], it is clear how this extension bears the apparition of a power b in both members. This allow for a procedure theoretically analogous to the unitary order situation. E and s are still found out by means of a suitable plot and their computations are independent on each other. Again, to an unknown b value, an optimization statistical problem corresponds. The method allows for some kind of variants: Muntoni and others [9] for this purpose start using a general order equation and a graphical estimation of E is attainable. Finally Maxia [10] postulated a singularity in the electron trap level and a multiplicity in the recombination centers.

The isothermal decay technique [11], apart from the details of the adopted thermal cycle, analyses in particular the phenomenon of trapped electron decay, that is to say of their rising to the conduction band. The magnitudes of physical interest are the temperature of the sample is stored at and the time elapsed; after these data, a graphical estimate of E and s may be carried out. The isothermal decay method be also appropriately extended to situations where the unitary order kinetics hypothesis, initially assumed, is no longer true; thus, the procedure can be applied to configurations where b

is both determined and unknown. In this later case a technique "by attempts" must be followed.

The Land's method [12] of inflection points, makes primarily use of two additional experimental parameters, defined as the temperature values where the glow-curve inflects. To their experimental determination, an analytical expression corresponds, computed by deriving twice the glow-curve equation, as defined for a first order kinetics, which this technique is applicable to. The accuracy available with this method is directly connected with the precision that may be reached in the experimental evaluation of the graphical variables of interest.

Several analytical procedures make use of the peak geometrical features. These parameters are derived by studying the glow-curve data, mainly as regard the total width, the left and right half-width and the maximum itself. The ratio between the two half-widths yields a measure of the degree of symmetry characterizing the peak on its whole. Lushchik [13] and Grossweiner [14] outline two procedures each furnishing estimate of E and s, based upon the experimental knowledge of the glow and half width temperatures, as well as their associate errors. On the other side, the Halperin and Braner technique [15] makes use of the maximum temperature, and both the half width ones. The relative theory starts from a delineated investigation about the two main phenomena which the electron-hole recombination is a result of. They assume that the recombination radiative event may occur both via the conduction band, or directly as a result of a tunneling between the electron trap and the recombination center under consideration. An analytical survey points out how the activation energy is connected to the glow temperature and to the above described geometrical parameters. These relationships show also the tie between the kinetic order and the curve symmetry or asymmetry; furthermore, it is remarkable that all the pertinent equations cam be elaborated only in an iterative way, because of the presence of an E-dependent term in the second members.

A more straightforward method, simplifying the E evaluation, has been outlined by Chen [16,17]. This method is not iterative and the evaluation of E is carried out by means of an expression, the form of which can be unified for various configurations differing from one another for the kinetic order and the kind of geometrical parameter involved.

A detailed critical review of the various expression based on the peak shape methods, giving the E/kT_M range of validity for each expression, is given in [18] and it is reported shortly here.

- the Lushchik's formula gives an error by 3.3% for $E/kT_M = 10$, reducing to 1.7% for $E/kT_M = 100$. However, in all cases the formula gives a higher value of E than the actual one.

- the Halperin and Braner's formula, based on τ value, underestimates E by 4.2% for $E/kT_M = 10$, is exact for $E/kT_M \approx 11$, over-estimates E by 12% for $E/kT_M = 20$ and by 17% for $E/kT_M = 100$.

- Grosswiener's expression overestimates E by 10.4% for $E/kT_M = 10$, by 7.1% for $E/kT_M = 20$ and by 4.1% for $E/kT_M = 100$.

- the Keating's expression is valid in the range $10 \leq E/kT_M \leq 18$ and it overestimates E by 3% at $E/kT_M = 10$, by a maximum 10% at $E/kT_M = 20$; it is exact at $E/kT_M = 60$ and underestimates E by 12.5% for $E/kT_M = 100$.

- the Chen's formula, based on ω, valid for E/kT_M between 14 and 42, underestimates E by 4% at $E/kT_M = 10$, by 1.6% at $E/kT_M = 14$; it is exact at $E/kT_M = 20$ and overestimates E by 1.6% at $E/kT_M = 40$ and by 2.4% at $E/kT_M = 100$. Chen also corrected the Lushchik's equation so that the errors being less than 0.5% for E/kT_M between 14 and 40 and less than 0.8% when E/kT_M is low as 10 or as high as 100. The Chen's formula based on τ underestimates E by 5.3% at $E/kT_M = 10$, by 2.5% at $E/kT_M = 14$; it is exact at $E/kT_M = 22$ and overestimates E by 2% at $E/kT_M = 43$ and by 3.2% at $E/kT_M = 100$.

Some Authors have also underlined the feasibility of computerized glow-peaks [19,20] analysis. A general program can be written: the input is given by the experimental data and by rough estimates of the physical parameters. These latter ones can be iteratively adjusted and each set of values gives a theoretical glow-curve. This plot can be statistically put in comparison to the experimental one, and so, the parameter optimization can kept on until a fair agreement is attained on the basis of statistical tests.

At the beginning of the '80 studies on computerized glow-curve deconvolution (CGCD) began to appear in the scientific literature. The CGCD programs are normally developed by each research group according to the particular needs and the material studied. A very useful review on this subject is appeared in 1995 [21].

In all the previous methods the hypothesis of constant s has been tacitly assumed. In some cases, however, there is evidence for a T-dependence of s and s'. From a mathematical standpoint, this temperature dependence affects the numerical solution of the integral comparing in all the equations.

Finally, it is to be noted how a convenient statistical treatment is of great practical interest, In the above discussed methods it has been often necessary to operate linear best fittings as well as to check their applicability. The procedure most commonly adopted is the last square method, by which, after the experimental data consideration, the slope and the intercept of the resulting line are computed, together with their errors. A first check on the actual linearity is given by the correlation coefficient; a more accurate way is the application of a so called "goodness of fit" statistical test, which the data are submitted to, and which can point out, within a given probability level, the opportunity to accept or to reject the linearity hypothesis.

Concerning the Moharil's methods [22-25], finally, it must be pointed out the quantity A/B, which varies from 0 to 1, which is physically more relevant than the general order of kinetics b.

References

1. Ilich B.M., Sov. Phys. Solid State **21** (1979) 1880
2. Randall J.T. and Wilkins M.H.F., Proc. Roy. Soc. **A184** (1945)366
3. Urbach F., Winer Ber. **IIa 139** (1930) 363
4. Christodoulides C., J. Phys. D: Appl. Phys. **18** (1985) 1501
5. Wintle A.G., J. Mater. Sci. **9** (1974) 2059
6. Hoogenstraaten W., Philips Res. Repts **13** (1958) 515
7. Chen R. and Winer S.A.A., J. Appl. Phys. **41** (1970) 5227
8. May C.E. and Partridge J.A., J. Chem. Phys. **40** (1964) 1401
9. Muntoni C., Rucci A. and Serpi A., Ricerca Scient. **38** (1968) 762
10. Maxia V., Onnis S. and Rucci A., J. Lumin. **3** (1971) 378
11. Garlick G.F.J. and Gibson A.F., Proc. Phys. Soc. **60** (1948) 574
12. Land P.L., J. Phys. Chem. Solids **30** (1969) 1681
13. Lushchik L.I., Soviet Phys. JEPT **3** (1956) 390
14. Grossweiner L.I., J. Appl. Phys. **24** (1953) 1306
15. Halperin A. and Braner A.A., Phys. Rev. **117** (1960) 408
16. Chen R., J. Appl. Phys. **40** (1969) 570
17. Chen R., J. Electrochem. Soc. **116** (1969) 1254
18. Christodoulides C., J. Phys. D: Appl. Phys. **18** (1985) 1501
19. Mohan N.S. and Chen R., J. Phys. D: Appl. Phys. **3** (1970) 243
20. Shenker D. and Chen R., J. Phys. D: Appl. Phys. **4** (1971) 287
21. Horowitz Y.S. and Yossian D., Rad. Prot. Dos. **60** (1995) 1
22. Moharil S.V., Phys. Stat. Sol. (a) **66** (1981) 767
23. Moharil S.V. and Kathurian S.P., J. Phys. D: Appl. Phys. **16** (1983) 425
24. Moharil S.V., Phys. Stat. Sol. (a) **73** (1982) 509
25. Moharil S.V. and Kathuria S.P., J. Phys. D: Appl. Phys. **16** (1983) 2017

Considerations on the symmetry factor, μ, and the order of kinetics, b

The order of kinetics, b, and the symmetry factor, $\mu=\delta/\omega$, are two important parameters. After the Chen's work [1], the graphical picture of dependence of the symmetry factor μ on b has been utilised to determine easily the order of kinetics. It has to be stressed that the order of kinetics b still remains a topic of controversy and matter of debate, even in the case of the most widely studied material, i.e., LiF [2-4]. Indeed it must be noted the fact that for a given value of b, the symmetry factor μ is not unique. Chen, in his work [1], has pointed out that μ is dependent of the thermal activation energy E and the frequency factor s, and for a given value of b and for extreme values of E and s, the maximum deviation in μ can be as high as ±7%. Therefore, without an a priori knowledge of E and s the absolute determination of b from the value of μ is not possible. The following mathematical treatment, as given in [5], allows to find a general expression for μ, considering any position selected on the glow-peak, in terms of the variable $u = E/kT$ and of its value at the peak temperature, $u_m = E/kT_M$.

The equation for a general order peak can be written as

$$I = sn_0 \exp\left(-\frac{E}{kT}\right)\left[1+\frac{(b-1)s}{\beta}\int_{T_0}^{T}\exp\left(-\frac{E}{kT'}\right)dT'\right]^{-\frac{b}{b-1}} \tag{1}$$

where $s = s'' n_0^{b-1}$ as usual. It has to be remind that the above equation is valid for $1<b\leq 2$.

Remembering the condition for the maximum intensity

$$1+\frac{(b-1)s}{\beta}\int_{T_0}^{T_M}\exp\left(-\frac{E}{kT'}\right)dT' = \left(\frac{sbkT_M^2}{\beta E}\right)\exp\left(-\frac{E}{kT_M}\right) \tag{2}$$

and replacing E/kT by u, E/kT_M by u_M and T_0 by 0, we get

$$I = sn_0 \exp(-u)\left[\frac{b\exp(-u_M)+(b-1)(J-J_M)u_M^2}{b\exp(u_M)-(b-1)J_M u_M^2}\right]^{-\frac{b}{b-1}} \tag{3}$$

where

$$J = \int_u^{\infty}\frac{\exp(-u')}{u'^2}du', \qquad J_M = \int_{u_M}^{\infty}\frac{\exp(-u')}{u'^2}du' \tag{4}$$

The intensity at the maximum is given then by

$$I_M = sn_0 \exp(-u_M)\left[\frac{b\exp(-u_M)}{b\exp(u_M)-(b-1)J_M u_M^2}\right]^{-\frac{b}{b-1}} \tag{5}$$

Expressing J and J_M in terms of second exponential integrals [6]:

$$E_2(u) = u\int_a^{\infty}\frac{\exp(-z)}{z^2}dz \tag{6}$$

one can write

$$\frac{I}{I_M} = \exp(u_M - u)\left[1-\frac{b-1}{b}F(u,u_M)\right]^{-\frac{b}{b-1}} \tag{7}$$

where

$$F(u,u_M)=u_M^2\exp(u_M)\left[\frac{E_2(u_M)}{u_M}-\frac{E_2(u)}{u}\right] \tag{8}$$

Equation (7) gives the TL signal I as a function of temperature when I_M and u_M are given. For a given value of u_M, the ratio I/I_M depends only on u.

Equation (7) can be transformed using any temperature value on the peak, i.e., T_x, for which $I/I_M = x$:

$$\begin{aligned}\ln\left(\frac{1}{x}\right)&=(u_x-u_M)+\frac{b}{b-1}\ln\left[1-\frac{b-1}{b}F\left(u_x,u_M\right)\right]=\\&=\left(u_x-u_M\right)+\frac{b}{b-1}\ln\left[1-\frac{b-1}{b}u_M^2\exp(u_M)\left\{\frac{E_2(u_M)}{u_M}-\frac{E_2(u_x)}{u_x}\right\}\right]\end{aligned} \tag{9}$$

In order to solve the last expression an iteration procedure is used, writing

a) $T_x=T_x^-$ for $T_x<T_M$

b) $T_x=T_x^+$ for $T_x>T_M$

For case a) the $(u_x - u_M)$ term dominates:

$$u_x^-=u_M+\ln\left(\frac{1}{x}\right)-\frac{b}{b-1}\ln\left[1-\frac{b-1}{b}u_M^2\exp(u_M)\left\{\frac{E_2(u_M)}{u_M}-\frac{E_2(u_x^-)}{u_x^-}\right\}\right] \tag{10}$$

For case b) the logarithm term in square brackets dominates:

$$u_x^+=u_M-\ln\left\{\frac{\left[\left(\frac{1}{x}\right)\exp\left(u_M-u_x^+\right)\right]^{\frac{b-1}{b}}-1+\frac{b-1}{b}u_M\exp(u_M)E_2(u_M)}{\frac{b-1}{b}u_M^2\exp(u_x^+)\frac{E_2(u_x^+)}{u_x^+}}\right\} \tag{11}$$

The above equations are valid for $b > 1$. For $b = 1$, the analogous expressions are given in **Christodoulides method**.

Knowing u and u_M it is possible to calculate the value of the symmetric factor as:

$$\mu_x(x)=\frac{T_x^+-T_M}{T_x^+-T_x^-}=\frac{u_x^-(u_M-u_x^+)}{u_x^--u_x^+} \tag{12}$$

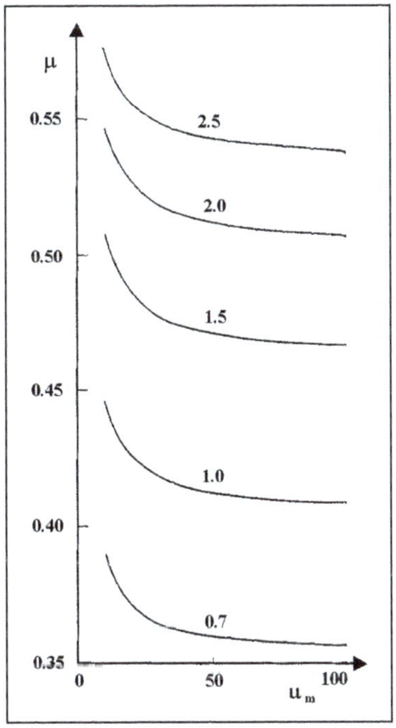

Fig. 1. Variation of $\mu(x)$ for $x = 0.5$ as a function of u_m for various order of kinetics.

order	u_M	$\mu(0.2)$	$\mu(0.5)$	$\mu(0.8)$
0.7	20	0.311	0.372	0.426
	30	0.302	0.365	0.422
	40	0.297	0.362	0.420
1.0	20	0.389	0.426	0.458
	30	0.378	0.418	0.453
	40	0.372	0.415	0.451
1.5	20	0.481	0.485	0.491
	30	0.468	0.477	0.487
	40	0.461	0.473	0.485
2.0	20	0.544	0.526	0.514
	30	0.531	0.518	0.510
	40	0.524	0.514	0.508
2.5	20	0.592	0.557	0.531
	30	0.579	0.549	0.527
	40	0.572	0.545	0.525

Table 1. Values of μ as a function of the kinetics order and u_m.

Fig. 1 shows the variation of $\mu(x)$, for $x = 0.5$, as a function of u_M for various order of kinetics. It is clear that to a given value of u_M not only one value of μ corresponds. This means that it is not possible to find out the true value of the order of kinetics. It is suggested to check the value of μ at various points on the glow-peak, x, to get an estimation of b. Table 1 gives the values of $\mu(x)$ for some particular values of u_M and various order of kinetics b, ranging from 0.7 to 2.5.

References

1. Chen R., J. Electrochem. Soc. **116** (1969) 1254
2. Kathuria S.P. and Sunta C.M., J. Phys. D: Appl. Phys. **15** (1982) 497
3. Kathuria S.P. and Moharil S.V., J. Phys. D: Appl. Phys. **16** (1983) 1331
4. Vana N. and Ritzinger G., Rad. Prot. Dos. **6** (1984) 29
5. Gartia R.K., Singh S.J. and Mazumdar P.S., Phys. Stat. Sol. (a) **106** (1988) 291
6. Abromowitz M. and Stegun I.A., Handbook of Mathematical Functions, Dover, N.Y. (1965)

Correction factor for the beam quality, F_{en} (general)

This factor must be evaluated when high atomic number thermoluminescent materials are used. In this case, the TL response at photon energies below about 100 keV becomes significantly greater than that, at the same dose, at higher energies.

The first step is then to calculate the effective atomic number of the dosimetric material, Z_{eff}, to check the possibility of an over estimation of the dose at low energies (see **Atomic number: calculation**).

The second step consists in the theoretical calculation of the energy response (see **Photon energy response: theory**) and, finally, the third step is the experimental determination of the energy response (see **Energy dependence: procedure**).

From the theoretical point of view, the absolute sensitivity, Σ, of a TLD, considered as the ratio between its net TL emission and the air absorbed dose D_t at which the dosemeter has been exposed, is defined, in the linear range of the TL response of the given TL material and for a given energy E of the radiation, as

$$\sum_E \propto \left(\frac{D_d}{D_a} \right)_E \tag{1}$$

where the index "d" is referred to the TL material, "a" stands for air. The same Eq. (1) can be referred to the tissue; in this case D_t substitutes D_a.

Taking into account that the absorbed dose in a material is a function of its mass energy absorption coefficient, the previous relation can be written as following

$$\sum_{E_\gamma} \propto \eta_E \left[\frac{\left(\frac{\mu_{en}}{\rho} \right)_d}{\left(\frac{\mu_{en}}{\rho} \right)_a} \right]_E \tag{2}$$

which is derived from the Bragg-Gray cavity theory applied to a large cavity.

Because $(\mu_{en}/\rho)_d$ is commonly referred to a compound of different elements, it must be substitutes by the expression

$$\left(\frac{\mu_{en}}{\rho} \right)_{TLD} = \sum_i \left(\frac{\mu_{en}}{\rho} \right)_i W_i \tag{3}$$

where W_i is the fraction by weight of the i-th element.

Considering the values relative to a reference energy E_0 (i.e. ^{60}Co or ^{137}Cs), one has the so called Relative Energy Response (RER):

$$RER = \frac{\Sigma_E}{\Sigma_{E_0}} = \frac{\left[\frac{\sum_i \left(\frac{\mu_{en}}{\rho} \right)_i W_i}{\left(\frac{\mu_{en}}{\rho} \right)_a} \right]_E}{\left[\frac{\sum_i \left(\frac{\mu_{en}}{\rho} \right)_i W_i}{\left(\frac{\mu_{en}}{\rho} \right)_a} \right]_{E_0}} \tag{4}$$

The behaviour of Σ_E/Σ_{E0} as a function of E gives the energy dependence of the TL response. As a consequence of this energy dependence, the calibration factor F_C also depends on the energy of the calibration source used for its determination.

The same calculation can be done for electrons, considering now the mass collision stopping power:

$$\Sigma'_{E_d} \propto \left[\frac{\left(\frac{S_{coll}}{\rho} \right)_d}{\left(\frac{S_{coll}}{\rho} \right)_t} \right]_E \tag{5}$$

Normally the calibration factor is determined using a ^{60}Co beam and in several situations also the TLDs used for applications are irradiated with gammas having the same energy. In this case the factor F_{en} is equal to unity. On the other hand, in much more practical situations the batch of TLDs is used in a radiation field having an energy different from the one used for calibration. Generally speaking, if we indicate F_c as the calibration factor obtained with a reference source and F_q the similar factor obtained with another beam quality, the F_{en} factor is defined as

$$F_{en} = \frac{F_c}{F_q} \tag{6}$$

Curve fitting method (Kirsh: general order)

Y.Kirsh proposed an alternative approach to the curve fitting method, trasforming the whole peak into a straight line. It may be regarded as an extension of the initial rise method and it can be applyed to the whole curve rather than to the initial part of the curve [1,2].

Starting from the **general order equation**

$$I = -\frac{dn}{dt} = n^b s' \exp\left(-\frac{E}{kT}\right) \tag{1}$$

and remembering that

$$s' = \frac{s}{n_0^{b-1}} \tag{2}$$

Eq. (1) can be rewritten as

$$I = \left(\frac{n}{n_0}\right)^b s n_0 \exp\left(-\frac{E}{kT}\right) \tag{3}$$

where n_0 is the initial value of n at $t = 0$.

Taking now the logarithm of both sides of Eq. (3), we obtain

$$\ln I = -\frac{E}{kT} + b \ln\left(\frac{n}{n_0}\right) + \ln(s n_0) \tag{4}$$

Taking now on the experimental glow curve any two points, i.e. (I_1, T_1, n_1) and (I_2, T_2, n_2), Eq. (4) can be written as

$$\ln I_1 = -\frac{E}{kT_1} + b\ln\left(\frac{n_1}{n_0}\right) + \ln(sn_0) \tag{5}$$

$$\ln I_2 = -\frac{E}{kT_2} + b\ln\left(\frac{n_2}{n_0}\right) + \ln(sn_0) \tag{6}$$

Subtracting Eq. (5) from Eq. (6) we obtain

$$\ln I_2 - \ln I_1 = b\left[\ln\left(\frac{n_2}{n_0}\right) - \ln\left(\frac{n_1}{n_0}\right)\right] - \left(\frac{E}{k}\right)\left(\frac{1}{T_2} - \frac{1}{T_1}\right)$$

which can be written as

$$\frac{\Delta \ln I}{\Delta \ln\left(\frac{n}{n_0}\right)} = b - \left(\frac{E}{k}\right)\left[\frac{\Delta\left(\frac{1}{T}\right)}{\Delta \ln\left(\frac{n}{n_0}\right)}\right] \tag{7}$$

where Δ represents the difference between any two points on the glow curve.

A plot of of the left hand side of Eq. (7) against the part in the square brakets should give a straight line with slope of $-\frac{E}{k}$ and an intercept of b at the y-axis.

Using this method one can simultaneously determine both the order of kinetics, b, and the activation energy, E. The frequency factor can be then determined by **the maximum condition**.

References

1. Kirsh Y., Phys. Stat. Sol. (a), **129** (1992) 15
2. Dorendrajit Singh S., Mazumdar P.S., Gartia R. and Deb N.C., J. Phys. D: Appl. Phys. **31** (1998) 231

CVD diamond

Chemical Vapour Deposition (CVD) diamond is a very interesting material as a thermoluminescent detector of ionizing radiations. Its atomic number is Z = 6 and then it can be considered a tissue equivalent material (effective atomic number of soft human tissue is $Z_{eff} = 7.4$). CVD diamond can be used in-vivo clinical dosimetry because it is no-toxic and chemically stable against all body fluids.

The growth technique for obtaining CVD diamond have been recently reviewed in [1]. The role plaied by the impurities atoms has been studied and reported in [2]. According to this paper, a boron concentration of 1 ppm is the optimum for obtaining a linear TL response Vs dose from 20 mGy to 10 Gy. More data about CVD TL properties are available in [3-9].

References

1. Sciortino S., Rivista del Nuovo Cimento **22** (1999) 3
2. Keddy R.J. and Nam T.L., Radiat. Phys. Chem. **41** (1993) 767
3. Avila O. and Buentil A.E., Rad. Prot. Dos. **58** (1995) 61
4. Biggeri U., Borchi E., Bruzzi M., Leroy C., Sciortino S., Bacci T., Ulivi L., Zoppi M. and Furetta C., Nuovo Cimento A, **109** (1996) 1277
5. Borchi E., Furetta C., Kitis G., Leroy C. and Sussmann R.S., Rad. Prot. Dos. **65** (1996) 291
6. Borchi E., Bruzzi M., Leroy C. and Sciortino S., J. Phys. D, **31** (1998) 1
7. Furetta C., Kitis G., Brambilla A, Jany C., Bergonzo P. and Foulon F., Rad. Prot. Dos. **84** (1999) 201
8. Furetta C., Kitis G. and Kuo C.H., Nucl. Instr. Meth. **B160** (2000) 65
9. Marczewska B., Furetta C., Bilski P. and Olko P., Phys. Stat. Sol. (a) **185** (2001) 183

D

Defects

Materials of interest in thermoluminescent dosimetry are principally insulators in which conduction electrons are entirely due to absorbed radiation energy. Examples of such insulators are the cubic structured alkali halides, such LiF and NaCl.

A crystal is an agglomerate of atoms or molecules characterized by a 3-fold periodicity. To describe completely a crystal one has to define the positions of atoms (or molecules) inside a unit cell, built a three-vector $\vec{a}_i$ (i = 1, 2, 3) of arbitrary origin. All the atoms of the crystal will be obtained from the atoms of the unit cell by all the translations $\vec{t}$:

$$\vec{t} = \sum_i \alpha_i a_i \tag{1}$$

where α_i represents all the positiv e and negative integers.

A crystal defined by Eq. (1) is termed ideal. Thermal vibrations disturb the periodicity and make it impossible to obey Eq. (1), so the crystal is now called imperfect. A further limitation to Eq. (1) is the finite crystal size. Crystals are limited by free surfaces which are the first type of crystal defect. A crystal which has free surfaces and probably other defects, is a real crystal.

Since alkali halides and their imperfections are particularly suitable for understanding luminescence phenomena, they will be used to discuss the behavior of a real crystal, all defects of which can potentially act as traps for the charge carriers created by secondary charged particles during irradiation.

The alkali halides structure consists of an orderly arrangement of alkali and halide ions, one after another, alternating in all three directions. Fig. 1 shows the structure of two ideal crystals.

At the contrary, a real crystal posses defects which are basically of three general type:

The intrinsic or native defects.

They can be:

a) vacancies or missing atoms (called Schottky defects). A vacancy is a defect obtained when one atom is extracted from its site and not replaced.

b) interstitial or Frenkel defect. It consists in an atom X inserted in a crystal X in a non-proper lattice site.

c) substitutional defects: for example, halide ions in alkali sites.

d) aggregate forms of previous defects.

Figure 2 depicts the previous mentioned defects.

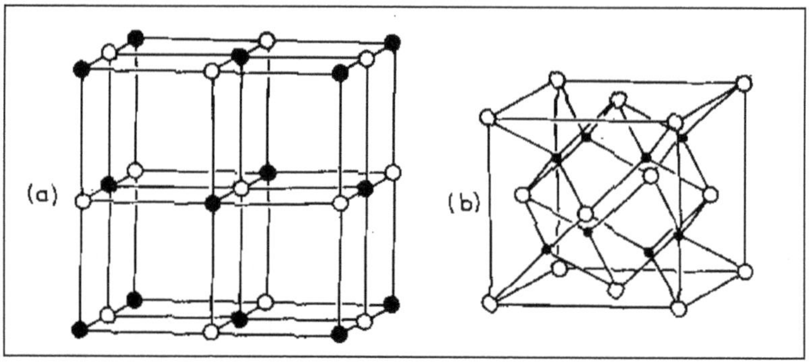

Fig. 1. The three-dimensional structure of an ideal crystal:
(a) structure of LiF (•Li, ○ F) ; (b) structure of CaF_2 (•Ca, ○ F)

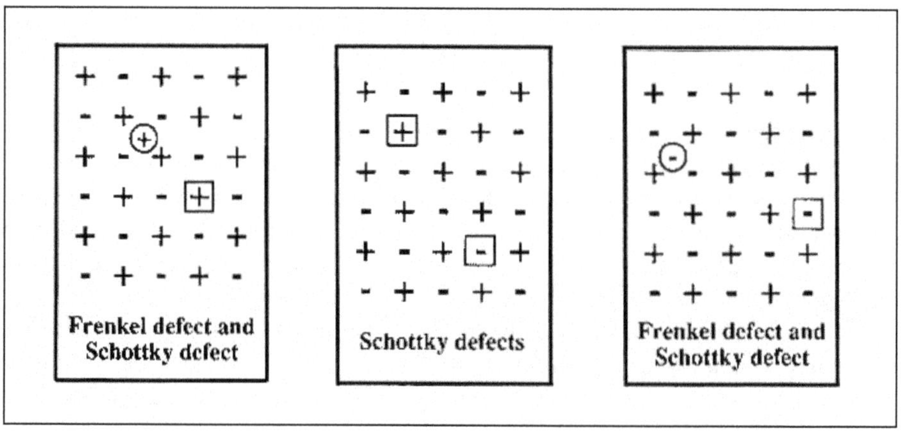

Fig. 2. Structures of a real crystal with intrinsic defects: i.e. LiF.
+ alkali ion (Li^+), - halide ion (F^-), $\boxed{+}$ alkali ion vacancy, $\boxed{-}$ halide ion vacancy,
$\oplus$ interstitial alkali ion, $\ominus$ interstitial halide ion

Extrinsic or impurity defects, like chemical impurities Y in a crystal X.

They can be:

a) substitutional impurity: an atom Y takes the place of an atom X.

b) interstitial impurity: an atom Y is inserted in an additional site not belonging to the perfect crystal.

These impurities are grown into the crystal structure from the melt, or diffused or implanted at a later stage. As an example, Fig. 3 shows the behavior of the divalent cation Mg^{2+} in LiF: it substitutes a Li^+ ion.

To understand the mechanism of chemical impurities, one can see the influence of a divalent ion on vacancy concentration, as shown in Fig. 4 (a): in order to compensate for the excess positive charge of impurity, an alkali ion must be omitted; furthermore, since the divalent cation impurity is a local positive charge and the cation vacancy is a local negative charge, the two attract each other giving rise to a complex as shown in Fig. 4 (b).

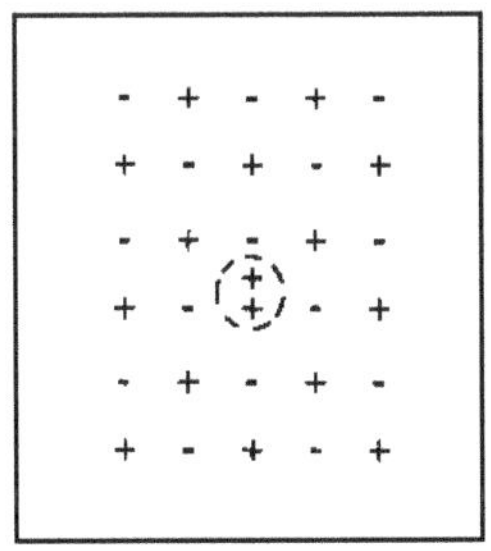

Fig. 3. Substitutional divalent cation impurity Mg^{2+}.

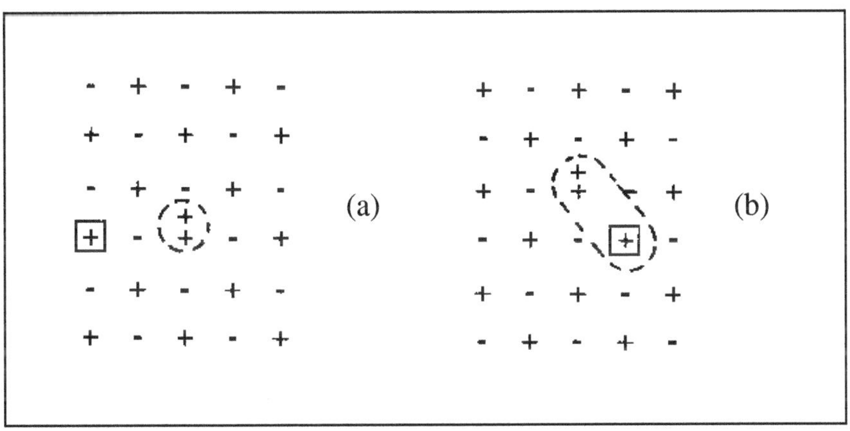

Fig. 4. (a) an alkali ion misssing; (b) attraction of ions to form a complex.

Ionizing radiation produces further defects in alkali halides.

These defects are called color centers which are absorption centers, coloring ionic crystals. For example, negative ion vacancies are regions of localized positive charge, because the negative ion which normally occupies the site is missing and the negative charges of the surrounding ions are not neutralized. As a result of ionizing radiation, an electron is free to wonder in the crystal and it can be attracted by a coulomb force to the localized positive charge and can be trapped in the vacancy. This system or center is called F center. Similarly, a positive ion vacancy represents a hole trap and the system is called V center, but no experimental data are known about it. Other types of hole centers are possible:

- the V_k center is obtained when a hole is trapped by a pair of negative ions,
- the V_3 center which consists in a neutral halogen molecule which occupies the site of a halogen ion: in effect two halide ions with two holes trapped.

All the previous defects are shown in Fig. 5.

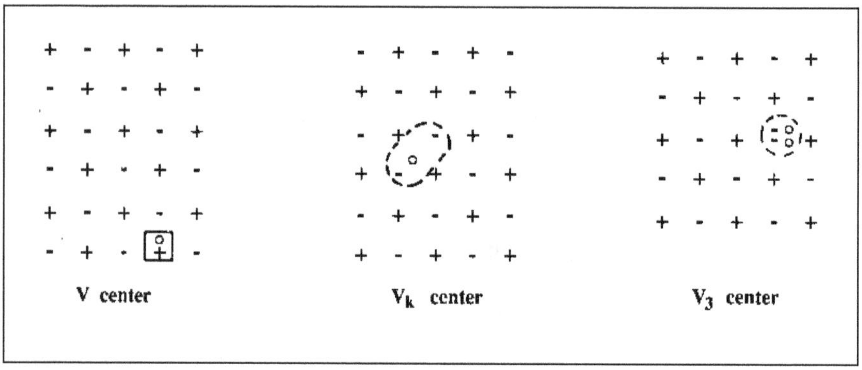

Fig. 5. V, V_k and V_3 centers in a real crystal.

We have to outline the importance of the defect production during irradiation because high dose levels can induce unwanted effects in TL materials, generally called radiation damage, which are important in the set up and maintenance of a thermoluminescent dosimetric system, i.e., lowering in sensitivity, saturation effects and so on. Furthermore, to study the color centers using various luminescence techniques, i.e., photoluminescence, can improve the knowledge of the thermoluminescent phenomena itself.

For this reason a phenomenological short feature of radiation damage in crystals is given below.

Photons, electrons, neutrons, charged and uncharged particles can create defects by displacement in the sense that the bombarding radiation displaces the crystal atoms from their normal position in the lattice, producing vacancies and interstitials. The number of defects produced is proportional to the flux of irradiation and to the irradiation time. However, during a long irradiation the number of defects produced will gradually decrease because the possibility of vacancy-interstitial recombination increases.

The irradiation can also produce negative ion vacancies by a process called ionization damage. This mechanism is related to the recombination of ionization electron and holes. During recombination a bound electron-hole pair (exciton) can be trapped on a negative lattice ion. The energy released during recombination is transferred to the negative ion which produces collisions leaving vacancies and interstitial atoms. The final result is the production of F centers and interstitial atoms.

Delocalized bands

Conduction band (CB) and valence band (VB).

Determination of the dose by thermoluminescence

The main algorithm which can be used to convert the light emission obtained during the readout of a thermoluminescent detector to the absorbed dose can be expressed by the following relationship

$$D = M \cdot F_c \tag{1}$$

where
M is the TL signal (integral light or peak height), and
F_c is the individual calibration factor of the detector.

The previous Eq. (1) can be generalised by inserting in it all the parameters which can influence the dose determination during the preparation of the detector, its irradiation, the possible period of time elapsed between the end of irradiation and readout and the readout itself.

A more general relation can then be written as following

$$D_m = M_{net} \cdot S_i \cdot F_c \cdot F_{st} \cdot F_{en} \cdot F_{lin} \cdot F_{fad} \tag{2}$$

where

- D_m is the absorbed dose in the mass m of the phosphor,
- M_{net} is the net TL signal (i.e., the TL signal corrected for the intrinsic background signal M_0: $M_{net} = M - M_0$),
- S_i is the relative intrinsic sensitivity factor or also called individual correction factor concerning the ith dosimeter,
- F_c is the individual calibration factor of the detector, relative to the beam quality, c, used for calibration purposes,
- F_{st} is the factor which takes into account the possible variations of F_c due to variations of the whole dosimetric system and of the experimental conditions (electronic instabilities of the reader, changes in the planchet reflectivity, changes in the light transmission efficiency of the filters interposed between the planchet and the PM tube, temperature instabilities of the annealing

ovens, variation of the environmental conditions in the laboratory, changes in the dose rate of the calibration source, etc.),

- F_{en} is the factor which allows for a correction for the beam quality, q, if the radiation beam used is different from the one used for the detector calibration,
- F_{lin} is the factor which takes into account for the non-linearity of the TL signal as a function of the dose,
- F_{fad} is the correction factor for fading which is a function of the temperature and the period of time between the end of irradiation and readout.

Dihalides phosphors

Dihalides have the general formula AXY, where A is an alkaline earth metal and X and Y are two halogens. Single crystals of dihalides are obtained by growth using different known techniques. From a melt containing a mixture of a metal halide and a dopant (i.e. rare earth of heavy metal ions).

The systems studied are BaFCl, BaFBr, SrFCl and SrFBr doped by Tl or Gd ions [1-3].

References

1. Somaiah K., Vuresham P., Prisad K.L.N. and Hari Babu V., Phys. Stat. Sol.(a) **56** (1979) 737
2. Somaiah K. and Hari Babu V., Phys. Stat Sol.(a) **79** (1984) 237
3. Somaiah K. and Hari Babu V., Phys. Stat Sol.(a) **82** (1984) 201

Dosimeter's background or zero dose reading (definition)

The dosimeter's background, also called **zero dose reading**, is obtained from repeated measurements carried out on unirradiated dosimeters. This quantity is particularly important when the dosimeters are used for low dose measurements. As the dose increases, the background and its variation become less important and then can be neglected at high doses.

The TL signal related to the background is due to various components:

- spurious signals from tribo- and chemi-luminescence,
- stimulation of the TL phosphor by UV and visible light,
- infrared emission of the heating element and its surroundings,
- dark current fluctuations of the PM tube,
- residual signals from the TL phosphor due to previous irradiations.

All the above given effects can be reduced or eliminated using appropriate procedures during handling and use of the TL dosemeters.

Dosemeter's background or zero dose reading (procedure)

The dosemeter's background, also called zero dose reading, is obtained from repeated measurements carried out on annealed and unirradiated dosemeters. This quantity is particularly important when the dosemeters are used for low dose measurements. As the dose increases, the background and its variation become less important and then can be neglected at high doses.

Quantitatively speaking, the zero dose reading is the result of two main components:

- reading without dosemeter: L_0 (dark current)
- reading of unexposed dosemeter: L_u

Several readings L_0 and L_u have to be performed for getting the average values $\overline{L_0}$ and $\overline{L_u}$. Then, the mean value of the zero dose reading is given by

$$\overline{L}_{BKG} = \overline{\overline{L}}_0 + \overline{L}_u \tag{1}$$

with the corresponding standard deviation, σ_{BKG}.

In the lower dose range, the mean value $\overline{L}_{BKG}$ given by Eq. (1) has to be subtracted from from the irradiated dosemeter readings.

Dosimetric peak

It indicates a very well resolved peak in the glow curve structure, having a high intensity and a good stability, i.e. it is not or almost not affected by fading. These characteristics allow an accurate determination of the given dose.

Dosimetric trap

It is usual to indicate as dosimetric trap the trapping center related to a particular peak in the glow curve and called dosimetric peak. This peak is used for dosimetric purposes.

E

Effect of temperature lag on trapping parameters

The effect of temperature lag on the determination of the trapping parameters can be determined using the Randall-Wilkins model for the first-order kinetics [1,2]. The equation of the TL intensity is given by

$$I(t) = ns\exp\left(-\frac{E}{kT_1}\right) \tag{1}$$

with the usual meaning of the symbols and where T_1, the temperature of the heating element, is given by

$$T_1 = T_0 + \beta \cdot t$$

Considering the temperature lag, Eq. (1) becomes

$$I(t) = ns\exp\left(-\frac{E}{k(T_1 - \Delta T)}\right) \tag{2}$$

The exponential in Eq. (2) can be developed around the temperature of the maximum TL intensity, T_M, into powers of $(T_1 - \Delta T - T_M)$. Thus, Eq. (2) can be approximated, as a first approximation, by

$$I(t) = ns\exp\left(\frac{E\Delta T}{kT_M^2}\right)\exp\left[-\frac{E\left(1+\frac{2\Delta T}{T_M}\right)}{kT_1}\right] \tag{3}$$

Comparing Eq. (1) with Eq. (3), it can be observed that, if the temperature lag is ignored, the activation energy and the logarithm of the frequency factor are overestimated by the quantities

$$\Delta E = \frac{2E\Delta T}{T_M} \qquad \Delta \ln s = \frac{\Delta E}{2kT_M}$$

According to the experimental results on the peak 4th of TLD-100, errors in E and $ln(s)$ can be of the order of 6% and 3% respectively if the temperature lag is neglected.

References

1. Piters T.M. and Bos A.J.J., J. Phys. D: Appl. Phys. **27** (1994) 1747
2. Piters T.M., A study into the mechanism of thermoluminescence in a LiF:Mg,Ti dosimetry material (Thesis, 1998), D.U.T.

Energy dependence (procedure)

- prepare *n* groups (as many points of energy as possible for one intends to use) of at least 6 TLDs each;
- each group of TLDs is inserted in plastic bags and the bags are irradiated in air, using the appropriate built-up thickness for each point of energy;
- irradiate each group with a reference dose at one energy;
- read each group;
- correct each reading by individual background and by individual sensitivity factor.

An example of data concerning the energy dependence of LiF:Mg,Ti (TLD-700) is reported in the following Table 1. The irradiations have been carried out in air with the appropriate built-up for ^{137}Cs and ^{60}Co. Six TLDs were inserted in a black plastic bag for each point of energy. The average TL readings were already corrected for background and sensitivity factor. Figure 1 shows the relative TL response as a function of the energy.

We have to mention that the procedures of irradiation for the energy dependence can be different according to the aim of the application, such as clinical, environmental and personal dosimetry.

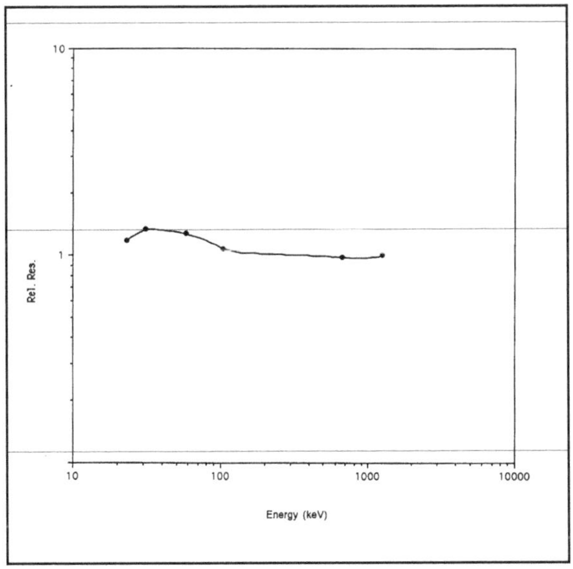

Fig. 1. Energy dependence for LiF :Mg,Ti (TLD-100).

Coming back to the correction factor F_{en}, if the calibration of the system was carried out using a ^{60}Co source and then the batch has been used at a different energy, i.e. 58 keV, the correction factor will be different from unity because at that energy the dose is overestimated. In this case we get the following correction factor

$$F_{en} = \frac{F_{1250}}{F_{58}} = \frac{1}{1.28} = 0.78$$

Energy (keV)	Dose (mGy)	Av_{cor}/mGy	σ (TL/mGy)	E_i/E_{Co}
23	7.84	146.4	14.5	1.19
31	6.36	164.8	14.7	1.34
58	6.77	158.0	14.9	1.28
104	9.15	133.0	17.9	1.08
662	10.36	120.5	24.4	0.98
1250	10.96	123.1	10.9	1.00

Table 1. Energy dependence of LiF:Mg,Ti.

It is better in principle and when it is possible, to determine the calibration factor with the same quality beam used for applications. This is easily done in clinical dosimetry, and in radiological or therapeutic monitoring. In situations like personal dosimetry the monitored radiation field is normally unknown and a different approach must be considered.

Environmental dose rate (calculation)

Considering the **escape probability rate per second** for electrons trapped in a trap

$$p = s\exp\left(-\frac{E}{kT}\right) \tag{1}$$

where
E = the trap depth (eV),
k = Boltzmann's constant ($8.6 \cdot 10^{13}$ eV/K),
T = the absolute temperature (K),
s = the frequency factor (s^{-1}) depending on the frequency of the number of hits in the trap which can be considered as a potential well.

The reciprocated p is the mean life of the trapped charges in their sites: therefore p itself is the fading factor related to the rate of fading rate, dn/dt, where n is the number of trapped charges, when the temperature is kept constant.

The fading factor can be determined using a fading experiment, under controlled environmental conditions.

In the present calculation only an isolated thermoluminescent peak is considered, without retrapping phenomenon (**first order kinetics**). In this case, the rate of release of electrons from the trap is given by

$$\frac{dn}{dt} = -p \cdot n\,. \tag{2}$$

In the assumption of constant temperature throughout the experimental period, the integration of Eq. (2) gives

$$n = n_0 \exp(-p \cdot t) \tag{3}$$

where n_0 is the initial number of the trapped charges.

Since the TL intensity is proportional to the release rate of the trapped charges

$$I(t) \propto -\frac{dn}{dt} \tag{4}$$

we obtain

$$I(t) = I_0 \exp(-p \cdot t) \tag{5}$$

where I_0 is the TL intensity at time $t = 0$.

From a read-out system the integral TL light is normally obtained; whence, introducing the function $\Phi(t)$, expressing the total TL light or the area below the TL peak, the function $\Phi(t)$ is related to $I(t)$ by the following relation:

$$\Phi(t) = \int_0^t I(t)dt \tag{6}$$

Then, $\Phi(t)$ coincides numerically with n and Eq. (3) can be rewritten as

$$\Phi(t) = \Phi_0 \exp(-p \cdot t) \tag{7}$$

from which the fading factor p is obtained as

$$p = -\frac{1}{t}\ln\frac{\Phi(t)}{\Phi_0} \tag{8}$$

The previous Eq. (8) gives then the fading factor in the case a single irradiation is performed at the beginning of the experimental period.

The experimental situation during the measurement of the environmental dose rate is described by a continuous irradiation of the dosimeter so, while the fading is equivalent to a progressive extinction of the stored information, the environmental contribution leads to a signal increasing. The two competing effects can be described modifying Eq. (2) as follows

$$\frac{dn}{dt} = \alpha \cdot \frac{dB}{dt} - p \cdot n \tag{9}$$

where α is the dosemeter sensitivity (i.e. reader units/dose) and dB/dt is the rate of increase of the background dose due to the environmental radiation.

The integration of Eq. (9) gives

$$n = c\exp(-p \cdot t) + \frac{\alpha}{p} \cdot \frac{dB}{dt} \tag{10}$$

Setting up the initial condition $n_0 = 0$ at the initial time of the environmental monitoring, the constant c is given by

$$c = -\frac{\alpha}{p} \cdot \frac{dB}{dt} \tag{11}$$

which can be substituted into Eq. (10):

$$n = \frac{\alpha}{p} \cdot \frac{dB}{dt} \left[1 - \exp(-p \cdot t) \right] \tag{12}$$

As already done before, the substitution of Φ instead of n can be operated, obtaining

$$\Phi(t) = \frac{\alpha}{p} \cdot \frac{dB}{dt} \left[1 - \exp(-p \cdot t) \right] \tag{13}$$

If t_W indicates the whole monitoring time during the environmental dose determination, the previous equation yields

$$\Phi_W(t_W) = \frac{\alpha}{p} \cdot \frac{dB}{dt} \left[1 - \exp(-p \cdot t_W) \right] \tag{14}$$

from which the environmental dose rate per day, corrected by fading, is obtained

$$\frac{dB}{dt} = p \frac{\Phi_W}{\alpha} \left[1 - \exp(-p \cdot t_W) \right]^{-1} \tag{15}$$

The accumulated environmental dose, B, is then

$$B = \frac{dB}{dt} \cdot t_W \tag{16}$$

The dose rate per hour is then given by

$$\left(\frac{dB}{dt} \right)_h = \frac{(\frac{dB}{dt})}{24} \tag{17}$$

The following Table 1 shows the numerical evaluation of p at different temperatures for LiF:Mg,Ti (TLD-100), using the data E = 1.36 eV and $s = 2.20 \cdot 10^{13}$ s^{-1}, corresponding to the dosimetric peak in LiF.

Temperature (K)	p (day)$^{-1}$	p% x 1 year	τ (year)
273	$1.3\cdot10^{-7}$	0.005	20693
275	$2.0\cdot10^{-7}$	0.007	13583
278	$3.8\cdot10^{-7}$	0.01	7304
280	$5.6\cdot10^{-7}$	0.02	4865
283	$1.0\cdot10^{-6}$	0.04	2673
285	$1.5\cdot10^{-6}$	0.1	1807
288	$2.7\cdot10^{-6}$	0.1	1014
290	$3.9\cdot10^{-6}$	0.2	694
293	$6.9\cdot10^{-6}$	0.3	397
295	$9.9\cdot10^{-6}$	0.4	275
298	$1.7\cdot10^{-5}$	0.6	161
300	$2.4\cdot10^{-5}$	0.9	113
303	$4.3\cdot10^{-5}$	1.6	64
308	$9.9\cdot10^{-5}$	3.6	28
313	$2.3\cdot10^{-4}$	8.4	12
318	$4.9\cdot10^{-4}$	17.9	6
323	$1.1\cdot10^{-3}$	40.2	3

Table 1. Calculated values of the fading factor at various temperatures.

Environmental dose rate (correction factors)

The environmental dose rate per day is given by

$$\frac{dB}{dt} = p\frac{\Phi_W}{\alpha}\left[1-\exp(-p\cdot t_W)\right]^{-1} \qquad (1)$$

according to the environmental dose rate calculation [see Eq. (15)].

The previous equation has to be corrected as follows.

The correction factor to be considered is the **zero dose reading** or **background** of the TL detector. We can denote this value as b. Then, the actual reading Φ, as well as the initial value Φ_0 have to be corrected by the background value b, subtracting it from both the previous values. This correction has to be done in both fading and environmental experiments.

Correction in the fading experiment

To take into account the zero dose reading in the fading experiment, a set of annealed and undosed dosimeters, called **control dosimeters**, have to be used. One group of these dosimeters has to be read out immediately after annealing to check the background. The second group will be read at the end of the fading experimental period to measure both the background and the possible environmental signals. Let us indicate with β this environmental signal. Then, the correct readings in the fading experiment will be:

$$\Psi_0 = \Phi_0 - b \tag{2}$$

and

$$\Psi = \Phi - (b + \beta) \tag{3}$$

The **fading factor** is given by

$$p = -\frac{1}{t}\ln\frac{\Psi}{\Psi_0} \tag{4}$$

<u>*Correction in the environmental measurement*</u>

The equation giving the TL reading after the environmental experiment is given by Eq. (1). In that expression, Φ_w represents the sum of the environmental as well as the dosimeter background signals:

$$\Phi_W = b + \Phi_{Wnet}$$

from which

$$\Phi_{Wnet} = \Phi_W - b \tag{5}$$

then Eq. (1) becomes

$$\frac{dB}{dt} = p\frac{\Phi_{Wnet}}{\alpha}\left[1 - \exp(-p \cdot t_W)\right]^{-1} \tag{6}$$

<u>*Correction of the sensitivity factor, α.*</u>

The sensitivity factor is obtained using a calibration dose, D_0. After irradiation of the calibration dosemeters with the calibration dose, the average reading will be S_0. The sensitivity is then given by

$$\alpha = \frac{S_0}{D_0} \tag{7}$$

Owing to the dosimeter background b, the previous equation has to be corrected as follows

$$\alpha = \frac{S_0 - b}{D_0} \tag{8}$$

Erasing treatment

The erasing treatment is the thermal procedure used to empty the traps of a phosphor. In some way it is different from the thermal annealing. More precisely, the annealing also has the function to stabilize the traps; the erasing procedure is just used to empty the traps and then it could be carried out in the reader.

Error sources in TLD measurements

There are many sources of error in a thermoluminescence dosimetry system and a considerable effort can be done to reduce the effects of uncertainty on the accuracy and precision of the system [1-3].

First of all we have to list the commonly encountered sources of error that affect the precision and accuracy of the system.

Both systematic and random sources of error can be originated from the characteristics of the thermoluminescent detector, or by the TL reader, or they come out by the incorrect heat treatment during readout or during the anneal process. In all cases it is essential to carry out the whole procedure in a very high reproducible manner.

Sources of error due to the dosimeter

They can be enumerated as follows:

- variation of transparency and other optical properties of the dosimeter;
- variation of the optical properties of the covering material of the TLD element if this material and the phosphor make a single body during readout (it is the case of some type of TLD cards);
- effects due to the artificial light and/or natural light (optical fading);
- effects due to the energy dependence of the thermoluminescent response;
- effects due to the directional dependence of the incident radiation on the thermoluminescent response;
- abnormal high values of the irradiation temperature;
- non-radioactive contaminations of the phosphor and/or the detector;
- non-efficient and non-reproducible procedure for cleaning the dosimeter;
- variations in the mass and size of the TL material;
- non-uniform distribution of the TL material on the reader tray when powder is used;
- variations in sensitivity owing to radiation damage of the TL material;

- loss of TL signal owing to thermal fading;
- increase of the TL background due to environmental radiations.

Several of the previous sources of error can be avoided by taking a considerable care during handling of the detectors. For instance, avoid any accidental contact between the TLD material and the fingers of the operator and/or the body of a patient during radiological inspections or therapy treatments.

The use of metal tweezers can provoke crashes on the TL element surface and/or detachment of fragment material; use vacuum tweezers.

Pay attention to the radiation history of each detector and reject the dosimeters which have received an abnormal high dose.

Take a considerable care in the annealing procedure and be sure that the set temperature is the correct and appropriate value for a given annealing treatment. Check also if the actual temperature matches the set temperature value. A particular attention must be paid to the temperature distribution inside the anneal oven. Inappropriate lower annealing temperatures can leave high residual TL signals due to previous irradiations. On the other hand, higher annealing temperatures can damage the crystal lattice and destroy traps and recombination centers.

Thermoluminescent materials are, in general, sensitive to light, especially to the ultraviolet component. The rate of fading can be increased substantially in the case of intense UV irradiations; in some cases the background can also be increased. It is always a good procedure to keep the TLDs away from any light sources, shielding them either during use or storage. A black plastic box is enough to avoid light effects.

Pay attention to any radiation sources which can occasionally be in the TLD laboratory. During storage after annealing, the TLDs must be located in appropriate lead box to avoid any radiation effects due to radioactive elements in some building materials (e.g. concrete) and/or from natural environmental radioactivity.

As it can be easily observed, many types of error can be avoided by making use of appropriate and accurate handling procedures.

One of them concerns the variations in sensitivity of TLDs within a batch. Variations of sensitivity within a batch of TLDs are quite inevitable even with a fresh batch of phosphors. These variations can increase with time due to loss of the phosphor material, changes in the optical properties and other damages, and the introduction of systematic errors in the measurements. Several methods are in use for limiting the effects of sensitivity variation in accuracy and precision for the measurements. The best approach is to divide the TLDs into batches each having similar sensitivity and then to use appropriate sensitivity factors, as it will be described later on. Further improvements can be obtained, if an individual calibration is carried out for each detector. For specific uses where an extreme accuracy is required, i.e. in clinical applications, the best procedure would be to calibrate the detectors before and after each measurement. The choice of a specific procedure depends strongly on the accuracy required. However, a check must be done frequently during the time of use of a given batch.

Large errors can be introduced in the dose determination when the dosimeters are exposed to photons of unknown energy, mainly in a range around 100 keV and below because in this region the photoelectric effect is predominant and then the dosimeters could overestimate the dose. This kind of error can be minimized by using the tissue equivalent phosphors which present a small variation in response to energy. Another method is to calibrate the detectors with a well-known beam quality and then use them with the same kind of beam. In some cases this procedure is not possible because the field dosimeters are used in personnel or environmental dosimetry where the energy field is not known. In these cases the errors can be minimized by using a combination of tissue equivalent materials and non-tissue equivalent materials so that information on the radiation energy can be obtained and corrections can be made.

The thickness of the dosimeter is another factor to be taken into consideration. For low energy photons and for beta irradiation a thick dosimeter can give an under-response owing to the self-absorption effect. On the contrary if the detector is too thin, it can give an under-response at high photon energies because of a lack of electron equilibrium.

Before using any TL material, it is necessary to perform an accurate thermal fading experiment simulating the real conditions of the field measurements. Fading depends on the depth of the trap corresponding to the dosimetric peak; the stability of the trap is a function of the annealing procedure which, in turn, depends on the characteristic of the anneal oven.

Errors generated in the reader

Errors associated with the reader can be generated by an unsuitable or instable readout cycle, as well as by non-reproducibility of the detector position in the reader tray.

In readers using planchet as heating element, an error is generated by a poor thermal contact between detector and heater.

If a built-in reference light source is used to check the stability and the background of the reader, attention should be paid to its performance which can change as a function of time and temperature.

The use of TL powder can provoke contamination of the PM tube or of the filter interposed between the PM and the tray and then their opacity. Irradiated powder lost in the reading chamber produces abnormal high background signals during successive use of the reader.

Concerning the background signal of the undosed TLDs or their zero dose reading, its effect on the dose evaluation is large when low doses have to be measured. It is very important to determine the standard deviation associated with the average background of the undosed detectors. It is easily observed that as the dose increases, the effect of the background and its variation becomes less and less important.

The light collection efficiency of the reader can change if the reflectance of the heater element changes; it is imperative to keep all parts of the reading chamber clean.

Another error can arise during the readout; the reading cycle must not include all the glow curve but only the dosimetric peak. Including low temperature peaks provoke errors due to their high fading rate. Use the pre-heat technique is necessary, both in oven or in the reader, to erase these low temperature peaks.

Errors due to the annealing procedures

It has well been demonstrated that the non-reproducibility of the annealing procedure can provoke large variations in the sensitivity of the TL materials.

It is recommended to carry out thermal erasing procedure in oven. An in-reader anneal can be done just in the case where very low irradiation doses have been detected and also in that case to be sure about the efficacy of the procedure in terms of reproducibility in the measurements.

The in-reader anneal procedure is normally done for some type of TLD cards where the phosphors are covered by plastic transparent materials and the covering material and the phosphors cannot be separated. In cases where the cards have received a high dose, the in-reader anneal is not efficient and the cards must be rejected.

For each TL material the proper annealing procedure must be determined and checked, both in temperature and time. The best combination of temperature and time will produce an effective depletion of the traps. Repeated cycles of irradiation and annealing-readout will show the precision of the thermal procedure as a function of the residual TL emission. Repeated cycles of annealing-irradiation-readout will show the precision of the TL response.

Another important factor which can introduce error in the dose determination concerns the cooling rate after annealing. As the cooling rate changes, the sensitivity changes dramatically. This effect is observed in any kind of TLDs. The best way is always to use the same procedure for cooling the TLDs. It must also be checked if a fast or low cooling rate is better for a given TLD material.

References

1. Busuoli G. in Applied Thermoluminescent Dosimetry, ISPRA Courses, Edited by M. Oberhofer and A. Scharmann. Adam Hilger Publisher (1981)
2. Marshall T.O. in Proc. of the Hospital Physicists' Association. Meeting on Practical Aspects of TLD. Edited by A.P. Hufton, University of Manchester, 29th March, 1984
3. Nambi K.S.V., Thermoluminescence: Its understanding and Applications. Instituto de Energia Atomica, São Paolo, Brasil, INF.IEA 54 (1977)

Exponential heating function (EHF)

The aim of the work of Kits et al. (1,2) is to derive the TL theory, for first and general order kinetics equations, describing single glow-peak shapes under an exponential heating function (EHF); this kind of heating system is widely employed in hot gas readers. The case of EHF was first proposed by Osada (3) and then applied by Dijk and Julius (4).

The temperature variation in EHF is given by the following expression:

$$T(t)=T_g-(T_g-T_0)\exp(-\alpha t) \tag{1}$$

where

T_g is the maximum hot gas temperature approached asymptotically with time t,

T_0 is the temperature at t = 0,

α (sec^{-1}) is given by

$$\alpha=\frac{\delta A}{mc_p} \tag{2}$$

where

δ is the heat transfer efficiency,

A is the heated area,

m is the mass of the TL sample,

c_p is the heat capacity

The heating rate is then, from Eq. (1):

$$\beta=\frac{dT}{dt}=\alpha(T_g-T) \tag{3}$$

Using Eq. (3), the first order kinetics equation becomes:

$$I(T)=n_0 s\exp\left(-\frac{E}{kT}\right)\exp\left[-\frac{s}{\alpha}\int_{T_0}^{T}\frac{\exp\left(-\frac{E}{kT'}\right)}{T_g-T'}dT'\right] \tag{4}$$

For the general order kinetics case the equation is:

$$I(T)=n_0 s''\exp\left(-\frac{E}{kT}\right)\left[1+\frac{(b-1)s''}{\alpha}\cdot\int_{T_0}^{T}\frac{\exp\left(-\frac{E}{kT'}\right)}{T_g-T'}dT'\right]^{-b(b-1)} \quad \ldots \tag{5}$$

with

$$s''=s'n_0^{b-1}$$

The second order kinetics equation is obtained by Eq. (5) by setting b=2 and s'=s/N

$$I(T)=n_0 s' \exp\left(-\frac{E}{kT}\right)\left[1+\frac{s'}{\alpha}\cdot\int_{T_0}^{T}\frac{\exp\left(-\frac{E}{kT'}\right)}{T_g-T'}dT'\right]^{-2} \tag{6}$$

All the previous TL intensity expressions contain the integral

$$F\left(T,T_g,E\right)=\int_0^T\frac{\exp\left(-E/kT'\right)}{T_g-T'}dT' \tag{7}$$

Eq. (7) can be solved replacing

$$\frac{1}{T'}=x,\qquad \frac{E}{k}=p,\qquad T_g=\alpha$$

and then it becomes

$$-\int_{\infty}^{u}\frac{\exp(-px)}{x(\alpha x-1)}dx=\int_{x=u}^{\infty}\frac{\exp(-px)}{x(\alpha x-1)}dx=$$

$$=-\int_u^{\infty}\frac{\exp(-px)}{x}dx-\int_u^{\infty}\frac{\exp(-px)}{\left(\frac{1}{\alpha}-x\right)}dx \tag{8}$$

The first term of the last expression is the exponential integral $E_i(u)$ defined in (5,6)

$$E_i\left(u\right)=-E_i\left(-u\right)=\int_u^{\infty}\frac{e^{-x}}{x}dx \tag{9}$$

and the second term has the form given in (6):
for p>0 and c<u it is

$$\int_u^{\infty}\frac{\exp(-px)}{c-x}dx=\exp(-pc)E_i\left(pc-pu\right) \tag{10}$$

According to Eq. (10) the right-hand side of Eq. (8) becomes

$$-\int_u^\infty \frac{\exp(-px)}{x}dx - \int_0^\infty \frac{\exp(-px)}{\left(\frac{1}{\alpha}-x\right)}dx = E_i(-px) - \exp\left(-\frac{p}{\alpha}\right)E_i\left(\frac{p}{\alpha}-px\right)$$

Replacing back the original quantities, the final form for *F* is then obtained:

$$F\left(T,T_g,E\right) = -\exp\left(-\frac{E}{kT_g}\right)E_i\left(\frac{E}{kT_g}-\frac{E}{kT}\right) + E_i\left(-\frac{E}{kT}\right) \quad (11)$$

and then Eq. (7) can be written as

$$\int_{T_0}^{T} \frac{\exp\left(-E/kT'\right)}{T_g - T'}dT' = F\left(T,T_g,E\right) - F\left(T_0,T_g,E\right) \quad (12)$$

As it has been shown in [6], the values of $F(T_0, T_g, E)$ are negligible relative to the values of $F(T, T_g, E)$ and then Eq. (12) becomes:

$$\int_{T_0}^{T} \frac{\exp\left(-E/kT\ \right)}{\left(T_g - T'\right)}dT' \approx F\left(T,T_g,E\right)$$

Using now the first two terms of the appropriate asymptotic approximation (ASA):

$$-E_i(-z) = \exp(-z)\sum_{n=0}^{N}\frac{(-1)^n n!}{z^{n+1}}$$

Eq. (11) becomes

$$F\left(T,T_g,E\right) \approx -\exp\left(-\frac{E}{kT_g}\right)\exp\left(\frac{E}{kT_g}-\frac{E}{kT}\right)\cdot$$

$$\cdot\frac{kTT_g}{E\left(T-T_g\right)} - \frac{kT}{E}\exp\left(-\frac{E}{kT}\right) \approx$$

$$\approx -\frac{kT^2}{E}\cdot\frac{1}{T-T_g}\exp\left[\frac{E}{kT}\left(1-\frac{T}{T_g}\right)\right] \quad (13)$$

Then we have

$$\frac{F\left(T_0,T_g,E\right)}{F\left(T,T_g,E\right)} \approx \frac{T_0^2\left(T-T_g\right)}{T^2\left(T_0-T_g\right)}\exp\left[\frac{E}{kT}\left(1-\frac{T}{T_0}\right)\right] \tag{14}$$

Coming back to Eqs. (4) and (5), the condition at the maximum is

$$\frac{d\ln\left[I(T)\right]}{dT}=0 \tag{15}$$

and then, for Eqs. (4) and (5), we get respectively

$$\frac{\alpha E\left(T_g-T_M\right)}{kT_M^2}=s\exp\left(-\frac{E}{kT_M}\right) \tag{16}$$

and

$$\frac{\alpha\left(T_g-T_M\right)E}{bkT_M^2}\left[1+\frac{(b-1)s}{\alpha}F\left(T_M,T_g,E\right)\right]=s''\exp\left(-\frac{E}{kT_M}\right) \;\cdots\cdots\cdots \tag{17}$$

Eqs. (4) and (5) are in the form $I(n_0, E, s, T)$ but they can be now expressed in the form $I(I_M, T_M, E, T)$ which is more useful for practical situations.

Eq. (4) of the first order kinetic, by substitution of the quantity s obtained from Eq. (16), becomes

$$\begin{aligned} I(T)= & \frac{n_0\alpha E\left(T_g-T\right)}{kT_M^2}\exp\left[-\frac{E\left(T_M-T\right)}{kTT_M}\right]\cdot \\ & \cdot\exp\left[-\frac{E\left(T_g-T\right)}{kT_M^2}\exp\left(\frac{E}{kT_M}\right)F\left(T,T_g,E\right)\right] \end{aligned} \tag{18}$$

Considering the value I_M of $I(T)$ at T_M, we get

$$\frac{n_0\alpha E\left(T_g-T\right)}{kT_M^2}=I_M\exp\left[\frac{E\left(T_g-T\right)}{kT_M^2}\exp\left(\frac{E}{kT_M}\right)F\left(T_M,T_g,E\right)\right] \tag{19}$$

Inserting this expression in Eq. (18) we finally get

$$I(T)=$$
$$=I_M\exp\left\{-\frac{E(T_M-T)}{kTT_M}+\frac{E(T_g-Y)}{kT_M^2}\exp\left(\frac{E}{kT_M}\right)\left[F(T_M,T_g,E)-F(T,T_g,E)\right]\right\}. \tag{20}$$

The expression for the second order is obtained in a similar manner as:

$$I(T)=I_M\exp\left[-\frac{E(T_M-T)}{kTT_M}\right]\cdot$$
$$\cdot\left\{1-\frac{(b-1)(T_g-T_M)}{bkT_M^2}\exp\left(\frac{E}{kT_M}\right)\cdot\left[F(T_M,T_g,E)-F(T,T_g,E)\right]\right\}^{-\frac{b}{b-1}} \tag{21}$$

Eqs. (20) and (21) represent single glow peaks depending on the experimental parameters I_M and T_M.

The next step is to calculate the appropriate values of F(T,T$_g$,E), i.e. the values of the exponential integrals appearing in Eq. (11). This can be obtained using an appropriate approximation for the exponential integral. Two kinds of approximation can be used:

(i) The asymptotic series approximation (ASA)
(ii) The convergent series approximation (CSA)

ASA

The ASA applied to both terms of $F(T,T_g,E)$ in Eq. (11), transforms Eqs. (4) and (5) in

$$I(T)=n_0 s\exp\left(-\frac{E}{kT}\right)\exp\left[-\frac{skT}{\alpha E}\exp\left(-\frac{E}{kT}\right)\cdot Z(T)\right] \tag{22}$$

for the first order, and

$$I(T)=n_0 s''\exp\left(-\frac{E}{kT}\right)\cdot\left[1+\frac{(b-1)s''kT}{\alpha E}\exp\left(-\frac{E}{kT}\right)Z(T)\right]^{-\frac{b}{b-1}} \tag{23}$$

for the general order case.

$Z(T)$ is given by

$$Z(T)=\sum_{n=0}^{N}(-1)^{n}\,n!D^{n}\left(B^{n+1}-1\right)+EC$$

where

$$B=\frac{T_g}{T_g-T},\qquad D=\frac{kT}{E}$$

EC is an error correction term which has to taken into account when ASA is used.

Eqs. (20) and (21) then become:

$$I(T)=I_M\exp\left\{-\frac{E(T_M-T)}{kTT_M}+\frac{T_g-T_M}{T_M}\cdot\right.$$
$$\left.\cdot\left(Z_M-\frac{T}{T_M}\exp\left[-\frac{E(T_M-T)}{kTT_M}\right]Z(T)\right)\right\} \tag{24}$$

for the first order, and

$$I(T)=I_M\exp\left[-\frac{E(T_MT)}{kTT_M}\right]\cdot\left\{1-\frac{(b-1)(T_g-T_M)}{bT_M}\cdot\right.$$
$$\left.\cdot\left(Z_M-\frac{T}{T_M}\exp\left[-\frac{E(T_M-T)}{kTT_M}\right]Z(T)\right)\right\}^{-\frac{b}{b-1}} \tag{25}$$

for the general order.

CSA

The CSA is now applied to the first term of Eq. (11) and the ASA to the second term. The first and general order equations are now respectively:

$$I(T)=n_0 s\exp\left(-\frac{E}{kT}\right)\cdot$$
$$\cdot\exp\left[-\frac{s}{\alpha}\exp\left(-\frac{E}{kT_g}\right)Z_1(T)-\frac{kT}{E}Z_2(T)\exp\left(-\frac{E}{kT}\right)\right] \tag{26}$$

$$I(T) = n_0 s'' \exp\left(-\frac{E}{kT}\right)\cdot$$
$$\cdot\left[1+\frac{(b-1)s''}{\alpha}\exp\left(-\frac{E}{kT_g}\right)Z_1(T)-\frac{kT}{E}Z_2(T)\exp\left(-\frac{E}{kT}\right)\right]^{-\frac{b}{b-1}} \tag{27}$$

where

$$Z_1(T) = \gamma + \ln\left|G(T)\right| + \sum_{n=1}^{50}\frac{G(T)^n}{n\cdot n!} \qquad \text{CSA}$$

$$Z_2(T) = \sum_{n=0}^{N}(-1)^n n! D^n + EC \qquad \text{ASA}$$

with

$$G(T) = \frac{E}{kT}\cdot\frac{T-T_g}{T_g} \qquad \text{and} \qquad D = \frac{kT}{E}$$

The new equations, expressed as $I(I_M,E,T_M,T)$, for the first order kinetics, and $I(I_M,E,T_M,b,T)$ for the general order, are then:

$$I(T) = I_M \exp\left[-\frac{E(T_m-T)}{kTT_M}-\frac{E(T_g-T_M)}{kT_M^2}\exp\left(\frac{E(T_g-T_M)}{kT_MT_g}\right)\cdot\right.$$
$$\left.\cdot(Z_1(T_M)-Z_1(T))-\frac{T_g-T_M}{T_M}\cdot\left(Z_2(T_M)-Z_2(T)\frac{T}{T_M}\exp\left(-\frac{E(T_M-T)}{kTT_M}\right)\right)\right]\cdot \tag{28}$$

and

$$I(T) = I_M \exp\left(-\frac{E(T_MT)}{kTT_M}\right)\cdot\left[1-\frac{(b-1)(T_g-T_M)}{bT_M}\left(\frac{E}{kT_M}\cdot\right.\right.$$
$$\cdot\exp\left(\frac{E(T_g-T_M)}{kT_gT_M}\right)\cdot(Z_1(T)-Z_1(T_M)) +$$
$$\left.\left.\frac{T}{T_M}Z_2(T)\exp\left(-\frac{E(T_M-T)}{kTT_M}\right)-Z_2(T_M)\right)\right]^{\frac{b}{b-1}} \tag{29}$$

References

1. Kitis G., Chen R., Pagonis V., Carinou E., Kamenopoulos V., J. Phys. D: Appl. Phys. **39** (2006) 1500
2. Kitis G., Chen R., Pagonis V., Carinou E., Kamenopoulos V., J. Phys. D: Appl. Phys. **39** (2006) 1508
3. Osada K., J. Phys. Soc. Japan **15** (1960) 145
4. Dijk Van J.W.E, Julius H.W, Rad. Prot. Dos. **47** (1993) 479
5. Chen R., McKeever S.W.S., "Theory of Thermoluminescence and Related Phenomenon" , World Scientific, Singapore (1997)
6. Chen R., J. Comput. Phys., **4** (1969) 415

F

Fading (theoretical aspects)

To study, theoretically, the fading effects in various situations simulating practical cases, it is possible to consider the simple TL system, shown in Fig. 1, in which only one kind of electron trap and one recombination center are present. In such a case, the system of equations describing the traffic of charges between the trapping levels and the conduction and valence bands is the following [1-5]:

$$\frac{dn}{dt} = -n\lambda + A_n n^*(N-n) \tag{1a}$$

$$\frac{dn^*}{dt} = n\lambda - A_n n^*(N-n) - A_m n^* m + \dot{n}_i \tag{1b}$$

$$\frac{dm}{dt} = -A_m n^* m + A_h m^*(M-m) - m\lambda_h \tag{1c}$$

$$\frac{dm^*}{dt} = -A_h m^*(M-m) + m\lambda_h + \dot{n}_i \tag{1d}$$

where

n	is the trapped electron concentration (cm^{-3}) at time t
n^*	is the electron concentration (cm^{-3}) in the CB at time t
m	is the trapped hole concentration (cm^{-3}) at time t
m^*	is the hole concentration (cm^{-3}) in the VB at time t
N	is the total density of electron traps (cm^{-3})
M	is the total density of recombination centers (cm^{-3})
A_n	is the probability factor for electron trapping (cm^{-3} sec^{-1})
A_m	is the recombination probability of electrons from the CB with holes in centers (cm^{-3} sec^{-1})
A_h	is the probability factor for hole capture (cm^{-3} sec^{-1})

$$\lambda = s\exp\left(-\frac{E}{kT}\right)$$

$$\lambda_h = s_h \exp\left(-\frac{E_h}{kT}\right)$$

s	is the frequency factor for electron traps (sec^{-1})

E	is the thermal activation energy for electron traps (eV)
s_h	is the frequency factor for recombination centers (sec^{-1})
E_h	is the thermal activation energy for recombination centers (eV)
k	is the Boltzmann's constant (8.6×10^{-5} eV K^{-1})
T	is the absolute temperature (K)
$\dot{n}_i$	is the rate of production of electron-hole pairs due to an applied external radiation field

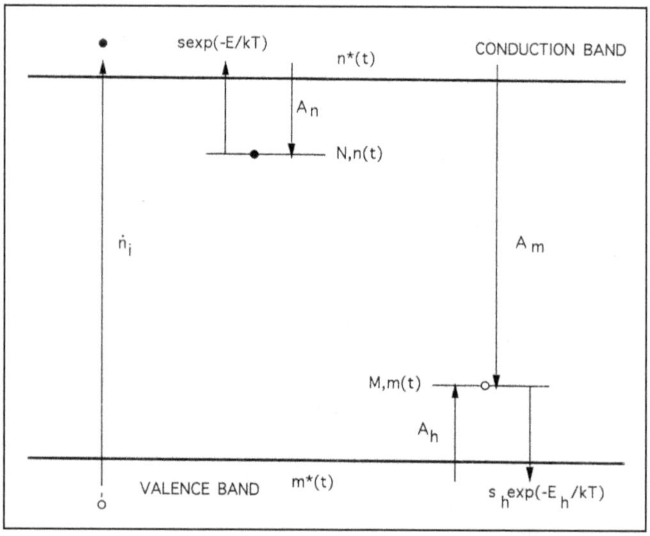

Fig. 1. Processes considered for fading simulation.

The process allowed in the system described by Eqs. (1) are:

- electron trapping and releasing from traps to the CB
- capture and releasing of holes from centers to the VB
- creation of free electron-holes pairs by the external radiation field
- recombination of free electrons with holes in recombination centers

The probability of direct band-to-band transitions and direct recombinations of trapped electrons and holes are both assumed negligible.

The evolution of the TL signal during storage it is considered at a constant temperature and the temperature dependence of the various parameters is not considered.

Eq. (1ab) can be rearranged with respect to n* :

$$n^* = \frac{\lambda n + \dot{n}_i - \dfrac{dn^*}{dt}}{A_n(N-n) + A_m m}$$

This expression can be inserted now in Eq. (1a) for obtaining the variation of the trapped electron density *n(t)* as

$$\frac{dn}{dt} = -n\lambda\left[\frac{A_m m}{A_n(N-n)+A_m m}\right] + \left[\frac{A_n(N-n)}{A_n(N-n)+A_m m}\right]\left(\dot{n}_i - \frac{dn^*}{dt}\right) \tag{2}$$

Eq. (2) can be transformed in an explicit form if the usual conditions for free carrier densities are considered:

$$\left|\frac{dn^*}{dt}\right| << \left|\frac{dn}{dt}\right|, \qquad n^* << n \tag{3}$$

$$\left|\frac{dm^*}{dt}\right| << \left|\frac{dm}{dt}\right|, \qquad m^* << m \tag{3'}$$

The conditions expressed by (3) and (3') mean that electrons and holes remain most of the time in localized states rather than in their respective bands.

Moreover, n, n^*, m and m^* are not independent functions but they are related by the charge neutrality equation:

$$\frac{dn}{dt} + \frac{dn^*}{dt} = \frac{dm}{dt} + \frac{dm^*}{dt} \tag{4}$$

which becomes, using conditions (3) (3'),

$$\frac{dn}{dt} = \frac{dm}{dt} \tag{5}$$

and then, by integration,

$$n + q = m \tag{6}$$

where q is a constant which could be different from zero. The value of q represents the net charge due to the presence of trapping centers not active at the considered temperature, i.e. disconnected traps.

Equation (2) can be then rearranged as following:

$$\frac{dn}{dt}+\left[\frac{\sigma(N-n)}{\sigma(N-n)+m}\right]\cdot\frac{dn^*}{dt}=-n\lambda\left[\frac{m}{\sigma(N-n)+m}\right]+\left[\frac{\sigma(N-n)}{\sigma(N-n)+m}\right]\cdot\dot{n}_i \tag{7}$$

where

$$\sigma=\frac{A_n}{A_m} \tag{8}$$

is the so-called retrapping-recombination cross-section ratio.

The term

$$\frac{\sigma(N-n)}{\sigma(N-n)+m}$$

comparing in Eq. (7), is always positive and lower than 1; so that one obtains for the left side of Eq. (7), assuming the conditions (3) (3'):

$$\frac{dn}{dt}+\left[\frac{\sigma(N-n)}{\sigma(N-n)+m}\right]\frac{dn^*}{dt}=\frac{dn}{dt}\left\{1+\left[\frac{\sigma(N-n)}{\sigma(N-n)+m}\right]\frac{\frac{dn^*}{dt}}{\frac{dn}{dt}}\right\}=\frac{dn}{dt} \tag{9}$$

Then, substituting $m=n+q$, one finally obtains from Eq. (7) and expression (9)

$$\frac{dn}{dt}=-n\lambda\left[\frac{n+q}{\sigma(N-n)+(n+q)}\right]+\left[\frac{\sigma(N-n)}{\sigma(N-n)+(n+q)}\right]\cdot\dot{n}_i \tag{10}$$

Equation (10) represents the form for a general order kinetics. The first and the second order kinetics are both particular cases of Eq. (10) in the case where no radiation-induced electron-hole pairs is present, i.e. $\dot{n}_i=0$. The first order kinetics is obtained for a small retrapping-recombination ratio and the second order when the ratio is high.

The two limit situations can be also obtained when $\dot{n}_i>0$.

- Assuming a strong recombination (first order kinetics), namely

$$\sigma(N-n)<<m \qquad \text{and} \qquad n(t)<<N$$

one obtains from Eq. (10)

$$\frac{dn}{dt} = -n\lambda + \frac{\sigma N \dot{n}_i}{n+q} \tag{11}$$

- On the other hand, if retrapping dominates (second order kinetics) this means

$$\sigma(N-n) >> m$$

Eq. (10) can be rearranged as following: considering both terms on its right side, they can be written as

(i) $$\frac{n+q}{\sigma(N-n)+(n+q)} = \frac{n+q}{\sigma(N-n)}\left[1+\frac{n+q}{\sigma(N-n)}\right]^{-1} =$$

$$= \frac{n+q}{\sigma(N-n)} - \left[\frac{n+q}{\sigma(N-n)}\right]^2 + \ldots\ldots \tag{12}$$

which is similar to the power expression of the type $x - x^2 + \ldots$.

(ii) $$\frac{\sigma(N-n)}{\sigma(N-n)+(n+q)} = \left[1+\frac{n+q}{\sigma(N-n)}\right]^{-1} = 1 - \frac{n+q}{\sigma(N-n)} + \ldots. \tag{13}$$

which is similar to the power expansion of the type $1 - x + \ldots.$
Now

$$\frac{n+q}{\sigma(N-n)} = \frac{m}{\sigma(N-n)}$$

but with the condition $\sigma(N-n) >> m$

$$\frac{m}{\sigma(N-n)} << 1 \tag{14}$$

so that the second order term can be rejected and finally one gets

$$\frac{n+q}{\sigma(N-n)+(n+q)} \approx \frac{n+q}{\sigma(N-n)} \tag{15}$$

Eq. (19) becomes then, using (12) and (15):

$$\frac{dn}{dt} = -n\lambda \cdot \frac{n+q}{\sigma(N-n)} + \left[1 - \frac{n+q}{\sigma(N-n)}\right] \cdot \dot{n}_i \tag{16}$$

If $n(t) << N$, far from saturation, Eq. (16) becomes

$$\frac{dn}{dt} = -n\lambda'(n+q) + \left[1 - \frac{n+q}{\sigma(N-n)}\right] \cdot \dot{n}_i \tag{17}$$

where

$$\lambda' = \frac{\lambda}{\sigma N} \tag{18}$$

If retrapping and recombination rates are equal, $\sigma = 1$, Eq. (10) simply becomes

$$\frac{dn}{dt} = -n\lambda \cdot \frac{n+q}{N+q} + \frac{N-n}{N+q} \cdot \dot{n}_i \tag{19}$$

or better

$$\frac{dn}{dt} = -n\lambda \cdot \frac{n+q}{N+q} + \left[1 - \frac{n+q}{N+q}\right] \cdot \dot{n}_i \tag{20}$$

and finally

$$\frac{dn}{dt} = -n\lambda''(n+q) + \left[1 - \frac{n+q}{N+q}\right] \cdot \dot{n}_i \tag{21}$$

where

$$\lambda'' = \frac{\lambda}{N+q} \tag{22}$$

If $\dot{n}_i = 0$ and $q = 0$, Eqs. (11), (17) and (21) reduce to the well known equations of first and second order decay processes.

Expressions related to different situations

- *Instantaneous irradiation*

A strong initial irradiation, very short compared with the storage duration, is now considered. The possible background radiation is neglected, which is equivalent to say $\dot{n}_i = 0$. Therefore, only the fading of an initial density of filled traps, i.e

$n(t=0)=n_0$ is taken into consideration. Assuming also that the net charge q is small enough compared with the density of the filled traps and then it can be considered equal to zero, Eq. (10) becomes

$$\frac{dn}{dt}=-n\lambda\left[\frac{n}{\sigma(N-n)+n}\right] \tag{23}$$

which can be rewritten as

$$\frac{dn}{dt}=-\frac{n^2\lambda}{n(1-\sigma)+\sigma N} \tag{24}$$

which can be integrated as following

$$(1-\sigma)\int_{n_0}^{n(t)}\frac{dn}{n}+\sigma N\int_{n_0}^{n(t)}\frac{dn}{n^2}=-\lambda\int_0^t dt$$

giving the solution

$$n(t)^{1-\sigma}\exp\left(-\frac{\sigma N}{n(t)}\right)=n_0^{1-\sigma}\exp\left(-\frac{\sigma N}{n_0}\right)\exp(-\lambda t) \tag{25}$$

If $\sigma=0$, Eq. (25) becomes

$$n(t)=n_0\exp(-\lambda\cdot t) \tag{26}$$

which is the equation describing the **first order kinetics** isothermal decay.
If $\sigma=1$, Eq. (25) transforms in

$$\exp\left(-\frac{N}{n}\right)=\exp\left(-\frac{N}{n_0}\right)\exp(-\lambda\cdot t)$$

and then in

$$n=n_0\left(1+\frac{n_0\lambda t}{N}\right)^{-1} \tag{27}$$

which is the **second order kinetics** equation.

- *Continuous irradiation*

Considering the more general Eq. (10), the density of trapped electrons $n(t)$ is the result of two competitive effects, acting simultaneously: the progressive storage of radiation-induced free electrons and the fading that leads to a progressive release of the trapped charges. As a consequence, the trapped electron concentration tends to a steady value when thermal raising exactly compensates the trapping of free electrons. The limit of Eq. (10), $dn/dt \to 0$, when $t \to +\infty$, gives the equilibrium value

$$\lambda n_\infty (n_\infty + q) = \sigma (N - n_\infty) \dot{n}_i$$

or better

$$n_\infty^2 = \frac{\sigma (N - n_\infty) \dot{n}_i}{\lambda q} \tag{28}$$

Considering that the system is far from saturation, i.e. $n_\infty << N$, and the net charge q is small compared with the equilibrium value at a given dose, the equilibrium density of the trapped electrons can be derived by Eq. (28) as:

$$n_\infty^2 = \frac{\sigma N \dot{n}_i}{\lambda} \tag{29}$$

In this case, it is possible to find an explict solution of Eq. (10). Indeed, taking into account expression (28) and the condition $q << n$, we get

$$\frac{dn}{dt} = -\lambda \left[\frac{n^2}{\sigma N + n(1-\sigma)} \right] + \lambda \left[\frac{(N-n) n_\infty^2}{\sigma N + n(1-\sigma)} \right] \cdot \frac{1}{N}$$

$$\frac{dn}{dt} = \left[\frac{\lambda}{\sigma N + n(1-\sigma)} \right] \left[n^2 - \frac{(N-n) n_\infty^2}{N} \right]$$

and then, considering we are far from saturation

$$\frac{dn}{dt} = \left[\frac{\lambda}{\sigma N + n(1-\sigma)} \right] \left[n^2 - n_\infty^2 \right] \tag{30}$$

After that we have

$$\int_{n_0}^{n(t)} \frac{\sigma N + (1-\sigma) n}{n^2 - n_\infty^2} dn = \sigma N \int_{n_0}^{n(t)} \frac{dn}{n^2 - n_\infty^2} + (1-\sigma) \int_{n_0}^{n(t)} \frac{n dn}{n^2 - n_\infty^2} = -\lambda t \tag{31}$$

The integrals in (31) have a singularity at $n = n_\infty$ and the $n(t)$ must be continuous; hence there are two possible solutions depending on whether n is greater or lower than n_∞.

The two possible solutions are:

- if $n_0 > n_\infty$, then $n(t)$ is always greater than n_∞ and $n(t)$ decreases with time according to the following expression

$$\left(n^2(t) - n_\infty^2\right)^{1-\sigma} \left(\frac{n(t) - n_\infty}{n(t) + n_\infty}\right)^{\frac{\sigma N}{n_\infty}} = \left(n_0^2 - n_\infty^2\right)^{1-\sigma} \left(\frac{n_0 - n_\infty}{n_0 + n_\infty}\right)^{\frac{\sigma N}{n_\infty}} \exp(-2\lambda t) \tag{32}$$

- if $n_0 < n_\infty$, then $n(t)$ is always lower than n_∞ and $n(t)$ increases with time according to

$$\left(n_\infty^2 - n^2(t)\right)^{1-\sigma} \left(\frac{n_\infty - n(t)}{n_\infty + n(t)}\right)^{\frac{\sigma N}{n_\infty}} = \left(n_\infty^2 - n_0^2\right)^{1-\sigma} \left(\frac{n_\infty - n_0}{n_\infty + n_0}\right)^{\frac{\sigma N}{n_\infty}} \exp(2\lambda t) \tag{33}$$

It has to be stressed that in both cases, $n(t)$ tends asymptotically to the steady value n_∞.

The previous equations can be now used for simulating various possible practical situations.

- *Instantaneous initial irradiation*

This case is depicted in Fig. 2.

The sample receives a short and strong initial irradiation, followed by a storage period at a constant temperature. The background irradiation is neglected in this case and so Eq. (25) can be applied.

Fig. 2. Instantaneous irradiation at the beginning of the storage period.

- *Continuous irradiation*

Fig. 3 shows the situation. The sample is irradiated during a long period, with a constant rate of trapping electrons $\dot{n}_\alpha$. During irradiation the fading acts as a competitive effect respect to the trapping. The conditions are $n(t=0)=0$, $\dot{n}_i=\dot{n}_\alpha$ and to be far from saturation. The equation to be used in this case is Eq. (32).

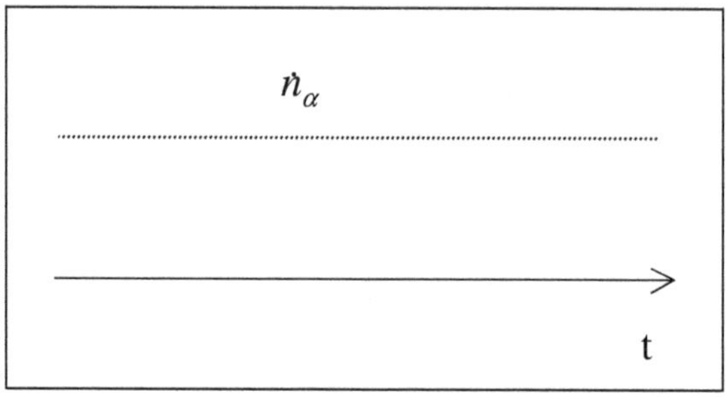

Fig. 3. Continuous irradiation during storage.

- *Instantaneous initial irradiation followed by a continuous background during storage*

Fig. 4 shows this case.

The conditions are $\dot{n}_i=\dot{n}_\alpha$, $n(t=0)=n_0$ and fading during the storage period. Eq. (31) can be applied if $n_0 > \left(\frac{\sigma N \dot{n}_\alpha}{\lambda}\right)^{1/2}$, and Eq. (33) if $n_0 < \left(\frac{\sigma N \dot{n}_\alpha}{\lambda}\right)^{1/2}$.

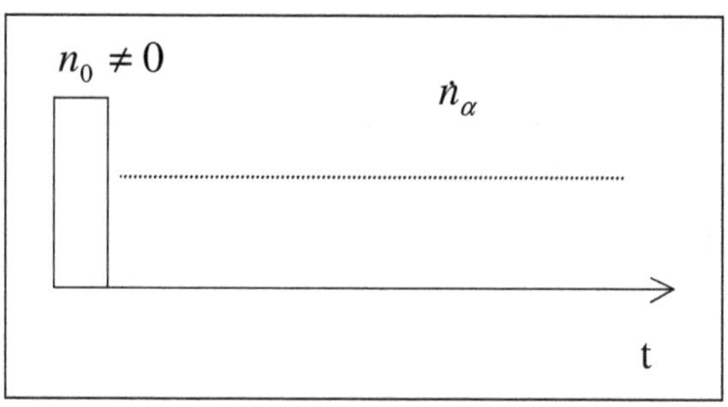

Fig. 4. Initial irradiation plus background.

- *Long irradiation during storage under continuous background irradiation*

This case is depicted in Fig. 5

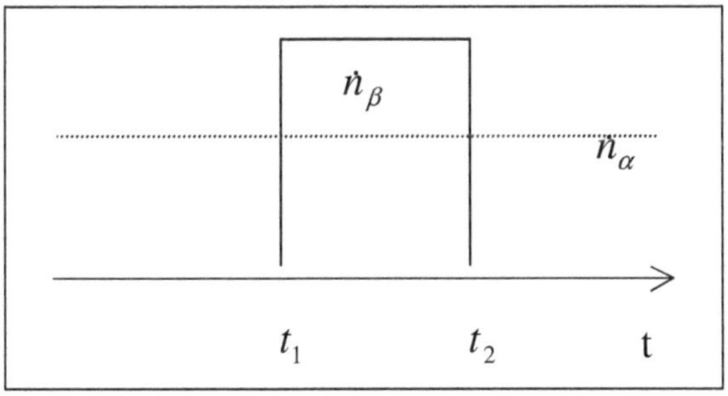

Fig. 5. Strong irradiation superposed on backgound irradiation.

The TL sample is continuously exposed to background, $\dot{n}_\alpha$, and for a given period of time bnetween t_1 and t_2, it undergoes a long and constant irradiation $\dot{n}_\alpha$. For $t<t_1$ the trapped electron density follows Eq. (33) and it will be $n_1=n(t=t_1)$ at the end of this period.

Between t_1 and t_2, the irradiation field is $\dot{n}_i=\dot{n}_\alpha+\dot{n}_\beta$. So, at time $t=t_1$ the condition

$$n(t_1)<\left[\frac{\sigma N(\dot{n}_\alpha+\dot{n}_\beta)}{\lambda}\right]^{1/2}$$

is fulfilled and Eq. (33) has to be again applied for calculation.

For $t>t_2$ two cases are possible:

$$\text{if}\quad n(t_2)>\left(\frac{\sigma N\dot{n}_\alpha}{\lambda}\right)^{1/2}, \text{ Eq. (32) is applied;}$$

$$\text{if}\quad n(t_2)<\left(\frac{\sigma N\dot{n}_\alpha}{\lambda}\right)^{1/2}, \text{ Eq. (33) is applied.}$$

References

1. Levy P.W., Nucl. Tracks Rad. Meas. **10**, 1985, 21
2. Furetta C., Nucl. Tracks Rad. Meas. **14**, 1988, 413
3. Delgado A. and Gomez Ros J.M., J. Phys. D: Appl. Phys. **23**, 1990, 571

4. Delgado A., Gomez Ros J.M. and Muñiz J.L., Rad. Prot. Dos. **45**, 1992, 101
5. Gomez Ros J.M., Delgado A., Furetta C. and Scacco A., Rad. Meas. **26**, 1996, 243

Fading factor

Starting from the **first order kinetics** equation

$$\frac{dn}{dt} = -ns\exp\left(-\frac{E}{kT}\right)$$

by integration one obtains

$$n = n_0 \exp\left[-st\exp\left(-\frac{E}{kT}\right)\right]$$

or more simply

$$n = n_0 \exp(-pt)$$

where n and n_0 are the trapped charges at time t and $t = 0$ respectively.

Considering that n is proportional to the TL emission, let us say the glow curve or peak area Φ, one gets

$$\Phi = \Phi_0 \exp(-pt)$$

and then

$$p = -\frac{1}{t}\ln\left(\frac{\Phi}{\Phi_0}\right) \quad (1)$$

Example: after irradiation of some TLDs, a part of them are immediately readout, giving an average TL reading of 1425 (reader units). The rest of the irradiated TLDs are stored in a lead box and readout after a period of 30 days, giving an average reading of 1285 (reader units). Using the previous equation one obtains

$$p = 3.45 \cdot 10^{-3} d^{-1}$$

which means a lost per day of 0.345%.

A fading based method for cheeking the presence of closely overlapping peaks in thermoluminescent (TL) materials

Several methods have been described along the years and applied to calculate the kinetics parameters [1-7]. The main problem arises from the fact that all the methods are concerning to TL glow peaks which are supposed to be originated from a single trapping level.

Some time these methods are difficult to apply when the TL peaks are closely overlapping and difficult to separate, or when a TL peak can be due to a continuous distribution of traps instead of a single level. Several experimental methods have been suggested to overcome these difficulties [8-13]. In particular, the McKeever's method [14] allows determining if the TL emission is due to closely overlapping peaks or to a quasi-continuous distribution of TL peaks. Then, as suggested by the Author, his method can be combined to the curve fitting procedure for calculating the kinetics parameters.

In this section, a simple method for cheeking the possibility that a TL peak could be due to closely overlapping peaks is presented. The method is based on fading experiments, at room temperature (RT) or at any other conveyable storage temperature, and then on the calculation of the activation energy, E, using the initial rise (IR) method [15] applied to each glow peak registered during the fading experiment, at different elapsed times from the sample irradiation.

Example

The samples are irradiated to a test dose giving a good TL signal according to the sensitivity of the material, but with no long irradiation time. The irradiation time should be chosen accordingly to the fading characteristics of the material. All the samples are irradiated at the same time. Dark condition is respected during irradiation, storage and readout of the samples

Immediately, after the end of the irradiation, some samples, i.e. 4, are read out: their TL readings represent the zero storage time point. The other experimental points are obtained at different times after the end of the irradiation. The total storage time, as well as the intervals among the experimental points, have to be chosen according to the stability of the material to avoid too little TL signals at the end of the storage time. For reading the sample, a heating rate of 2°C/sec can be used to obtain a good thermal contact between the TL sample and the heating strip.

At each experimental point after irradiation, the corresponding glow curve is registered and the IR method is applied to calculate the activation energy.

In the following, two examples are given. The thermoluminescent materials used are two kinds of spices from which the minerals contain has been separated from the organic materials. The herbs are Oregano and Nopal. The mineral contain is given in Table 1. Indeed, because the mineral composition, the shape of the glow curves suggests a trap distribution instead of a single trapping level. Indeed, the glow curves generated by the thermoluminescence of the herbs are not the ones generated by a pure or single material. The herb dust consists in many different materials, each one giving its own TL emission; so, the glow curve from herb is really the result of a sum of all glow peaks generated by the minerals contain. In this sense, the registered glow curve has the shape of a continuous which, however, it is not a real continuous but the sum of overlapped glow peaks.

Table 1. Mineral composition of the spices.

Nopal	Whewellite ($CaC_2O_4(H_2O)$) 98.9%	Weddellite ($CaC_2O_4(H_2O)_{2.375}$ 1.1%	Anorthoclase ($Na_{0.75}K_{0.25}AlSi_3O_8$) 76.8%	Quartz (SiO_2) 23.2%
Oregano	Quartz (SiO_2) 50%	Calcite ($CaCO_3$) 20%	Albite ($NaAlSi_3O_8$) 20%	Clay 10%

Fig. 6 shows the glow curves of Nopal, registered at different periods of time during the fading experiment. Fig. 7 shows the activation energy values, E, as function of the duration of the fading experiment.

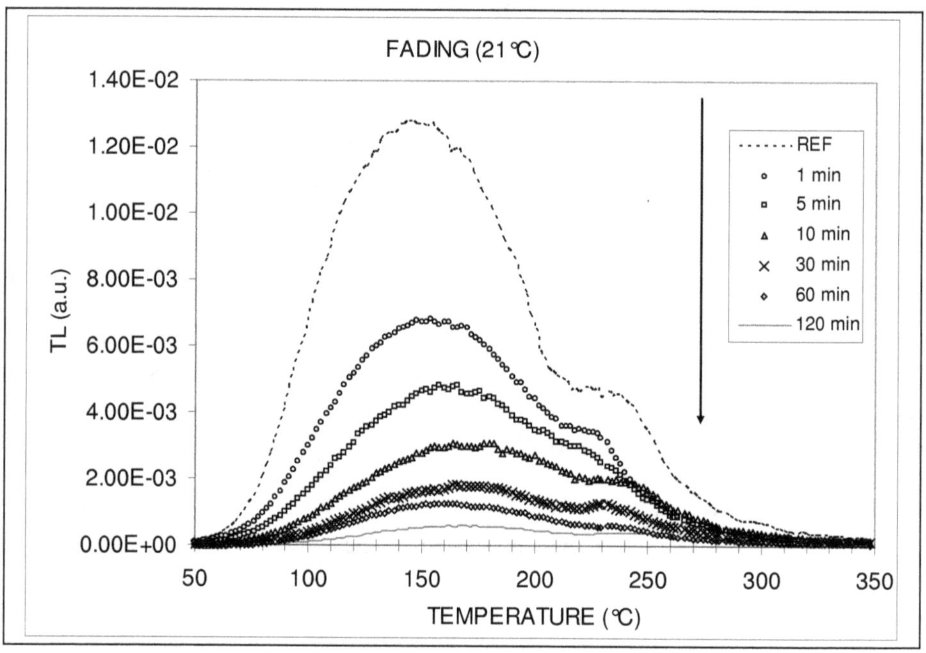

Fig. 6. Glow curves of Nopal registered during the fading experiment at R.T. and in dark conditions.

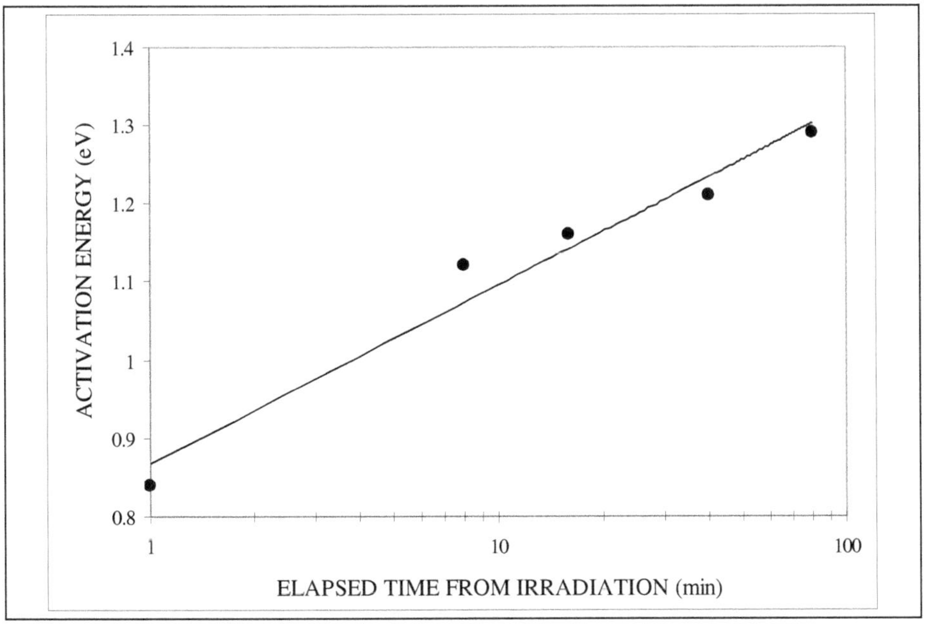

Fig. 7. Behavior of the activation energy of Nopal during the fading experiment.

Figs. 8 and 9 show, respectively, the glow curves of Oregano and the behaviour of the activation energy E.

The result concerning the increase of E, as a function of the fading time, is not expected from the kinetics models concerning with a single trap level. This result may be only explained considering a series of overlapped glow peaks.

The result concerning the increase of E, as a function of the fading time, is not expected from the kinetics models concerned with a single trap level. This result may be only explained considering a continuous distribution of traps. According to this model, during the fading for a fixed time, some of the sub levels, from the shallowest, are emptied. Increasing the fading time more sub levels are emptied. The activation energy, evaluated at the end of each time, should correspond to a value averaged over the values of each single sub-level taken into consideration. As the fading time increases, more deep sub-levels are emptied, in turn resulting in an increase of the activation energy value.

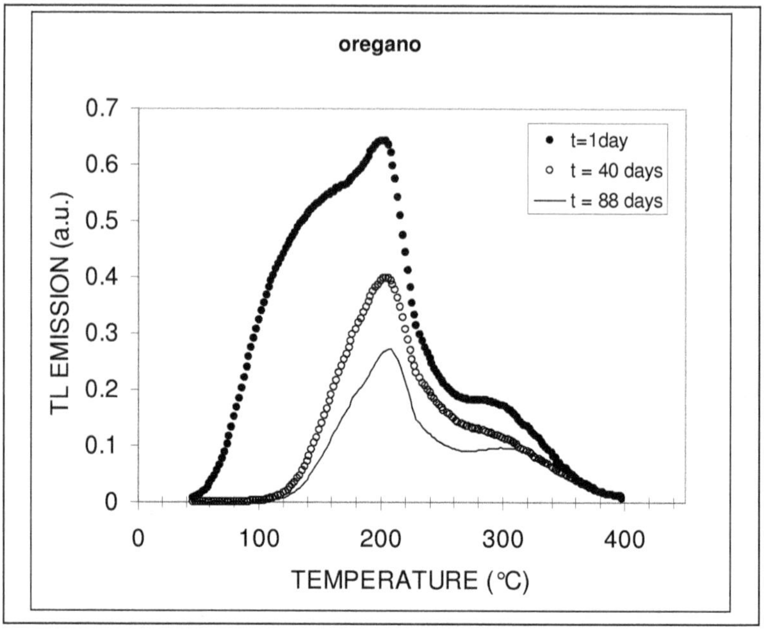

Fig. 8. Glow curves of Orégano registered during fading experiment at RT and in dark conditions.

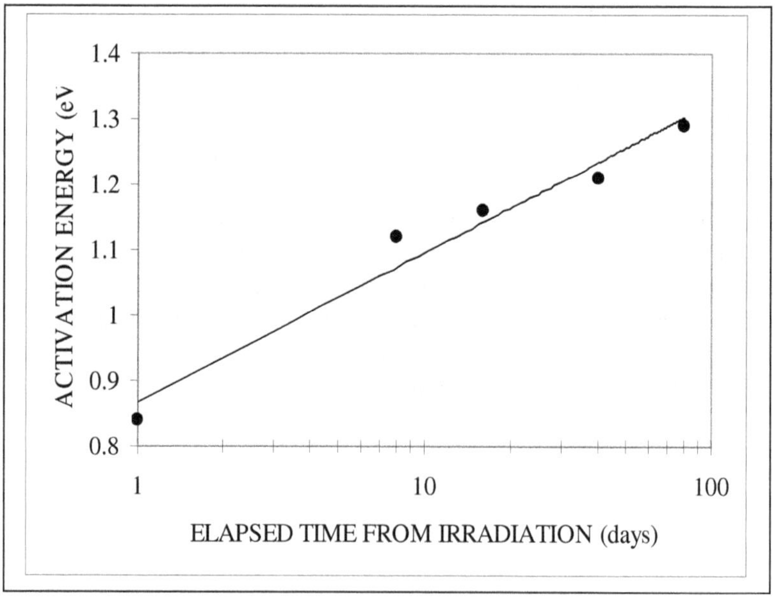

Fig. 9. Behavior of the activation energy of Oregano during the fading experiment.

References

1. R. Chen, J. Mater. Sci. **11**, 1521 (1976)
2. P. Kivits and H.J.L. Hagebeuk, J. Lum. **15**, 1, (1977)
3. K.H. Nicholas and J. Woods, Brit. J. Appl. Phys. **15**, 783 (1964)
4. S.W.S. McKeever, Thermoluminescence of Solids, Cambridge University Press, Cambridge, UK (1985)
5. C. Furetta and P.S. Weng, Operational Thermoluminescence Dosimetry, World Scientific, Singapore (1998)
6. C. Furetta, Handbook of Thermoluminescence, World Scientific, Singapore (2003)
7. J. Azorin, C. Furetta and A. Scacco, Phys. Stat. Sol. (a) **138**, 9 (1993)
8. H. Gobrecht and D. Hofmann, J. Phys. Chem. Solids **27**, 509 (1966)
9. D.R. Rao, Phys. Stat. Sol. (a) **22**, 337 (1974)
10. R.P. Khare and J.D. Ranade, Phys. Stat. Sol. (a) **32**, 221 (1975)
11. G.C. Taylor and E. Lilley, J. Phys. D **11**, 567 (1978)
12. G. Rudlof, J. Becherer and H. Glaefeke, Phys. Stat. Sol. (a) **49**, K121 (1978)
13. G. Rudlof, J. Becherer and H. Glaefeke, Phys. Stat. Sol. (a) **52**, K137 (1979)
14. S.W.S. McKeever, Phys. Stat. Sol. (a) **62**, 331 (1980)
15. G.F.J. Garlick and A.F. Gibson, Proc. Phys. Soc. **60**, 574 (1948)

Fading: useful expressions

In the following, some expressions for fading correction in practical situations will be given. They are based considering the first order process and the general case in which, during the experimental period of time, two effects are in competition between them: one is the trapping rate due to a continuous irradiation over all the experimental period, i.e. environmental background irradiation; the second one is the detrapping rate which takes place at the same time, i.e. thermal fading. Such a situation can be described by the following first order differential equation:

$$\frac{d\Phi}{dt} = -\lambda\Phi + \frac{\dot{D}}{F_C} \tag{1}$$

where

- Φ is the total TL light of a given peak in the glow curve;
- λ is the fading factor and it is constant for a constant temperature. In case of the kinetics parameters of the considered peak are known, i.e. E and s, it can be expressed by $s\exp\left(-\frac{E}{kT}\right)$;
- $\dot{D}$ is the dose rate of the irradiation field;
- F_C is the calibration factor of the thermoluminescence system, expressed in *dose/TL*.

Eq. (1) represents a dynamic situation where two competing effects are taken into account. This equation tends to an asymptotical limit as the fading produces a progressive extinction of the accumulated charges, whereas the continuous irradiation leads to an increase of them. Eq. (1) only holds in the case we are far from saturation.

The solution of Eq. (1) ia then obtained as following:

$$\frac{d\Phi}{-\lambda\Phi + \frac{\dot{D}}{F_C}} = dt$$

Using the substitutions

$$x = -\lambda\Phi + \frac{\dot{D}}{F_C}, \qquad dx = -\lambda d\Phi$$

we get

$$-\frac{1}{\lambda}\int\frac{dx}{x} = -\frac{1}{\lambda}[\ln x]_{\Phi_0}^{\Phi}$$

and then

$$-\frac{1}{\lambda}\ln\frac{-\lambda\Phi + \frac{\dot{D}}{F_C}}{-\lambda\Phi_0 + \frac{\dot{D}}{F_C}} = t$$

from which

$$\Phi = \Phi_0 \exp(-\lambda t) + \frac{\dot{D}}{\lambda F_C}\left[1 - \exp(-\lambda t)\right] \tag{2}$$

Eq. (2) depicts a situation where a non-zero charge population is already trapped at the beginning of the experimental time, i.e. $\Phi_0 \neq 0$.

Considering the practical situation where the TLDs are annealed before use, all the traps are empty at the beginning of the experimental period. In such a case Eq. (2) becomes

$$\Phi = \frac{\dot{D}}{\lambda F_C}\left[1 - \exp(-\lambda t)\right] \tag{3}$$

When a very long time has elapsed, i.e. $t \to \infty$, Φ gets more and more similar to the asymptotical value

$$\Phi_\infty = \frac{\dot{D}}{\lambda F_C} \tag{4}$$

Such a value grows larger as the dose rate and /or the sensitivity ($1/F_C$) increases, or as the fading effect decreases. The asymptotical value given by Eq. (4) may be explained assuming that, at infinity, a dynamical equilibrium is attained, providing the trapped charges to compensate at each instant those escaping owing the fading phenomenon.

Discussion of some practical situations

1. Initial and instantaneous irradiation followed by fading at room temperature

Fig. 10 depicts the situation. In this case the irradiation is delivered to the dosimeters at the beginning of the experimental period and the duration of irradiation, t_i , is very short so that any fading effect during irradiation can be neglected. After irradiation the irradiated samples are stored, at room temperature or at any other controlled temperature, for a time $t_S >> t_i$.

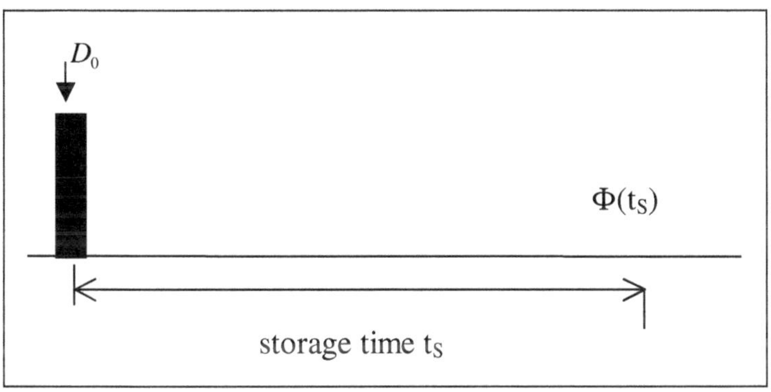

Fig. 10. Case 1. Initial irradiation followed by storage at R.T.

The situation depicted in Fig. 10 is the usual for fading studies. Eq. (2) reduces to the simply expression

$$\Phi(t_S) = \Phi_0 \exp(-\lambda t_S) \tag{5}$$

from which

$$\Phi_0 = \Phi(t_S) \exp(\lambda t_S) \tag{6}$$

Through the calibration factor F_C, the initial deliverd dose is then obtained:

$$D_0 = F_C \Phi(t_S) \exp(\lambda t_S) \tag{7}$$

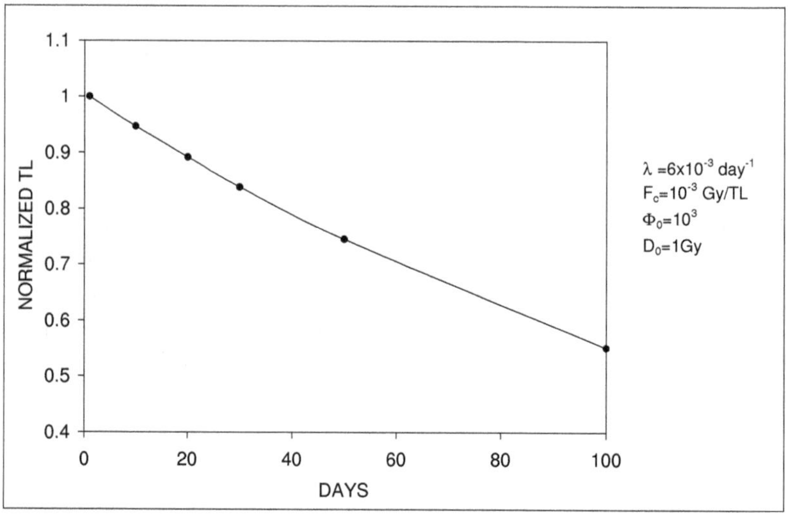

Fig. 11. Case 1. Plot of Eq. (7). The imput data are given in the same figure.

<u>*2. Initial but not instantaneous irradiation, followed by fading at room temperature*</u>

An initial irradiation is delivered at the beginning of the experimental period, but the irradiation time, t_i , is long enough so that a fading effect is not any more negligeble during the period of irradiation.

After the irradiation the samples are stored for the time t_S. Fig. 12 depicts the situation.

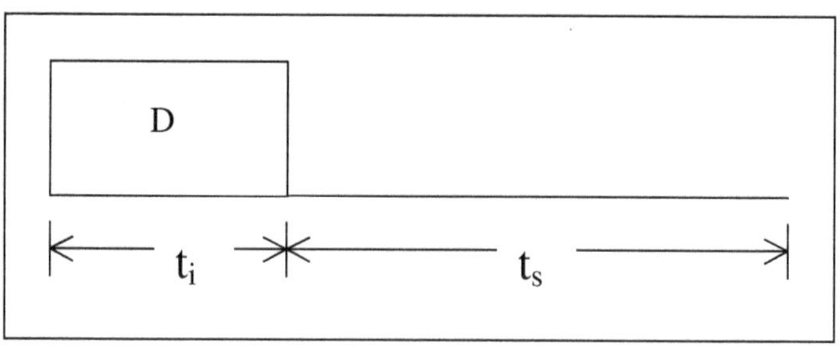

Fig. 12. Case 2. Long irradiation followed by storage at RM.

During the irradiation time, Eq. (2) reduces to the following expression

$$\Phi(t_i)=\frac{\dot{D}}{\lambda F_C}\left[1-\exp(-\lambda t_i)\right] \tag{8}$$

which gives the TL emission at the end of the irradaition time. As the irradiation stops, the samples are only subjected to fading at room temperature, so that

$$\Phi(t_S)=\Phi(t_i)\exp(-\lambda t_S) \tag{9}$$

Combining Eq. (8) and Eq. (9), we get

$$\Phi(t_S)=\frac{\dot{D}}{\lambda F_C}\left[1-\exp(-\lambda t_i)\right]\exp(-\lambda t_S) \tag{10}$$

from which the true deliverd dose is obtained, taking into account that $D=\dot{D}\cdot t_i$,

$$D=\lambda F_C\Phi(t_S)t_i\exp(\lambda t_S)\left[1-\exp(-\lambda t_i)\right]^{-1} \tag{11}$$

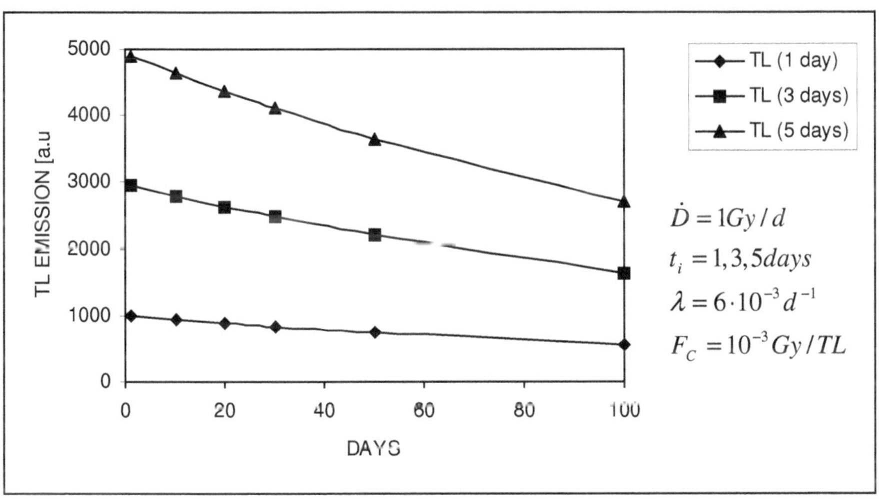

Fig. 13. Case 2. Plot of Eq. (10).

3. The irradiation is carried out over all the experimental period.

This is the case of environmental background measurements or self-dose irradiation. See Fig. 14.

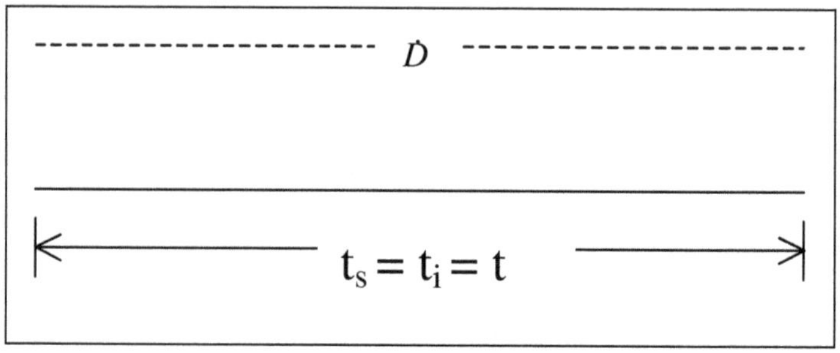

Fig. 14. Case 3. The irradiation is carried out over all the storage time.

The irradiation time, t_i, is now equal to the storage time t_S. The TLD samples are prepared and exposed to the irradiation field, then the initial condition is $\Phi_0 = 0$ and Eq. (2) reduces to the following expression:

$$\Phi = \frac{\dot{D}}{\lambda F_C}\left[1 - \exp(-\lambda t)\right] \tag{12}$$

having indicated $t_S = t_i = t$.

If $\dot{D}$ is the environmental background dose rate or the internal dose rate due to the self irradiation of the samples, the total dose is obtained as

$$D = \Phi \lambda F_C t\left[1 - \exp(-\lambda t)\right]^{-1} \tag{13}$$

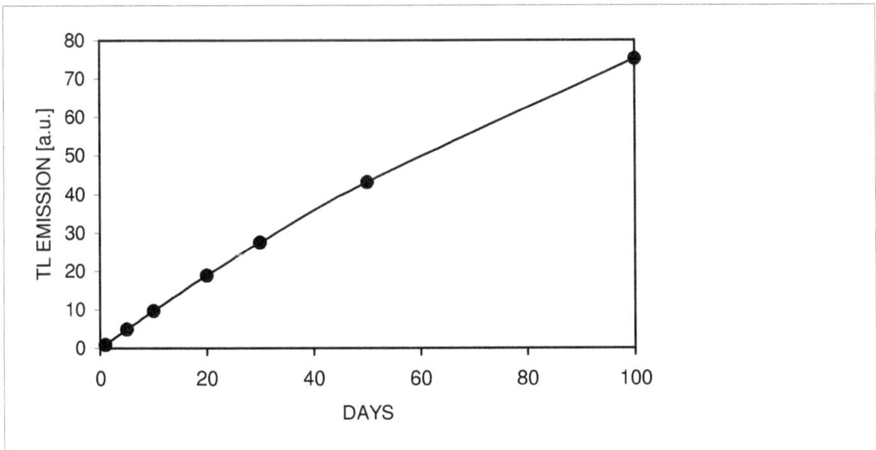

Fig. 15. Case 3. Plot of Eq. (13). Input data are: $\dot{D} = 10^{-3} Gy/d, \lambda = 6 \cdot 10^{-3} d^{-1}$, $F_C = 10^{-3} Gy/TL$.

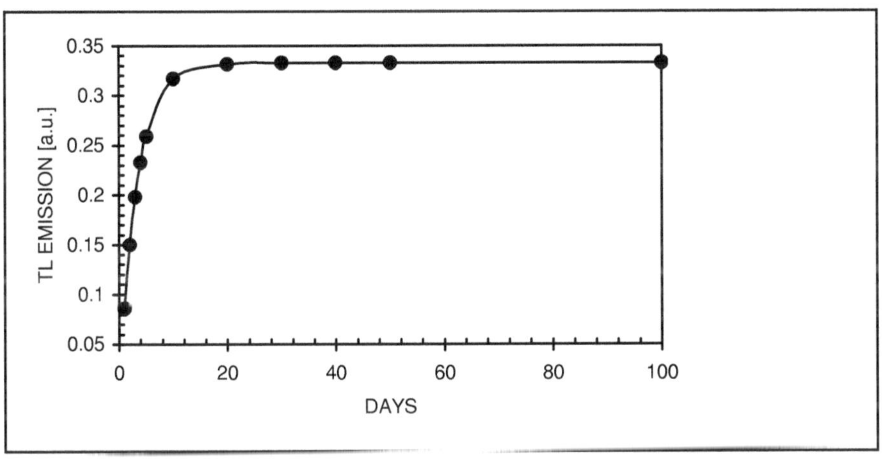

Fig. 16. Case 3. This figure depicts the case of saturation after few days. The numerical values are the following:
$\dot{D} = 10^{-4} Gy/d, \lambda = 3 \cdot 10^{-1} d^{-1}, F_C = 10^{-3} Gy/TL$

4. An initial and short irradiation is superposed to a background irradiation

Let us indicate with $\dot{D}_B$ the background irradiation, which acts over all the period of storage, t_S. Fig. 8 shows the situation. The fading during the short irradiation is neglected. The equation simulating this case is always Eq. (2), written in the following way:

$$\Phi = \Phi_0 \exp(-\lambda t_S) + \frac{\dot{D}_B}{\lambda F_C}\left[1 - \exp(-\lambda t_S)\right] \tag{14}$$

which gives, in explicit form

$$\Phi_0 = \left\{\Phi - \frac{\dot{D}_B}{\lambda F_C}\left[1 - \exp(-\lambda t_S)\right]\right\}\exp(\lambda t_S) \tag{15}$$

from which the initial deliverd dose is obtained:

$$D_0 = F_C \Phi_0 \tag{16}$$

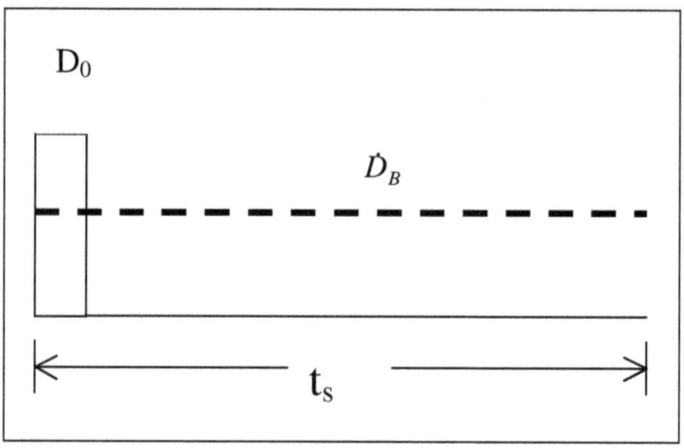

Fig. 17. Case 4. Initial and short irradiationsuperposed to background irradiation.

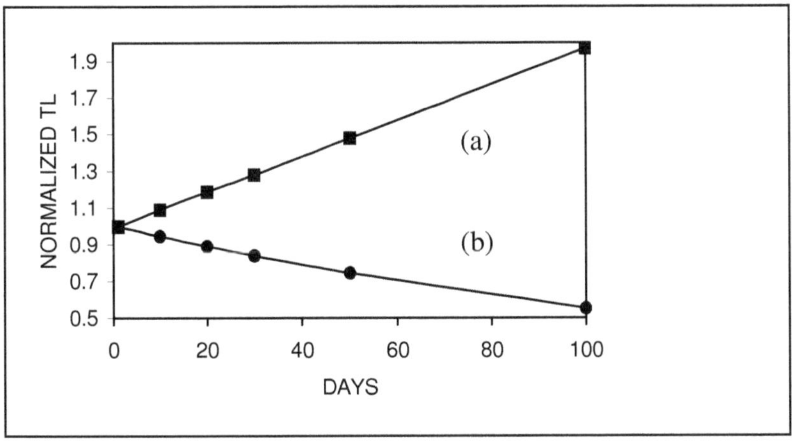

Fig. 18. Case 4. Eq. (16) has been computed for two different set of imput data.
(a): $\dot{D}_B = 10^{-1} Gy/d, D_0 = 10Gy, F_C = 10^{-3} Gy/TL, \lambda = 10^{-4} d^{-1}$
(b): $\dot{D}_B = 10^{-3} Gy/d, D_0 = 100Gy, F_C = 10^{-3} Gy/TL, \lambda = 6 \cdot 10^{-3} d^{-1}$

First-order kinetics when $s = s(T)$

The frequency factor s may be considered in some cases to be dependent on temperature, and proportional to T^{α}, where α has various values in the range $-2 \leq \alpha \leq 2$ [1-3]

Let us suppose that s depends on temperature according to the following relation

$$s = s_0 T^{\alpha} \tag{1}$$

The detrapping rate is now

$$\frac{dn}{dt} = -n s_0 T^{\alpha} \exp(-\frac{E}{kT}) \tag{2}$$

Using a linear heating rate $\beta = dT/dt$, the solution of Eq. (2) is

$$n = n_0 \exp\left[-\frac{s_0}{\beta}\int_{T_0}^{T} T^{\alpha} \exp(-\frac{E}{kT}) dT\right] \tag{3}$$

and the TL intensity $I(T)$ will be expressed by

$$I(T) = n_0 s_0 T^{\alpha} \exp(-\frac{E}{kT}) \exp\left[-\frac{s_0}{\beta}\int_{T_0}^{T} T^{\alpha} \exp(-\frac{E}{kT}) dT\right] \tag{4}$$

References

1. Chen R. and Kirsh Y., Analysis of Thermally Stimulated Processes (Pergamon Press, 1981)
2. McKeever S.W.S., Thermoluminescence of Solids (Cambridge University Press, 1985)
3. Furetta C. and Weng P.S., Operational Thermoluminescence Dosimetry (World Scientific, 1998)

Fluorescence

Fluorescence is a luminescence effect occurring during excitation. The light is emitted at a time less than 10^{-8} s after the absorption of the radiation. This means that fluorescence is a luminescent process that persists only as long as the excitation is continued. The decay time of fluorescence is independent on temperature: it is determined by the transition probability of the transition from an excited level E_e to the ground state E_0. The process is shown in the following Figure 19.

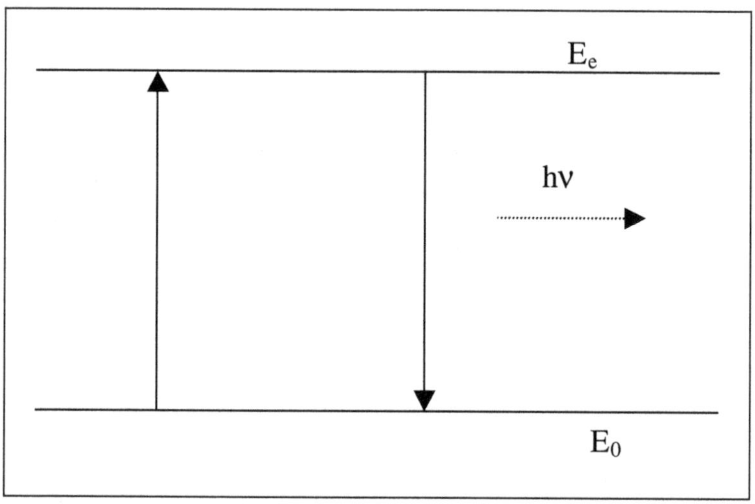

Fig. 19. Fluorescence process.

Fluorapatite ($Ca_5F(PO_4)_3$)

This material belongs to a class of compounds, mineralogically known as apatites. The effective atomic number is about 14.

Fluorapatite is prepared by synthesis from CaF_2 and $CaHPO_4$ through elimination of hydrofluoric acid. About 160 mg of the resulting powder, covered with thin LiF crystals and contained im silver boats, are typical samples for thermoluminescence investigation.

The TL glow curve of synthetic fluorapatite powder exhibits peaks at 145, 185, 260 and 395°C.

Reference

Ratnam V.V., Jayaprakash R. and Daw N.P., J. Lum. **21** (1980) 417

Frequency factor, *s*

The frequency factor, *s*, is known as the attempt-to-escape frequency and is interpreted as the number of times per second, ν, that an electron interacts with the crystal lattice of a solid, multiplied by a transition probability κ, multiplied by a term which accounts for the change in entropy ΔS associated with the transition from a trap to the delocalised band. s may be written as

$$s = \nu \cdot \kappa \cdot \exp\left(\frac{\Delta S}{k}\right)$$

where *k* is Boltzmann's constant [1,2].

The expected maximum value of *s* should be similar to the lattice vibrational frequency (Debye frequency), i.e. $10^{12} - 10^{14}\ s^{-1}$. According to Chen, the possible range for *s* is from 10^5 to $10^{13}\ s^{-1}$ [3].

Randall and Wilkins gave the following meaning to the frequency factor: they described the trap as a potential well and *s* should be the product between the frequency with which the trapped electrons strike the wells of the potential barrier and the reflection coefficient. According to this definition, *s* should be expected to be about of the order of the vibrational frequency of the crystal, i.e. $10^{12}\ s^{-1}$.

Alternatively, *s* should be considered as connected with the capture cross-section, σ, of the trap by the following relation

$$s = v_e N_c \sigma$$

where v_e is the thermal velocity of the lectrons in the conduction band, N_c is the density of states (available electron levels) near the bottom of the conduction band and σ the capture cross-section of the trap. In this case the values of *s* are ranging between 10^8 and 10^{14} s-1.

In some cases, the frequency factor, as well as the pre-exponential factors, may be considered temperature dependent and proportinal to T^{α} with α ranging from -2 to +2.

References

1. Glasstone S., Laidler K.J. and Eyring H., The theory of rate processes McGraww-Hill, New York, 1941
2. Curie D., Luminescence in Crystals, Methuen, London, 1960
3. Chen R. and Kirsh Y., Analysis of Thermally Stimulated Processes, Pergamon Press, 1981

Frequency factor, *s* (errors in its determination)

It must be stressed that any error in the evaluation of the activation energy and of temperature introduces very high error in the determination of *s*. Making reference to the expression of the condition at the maximum for the first order kinetics

$$s = \frac{\beta E}{kT_M^2} \exp\left(\frac{E}{kT_M}\right) \tag{1}$$

its logarithm is

$$\ln s = \ln \beta + \ln E - \ln k - 2\ln T_M + \frac{E}{kT_M} \tag{2}$$

Assuming that the heating rate β has no error, we can differentiate Eq. (2) to give

$$\left|\frac{\Delta s}{s}\right| = \left|\frac{\Delta E}{E}\right| + 2\left|\frac{\Delta T_M}{T_M}\right| + \frac{1}{k}\left|\frac{T_M \Delta E + E\Delta T_M}{T_M^2}\right| \tag{3}$$

Supposing an error of 2% on the evaluation of *E* and the same error on the temperature determination, it is easy to see that the error on *s* is very large. Making the assumption that no error is done on the temperature measurement, Eq. (1) gives

$$\left|\frac{\Delta s}{s}\right| = \left|\frac{\Delta E}{E}\right| + \frac{1}{kT_M}|\Delta E| \tag{4}$$

which still remains a large error.

Frequency factor and pre-exponential factor expressions

- Frequency factor : 1st order of kinetics

$$s = \frac{\beta \cdot E}{k \cdot T_M^2} \exp\left(\frac{E}{kT_M}\right) \qquad [\text{sec}^{-1}] \tag{1}$$

- Pre-exponential factor: 2nd order of kinetics

$$s' = \frac{\beta E \exp\left(\frac{E}{kT_M}\right)}{kn_0 T_M^2}\left[1+\frac{2kT_M}{E}\right]^{-1} \tag{2}$$

which becomes, introducing $s = s'n_0$,

$$s = \frac{\beta E \exp\left(\frac{E}{kT_M}\right)}{kT_M^2}\left[1+\frac{2kT_M}{E}\right]^{-1} \qquad [\text{sec}^{-1}] \tag{3}$$

- Pre-exponential factor: general kinetics order $(1 < b \leq 2)$

$$s = \left[\frac{kT_M^2 \exp\left(-\frac{E}{kT_M}\right)}{\beta E}\left(1+\frac{2kT_M(b-1)}{E}\right)\right]^{-1} \qquad [\text{sec}^{-1}] \tag{4}$$

or

$$s'' = \frac{\beta E \exp\left(\frac{E}{kT_M}\right)}{n_o^{b-1} kT_M^2}\left[1+\frac{2kT_M(b-1)}{E}\right]^{-1} \qquad [\text{cm}^{3(b-1)}\ \text{sec}^{-1}] \tag{5}$$

The frequency factor given by Eq. (5) acts as an effective frequency factor expressed by the following equation

$$s_{eff} = s'' n_0^{b-1} \tag{6}$$

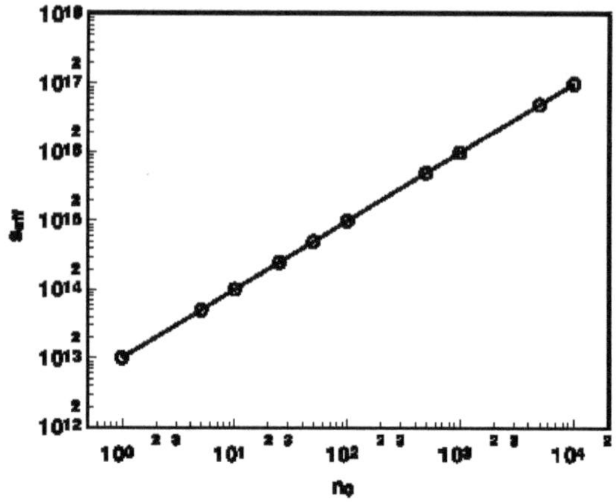

Fig. 20. Plot of Eq. (6), [1].

- Frequency factor vs temperature: 1st order kinetics

$$s_0 = \frac{\beta E}{kT_M^{\alpha+2}}\left(1+\frac{\alpha}{2}\Delta_M\right)\exp\left(\frac{E}{kT_M}\right) \tag{7}$$

where $\Delta_M=2kT_M/E$ and $-2 \le \alpha \le 2$.

- Pre-exponential factor vs temperature: 2nd order kinetics

$$s_0' = \frac{\beta E\left(1+\frac{\alpha}{2}\Delta_M\right)}{n_0 kT_M^{\alpha+2}}\exp\left(\frac{E}{kT_M}\right) \tag{8}$$

- Pre-exponential factor vs temperature: general kinetics order $(1 < b \le 2)$

$$s_0'' = \frac{\beta E}{kT_M^{\alpha+2}}\left[\frac{1-\Delta_M(1+\alpha)(1-b)}{1+\frac{\alpha}{2}\Delta_M}\right]^{-1}\exp\left(\frac{E}{kT_M}\right) \tag{9}$$

- Remarks

It is easy to note that, except in the first order case, *s*' and *s*'' are constant for a given TL sample and dose but would vary in the same sample as the dose is varied.

In order to overcome this difficulty, it has been suggested in [2] to rewrite the **general-order equation** in the form

$$\frac{dn}{dt} = -s\frac{n^b}{N^{b-1}}\exp\left(-\frac{E}{kT}\right) \tag{10}$$

taking into account that the equations for the first- and second-order can be written, respectively, as

$$\frac{dn}{dt} = -n\frac{s}{N^0}\exp\left(-\frac{E}{kT}\right) \tag{11}$$

$$\frac{dn}{dt} = -n^2\frac{s}{N}\exp\left(-\frac{E}{kT}\right) \tag{12}$$

In all cases s has units of sec^{-1}, having eliminated any difficulty related to the dimension problem of s' and s".

According to [1], the empirical expression (10) should also eliminate the variation of s with respect to the variation of the absorbed dose. Indeed, this point does not seem correct. Infact, the TL intensity obtained from the new expression (10), is

$$I(T) = \frac{sn_0^b\exp\left(-\frac{E}{kT}\right)}{N^{b-1}}\left[1+\frac{sn_0^{b-1}(b-1)}{\beta N^{b-1}}\int_{T_0}^{T}\exp\left(-\frac{E}{kT'}\right)dT'\right]^{-\frac{b}{b-1}} \tag{13}$$

Using the **condition at the maximum** and the **integral approximation**, the new expression is

$$s = \left(\frac{N}{n_0}\right)^{b-1}\left[\frac{kT_M^2\exp\left(-\frac{E}{kT_M}\right)}{\beta E}\right]^{-1}\left[1+\frac{2kT_M(b-1)}{E}\right]^{-1} \tag{14}$$

from which the expression for the second-order process (b=2) is easily obtained.

Although in Eq. (14) s is now expressed in sec^{-1}, as in the first-order kinetics, the dependence of s on the initial trapped charges, n_0, still remains. Only in the saturation case, i.e. $n_0 = N$, the s values results to be independent of the value of the initial trapped charges. Furthermore, Eq. (14) includes the parameter N which cannot be easily determined.

It is then evident that the suggested way to rewrite the rate equations does not eliminate the dependence on the dose. The new formulation only allows to express s in units of time in all cases.

References

1. Kitis G., private communication.
2. Rasheedy M.S., J. Phys.: Condens. Matter. **5**, 1993, 633

G

Garlick-Gibson model (second-order kinetics)

In 1948 Garlick and Gibson, in their studies on phosphorescence, considered the case when a free charge carrier has probability of either being trapped or recombining within a recombination center. The term second order kinetics is used to describe a situation in which retrapping is present.

They assumed that the escaping electron from the trap has equal probability of either being retrapped or of recombining with hole in a recombination centre.

Let us indicate:

N = concentration of traps,
n = electrons in N,
m = concentration of recombination centres,
$n = m$ for charge neutrality condition.

The probability that an electron escapes from the trap and recombine in a recombination centre is

$$\frac{m}{(N-n)+m} = \frac{n}{N} \tag{1}$$

So, the intensity of phosphorescence, I, is given by the rate of decrease of the occupied trap density, resulting in the recombination of the released electrons with hole in the recombination centres:

$$I(t) = -\frac{dn}{dt} = c\left(\frac{n}{N}\right)\cdot\left(\frac{n}{\tau}\right) = c\frac{n^2}{N}s\exp\left(-\frac{E}{kT}\right) \tag{2}$$

where τ is the mean trap lifetime.
Eq. (2) can be rewritten as

$$\frac{dn}{dt} = -n^2 s' \exp\left(-\frac{E}{kT}\right) \tag{3}$$

The quantity $s' = s/N$ is called pre-exponential factor and it is a constant having dimensions of cm^3sec^{-1}. Eq. (3) is different from that one obtained in the case of **first order kinetics**, where the recombination probability was equal to 1, since no retrapping is possible.

From Eq. (3), by integration with constant temperature T, we obtain:

$$\int_{n_0}^{n} \frac{dn}{n^2} = -s' \exp\left(-\frac{E}{kT}\right) \int_0^t dt$$

$$\frac{1}{n_0} - \frac{1}{n} = -s't \exp\left(-\frac{E}{kT}\right)$$

$$n = n_0 \left[1 + s'n_0 t \exp\left(-\frac{E}{kT}\right)\right]^{-1} \tag{4}$$

and then, the intensity I(t) is:

$$I(t) = -\frac{dn}{dt} = n^2 s' \exp\left(-\frac{E}{kT}\right) = \frac{n_0^2 s' \exp\left(-\frac{E}{kT}\right)}{\left[1 + s'tn_0 \exp\left(-\frac{E}{kT}\right)\right]^2} \tag{5}$$

which describes the hyperbolic decay of phosphorescence.

Otherwise, the luminescence intensity of an irradiated phosphor under rising temperature, i.e. thermoluminescence, taking into account that $dt=dT/\beta$, is obtaining as following:

$$\frac{dn}{n^2} = -\frac{s'}{\beta} \exp\left(-\frac{E}{kT}\right) dT$$

therefore

$$\int_{n_0}^{n} \frac{dn}{n^2} = -\frac{s'}{\beta} \int_{T_0}^{T} \exp\left(-\frac{E}{kT}\right) dT$$

and then

$$\frac{1}{n_0} - \frac{1}{n} = -\frac{s'}{\beta} \int_{T_0}^{T} \exp\left(-\frac{E}{kT'}\right) dT'$$

$$n = n_0 \left[1 + \left(\frac{s'n_0}{\beta}\right) \int_{T_0}^{T} \exp\left(-\frac{E}{kT'}\right) dT'\right]^{-1} \tag{6}$$

The intensity $I(T)$ is then

$$I(T) = -\frac{dn}{dt} = n^2 s' \exp\left(-\frac{E}{kT}\right) = \frac{n_0^2 s' \exp\left(-\frac{E}{kT}\right)}{\left[1 + \frac{s' n_0}{\beta} \int_{T_0}^{T} \exp\left(-\frac{E}{kT'}\right) dT'\right]^2} \tag{7}$$

Eq. (7) can be rewritten as

$$I(T) = \frac{n_0 s \exp\left(-\frac{E}{kT}\right)}{\left[1 + \frac{s}{\beta} \int_{T_0}^{T} \exp\left(-\frac{E}{kT'}\right) dT'\right]^2} \tag{8}$$

where $s = s' n_0$. In this case s has units of s^{-1} like the frequency factor in the first-order kinetics, but it depends on n_0.

Reference

Garlick G.F.J. and Gibson A.F., Proc. Phys. Soc. **60** (1948) 574

General characteristics of first and second order glow-peaks

Some general characteristics can be listed to distinguish between first and second order glow-peaks when a linear heating rate function is used.

First order peaks

- The first order peaks are asymmetrical and $\tau = T_M - T_1$, the half-width at the low temperature side of the peak, is almost 50% larger than $\delta = T_2 - T_M$, the half-width towards the fall-off of the glow-peak $\tau \cong 1.5\delta$).
- The shape and the peak temperature depend on the heating rate.
- For a fixed heating rate, both peak temperature and shape are independent of the initial trapped electron concentration n_0, as it can be observed from the condition at the maximum

$$\frac{\beta E}{kT_M^2} = s \exp\left(-\frac{E}{kT_M}\right)$$

- The value of n_0 depends on the pre-measurement dose.

- The TL glow-curve obtained for any n_0 value can be superimposed out the curve obtained for a different n_0 by multiplying by an appropriate factor (Fig. 1).
- A first order peak is characterized by a geometrical factor $\mu = \delta/\omega = (T_2 - T_M)/(T_2 - T_1)$ equal to about 0.423.
- For fixed values of dose and heating rate, the ω value increases as E decreases (Fig. 2).
- The decay at constant temperature of a first order peak is exponential.

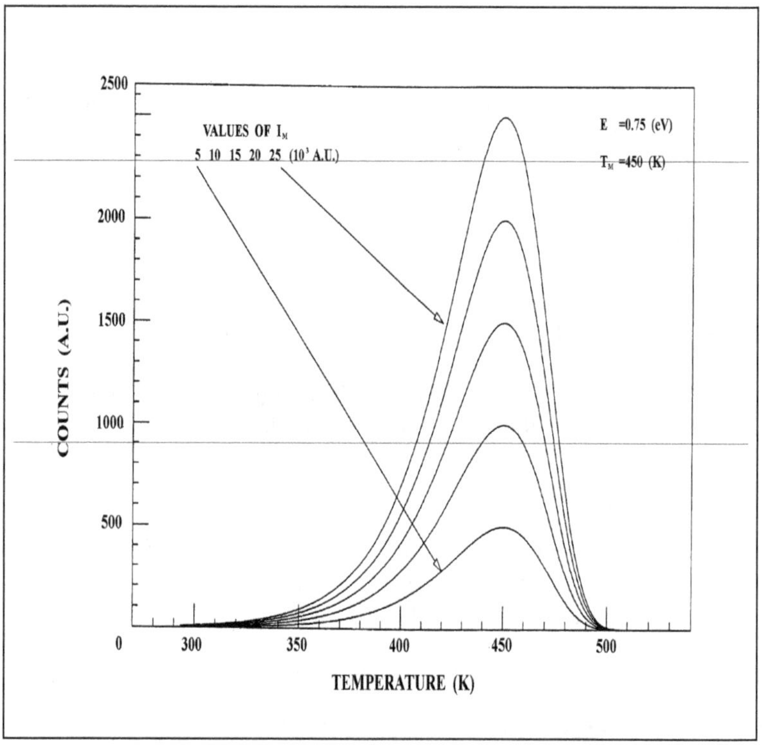

Fig. 1. A computed first order glow-peak showing the linear increase of I_M as a function of dose.

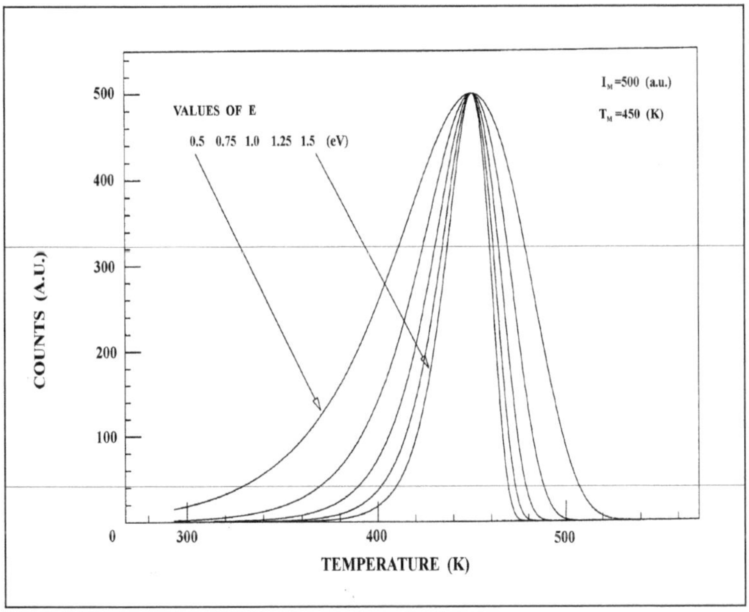

Fig. 2. A computed first order glow-peak showing the increase of ω as E decreases.

<u>*Second order peaks*</u>

- A second order peak is practically symmetrical ($\delta \approx \tau$).
- For all other parameters kept constant, the shape and the peak temperature depend on the heating rate.
- For a fixed heating rate, the peak temperature and shape are strongly dependent on the initial trapped charge concentration n_0. Peaks obtained for different initial trapped charge concentrations cannot be superimposed by multiplying by a factor.
- The glow-peaks obtained for different n_0 values tend to superimpose at the high temperature extremity of the glow-peak.
- An increase of n_0 produces a decrease in the temperature of the peak, according to the maximum condition

$$\frac{\beta E}{2kT_M^2}\left[1+\frac{s'n_0}{\beta}\int_{T_0}^{T_M}\exp\left(-\frac{E}{kT'}\right)dT'\right]=s'n_0\exp\left(-\frac{E}{kT_M}\right)$$

- The isothermal decay of a second order peak is hyperbolic.

- A second order peak is characterized by a geometrical factor $\mu \cong 0.524$.
- Furthermore, a decrease in the temperature of the peak, T_M, is observed as a function of the kinetics order changing from 1 to 2. This effect is illustrated in Fig. 3.

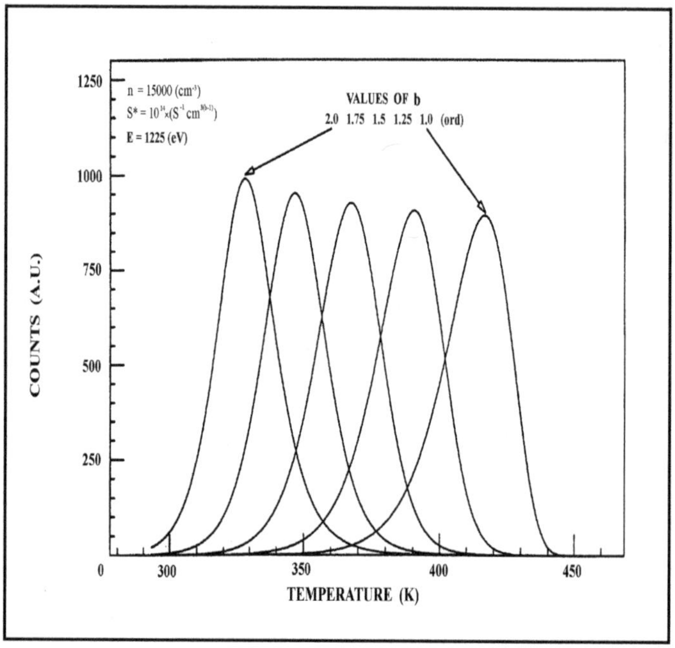

Fig. 3. Computerized glow-peaks showing the effect of the kinetics order on the position of the peak temperature.

Reference

Bacci C., Bernardini P., Di Domenico A., Furetta C. and Rispoli B., Nucl. Instr. Meth. A **286** (1990) 295

General-order kinetics when $s'' = s''(T)$

In the case of general order kinetics, b, the TL intensity equation $I=I(T)$ has to be modified by substituting s'' with the following expression [1-3]:

$$s'' = s''_0 T^{\alpha} \tag{1}$$

In this case the TL intensity, $I(T)$, becomes

$$I(T) =$$

$$= s_0'' n_0 T^{\alpha} \exp(-\frac{E}{kT}) \cdot \left[1 + \frac{s_0''(b-1)}{\beta} \int_{T_0}^{T} T^{\alpha} \exp\left(-\frac{E}{kT} \right) dT \right]^{\frac{-b}{b-1}} \quad (2)$$

References

1. Chen R. and Kirsh Y., Analysis of Thermally Stimulated Processes (Pergamon Press, 1981)
2. McKeever S.W.S., Thermoluminescence of Solids (Cambridge University Press, 1985)
3. Furetta C. and Weng P.S., Operational Thermoluminescence Dosimetry (World Scientific Publ., 1998)

Geometrical characteristics of a single peak

Fig. 1 shows the characteristics of a single glow-peak [1].

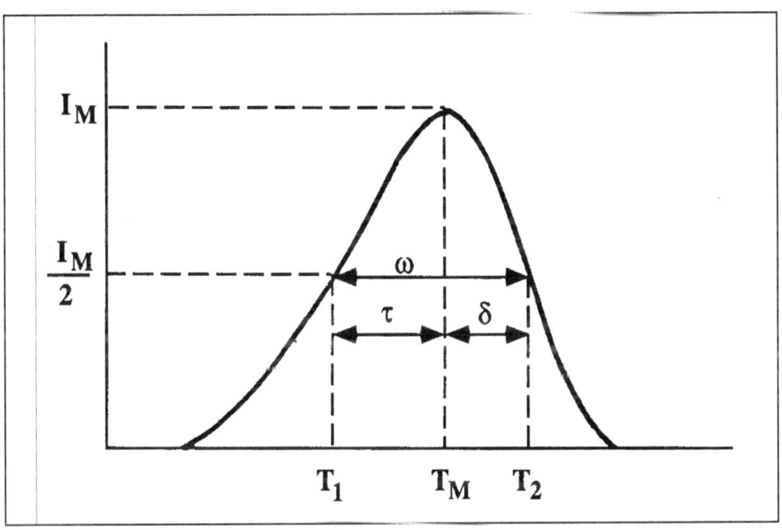

Fig. 1

They are:

T_M = the peak maximum temperature

T_1 and T_2 = the temperatures at half maximum TL intensity at the low and higher temperature side of the glow-peak respectively

$\omega = T_2 - T_1$

$\delta = T_2 - T_M$

$\tau = T_M - T_1$
$\mu = \delta/\omega$ the symmetry factor of the glow-peak

The general order kinetics equation is

$$I(T) = -\frac{dn}{dt} = s'n^b \exp\left(-\frac{E}{kT}\right) \tag{1}$$

where
n (cm^{-3}) = concentration of electrons in traps
E (eV) = activation energy
b = kinetic order
s' ($m^{3(b-1)}$ s^{-1}) = pre-exponential factor

The solution of Eq. (1), using a linear heating rate β, is

$$I = n_0^b s' \exp\left(-\frac{E}{kT}\right)\left[1 + \frac{(b-1)s'n_0^{b-1}}{\beta}\int_{T_0}^{T} \exp\left(-\frac{E}{kT'}\right)dT'\right]^{-\frac{b}{b-1}} \tag{2}$$

The condition for maximum TL intensity is obtained by equating the derivative of Eq. (2) to zero:

$$1 + \frac{(b-1)s'n_0^{b-1}}{\beta}\int_{T_0}^{T_M} \exp\left(-\frac{E}{kT}\right)dT = \frac{bs'n_0^{b-1}kT_M^2}{\beta E}\exp\left(-\frac{E}{kT_M}\right) \tag{3}$$

The integral on the left-hand side of Eq. (3) can be approximated using the first two terms of the asymptotic series

$$\int_{T_0}^{T} \exp\left(-\frac{E}{kT'}\right)dT' \approx T\exp\left(-\frac{E}{kT}\right)\sum_{n=1}\left(\frac{kT}{E}\right)^n (-1)^{n-1} n! \tag{4}$$

From Eq. (3) we then obtain

$$\frac{\beta E}{kT_M^2} = n_0^{b-1} s' \exp\left(-\frac{E}{kT}\right)\left[1 + (b-1)\Delta_M\right] \tag{5}$$

where $\Delta_M = \frac{2kT_M}{E}$

The geometrical characteristics are the followings:

- $$C_\omega = \frac{\omega \cdot I_M}{\beta \cdot n_0} \tag{6}$$
- $$C_\delta = \frac{\delta \cdot I_M}{\beta \cdot n_M} \tag{7}$$
- $$C_\tau = \frac{\tau \cdot I_M}{\beta \cdot (n_0 - n_M)} \tag{8}$$

Furthermore

- $$n_M = \int_{t_M}^{\infty} I dt \tag{9}$$

- $\frac{I_M}{n_0}$ is obtained from Eq. (2) considering $I(T_M) = I_M$

$$\frac{I_M}{n_0} = n_0^{b-1} s' \exp\left(-\frac{E}{kT_M}\right)\left[1 + \frac{(b-1)s' n_0^{b-1}}{\beta}\int_{T_0}^{T_M} \exp\left(-\frac{E}{kT'}\right) dT'\right]^{-\frac{b}{b-1}}$$

Using now Eqs. (3) and (5) we obtain the final expression

$$\frac{I_M}{n_0} = b^{-1}\left[\frac{b}{1+(b-1)\Delta_M}\right]^{-\frac{1}{b-1}}\left(\frac{\beta E}{kT_M^2}\right) \tag{10}$$

- $\frac{n_M}{n_0}$ is obtained as following:

Solving Eq. (1) for n_M we have

$$-\int_{n_0}^{n_M} \frac{dn}{dt} = \frac{s'}{\beta}\int_{T_0}^{T_M} \exp\left(-\frac{E}{kT'}\right) dT'$$

and then

$$n_M = n_0\left[1+\frac{(b-1)s'n_0^{b-1}}{\beta}\int_{T_0}^{T_M}\exp\left(-\frac{E}{kT'}\right)dT'\right]^{\frac{1}{1-b}}$$

From which, using the condition at the maximum Eq. (3), we get

$$\frac{n_M}{n_0} = \left[\frac{bs'n_0^{b-1}kT_M^2}{\beta E}\exp\left(-\frac{E}{kT_M}\right)\right]^{-\frac{1}{b-1}}$$

and finally, using Eq. (5), the approximate condition for the maximum, we obtain

$$\frac{n_M}{n_0} = \left[\frac{b}{1+(b-1)\Delta_M}\right]^{-\frac{1}{b-1}} \tag{11}$$

- the quantity $\frac{I_M}{n_M}$ is simply obtained dividing Eq. (10) by Eq. (11):

$$\frac{I_M}{n_M} = \frac{\beta E}{bkT_M^2} \tag{12}$$

The previous quantities can be used for finding explicit forms for the activation energy in different cases.

- Method of the total peak width (ω)

Eq. (6) can be re-written as

$$\frac{I_M}{\beta n_0} = \frac{C_\omega}{\omega} \tag{13}$$

and using Eq. (10) we obtain

$$\frac{I_M}{\beta n_0} = (b^{-1})\cdot\left[\frac{b}{1+(b-1)\Delta_M}\right]^{-\frac{1}{b-1}}\cdot\left(\frac{E}{kT_M^2}\right) \tag{14}$$

Combining Eqs. (13) and (14) we get

$$E_\omega = (C_\omega \cdot b) \cdot \left[\frac{b}{1+(b-1)\Delta_M} \right]^{\frac{1}{b-1}} \cdot \left(\frac{kT_M^2}{\omega} \right) \tag{15}$$

Because the quantity $(b-1)\Delta_M \prec\prec 1$, the following approximation can be used:

$$\left[1+(b-1)\Delta_M\right]^{\frac{1}{b-1}} \approx 1 + \frac{b-1}{b-1}\Delta_M = 1 + \Delta_M \tag{16}$$

So that Eq. (15) becomes

$$E_\omega = C_\omega \cdot b^{\frac{b}{b-1}} \cdot \left(\frac{1}{1+\Delta_M} \right) \cdot \frac{kT_M^2}{\omega}$$

and then

$$E_\omega = C_\omega \cdot b^{\frac{b}{b-1}} \cdot \frac{kT_M^2}{\omega} - 2kT_M \tag{17}$$

This equation is similar to the Chen equation for the peak shape method, but it contains the kinetic order *b*.

The following table lists the values of μ, C_ω and $b^{\frac{b}{b-1}}$ as a function of the kinetic order.

b	μ_g	$b^{\frac{b}{b-1}}$	C_ω
1.0	0.419	2.72	0.903
1.1	0.431	2.85	0.912
1.2	0.444	2.99	0.909
1.3	0.456	3.12	0.906
1.4	0.467	3.25	0.903
1.5	0.476	3.38	0.899
1.6	0.485	3.50	0.896
1.7	0.494	3.63	0.892
1.8	0.502	3.75	0.888
1.9	0.509	3.88	0.884
2.0	0.517	4.00	0.880

- Method of the high temperature half-width (δ)

Eq. (7) can be re-written as

$$\frac{I_M}{\beta \cdot n_M} = \frac{C_\delta}{\delta} \tag{18}$$

and using Eq. (12) one obtains

$$E_\delta = \frac{C_\delta \cdot b \cdot k \cdot T_M^2}{\delta} \tag{19}$$

b	μ_g	C_δ
1.0	0.419	0.981
1.1	0.431	0.967
1.2	0.444	0.951
1.3	0.456	0.937
1.4	0.467	0.925
1.5	0.476	0.913
1.6	0.485	0.902
1.7	0.494	0.892
1.8	0.502	0.882
1.9	0.509	0.872
2.0	0.517	0.864

- Method of the low temperature half-width (τ)

Eq. (8) can be re-written as

$$\frac{n}{n_M} - 1 = \frac{\tau \cdot I_M}{C_\tau \cdot \beta \cdot n_M}$$

Using Eqs. (11) and (12), one obtains

$$E_\tau = \frac{C_\tau \cdot b \cdot k \cdot T_M^2}{\tau}\left[\left(\frac{b}{1+(b-1)\Delta_M}\right)^{\frac{1}{b-1}} - 1\right]$$

which becomes, using Eqs. (16):

$$E_\tau = \frac{C_\tau \cdot b \cdot k \cdot T_M^2}{\tau}\left[\left(\frac{b^{\frac{1}{b-1}}}{1+\Delta_M}\right) - 1\right] \tag{20}$$

This equation can be simplified using the following approximation:

$$\frac{b^{\frac{1}{b-1}}}{1+\Delta_M} \approx b^{\frac{1}{b-1}} \cdot (1-\Delta_M)$$

So that Eq. (20) becomes

$$E_\tau = C_\tau \cdot b \cdot \frac{kT_M^2}{\tau} \cdot \left(b^{\frac{1}{b-1}} - 1\right) - C_\tau \cdot b \cdot \frac{kT_M^2}{\tau} \cdot \left(b^{\frac{1}{b-1}} - 1\right) \cdot \frac{b^{\frac{1}{b-1}}}{b^{\frac{1}{b-1}} - 1} \cdot \frac{2kT_M}{E_\tau} \tag{21}$$

The first term of Eq. (21) is the dominant term and the second term represents a very small correction, so that we have the final form of the equation as:

$$E_\tau = C_\tau \cdot b \cdot \left(b^{\frac{1}{b-1}} - 1\right) \cdot \frac{kT_M^2}{\tau} - \left(\frac{b^{\frac{1}{b-1}}}{b^{\frac{1}{b-1}} - 1}\right) \cdot (2kT_M) \tag{22}$$

b	μ_g	$b \cdot \left(b^{\frac{1}{b-1}} - 1\right)$	$\frac{b^{\frac{1}{b-1}}}{b^{\frac{1}{b-1}} - 1}$	C_τ
1.0	0.419	1.778	1.582	0.871
1.1	0.431	1.753	1.628	0.875
1.2	0.444	1.786	1.672	0.878
1.3	0.456	1.817	1.715	0.881
1.4	0.467	1.847	1.758	0.884
1.5	0.476	1.875	1.800	0.887
1.6	0.485	1.902	1.841	0.889
1.7	0.494	1.928	1.882	0.892
1.8	0.502	1.953	1.922	0.894
1.9	0.509	1.970	1.960	0.896
2.0	0.517	2.000	2.000	0.895

Reference

Kitis, G. and Pagonis V., NIM-B **262** (2007) 313

Glow curve

It is the plot of the thermoluminescence intensity, *I*, as a function of the sample temperature during read out. Each trapping level in the material gives rise to an associated glow peak; so, a glow curve may be formed by several peaks, each one related to different trapping levels. These peaks may or may not be resolved in the glow curve.

Considering the basic equation

$$I = -c\frac{dn}{dt} \tag{1}$$

where the TL intensity is proportional to the detrapping rate, by its integration wer get

$$n(t) - n_{\infty} = \int_{t}^{\infty} I \cdot dt' \tag{2}$$

Dealing with only one peak, $n_{\infty} = 0$ and therefore

$$n(t) = \int_{t}^{\infty} I \cdot dt' \tag{3}$$

Furthermore, using a linear heating rate, Eq. (3) transforms in

$$n(T) = \frac{1}{\beta}\int_{T}^{\infty} I(T')dT' \tag{4}$$

From a practical point of view, *n(T)* can be evaluated from the area under the peak, from a value $T=T_i$, initial rise region of the peak, to a temperature T_f, end of thwe peak (i.e., when the glow intensity is at its minimum. So, Eqs. (3) and (4) can be rewritten as

$$n = \int_{t_i}^{t_f} Idt' = \frac{1}{\beta}\int_{T_i}^{T_f} IdT' \tag{5}$$

Because the trapped charge concentration, n, is proportional to the dose delivered to the TL sample, the concept expressed by Eq. (5) is of great importance in radiation dosimetry.

Glow peak parameters

The following parameters have been obtained by glow peak simulation [1]

b	Tmax	ω	τ	δ	μ
1	384.58	29.825	17.335	12.49	0.418776
1.1	384.51	31.235	17.72	13.515	0.432688
1.2	384.435	32.595	18.07	14.525	0.44562
1.3	384.365	33.9	18.4	15.5	0.457227
1.4	384.295	35.17	18.71	16.46	0.468013
1.5	384.225	36.4	19.005	17.395	0.477885
1.6	384.155	37.605	19.285	18.32	0.487169
1.7	384.085	38.77	19.545	19.225	0.495873
1.8	384.015	39.91	19.795	20.115	0.504009
1.9	383.95	41.025	20.04	20.985	0.511517
2	383.88	42.11	20.265	21.845	0.51876

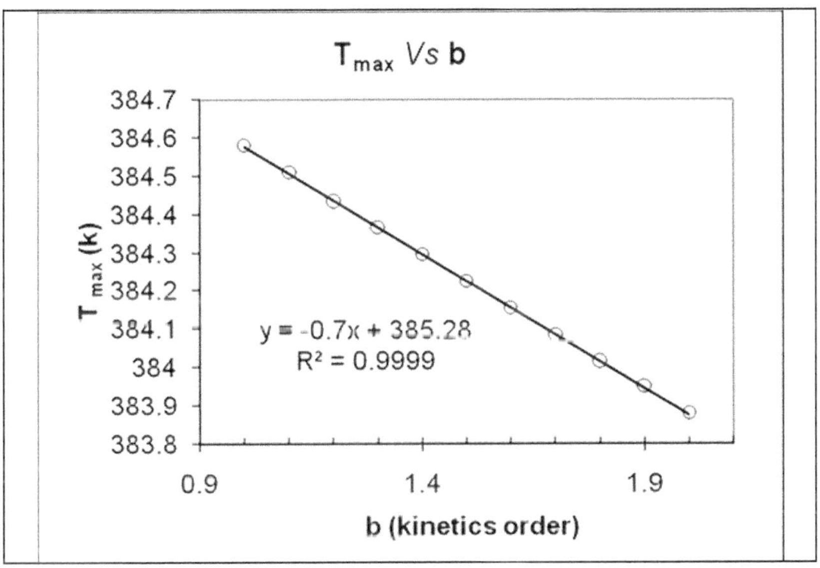

Fig. 1. Peak temperature at the maximum Vs kinetics order.

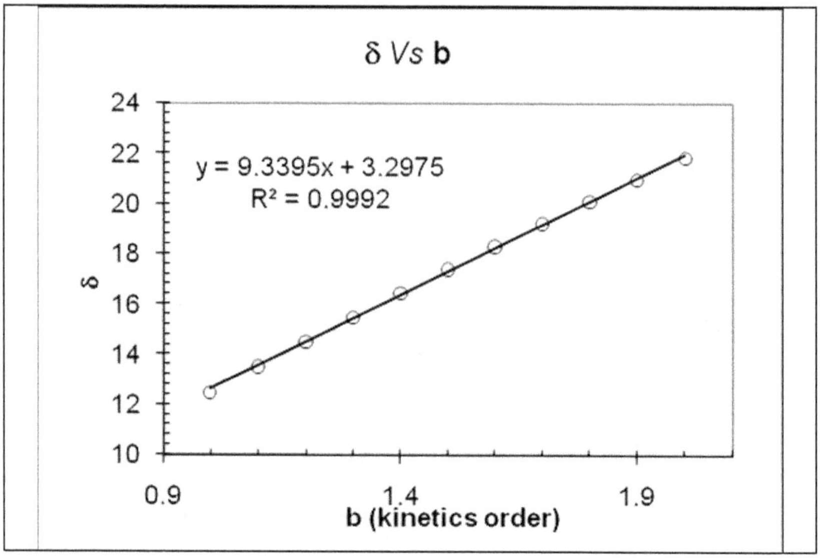

Fig. 2. $\delta = T_2 - T_M$ Vs kinetics order.

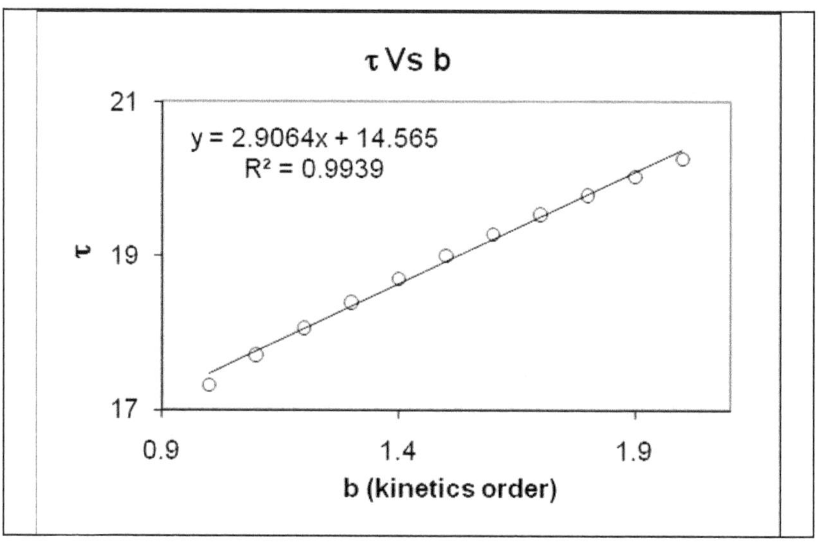

Fig. 3. $\tau = T_M - T_1$ Vs kinetics order.

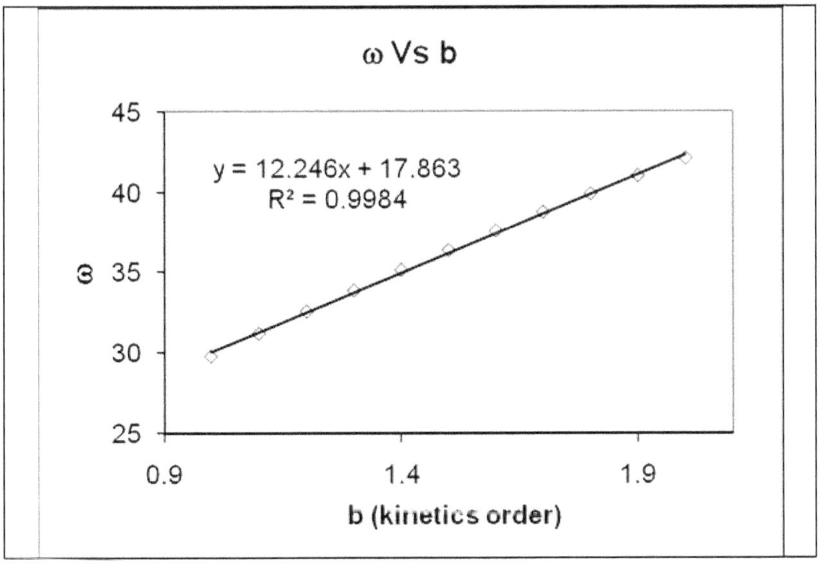

Fig. 4. $\omega = T_2 - T_1$ Vs kinetics order.

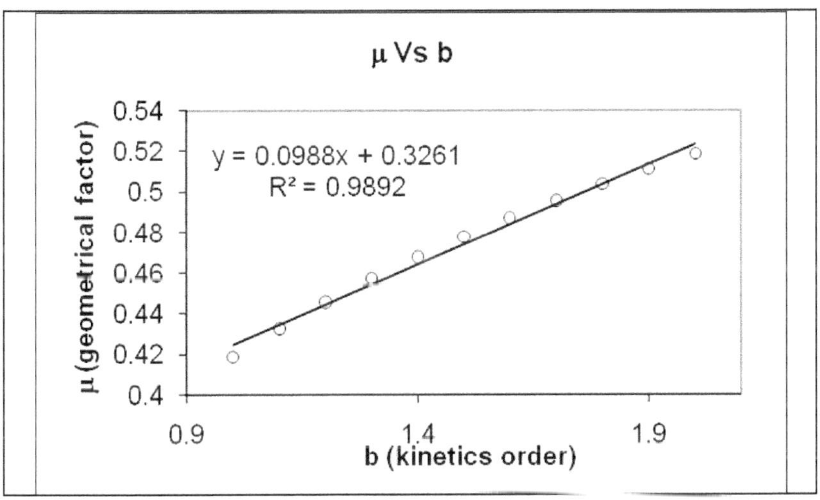

Fig. 5. $\mu=\delta/\omega$ Vs kinetics order.

- Pre-exponential factor Vs Kinetics order

kinetics order	S" numerical	S" approximated
1.02	9.94E+12	9.94E+12
1.05	9.94E+12	9.94E+12
1.1	9.94E+12	9.94E+12
1.15	9.94E+12	9.94E+12
1.2	9.95E+12	9.94E+12
1.25	9.95E+12	9.93E+12
1.3	9.94E+12	9.92E+12
1.35	9.94E+12	9.92E+12
1.4	9.94E+12	9.92E+12
1.45	9.94E+12	9.92E+12
1.5	9.94E+12	9.92E+12
1.55	9.94E+12	9.92E+12
1.6	9.94E+12	9.91E+12
1.65	9.95E+12	9.91E+12
1.7	9.94E+12	9.90E+12
1.75	9.94E+12	9.90E+12
1.8	*9.94E+12*	9.90E+12
1.85	9.94E+12	9.90E+12
1.9	9.94E+12	9.90E+12
1.95	9.94E+12	9.89E+12
2	9.94E+12	9.89E+12

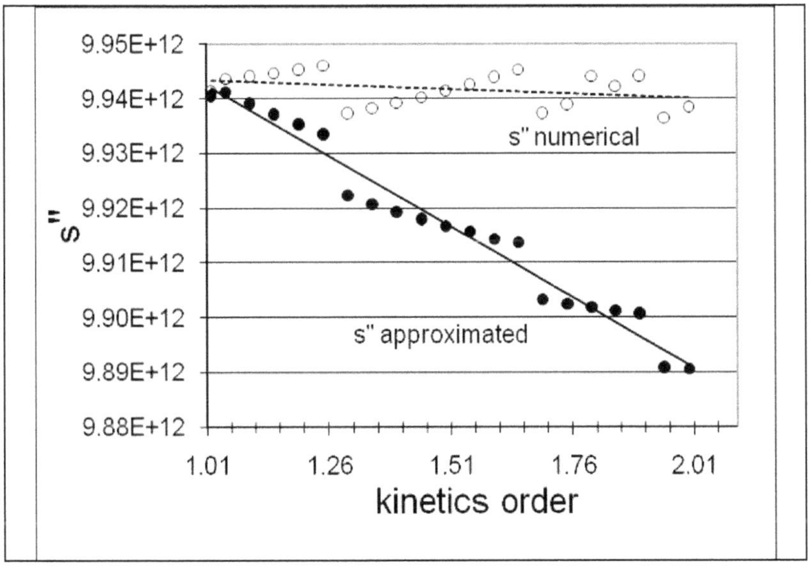

Pre-exponential factor Vs kinetics order.

Reference

1. Kitis, G., private communication

H

Heating rate effects

The various heating rate method for the calculation of the activation energy has been proposed by several authors [1-5]. The results from many TL phosphors showed, if the integral of the TL emission, the glow curve, is calculated using a temperature scale, a shift of the peak maximum temperature toward high temperature values. Furthermore, a second effect appears as a function of the heating rate: a drastic decrease of the TL response, in some materials, as the heating rate increases. This behaviour has been attributed to a thermal quenching effect.

According to M.S. Rasheedy and E.M. Zaharan [6], the decrease in the TL intensity with increasing heating rate cannot be explained by using the kinetics equations in their usual time-dependent forms.

A first observation has done in [7]: the initial concentration of trapped carriers, n_0, is proportional to the given dose and the glow curve area is proportional to n_0. As a consequence, for a given dose, n_0 is conserved and then also the area is conserved with changing hating rate. However, the glow curve can move and change according to its plot:

1) If the glow curve plot is given as TL intensity expressed in (cm^{-3} s^{-1}) Vs time (s), at same values of n_0 the glow curve moves to lower times as the H.R. increases.
2) If the glow curve plot is given as TL intensity expressed in (cm^{-3} k^{-1}) Vs temperature (k), the glow curve moves to higher values of temperature as the H.R. increases.

In both cases the total area, proportional to n_0, is conserved.

In case 1) the TL intensity is given by

$$I(t) = -\frac{dn}{dt} \tag{1}$$

In case 2), the TL intensity is given by

$$I(T) = -\frac{dn}{dT} = \frac{I(t)}{\beta} = \frac{ns}{\beta}\exp\left(-\frac{E}{kT}\right) \tag{2}$$

for the first order, and by

$$I(T) = -\frac{dn}{dT} = \frac{n^2}{N\beta}s\exp\left(-\frac{E}{kT}\right) \tag{3}$$

and

$$I(T) = -\frac{dn}{dT} = \frac{n^b}{\beta N^{b-1}} s \exp\left(-\frac{E}{kT}\right) \tag{4}$$

for second and general order respectively.

By solving the differential equation (2), assuming a linear heating rate β, one has, for the first order kinetics

$$I(T) = \frac{n_0 s}{\beta} \exp\left(-\frac{E}{kT}\right) \exp\left[-\frac{s}{\beta}\int_{T_0}^{T} \exp\left(-\frac{E}{kT'}\right) dT'\right] \tag{5}$$

The solution of Eqs. (3) and (4) gives

$$I(T) = \frac{n_0^2 s \exp\left(-\frac{E}{kT}\right)}{N\beta\left[1 + \left(\frac{n_0 s}{N\beta}\right)\int_{T_0}^{T} \exp\left(-\frac{E}{kT'}\right) dT'\right]^2} \tag{6}$$

for the second order kinetics, and

$$I(T) = \frac{n_0^b s N^{1-b} \exp\left(-\frac{E}{kT}\right)}{\beta\left\{1 + \left[\frac{s}{\beta}(b-1)\left(\frac{n_0}{N}\right)^{b-1}\right]\int_{T_0}^{T} \exp\left(-\frac{E}{kT'}\right) dT'\right\}^{\frac{b}{b-1}}} \tag{7}$$

for the general order kinetics.

The same problem has been also discussed by G. Kitis et al. in [8] The Authors reported the following interesting expression which relates the heating rate β, the FWHM of the glow peak, in terms of temperature, and the half-life at T_{MAX}

$$\frac{[FWHM]_T}{\beta \times [t_{1/2}]_{T_{MAX}}} = c \tag{8}$$

where c =2.3472±0.001 for a first order kinetics, and it is equal to 3.1067±0.001 for a second order kinetics.

References

1. Booth A.H., Can. J. Chem., **32** (1954) 214
2. Bohum A., Czech, J. Phys., **4** (1954) 91
3. Hoogenstraaten W., Philips Res. Rev., **13** (1958) 515

4. Chen R. and Winer S.A.A., J. Appl. Phys., **41** (1970) 5227
5. Rasheedy M.S., Int. J. Mod. Phys. B, **18** (2004) 2877
6. Rasheedy M.S. and Zahram E.M., Phys. Scr. **73** (2006) 98
7. Favalli A., Private communication
8. Kitis G., Spiropulu M., Papadopoulos J. and Charalambous Stef., N.I.M. **73** (1993) 367

Heating rate effects on thermoluminescence of $BaSO_4$:Eu+PTFE

The crystals of $BaSO_4$, activated by Eu, have been prepared using the crystallization technique of $Ba(NO_3)_2$ with concentrated H_2SO_4 (95 to 98%) in stoichiometric mixture [1]. The highest sensitivity has been obtained for $BaSO_4$:Eu, with an Eu concentration of 0.5 mol%. The thermoluminescent powder was then mixed with PTFE resin powder in the ratio 2:3. The samples, having an average mass of 30 mg, a diameter of 4.5 mm and a thickness of 0.8 mm, were carefully selected in order to obtain a good consistency in the TL response.

The TL reading system was a Harshaw TL reader model 4000, with constant heating rate from 2 to 12 K/s, i.e. 2, 4, 6, 8 and 12 K/s with read out times of 120, 60, 40, 30 an 20 s respectively. In all cases the initial read out temperature was 333 K and the final one 573 K. Nitrogen gas was allowed to flow into the reader during read out to avoid any spurious signals; to avoid any contribution of mass and sensitivity variation in all cases we used the same dosimeter. A PC connected to the TL reader allowed to record and analyzes the received glow curves. Before any irradiation, the sample was subjected to a thermal procedure for erasing any previous irradiation effects. The annealing was performed in an oven at 573 K for a period of 30 min. The irradiation dose was 0.1 Gy from ^{60}Co. The material presents a glow curve having, apparently, only one peak.

Table 1 shows the TL readings at each heating rate used. The reported TL integral values have been obtained using a time scale and then, as a consequence, the integral values result to be constant and independent by the heating rate. The average value has a S.D. of 0.85%.

Figs. 1 and 2 show, respectively, the glow peak and the TL intensity behaviors as a function of the heating rate.

Fig. 3, as an example, shows the glow curve deconvolution applied to an experimental glow curve obtained with an heating rate of 2 K/s. The best FOM (Figure Of Merit) was obtained using two peaks, both following a second order kinetics, i.e. $b = 2$.

Figs. 4, 5, 6 and 7 show for both peaks, as a function of the heating rate, the peak temperature at the maximum, T_M, the peak intensity at the maximum, I_M, the activation energy, E, the frequency factor, s. Fig. 8 shows the full width at half maximum (FWHM) as a function of the H.R. too.

The most important experimental result is concerning to the glow curve behavior; indeed, as it is shown in Fig. 1, the glow peak intensity increases as the H.R. increases and the peak temperature at the maximum moves towards higher temperature values. Accordingly to the thermoluminescent theory, the following glow curve characteristics should be fulfilled as a function of increasing heating rate [2,3]:

a) the peak temperature at the maximum moves towards higher temperature values
b) the TL integral of the glow curve remains constant
c) the FWHM increases
d) the TL intensity at the maximum, I_M, decreases.

According to the experimental results, the property d) is not verified

Table 1. TL readings as a function of the H.R.

H.R (K/s)	TL intensity (nC)
2	1665.83
4	1666.09
6	1638.37
8	1676.42
12	1665.80

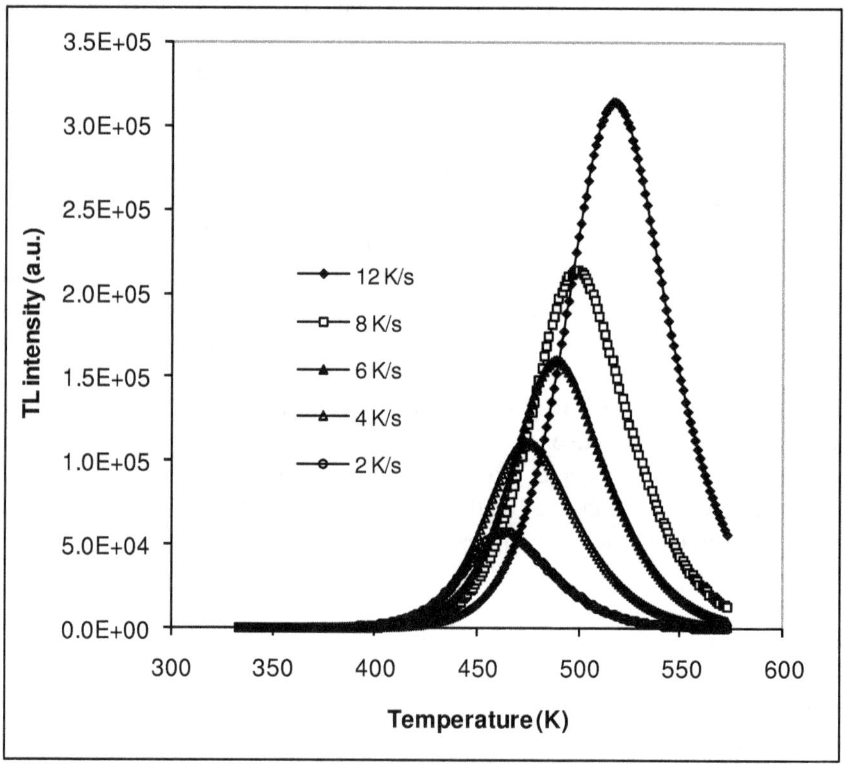

Fig. 1. Total Glow Curve read out from 2 to 12 K/s H.R.

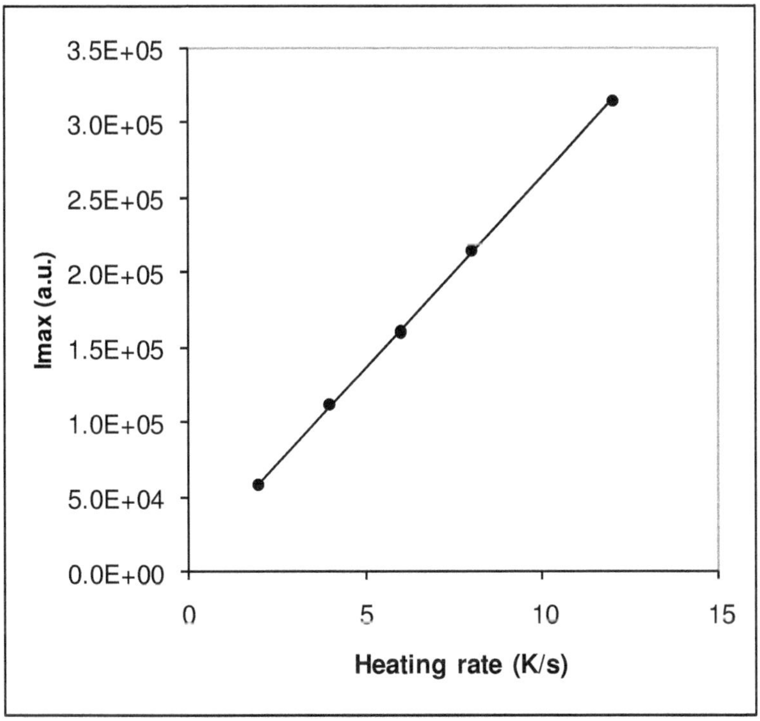

Fig. 2. Maximum intensity, Im, as a function of the H.R.

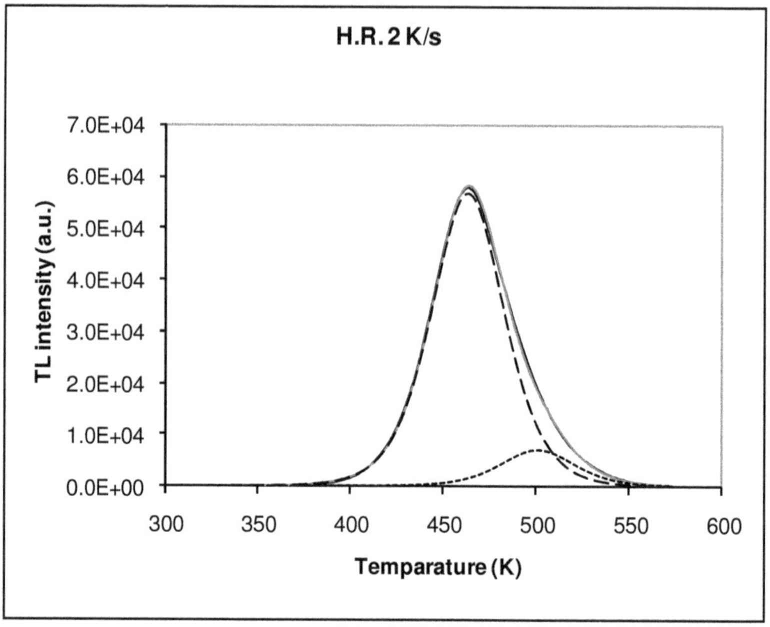

Fig. 3. Glow Curve deconvolution at 2 K/s H.R (FOM = 1.35 %).

Table 2. Kinetic parameters of $BaSO_4$:Eu+PTFE.

H.R. (K/s)	Peak	Tm	Im	b	E(eV)	$s(s^{-1})$	n_0 (cm^{-3})	s* ($cm^{3(b-1)}s^{-1}$)	FOM (%)
2	**1**	463.4	56948	2.00	1.42	4.47E+14	1.40E+06	3.19E+08	1.35
	2	500.9	7026	2.00	1.63	3.49E+15	1.77E+05	1.97E+10	
4	**1**	473.9	108255	2.00	1.46	8.45E+14	1.36E+06	6.20E+08	1.05
	2	510.9	16956	2.00	1.66	5.90E+15	2.19E+05	2.70E+10	
6	**1**	487.8	155311	2.00	1.50	1.23E+15	1.34E+06	9.17E+08	0.89
	2	525.0	24737	2.00	1.70	8.07E+15	2.19E+05	3.69E+10	
8	**1**	497.4	205689	2.00	1.52	1.34E+15	1.37E+06	9.83E+08	0.96
	2	533.9	35696	2.00	1.72	9.06E+15	2.42E+05	3.74E+10	
12	**1**	514.6	290991	2.00	1.56	1.61E+15	1.34E+06	1.21E+09	1.09
	2	547.3	65829	2.00	1.76	1.36E+16	3.05E+05	4.46E+10	

1 is the dosimetric peak

Figure 4 shows the behavior of the maximum temperature, T_m, of the glow peaks of $BaSO_4$:Eu+PTFE as a function of the heating rate. Figure 5 shows the maximum TL intensity, I_m, as a function of the heating rate.

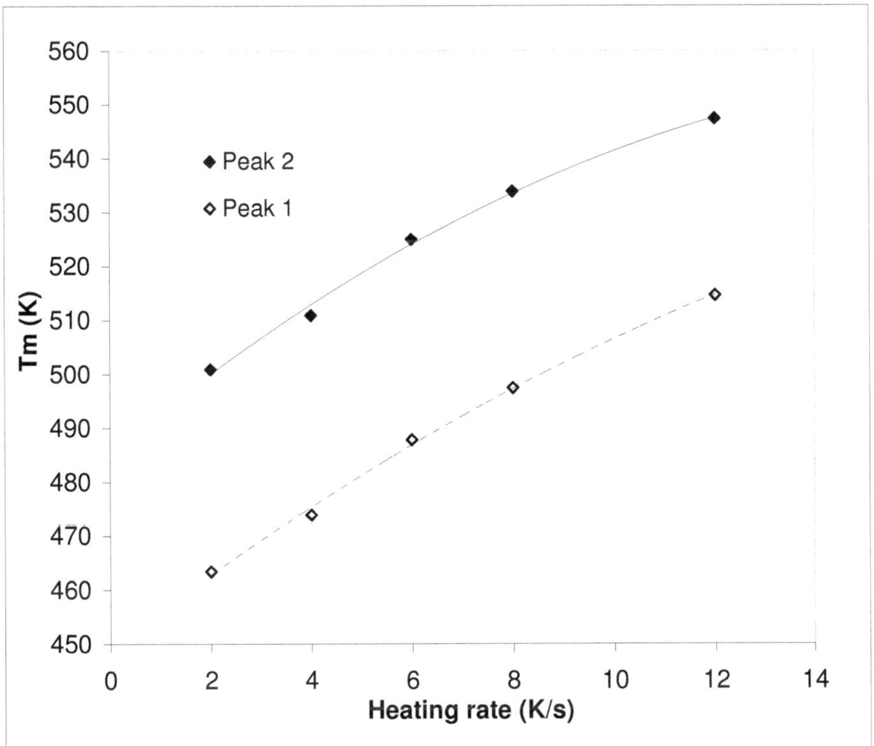

Fig. 4. The maximum temperature, T_m, of the glow peaks of $BaSO_4$:Eu+PTFE as a function of the heating rate.

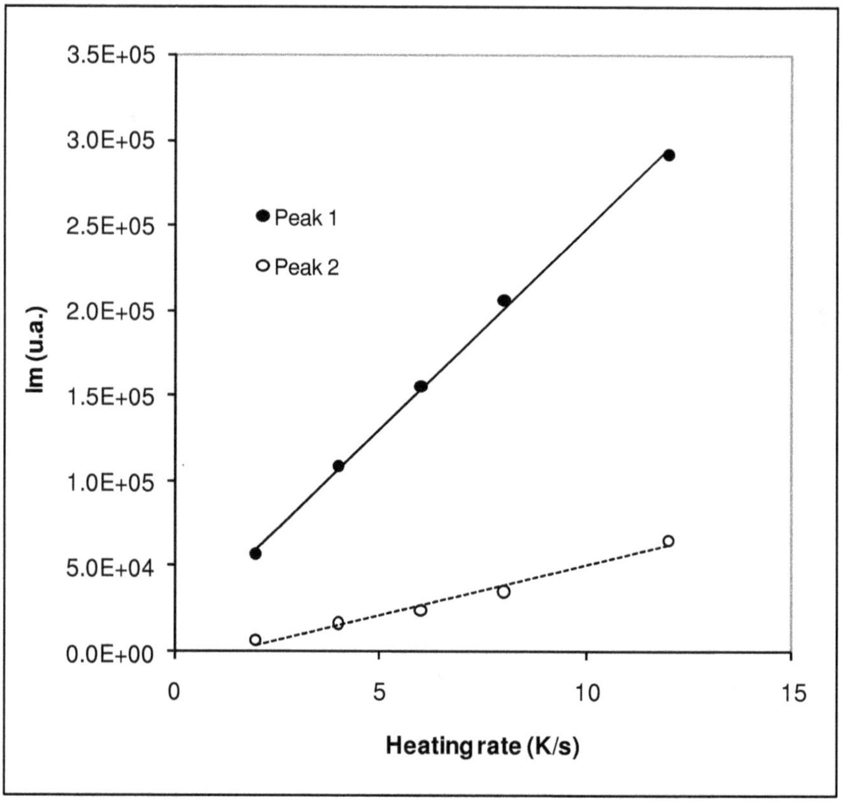

Fig. 5. The maximum TL intensity, I_m, as a function of the heating rate.

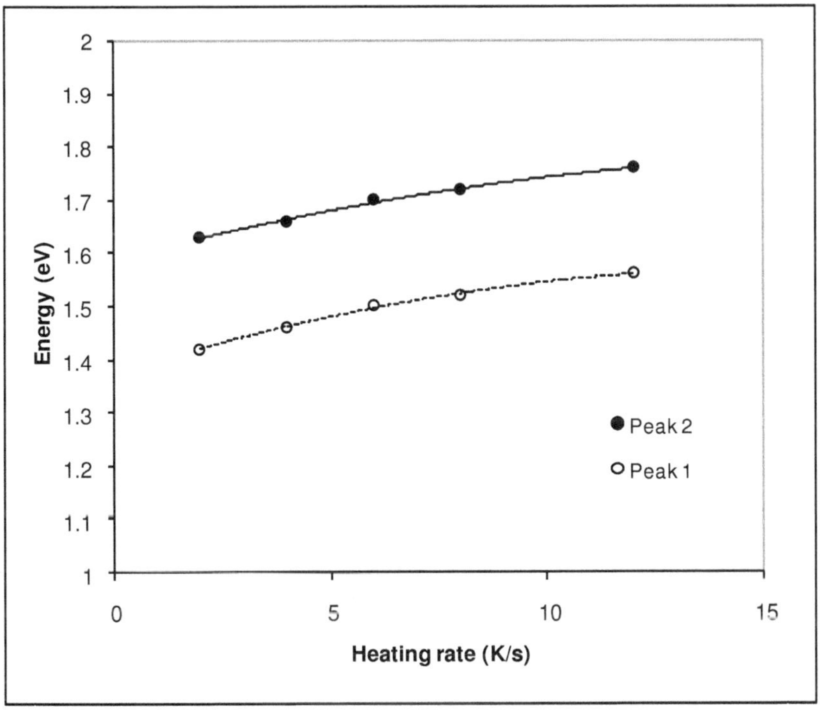

Fig. 6. Energy of the TL peaks as a function of H.R.

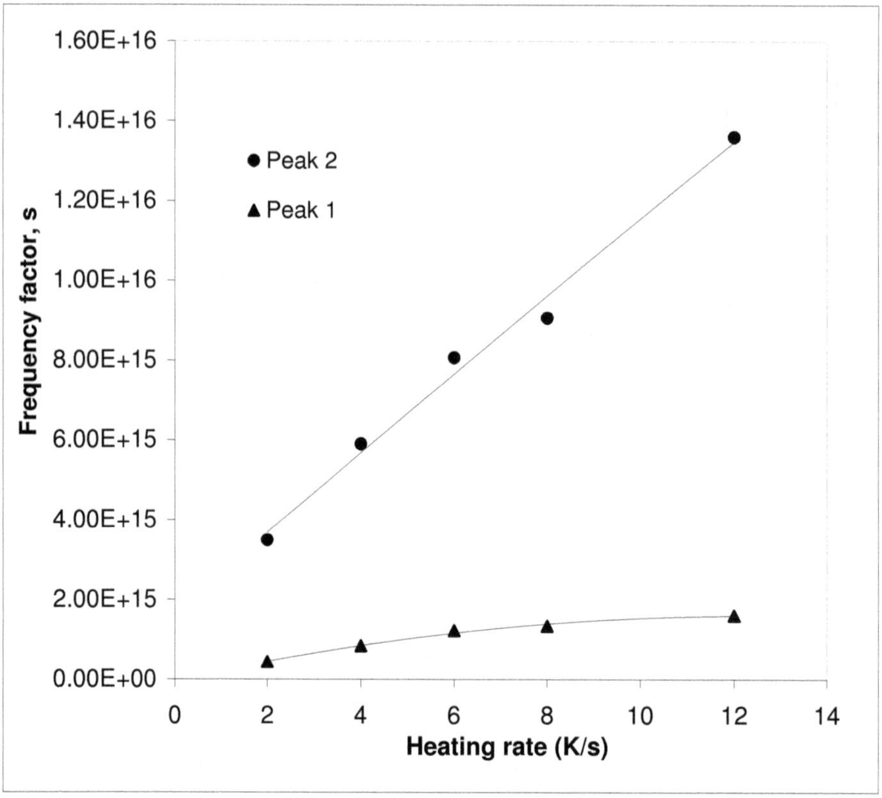

Fig. 7. Frequency factor, *s*, as a function of H.R.

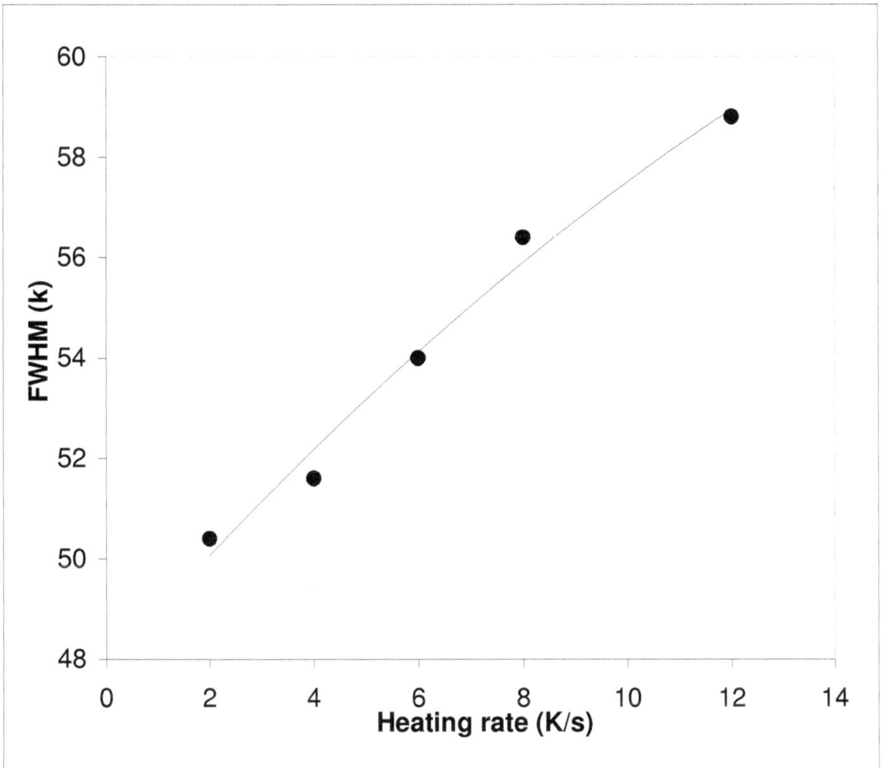

Fig. 8. The FWHM of the main glow peak as a function of the H.R.

References

1. P.R. González, et al., NIMB, in press
2. S.G. Gorbics, et al., J. Appl. Rad. Isot. **20**, (1969) 829
3. G. Kitis, et al., NIM Phys. Res., **73** (1993), 367

I

In-vivo dosimetry (dose calibration factors)

It is strongly recommended to perform a separate calibration for each radiation beam quality. If the TLDs can be identified, a calibration factor could be given to each dosemeter and it is necessary to monitor the individual factors from time to time.

In practice, having a large number of dosemeters, it is possible to save a part of them for the purpose of calibration. The readings of the patient dosemeters can be then converted in dose by comparison their response to the ones of the calibrated dosemeters.

Entrance dose calibration factor

The entrance dose calibration factor, F_{IN}, is defined as the factor with which the TL signal, TL_{IN}, of the TLD, positioned on the skin of the patient at the entrance surface, with the right build-up cap, must be multiplied to obtain the entrance dose, D_{IN}:

$$F_{IN} = \frac{D_{IN}}{TL_{IN}}$$

F_{IN} is determined positioning the TLD on the surface of a flat phantom, at the entrance side of the beam. The TLD response (TL_{IN}) is then compared to the response of a calibrated ionization chamber (D_{IN}), positioned at depth $d_{IN,\max}$.

Exit dose calibration factor

The exit dose calibration factor, F_{OUT}, is determined in a very similar way. The TLD is positioned on the exit surface of the beam and its signal is compared to the response of the calibrated ionization chamber positioned in the phantom at $d_{OUT,\max}$ from the exit surface:

$$F_{OUT} = \frac{D_{OUT}}{TL_{OUT}}$$

The phantom thickness should be variable for matching must of the various thickness of the patients

It is also suggested to determine the calibration factors for each particular kind of treatment.

The following Fig. 1 shows the experimental set up for the calibration factors determination.

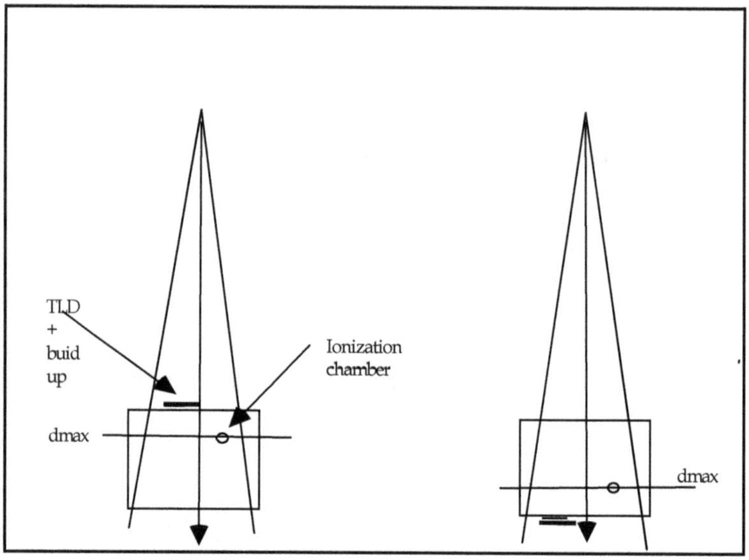

Fig. 1. Experimental set up for determing the entrance and exit calibration factors.

Inflection points method (Land: first order)

This method, proposed and applied by Land, uses, in addition to the temperature at the maximum, T_M, the two inflection points in the curve of the TL emission. Using the Randall and Wilkins expression the second derivative of the TL intensity can be written as

$$\frac{d^2I}{dT^2}=\frac{d}{dT}\left(\frac{dI}{dT}\right)=\frac{d}{dT}\left(I\frac{1}{I}\frac{dI}{dT}\right)=\frac{d}{dT}\left[I\frac{d}{dT}(\ln I)\right]$$

$$=\frac{dI}{dT}\frac{d}{dT}(\ln I)+I\frac{d^2}{dT^2}(\ln I)=I\left[\frac{d}{dT}(\ln I)\right]^2+I\frac{d^2}{dT^2}(\ln I) \tag{1}$$

The temperature values T_i, corresponding to the inflection points, are obtained from Eq. (1) by quoting it to zero, for $T=T_i$:

$$\left[\frac{d}{dT}(\ln I)\right]^2+\frac{d^2}{dT^2}(\ln I)=0 \tag{2}$$

Using now the logarithm of the intensity $I(T)$ one obtains

$$\left[\frac{d}{dT}(\ln I)\right]^2_{T=T_i} = \left[\frac{E}{kT_i^2} - \left(\frac{s}{\beta}\right)\exp(-\frac{E}{kT_i})\right]^2 \tag{3}$$

$$\frac{d^2}{dT^2}(\ln I) = -\left(\frac{2E}{kT_i^3}\right) - \left(\frac{E}{kT_i}\right)\left(\frac{s}{\beta}\right)\exp(-\frac{E}{kT_i}) \tag{4}$$

Inserting Eq. (4) into Eq. (2) and using the condition at the maximum, one obtains

$$\frac{E}{kT_M^4}\exp\left[\frac{2E}{k}\left(\frac{1}{T_M}-\frac{1}{T_i}\right)\right] - \frac{3E}{kT_i^2T_M^2}\exp\left[\frac{E}{k}\left(\frac{1}{T_M}-\frac{1}{T_i}\right)\right] + \frac{E}{kT_i^4} - \frac{2}{T_i^3} = 0 \tag{5}$$

Using now the substitutions

$$Q = \left(\frac{T_i}{T_M}\right)^2 \tag{6}$$

the following final form is obtained

$$E = \frac{kT_M T_i}{|T_i - T_M|}\ln\left(\frac{A}{Q}\right) \tag{7}$$

with $A = 0.77$ if $T_i < T_M$, or, $A = 2.66$ if $T_i > T_M$.

The frequency factor is then obtained from the condition at the maximum.

This method is useful even in case of closed peaks and E and s can be obtained for all peaks from a single glow-curve.

Reference

Land P.L., J. Phys. Chem. Solids **30** (1969) 1681

Inflection points method (Singh et al.: general order)

Singh et al. presented the **method of Land** in a more simple form.

Considering the equation

$$I(t) = sn_0\exp\left(-\frac{E}{kT}\right)\left[1 + s(b-1)t\exp\left(-\frac{E}{kT}\right)\right]^{\frac{1}{1-b}} \tag{1}$$

which gives the TL intensity function $I(T)$ for a general order peak, the first and the second derivatives of $I(T)$ with respect to T are expressed by

$$\frac{dI}{dT} = I \cdot f(T) \tag{2}$$

$$\frac{d^2 I}{dT^2} = \frac{dI}{dT} \cdot f(T) + I \cdot \frac{df(T)}{dT} \tag{3}$$

where

$$f(T) = \frac{E}{kT^2} - \frac{bs\exp(-\frac{E}{kT})}{\beta[1+(b-1)\frac{s}{\beta}\int_{T_0}^{T}\exp(-\frac{E}{kT'})dT']} \tag{4}$$

if b ≠ 1; and

$$f(T) = \frac{E}{kT^2} - \frac{s}{\beta}\exp(-\frac{E}{kT}) \tag{5}$$

if b = 1.

$dI/dT = 0$ gives the peak temperature at the maximum, T_M , and $d^2I/dT^2 = 0$ gives the inflection pints T_i of the glow-curve. Furthermore, $T_i = T_i^-$ corresponds to the inflection point on the raising side of the glow-curve and $T_i = T_i^+$ is the inflection point on the falling side.

According to Land, one can write

$$x_i^{\pm} = \frac{E}{kT_i^{\pm}}, \qquad x_M = \frac{E}{kT_M} \tag{6}$$

Because a good linear correlation exists between the following pairs of variables:

$$[x_M, \frac{x_i^-}{(x_i^- - x_M)}]; \; [x_M, \frac{x_i^+}{(x_M - x_i^+)}]; \; [x_M, \frac{x_i^+ x_i^-}{x_M(x_i^- - x_i^+)}] \tag{7}$$

one can write

$$x_M = A_1 \frac{x_i^-}{x_i^- - x_M} + B_1$$

$$x_M = A_2 \frac{x_i^+}{x_M - x_i^+} + B_2 \tag{8}$$

$$x_M = A_3 \frac{x_i^+ x_i^-}{x_M (x_i^- - x_i^+)} + B_3$$

where the coefficients A_j and B_j depend on the order of kinetics b.

The previous equations can be rewritten in the following explicit form

$$E = \frac{A_1 k T_M^2}{\left(T_M - T_i^-\right)} + B_1 k T_M$$

$$E = \frac{A_2 k T_M^2}{\left(T_i^+ - T_M\right)} + B_2 k T_M \tag{9}$$

$$E = \frac{A_3 k T_M^2}{\left(T_i^+ - T_i^-\right)} + B_3 k T_M$$

By using the method of non-linear least-square regression, the coefficients A_j and B_j can be expressed as a quadratic function of the kinetics order, for b ranging from 0.7 to 2.5:

$$\begin{aligned} A_j &= a_{0j} + a_{1j} b + a_{2j} b^2 \\ B_j &= c_{0j} + c_{1j} b + c_{2j} b^2 \end{aligned} \tag{10}$$

The following Table 1 shows the numerical values of the coefficients comparing in Eqs. (10):

J	a_{0j}	a_{1j}	a_{2j}	c_{0j}	c_{1j}	c_{2i}
1	0.8730	-1.5619	0.1334	0.4489	0.5853	-0.0751
2	0.6676	-1.8493	0.1499	0.4479	0.5866	-0.0756
3	0.9394	-1.7055	0.1422	0.8967	1.1721	-0.1507

Table 1. Values of the coefficients a_{kj} and c_{kj} in Eq. (10).

Fig. 1 shows the behaviour of *I*, *dI/dT* and d^2I/dT^2 as a function of temperature for an isolated peak at 320°C in $KAlSi_3O_8$ following a second order kinetics.

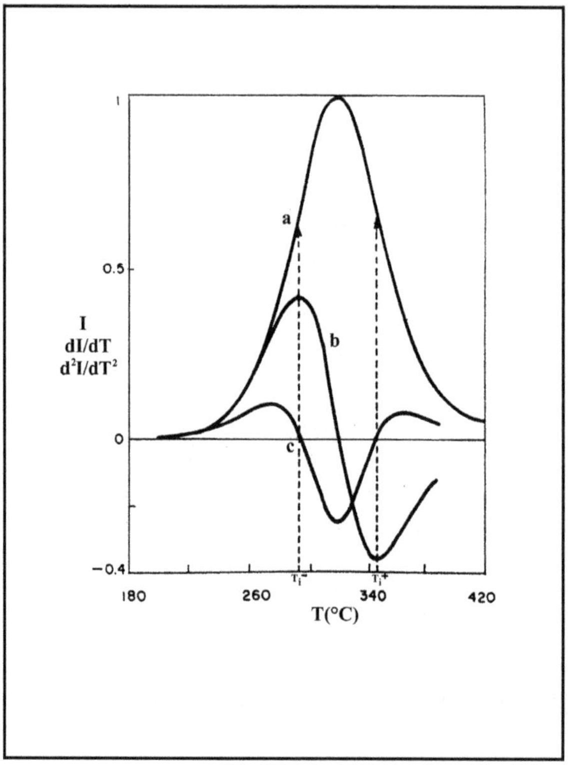

Fig. 1. Behaviour of I, (a), dI/dT, (b) and d^2I/dT^2 (c) as a function of the temperature T.

Reference

Singh T.C.S., Mazumdar P.S. and Gartia R.K., J. Phys. D: Appl. Phys. **23** (1990) 562

Initial rise method

The initial rise (IR) method is based on the analysis of the low temperature interval of a peak. This method can be applied when a thermoluminescent sample has a single, well defined peak or, if more than one peak, the peaks are not overlapped. The method was formerly suggested by Garlick and Gibson (1). The IR method is independent of kinetics order.

In the low temperature tail of a peak the amount of trapped electrons can be assumed as a constant, the dependence on temperature being neglectable. Making reference to the following figure, increasing the temperature up to a value $T_C < T_M$, for which the corresponding intensity I_C should be not larger than 15% of I_M, the thermoluminescent emission may be expressed by

$$I(T) \propto \exp\left(-\frac{E}{kT}\right) \tag{1}$$

if the frequency factor is assumed to be independent of temperature.

From Eq. (1) we get

$$\ln I(T) \propto -\frac{E}{kT}$$

A plot of $\ln I(T)$ as a function of $\frac{1}{T}$ should produce a straight line. From the slope $-\frac{E}{kT}$ the value of E is evaluated.

Reference

1. Garlick G.F.J and Gibson A.F., Proc. Phys. Soc. **60** (1948) 574

Initial rise method when $s = s(T)$ (Aramu et al.)

Aramu and his colleagues applaied the **initial rise method** in the case of the **frequency factor** *s* is temperature dependent. In this case, the intensity *I* is proportinal to the first exponential only:

$$I \propto T^{\alpha} \exp\left(-\frac{E}{kT}\right) \tag{1}$$

from which

$$\ln I = \alpha \ln T - \frac{E}{kT}$$

$$\frac{d}{dT}(\ln I) = \frac{\alpha}{T} + \frac{E}{kT^2} \tag{2}$$

Comparing Eq. (2) with the following equation

$$I \propto \exp\left(-\frac{E_{ir}}{kT}\right)$$

or better with

$$\frac{d}{dT}(\ln I) = \frac{E_{ir}}{kT} \tag{3}$$

one obtains

$$\frac{E_{ir}}{kT^2} = \frac{\alpha}{T} + \frac{E}{kT^2}$$

form which

$$E = E_{ir} - \alpha kT \tag{4}$$

This means the need to correct E for a few percent.

Reference

Aramu F., Brovetto P. and Rucci A., Phys. Letters **23** (1966) 308

Initialisation procedure

The initialisation procedure on a new batch of TLDs is recommended to reduce the possibility of variations in dosimeter performance characteristics during usage [1,2].

The first stage of the procedure involves heating dosimeters inside a furnace using the optimum annealing parameters (temperature and time) indicated for the TL material under test. In another section of this book all the annealing procedures used for different materials are listed. The dosimeters are placed in lidded crucible or in suitable annealing stacks (such as those made from quality stainless steel or electroplated copper). Annealing stacks allow separation and identification of dosimeter elements and are particularly useful if these elements are to be calibrated individually rather than in batches. The annealing stack containing the dosimeters is placed in the furnace, preheated to the required temperature. The actual duration of annealing will be longer than the required annealing time in order to attain thermal equilibrium at the required temperature. This additional time should be determined before all the setting up procedures as it will be indicated in the section concerning the quality control of the furnaces. After annealing, the dosimeters are cooled in their containers in a reproducible manner. It is imperative to use always the same cooling procedure and that this is reproducible because the glow-curve of the material is strongly affected by the cooling. The cooling may be accomplished by keeping the furnace door open after the heating has been switched off. In this manner the cooling will be more or less long, depending on the starting temperature. Alternatively, the crucible or annealing stack may be removed from the furnace immediately after the thermal treatment in order to allow the dosimeters to be cooled much faster to room temperature. This can be obtained lying the annealing container on a metal plate. Tests should be made before initialisation to find the most suitable means of cooling for the user's particular requirements. It is not recommended to switch to other methods once a cooling procedure has been adopted. In some case the annealing procedure consists of two subsequent annealing (see the annealing section): the first is carried out at high temperature and the second at low temperature. An example is given by LiF:Mg,Ti in the form of TLD-100, 600 or 700, which needs a first annealing at 400°C during 1 hour followed by 2 hours at 100°C (or 24 hours at 80°C). In all the cases where the annealing procedure is formed by two thermal treatments, the first at high temperature followed by one at low temperature, the dosimeters have to be cooled to room temperature at the end of the first annealing and then placed in the preheated oven for the second annealing. There are now several commercial programmed ovens in which the thermal cycles can be programmed at the beginning of the treatment; in this case the low temperature annealing is switched on when the high temperature of the first annealing decreases until the lower temperature of the second one. However, the procedures of heating and cooling have to be always in the same manner. At the end of the annealing procedure the dosimeters are read to check the **background signal**. The background depends on the H.V. applied to the P.M. tube, on its age and on the room temperature: the stability of the TL reader must be checked before and after any reading session. The initialisation procedure is repeated over three cycles. If the backgrounds on the dosimeters have remained low over these cycles, the initialisation is terminated and the dosimeters are ready for the subsequent tests. If backgrounds on the dosimeters are variable, the initialisation can be continued for further two cycles of treatment. If backgrounds continue to remain high or variable the efficiency of the readout system should be checked and/or the dosimeters rejected.

An example of the above initialisation procedure is given for 10 TLD-100. The TLD reader was an Harshaw Mod. 2000 A+B with a heating rate of 5°C/s. No nitrogen flux was used. The following table shows the results obtained. Considering the negligible changes in the average values obtained through the three subsequent cycles (annealing + readout) one can consider the background to be stable and the initialisation ended. The background values determined for each dosimeter have to be collected (i.e. memorised in a file concerned the batch under test) so that they can be used for the successive tests. In many cases an average background value is considered for the whole batch and then subtracted from each individual reading of the irradiated TLDs. This procedure is valid when the background is very low and constant for the whole batch. In other specific situations, as in radiotherapy where a high accuracy is necessary, an individual background is used and checked periodically to avoid any possible mistakes in the dose determination owing to large variations of the background. The following Table 1 shows an example of initialisation procedure.

TLD No.	1st BKG	2nd BKG	3rd BKG	$\overline{\overline{B}}$ (a.u.)	%CV
1	0.091	0.087	0.090		
2	0.099	0.101	0.098		
3	0.101	0.098	0.099		
4	0.087	0.091	0.090		
5	0.095	0.087	0.091		
6	0.107	0.095	0.097		
7	0.085	0.090	0.088		
8	0.083	0.087	0.085		
9	0.085	0.088	0.091		
10	0.093	0.091	0.089		
$\overline{B}$	0.093	0.092	0.092	0.092	0.60
%CV	8.60	5.40	5.10		

Table 1. Example of initialisation procedure (BKG = background).

References

1. Driscoll C.M.H., National Radiological Protection Board, Tech. Mem. 5(82)
2. Scarpa G. in Corso sulla Termoluminescenza Applicata alla Dosimetria, Universita' di Roma "La Sapienza", 15-17 Febbraio 1994

Integral approximation

The integral comparing the thermoluminescence theory

$$\int_{T_0}^{T} \exp\left(-\frac{E}{kT'}\right) dT' \tag{1}$$

cannot be solved in an analytical form.

A method which is usually followed for evaluating the value of the integral is by integration in parts, when the lower limit of integration is 0 instead of T_0. So, a good approximation is provided by the asymptotic series

$$F(T,E) = \int_{0}^{T} \exp\left(-\frac{E}{kT'}\right) dT' = T\exp\left(-\frac{E}{kT}\right) \sum_{n=1} \left(\frac{kT}{E}\right)^{n} (-1)^{n-1} n! \tag{2}$$

The value of (1) is then given by

$$\int_{T_0}^{T} \exp\left(-\frac{E}{kT'}\right) dT' = F(T,E) - F(T_0,E)$$

Since $F(T,E)$ is a very strong increasing function of T, $F(T_0,E)$ is negligible compared to $F(T,E)$, the right hand side of Eq. (2) can be considered to represent the integral value from T_0 as well.

In the practical case, a good approximation of the integral is given by the second order approximation of Eq. (2):

$$F(T,E)=\frac{kT^2}{E}\exp\left(-\frac{E}{kT}\right)\left(1-\frac{2kT}{E}\right) \tag{3}$$

Integral approximation when $s = s(T)$

$$\int_{T_0}^{T} T^{\alpha}\exp\left(-\frac{E}{kT'}\right)dT' \approx$$

$$\approx \frac{kT^{\alpha+2}}{E}\left[1-(\alpha+2)\frac{kT}{E}\right]\exp\left(-\frac{E}{kT}\right) \tag{1}$$

if $T=T_M$, expression (1) becomes

$$\approx \frac{kT_M^{\alpha+2}}{E}\left[1-(\alpha+2)\frac{kT_M}{E}\right]\exp\left(-\frac{E}{kT_M}\right) \tag{2}$$

Interactive traps

Electrons released by a shallow trap may be captured by a deep trap (**thermally disconnected traps**): in this way the traps are called interactive. The deep traps are in competition with the recombination centers for capturing electrons released by the shallow traps.

Isothermal decay method (Garlick-Gibson: first order)

Formerly the isothermal decay method was illustrated for the first-order kinetics by Garlick and Gibson.

Let the initial integral light be S_o, while S_{t_i} will be the integral light at time t_i:

$$S_0 = n_0$$
$$S_{t_i} = n_1 = n_0 \exp(-pt_1)$$
$$\ldots\ldots\ldots\ldots\ldots\ldots\ldots\ldots\ldots$$
$$S_{t_n} = n_n = n_0 \exp(-pt_n) \qquad at \quad T = const \tag{1}$$

Making the ratios

$$\ln(\frac{S_{t_1}}{S_0}) = -pt_1 . \quad . \quad . \quad .\ln(\frac{S_{t_n}}{S_0}) = -pt_n \tag{2}$$

the graphs of $\ln(S_{t_i}/S_0)$ versus t_i is then plotted for data obtained at a given storage temperature T. Using different storage temperatures (T_i) one can obtain a set of straight line of slopes

$$m_i = -s\exp(-\frac{E}{kT_i}) \tag{3}$$

from which

$$\ln(m_i) = \ln(-s) - \frac{E}{kT_i} \tag{4}$$

Therefore a plot of ln(m) versus $1/T$ yields a straight line of slope $-E/k$ and intercept ln($-s$) on the ordinate axis.

If the experiment is carried out with two different constant storage temperatures, T_1 and T_2, two different slopes, m_1 and m_2, are obtained and then from them

$$\ln\left(\frac{m_1}{m_2}\right) = \frac{E}{k}\left(\frac{1}{T_2} - \frac{1}{T_2}\right) \tag{5}$$

The last equation allows to calculate E. The frequency factor s is derived by substitution of the E value in Eq. (3).

Reference

Garlick G.F.J. and Gibson A.F., Proc. Phys. Soc. **60** (1948) 574

Isothermal decay method (general)

Isothermal decay of the thermoluminescence emission does not employ any particular heating. Strictly speaking, the isothermal decay technique is not a TL based method but nevertheless is a general method to determine E and s. The experimental steps consist in quickly heating the sample, after irradiation, to a specific temperature just below of the maximum temperature of the peak under study, and keeping it at this constant temperature during a given time. The light output (phosphorescence decay) is measured and so it is possible to evaluate the decay rate of trapped electrons.

Isothermal decay method (May-Partridge: (a) general order)

May and Partridge suggested to apply the isothermal decay method in the general case of any order. In this case it is possible to find the order b also. The TL intensity, at any temperature, is given by the equation

$$\frac{dn}{dt} = -s'' n^b \exp\left(-\frac{E}{kT}\right) \tag{1}$$

whence

$$\int_{n_0}^{n} \frac{dn}{n^b} = -\int_0^t s'' \cdot \exp(-\frac{E}{kT})dt \tag{2}$$

By integration, the following expression is obtained

$$\frac{n^{1-b} - n_0^{1-b}}{1-b} = -s'' \cdot \exp(-\frac{E}{kT}) \tag{3}$$

which, with the substitutions

$$a = n_0^{1-b}$$

$$c = -(1-b)s'' \exp(-\frac{E}{kT}) \tag{4}$$

reduces to

$$n = (a+ct)^{\frac{1}{1-b}} \tag{5}$$

Executing the derivative of Eq. (5) one gets

$$\frac{dn}{dt} = c\frac{1}{1-b}(a+ct)^{\frac{b}{1-b}} \tag{6}$$

Because

$$I = -\frac{dn}{dt}$$

we obtain

$$I = -\frac{c}{1-b}(a+ct)^{\frac{b}{1-b}}$$

that is

$$I^{\frac{1-b}{b}} = (a+ct)\left[s''\exp\left(-\frac{E}{kT}\right)\right]^{\frac{1-b}{b}}$$

which becomes

$$I^{\frac{1-b}{b}} = A + B\cdot t \tag{7}$$

where

$$A = a\left[s''\exp(-\frac{E}{kT})\right]^{\frac{1-b}{b}} \tag{8}$$

$$B = c\left[s''\exp(-\frac{E}{kT})\right]^{\frac{1-b}{b}} \tag{9}$$

The *I(t)* function given by Eq. (7) is a linear function of the time; thus a plot of the left side versus time yields a straight line when by iterative procedure using different values of *b* the best *b* value is determined to fit Eq. (7).

Reference

May C.E. and Partridge J.A., J. Chem. Phys. **40** (1964) 1401

Isothermal decay method (May and Partridge: (b) general order)

May and Partridge gave an alternative method to the one proposed by them for the **(a) general order** case. Their method can be explained as follows.

By differentiation of the Eq. (1) (see **Isothermal decay method (May-Partridge: (a) general order)** at constant temperature:

$$I^{\frac{1-b}{b}} = A + B \cdot t \tag{1}$$

one obtains

$$\frac{1-b}{b} I^{\frac{1-2b}{b}} \frac{dI}{dt} = B \tag{2}$$

from which

$$\frac{dI}{dt} = CI^{\frac{2b-1}{b}} \tag{3}$$

The logarithm of Eq. (3) yields

$$\ln(\frac{dI}{dt}) = \ln C + \frac{2b-1}{b} \ln(I) \tag{4}$$

thus a plot of ln(*dI/dt*) versus ln(*I*) gives a straight line having a slope $m=(2b-1)/b$ from which b can be evaluated.

Reference

May C.E. and Partridge J.A., J. Chem. Phys. **40** (1964) 1401

Isothermal decay method (Moharil: general order)

Moharil suggests the isothermal decay technique for obtaining a parameter which is physically more relevant than the order of the kinetics [1]. The theory is based on the **Antonov-Romanovskii equation** [2]:

$$I = -\frac{dn}{dt} = s \frac{Bnm}{Bn + A(N-n)} \exp(-\frac{E}{kT}) \tag{1}$$

where
B = probability of recombination, A = probability of retrapping, m = number of recombination centres at time t, n = number of filled traps at time t, N – total number of traps and the usual meaning for the other quantities. If $n = m$, Eq. (1) becomes

$$I = -\frac{dn}{dt} = \frac{Bn^2}{Bn + A(N-n)} s \exp(\frac{E}{kT}) \tag{2}$$

which reduces to the first-order equation if $A<<B$, and to the second-order equation for $A = B$. When neither of the two conditions are satisfied, one has the general order

kinetics. In this case, the general order equation cannot be derived from Eq. (2) and the kinetics order b cannot be related to the physical quantities A and B.

As it is suggested by Moharil, the ratio A/B can be obtained from isothermal decay experiment and using Eq. (2). Rearranging this equation one has

$$\frac{Bn + A(N-n)}{Bn^2}dn = -s\exp\left(-\frac{E}{kT}\right)dt \tag{3}$$

Integrating between 0 and t, with the condition $n = n_0$ for $t = 0$, we obtain:

$$\int_{n_0}^{n}\frac{B-A}{Bn}dn + \int_{n_0}^{n}\frac{AN}{Bn^2}dn = -\int_0^t s\exp\left(-\frac{E}{kT}\right)dt$$

$$\left(1-\frac{A}{B}\right)\ln\frac{n_0}{n} + \frac{AN}{B}\left(\frac{n_0-n}{nn_0}\right) = st\exp\left(-\frac{E}{kT}\right) \tag{4}$$

The following hipothesis are now assumed:
n_0 is proportional to the area under the isothermal decay curve ($= A_0$); n is proportional to the remaining area under the decay curve after time t (A_t). If $n_0 = N$, saturation case, A_0 is proportional to N: in this case the area is denoted A_S instead of A_0. Equation (4) can be written as

$$\left(1-\frac{A}{B}\right)\ln\frac{A_0}{A_t} + \frac{AA_S}{B}\left(\frac{A_0-A_t}{A_0A_t}\right) = st\exp\left(-\frac{E}{kT}\right) \tag{5}$$

A graph of the left-hand side of Eq. (5) against time should give a straight line when the best value of A/B is chosen.

References

1. Moharil S.V. and Kathuria P.S., J. Phys. D: Appl. Phys **16** (1983) 425
2. Atonov-Romanoskii V.V., Bull. Acad. Sci. USSR Phys. Res. **15** (1951) 673

Isothermal decay method (Takeuchi et al.: general order)

Takeuchi et al. reported a method slightly different from the one described by May and Partridge. From the equations for general order:

$$I = s''n^b\exp\left(-\frac{E}{kT}\right) \tag{1}$$

$$I(t) = s''n_0^b \exp\left(-\frac{E}{kT}\right)\left[1 + s''n_0^{b-1}(b-1)t\exp\left(-\frac{E}{kT}\right)\right]^{\frac{1}{1-b}} \tag{2}$$

keeping constant the temperature, one obtains:

$$I_0 = s''n_0^b \exp(-\frac{E}{kT})$$

$$I_t = s''n_0^b \left[1 + s''n_0^{b-1}(b-1)t\exp(-\frac{E}{kT})\right]^{\frac{b}{1-b}} \cdot \exp(-\frac{E}{kT})$$

where I_o and n_o are respectively the initial intensity and the initial concentration of trapped charges and I_t is the intensity at time t. The ratio of the two equations gives

$$\left(\frac{I_0}{I_t}\right)^{\frac{b}{b-1}} = \left[1 + s(b-1)t\exp(-\frac{E}{kT})\right] \tag{3}$$

with $s = s''n_0^{b-1}$

The plot of the left side term versus time should then be a straight line when a suitable value of b is found.

Using different decay temperatures, a set of straight lines of slopes

$$m = s(b-1)\exp(-\frac{E}{kT}) \tag{4}$$

are obtained and the activation energy E will be determined from the plot of $\ln(m)$ versus $1/T$:

$$\ln(m) = \ln s(b-1) - \frac{E}{kT} \tag{5}$$

Reference

Takeuchi M., Inabe K. and Nanto H., J. Mater. Sci. **10** (1975) 159

Isothermal decay method: a new approach

The isothermal decay technique is a very well know method for the activation energy determination in thermoluminescence experiments.

The aim of this paragraph is to present a variant of the method. Expressions for first, second and general order kinetics are developed on the basis of the original equations of Randall-Wilkins, Garlick-Gibson and May-Partridge.

The kinetic equations allow to characterise the TL decay after irradiation, i.e. exponential for a first order, hyperbolic for the second order and intermediate for any order between first and second.

The equations have been modified for allowing a more practical and easy application.

Let us start considering a first order TL process which is described by the Randall – Wilkins equation

$$\frac{dn}{dt} = -n\lambda \tag{1}$$

Eq. (1) gives the charge change rate, *dn/dt*, as a function of the detrapping rate, $n\lambda$.

n is the trapped electron concentration (cm^{-3}) at time t, λ is the escape probability per unit of time, i.e. $\lambda = s\exp\left(-\frac{E}{kT}\right)$ where s is the frequency factor (sec^{-1}), E (eV) is the activation energy, i.e. the energy associated to the trapping state located in the band gap of the crystal, k the Boltzmann's constant (0.862×10^{-4} eV/K) and T the absolute temperature (K). λ is also called fading factor.

Integration of Eq. (1), at constant temperature, gives

$$n(t) = n_0 \exp(-\lambda \cdot t) \tag{2}$$

where n_0 is the concentration of the initial trapped charges at t = 0.

The TL intensity, I(t), is given by

$$I(t) = -\frac{dn}{dt} = n \cdot \lambda \tag{3}$$

or better

$$I(t) = n_0 \lambda \exp(-\lambda \cdot t) \tag{4}$$

By integration of the basic Eq. (3), we get

$$n(t) - n(\infty) = \int_t^{\infty} I dt' \tag{5}$$

Dealing with only one single peak, $n(\infty) = 0$ and therefore

$$n(t) = \int_t^{\infty} I dt' \tag{6}$$

The integral in Eq. (6) gives the total area, *Φ*, of the peak. Then, it turns out that the peak area coincides numerically with n. In this way, Eqs. (1) and (2) can be rewritten as

$$\frac{d\Phi}{dt} = -\Phi \cdot \lambda \tag{7}$$

$$\Phi(t) = \Phi_0 \exp(-\lambda \cdot t) \tag{8}$$

This kind of writing is much more useful for the following development of this work because the peak area is a quantity easily available from the TL reading.

In a very similar way, the equation of Garlick and Gibson, dealing with a second order kinetics process,

$$\frac{dn}{dt} = -n^2 \cdot \lambda^* \tag{9}$$

can be modified in the same way, giving

$$\frac{d\Phi}{dt} = -\Phi^2 \cdot \lambda^* \tag{10}$$

where $\lambda^* = s' \exp\left(-\frac{E}{kT}\right)$ and *s'* is the so-called pre-exponential factor.

Integration of Eq. (10) leads to the expression

$$\Phi = \Phi_0 \left[1 + \Phi_0 \cdot t \cdot \lambda^*\right]^{-1} \tag{11}$$

Both Eqs. (8) and (11) can be further modified introducing the calibration factor of the thermoluminescent system, i.e. F_C, expressed in dose/TL. F_C is related to the calibration dose, D_0, as well as to the corresponding TL reading, Φ_0, so that

$$F_C = \frac{D_0}{\Phi_0} \tag{12}$$

Eqs. (8) and (11) become then

$$\Phi(t\ ,T)=\frac{D_0}{F_C}\exp\left[s\cdot t\ \cdot\exp\left(-\frac{E}{kT}\right)\right] \tag{13}$$

for the first order, and

$$\Phi\left(t\ ,T\right)=\frac{D_0}{F_C}\left[1+\frac{D_0}{F_C}\cdot t\ \cdot s'\cdot\exp\left(-\frac{E}{kT}\right)\right]^{-1} \tag{14}$$

for the second order.

In both cases t is the period of storage at a temperature T.

The general order equation can be written, as already done for first and second order, in the following way

$$\frac{d\Phi}{dt}=-\lambda''\Phi^b \tag{15}$$

with

$$\lambda''=s''\exp\left(-\frac{E}{kT}\right)$$

The analytical solution of Eq. (15) is

$$\Phi=\Phi_0\left[1+s''\cdot\Phi_0^{b-1}\left(b-1\right)\cdot t\cdot\exp\left(-\frac{E}{kT}\right)\right]^{\frac{1}{1-b}} \tag{16}$$

Eq. (13) may be written for two different storage temperature at the same time of storage t_S as

$$\Phi_1(t_S,T_1)=\frac{D_0}{F_C}\exp\left[s\cdot t_S\cdot\exp\left(-\frac{E}{kT_1}\right)\right] \tag{17}$$

$$\Phi_2(t_S,T_2)=\frac{D_0}{F_C}\exp\left[s\cdot t_S\cdot\exp\left(-\frac{E}{kT_2}\right)\right] \tag{18}$$

The quantity

$$\exp\left[s\cdot t_S\cdot\exp\left(-\frac{E}{kT}\right)\right]$$

can be solved using the approximation $e^x = 1 + x$ and then Eqs. (17) and (18) become

$$1 + s \cdot t_S \exp\left(-\frac{E}{kT_1}\right) = \frac{F_C \cdot \Phi_1}{D_0}$$

and

$$1 + s \cdot t_S \exp\left(-\frac{E}{kT_2}\right) = \frac{F_C \cdot \Phi_2}{D_0}$$

The ratio of the two last expressions gives

$$\exp\left(-\frac{E}{kT_1} + \frac{E}{kT_2}\right) = \frac{\frac{F_C \cdot \Phi_1}{D_0} - 1}{\frac{F_C \cdot \Phi_2}{D_0} - 1}$$

from which the activation energy is easily found as

$$E = k\frac{T_2 \cdot T_1}{T_1 - T_2} \ln\left(\frac{F_C\Phi_1 - D_0}{F_C\Phi_2 - D_0}\right) \tag{19}$$

Starting from the 2nd order decay expressions, for two different temperatures T_1 and T_2 of storage at the same time t_S:

$$\Phi_1 = \Phi_0\left[1 + \Phi_0 \cdot t_S \cdot s' \cdot \exp\left(-\frac{E}{kT_1}\right)\right]^{-1} \tag{20}$$

$$\Phi_2 = \Phi_0\left[1 + \Phi_0 \cdot t_S \cdot s' \cdot \exp\left(-\frac{E}{kT_2}\right)\right]^{-1} \tag{21}$$

where

- Φ_1 and Φ_2 are the peak area integral of the peak under investigations after its decay at temperature T_1 and T_2 over a time t_S
- Φ_0 is the peak area integral of the same peak of the prompt read out after irradiation.

From the previous equations we obtain

$$E = k\left[\frac{T_2 T_1}{T_1 - T_2}\right] \cdot \ln\left[\frac{\Phi_2 \cdot (\Phi_0 - \Phi_1)}{\Phi_1 \cdot (\Phi_0 - \Phi_2)}\right] \tag{22}$$

For the general order decay, from Eq. (16) and for two different storage temperatures at the same time of storage, we get:

$$\Phi_1 = \Phi_0\left[1 + s'' \cdot \Phi_0^{b-1}(b-1) \cdot t_S \cdot \exp\left(-\frac{E}{kT}\right)\right]^{\frac{1}{1-b}} \tag{23}$$

$$\Phi_2 = \Phi_0\left[1 + s'' \cdot \Phi_0^{b-1}(b-1) \cdot t_S \cdot \exp\left(-\frac{E}{kT}\right)\right]^{\frac{1}{1-b}} \tag{24}$$

from which we obtain

$$E = k\frac{T_1 T_2}{T_1 - T_2} \ln \frac{\left(\dfrac{\Phi_1}{\Phi_0}\right)^{1-b} - 1}{\left(\dfrac{\Phi_2}{\Phi_0}\right)^{1-b} - 1} \tag{25}$$

In all cases the kinetics order is not known a-priori: then, it is necessary to determine it using the geometrical factor of the peak, μ.

K

Keating method (first-order, $s = s(T)$)

Keating has proposed a method to determine E, for the first-order, when s is supposed to be temperature dependent [1].

The equation giving the TL intensity, when the frequency factor is temperature dependent, is the following:

$$I(T) = n_0 s_0 T^{\alpha} \exp(-\frac{E}{kT}) \left[-\frac{s_0}{\beta} \int_{T_0}^{T} T^{\alpha} \exp(-\frac{E}{kT'}) dT' \right] \tag{1}$$

Putting $n_o s_o = I_o$ and making the logarithm of Eq. (1), one gets

$$\ln\left(\frac{I}{I_0}\right) = \alpha \ln T - \frac{E}{kT} - \left(\frac{s_0}{\beta}\right) \int_{T_0}^{T} T^{\alpha} \exp(-\frac{E}{kT}) dT \tag{2}$$

Differentiation of this equation with respect to T, and setting the derivative at the maximum equal to zero, yields

$$\frac{\alpha}{T_M} + \frac{E}{kT_M^2} - \left(\frac{s_0}{\beta}\right) T_M^2 \exp\left(-\frac{E}{kT_M}\right) = 0$$

from which

$$\frac{s_0}{\beta} = \left(\frac{\alpha}{T_M^{\alpha+1}} + \frac{E}{kT_M^{\alpha+2}}\right) \exp(\frac{E}{kT_M}) \tag{3}$$

Remembering that the integral in Eq. (1) can be evaluated by an asymptotic series, in this case we have

$$\int_{T_0}^{T} T^{\alpha} \exp(-\frac{E}{kT}) dT \cong \frac{kT^{\alpha+2}}{E} (1 - \Delta) \exp(-\frac{E}{kT}) \tag{4}$$

with

$$\Delta = (\alpha + 2) \frac{kT}{E}$$

Thus, Eq. (2) becomes

$$-\ln(\frac{I}{I_0}) = -\alpha \ln T + \frac{E}{kT} + \frac{s_0}{\beta}\frac{kT^{\alpha+2}}{E}(1-\Delta)\exp(-\frac{E}{kT}) \tag{5}$$

Inserting in Eq. (5) the expression (3), we get

$$-\ln(\frac{I}{I_0}) =$$

$$= -\alpha \ln T + \frac{E}{kT} + \left(\frac{T}{T_M}\right)^{\alpha+2}\left(\frac{\alpha k T_M}{E}+1\right)(1-\Delta)\exp\left(\frac{E}{kT_M}-\frac{E}{kT}\right) \tag{6}$$

Using now the temperature T_1 and T_2 when $I=I_M/2$, the following parameters are defined:

$$\Gamma_1 = \frac{T_M - T_1}{T_M}$$

$$\Gamma_2 = \frac{T_2 - T_M}{T_M} \tag{7}$$

Hence, the following expressions, with T_1 and T_2 respectively, can be obtained

$$\left(1+\frac{\alpha k T_M}{E}\right)(1-\Delta) + \ln 2 =$$

$$= -\Psi_1 - \alpha \ln(1-\Gamma_1) + (1-\Gamma_1)^{\alpha+2}(1+\frac{\alpha k T_M}{E})(1-\Delta)\exp(\Psi_1) \tag{8a}$$

$$\left(1+\frac{\alpha k T_M}{E}\right)(1-\Delta) + \ln 2 = \tag{8b}$$

$$= -\Psi_2 - \alpha \ln(1-\Gamma_2) + (1-\Gamma_2)^{\alpha+2}(1+\frac{\alpha k T_M}{E})(1-\Delta)\exp(\Psi_2)$$

with

$$\Psi_1 = \left(\frac{E}{kT_M}\right)\left[\frac{\Gamma_1}{\Gamma_1 - 1}\right] \tag{9a}$$

$$\Psi_2 = \left(\frac{E}{kT_M}\right)\left[\frac{\Gamma_2}{\Gamma_2 - 1}\right] \tag{9b}$$

Since $\Delta \prec\prec 1$ for $E/kT>10$ the expressions $(\alpha+2)kT_1/E$ and $(\alpha+2)kT_2/E$ have been taken equal to $\Delta=(\alpha+2)kT_M/E$.

Equations (8a,b) can be resolved numerically for Γ_1 and Γ_2 for values of $a = 0$, ± 2 and E/kT_M between 10 and 35. Analysis of the data shows that E can be found by the following linear equation

$$E = kT_M \left[\gamma(1.2\Gamma - 0.54) + 5.5\cdot 10^{-3} - \left(\frac{\Gamma - 0.75}{2} \right)^2 \right]^{-1} \tag{10}$$

with

$\gamma = \Gamma_1 + \Gamma_2$ and $\Gamma = \delta/\tau$

Nicholas and Woods have found that Eq. (10) holds true for $0.75 < \Gamma < 0.90$ [2].

References

1. Keating P.N., Proc. Phys. Soc. **78** (1964) 1408
2. Nicholas K.H. and Woods J., Br. J. Appl. Phys. **15** (1964) 783

Killer centres

The killer centres have been introduced by Schön and Klasens to explain the thermal quenching of luminescence. At high enough temperatures, holes may be released from luminescence centres and migrate to other centres called "killers", in which the recombination between free electrons with the trapped holes is not accompanied by emission of light due to phonon interaction. An increase in the concentration of the killer centres provokes the decrease of the luminescence efficiency.

Kinetic parameters determination: observations

The glow-curve computerized deconvolution analysis (GCD) is the most recent and widely used technique for determining the kinetics parameters. Any way, it has to be stressed that it is possible, in principle, to deconvolute a complex, and even a single peak, in a very large number of different configurations and to chose that one or those which give the best figure of merit (FOM). Indeed, even in this case many configurations may be obtained, each one with a different set of the trapping parameters. Of course, this kind of result is not physically acceptable.

For this reason, trapping parameters obtained just using the GCD are not acceptable and some suggestions how to proceed are given below:

- Start the analysis using at least two classical methods which are independent of the shape of the peak. The GCD depends, on the contrary, on the shape. The **initial rise** and the **various heating rate** methods may be used for this purpose.

- Use now the GCD and compare the trapping data to the ones obtained in the first point.

As an example, the following table reports the values of the activation energy of two different kind of lithium borate. The data are referred to the very intense peak only [1].

Material	Initial Rise (IR)	Various Heating Rates (VHR)	GCD
$Li_2B_4O_7$:Cu	1.56 ± 0.04	1.57 ± 0.02	1.37 ± 0.03
$Li_2B_4O_7$:Cu, In	1.61 ± 0.03	1.66 ± 0.02	1.35 ± 0.03

Table 1. Activation energy (eV).

Figs. 1 and 2 show the glow curves of both materials. The experimental data are given by the open circles. In the same figure it is also shown the deconvolution performed.

From Table 1 it is evident the discrepancy between the data obtained by IR and VHR methods and the results of the deconvolution. The values obtained in the last case are lower in comparison to the data resulting from IR and VHR.

- Check which of the results should be the more realistic and physically acceptable. For this, one should apply a method which depends, as the GCD is, by the quantities characterizing the shape of the peak: i.e. one of the peak shape methods (PS), for instance the Chen's method. This method should give results very similar to those obtained by GCD. Table 2 shows the results obtained using the PS method.

Material	E_τ (eV)	E_δ (eV)	E_ω (eV)
$Li_2B_4O_7$:Cu	1.38 ± 0.03	1.39 ± 0.02	1.40 ± 0.03
$Li_2B_4O_7$:Cu, In	1.38 ± 0.04	1.40 ± 0.04	1.40 ± 0.04

Table 2. Activation energy values obtained by PS method.

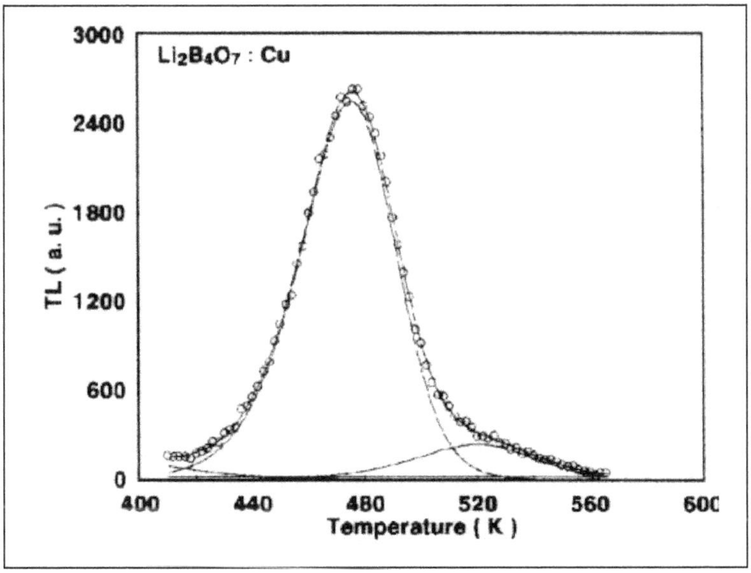

Fig. 1. Glow curve of $Li_2B_4O_7$:Cu. The open circles indicate the experimental data. The performed deconvolution is also shown.

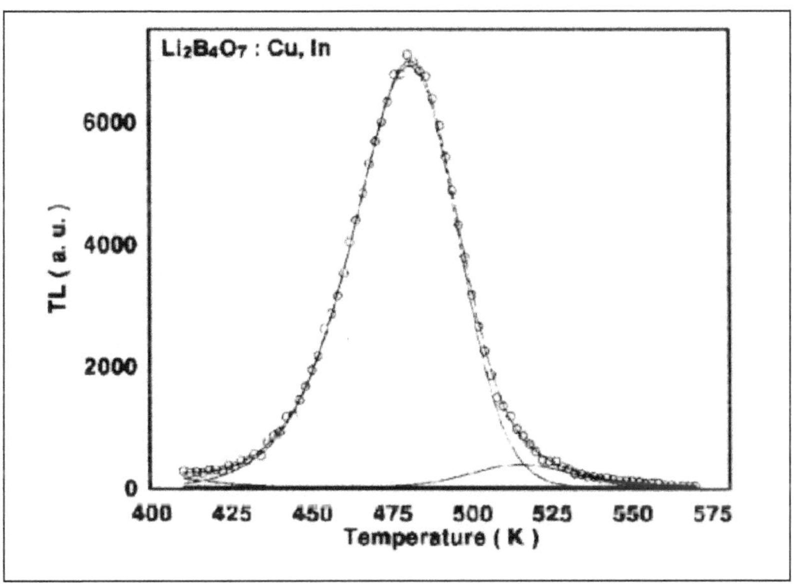

Fig. 2. Glow curve of $Li_2B_4O_7$:Cu,In. The open circles indicate the experimental data. The performed deconvolution is also shown.

From Table 2 results that the data obtained by the PS method are very similar to the data resulting from GCD.

- Make the following assumption: it could be possible that the peaks under investigation are not single peaks but rather there is some satellite peak/peaks that made their shape broader than a pure single peak. In turn, this should cause the activation energy to be lower than the real one in both PS and GCD methods.
- Look for a method which is again independent of the glow shape and, furthermore, which should allow to estimate number and position of individual, not resolved peaks within the glow peak appearing as a single peak. This method is the modified IR method introduced by McKeever [2].
- Perform a second deconvolution according to the results obtained above. Figs. 5 and 6 show the new deconvolution and Table 3 the new data.

The application of the McKeever method allows us to obtain the following plots shown in Figs. 3 and 4.

For $Li_2B_4O_7$:Cu three distinct plateau can be observed, the first corresponding to the main peak, the second and third indicate the presence of two high-temperature peaks. $Li_2B_4O_7$:Cu,In analysis shows the main peak, corresponding to the first plateau, and a possible second peak at higher temperature.

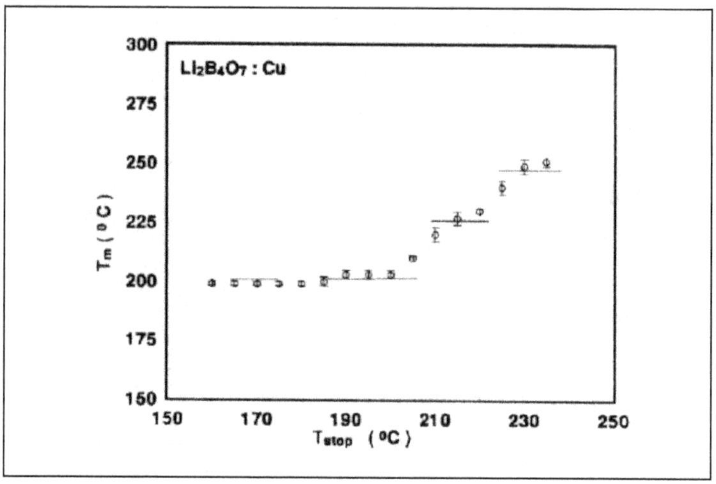

Fig. 3. I.R. plot for $Li_2B_4O_7$:Cu.

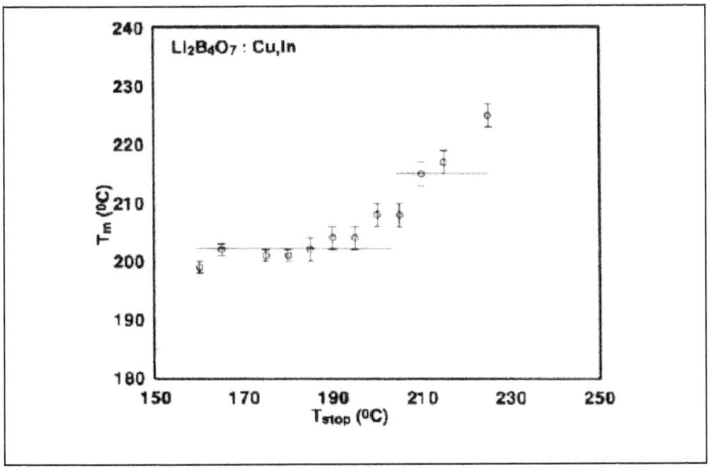

Fig. 4. I.R. plot for $Li_2B_4O_7$:Cu,In.

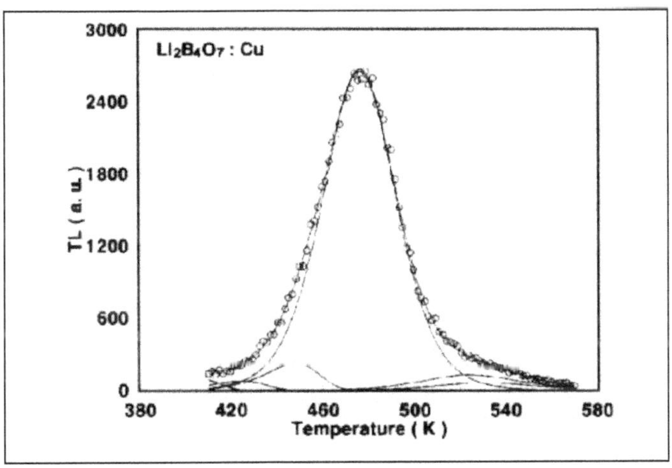

Fig. 5. The new deconvolution performed for $Li_2B_4O_7$:Cu.

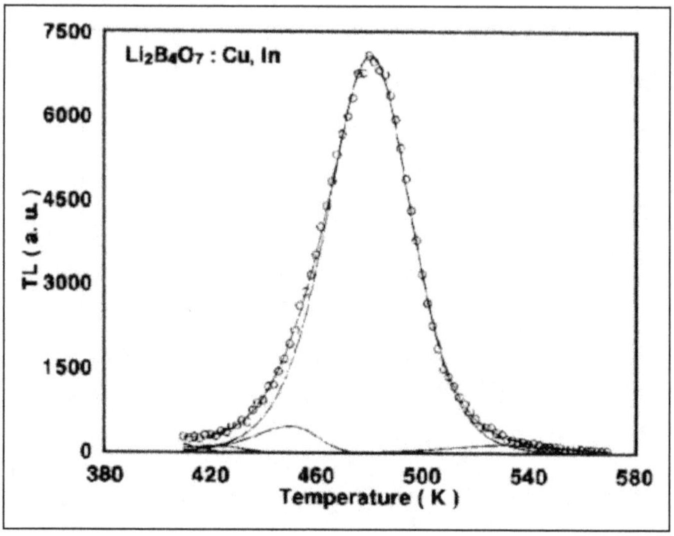

Fig. 6. The new deconvolution performed for $Li_2B_4O_7$:Cu,In.

Material	E (eV)
$Li_2B_4O_7$:Cu	1.61 ± 0.03
$Li_2B_4O_7$:Cu, In	1.62 ± 0.02

Table 3. New GCD data.

Table 3 shows that the new data are now in a very good agreement with the data obtained by IR and VHR methods.

The discrepancies observed before are now disappeared and it is possible to trust in the second deconvolution performed taking into account a more complex glow peak structure.

References

1. Kitis G., Furetta C., Prokic M. and Prokic V., J. Phys. D: Appl. Phys. **33** (2000) 1252
2. McKeever S.W.S., Phys. Stat. Sol. (a) **62** (1980) 331

Kinetics order: effects on the glow-curve shape

The practical effect of the order of kinetics on the glow-peak shape is illustrated in Fig. 1, in which two glow-curves from a single type of trap are compared.

In the case of second order kinetics T_M increases by the order of 1% respect to the temperature at the maximum of a first order peak. The main difference is that the light is produced at temperatures above T_M because the trapping delays the release of the electrons. Furthermore, for a fixed value of E, T_M increases as β increases or s' decrease; for a fixed value of β, T_M results to be directly proportional to E.

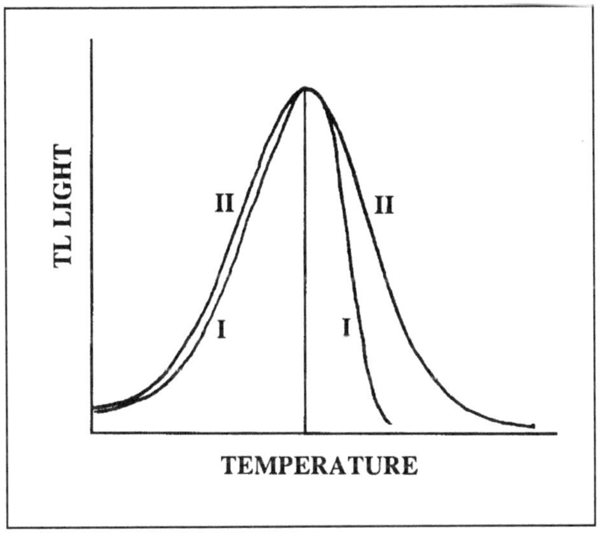

Fig. 1. Glow-peak shapes for a first order (I) and a second order (II). The largest difference is related to the descending part of the curve.

L

Light transmission of semi-transparent dosemeter

The aim of this paragraph is to give a simple method of measuring the light absorption coefficient of a dosemeter, $A(\lambda)$. This means a knowledge of the quantity $a = cA(\lambda)$ (see the section Theory of thermoluminescent dose measurement) for the read out instrument.

The method is the following:
A thin solid-state light source, having an emission spectrum similar to that of the dosemeter, is placed on the dosemeter tray and a reading R_1 is taken. A dosemeter is now placed on the top of the light source and the reading R_2 is taken. The effective light transmission, t_{eff}, is then defined as the ratio of the second reading over the first one; hence

$$t_{eff} = \frac{R_2}{R_1} = \exp\left[-cA(\lambda) \cdot T\right]$$

or

$$t_{eff} = \exp(a \cdot T)$$

From which the quantity $cA(\lambda)$, the effective coefficient of attenuation of the dosemeter internally generated light, can be found.

Linearisation factor, F_{lin} (general requirements for linearity)

Let us define, at first, the yield or efficiency of the thermoluminescent emission, η, from a material having a mass m, as the ratio between the energy, ε, released as light from the material itself, and the mass m multiplied by the absorbed dose D [1]:

$$\eta = \frac{\varepsilon}{m \cdot D} \tag{1}$$

In the range where the efficiency η is constant, there is a linear relationship between the TL signal, M, and the absorbed dose, D:

$$M = k \cdot D \tag{2}$$

where k is a constant.

It is important in any thermoluminescent dosimetric application to have, if it is possible, a linear relationship between the TL emission and the absorbed dose. The linearity zone, if existing, is more or less depending on the material as well as on the reader.

A typical first-order relationship can be written as [2]

$$y = ax + b \tag{3}$$

The linearity range, as already mentioned, depends on the particular thermoluminescent material. The plot of Eq. (3) is a straight line with slope "a" and intercept "b" on the Y-axis.

The physical meaning of the x and y variables, when using the Eq. (3) to describe the TL yield as a function of dose, are:

- the independent variable x represents the absorbed dose D received by the TL dosemeter,
- the depending variable y is the TL light emitted by the dosemeters irradiated at the dose D; it is expressed in reader units,
- the slope "a" identifies itself with the absolute sensitivity of the dosemeter (expressed in terms of reader units per dose), or, with the inverse of the calibration factor F_c (expressed in terms of dose per reader units),
- the intercept on the Y axis, "b", is the TL reading due to the intrinsic background for the same dosemeter just annealed and not irradiated.

Equation (3) can then be rewritten according to the symbols used previously

$$M = \frac{1}{F_c} D + M_0 \tag{4}$$

where M is the TL signal at a given dose and M_0 is the intrinsic background of the dosemeter.

Equation (4) is strictly valid only for a material having a relative intrinsic sensitivity factor (individual correction factor) equal to 1; if this is not true, the TL reading must be corrected consequently. In the following discussion the case of $S_i \neq 1$ is omitted to avoid a heavy formalism.

Considering the net TL response, Eq. (4) becomes

$$M - M_0 = \frac{1}{F_c} D \tag{5}$$

In this form Eq. (5) can be better defined as a proportionality relationship between the TL emission and the dose.

Figure 1 shows the plots of both Eq. (4) and (5), where

$$\tan\alpha = \frac{1}{F_c} \tag{6}$$

For practical reasons, the data concerning the TL emission vs. dose are normally plotted on a log-log paper. In this way Eq. (5) becomes

$$\log(M - M_0) = \log D + \log\left(\frac{1}{F_c}\right) \tag{7}$$

which is still the equation of a straight line having now a slope equal to unity.

Figure 2 shows, schematically, the behaviour of the TL vs. dose for three different materials. The dotted line represents the proportionality as indicated by Eq. (6).

An unfortunate use of terminology has crept into the literature on thermoluminescence dosimetry which may easily mislead the uninitiated.

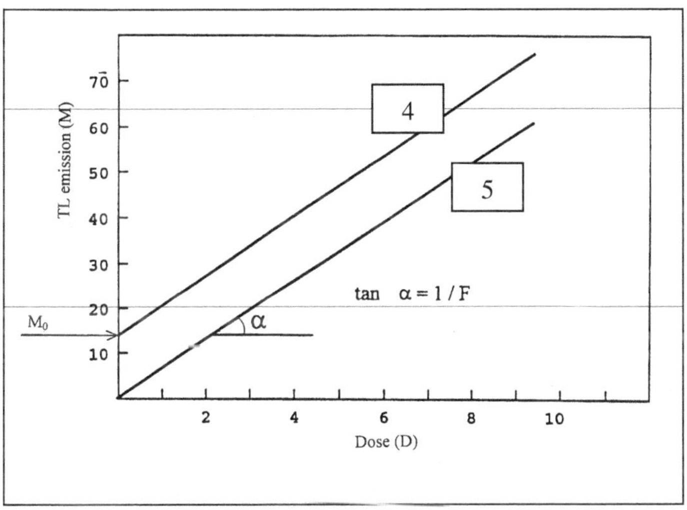

Fig. 1. Plots of Eq. (4) and Eq. (5).

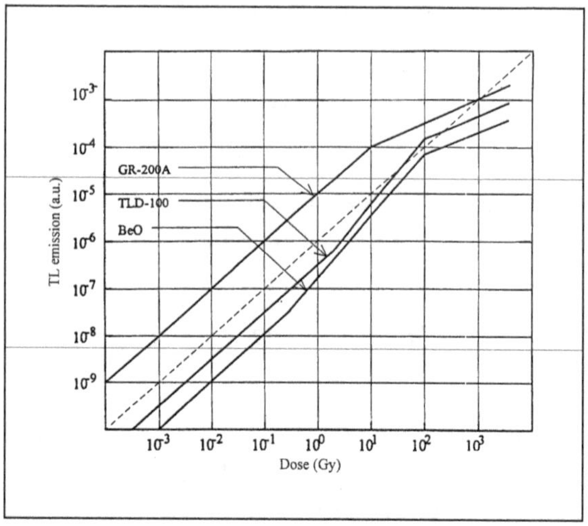

Fig. 2. TL response as a function of dose for three different types of TLDs.

Calibration data for various dosemeter materials are usually presented, as already stated, as a plot of the logarithm of thermoluminescence response vs. the logarithm of the absorbed dose. It must be stressed that a straight line on full log paper implies linearity only in the special case when it makes an angle of 45° with the logarithm axes. Other straight lines imply some power relationship between the variables. Then remember that a straight line on full log paper is not necessary linear.

References

1. Busuoli G. in Applied Thermoluminescence Dosimetry, ISPRA Courses, Edited by M. Oberhofer and A. Scharmann, Adam Hilger Publ. (1981)
2. Scarpa G. in Corso sulla Termoluminescenza Applicata alla Dosimetria, 15-17 February 1994, Rome University La Sapienza (I)

Linearity (procedure)

- prepare a group of N dosemeters. For each detector one must know the individual background and the intrinsic sensitivity factor.
- the N dosemeters are divided in n subgroups (n = 1, 2, …, i), each subgroup corresponding to a dose level. Each subgroup has a number m of detectors (m = 1, 2, …, j ≥ 5).
- each subgroup is then irradiated using a calibration source possibly of the same quality of the radiation used for the applications so that no correction factor for energy will be necessary.

- the range of doses delivered to the dosemeters has to be chosen according to the needs. In any case it is good to give increasing doses following a logarithm scale (i.e., ...0.1, 1, 10, 100 Gy, ...).
- read all the dosemeters in only one session.
- correct the readings by background and sensitivity factors.
- for each subgroup calculate the average value

$$\overline{M}_i = \sum_{j=1}^{m} \frac{M_j}{m} \tag{1}$$

where $\overline{M}_i$ stands for the average value of the ith subgroup and M_j stands for the reading of the jth dosemeter already corrected by background and sensitivity factor.

- plot on a log-log paper the $\overline{M}_i$ values as a function of the doses.
- test the linear behaviour using a statistical methods.

The following Table 1 lists the data obtained after irradiation of CaF_2:Dy (TLD-200) samples to ^{60}Co gamma rays at various doses in the range from 25 to 300 μGy. Each experimental point is the average of the readings of five TLDs. The data are corrected by subtraction of the individual background and by the intrinsic sensitivity factor. For simplicity, Table 1 reports only the average values and the corresponding standard deviations. The plot is shown in Fig. 1.

Dose (μGy)	Average reading (a.u.)	σ% (a.u.)
25	0.340	2.4
50	0.644	2.1
75	0.980	2.2
100	1.327	1.8
125	1.605	1.6
150	1.977	2.0
200	2.695	1.5
250	3.215	1.3
300	3.972	1.1

Table 1. Example of TL response vs. dose.

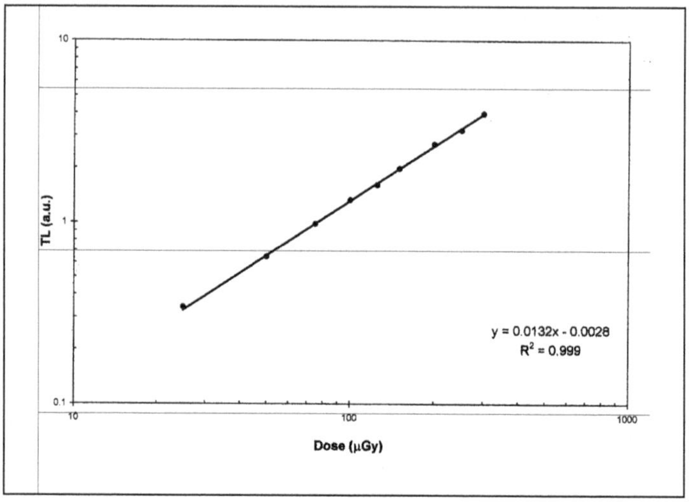

Fig. 1. Linearity plot for TLD-200.

Reference

Furetta C. and Weng P.S., Operational Thermoluminescence Dosimetry, World Scientific, 1998

Linearity test (procedure)

The aim of this test is to verify if a TL system is linear as a function of the doses used or, in other words if, within the reproducibility characteristics of the system, the net reading is proportional to the given dose.

- select a randomgroup of 10 TLDs from the batch,
- anneal the TLD samples according to the appropriate standard annealing,
- irradiated the samples at a test dose of 0.1 mGy,
- read out the irradiated samples,
- second read out for background determination
- repeat points 2 to 5 for different doses, i.e. 1 mGy, 10 mGy and 100 mGy,
- create the following table, using the net readings and correct them according the intrinsic sensitivity factors

TLD N.o	readings				S_i	corrected readings			
	Dose (D_c) mGy					Dose (D_c) mGy			
	0.12	1	10	100		0.12	1	10	100
1	242341	1943571	20227100	196833833	1.24	192211	1564170	16308952	158733736
2	204128	1654501	16578730	164846367	1.02	196203	1618138	16253657	161610164
3	208551	1584265	16833900	164535400	1.03	198593	1534238	16343592	159739223
4	231101	1861617	18020780	183750000	1.11	204596	1673529	16234937	165536937
5	226310	1888630	19638860	188766300	1.20	185258	1570525	16365717	157301917
6	184280	1436054	15034580	147125067	0.92	195957	1556580	16341935	159914203
7	155635	1278430	12628450	126211967	0.78	194404	1633885	16190321	161805086
8	159326	1313236	13000080	129595100	0.80	194158	1636545	16250100	161988875
9	160594	1234075	12814190	126674767	0.79	198220	1557057	16220494	160342743
10	229475	1926944	18099080	185755033	1.12	201317	1716914	16159893	165849137
Average						196092	1606158	16266960	161282202
σ						5265	59363	69915	2734799
m/D_c						1634097	1606158	1626696	1612822
Φ						1619943	1619943	1619943	1619943
D_{ev}						0.121	0.991	10.042	99.560
σ_{ev}						0.003	0.037	0.043	1.688
$D_{ev}+0.7\sigma_{ev}/D_c$						1.028	1.018	1.007	1.008
$D_{ev}-0.7\sigma_{ev}/D_c$						0.989	0.965	1.001	0.984

Where;

- D_c is the given dose
- S_i is the relative intrinsic sensitivity factor, defined as

$$S_i = \frac{M_i - M_{i0}}{\bar{M}}$$

- m is the average of the corrected readings over the ten dosimeters at each dose
- σ is the corresponding standard deviation
- $\Phi = \frac{1}{4}\sum_{i=1}^{4}\frac{m_i}{D_C}$ is the slope of the best fit straight line crossing the origin of the axis in the plot of TL emission vs dose
- D_{ev} is the evaluated dose $= \frac{m_i}{\Phi}$

- σ_{ev} is the standard deviation of the evaluated dose $\frac{\sigma}{\Phi}$

The acceptability limit at each level of dose is given by

$$0.9 \leq \frac{(D_{ev} \pm 0.7\sigma_{ev})}{D_c} \leq 1.1$$

where the coefficient 0.7 is given by the ratio

$$\frac{t}{\sqrt{n}}$$

being n the number of measurements at each level of dose (10 in the example) and t is the t-Student value (equal to 2.26 for 10 measurements and a confidence level of 95%).

Reference

Scarpa G. in Corso sulla Termoluminescenza Applicata alla Dosimetria. Rome University "La Sapienza", Rome (Italy)

Lithium borate ($Li_2B_4O_7$)

The effective atomic number 7.3 makes $Li_2B_40_7$ a tissue equivalent material and encourages studies on its TL properties for radiation dosimetry in general and personnel monitoring in particular. This is the reason why repeated investigations were devoted in the last two decades to identify effective TL activators and to optimize the method of preparation of lithium borate [1-6].

Two different methods can be adopted to prepare $Li_2B_40_7$ phosphors.

- melting method: lithium carbonate Li_2CO_3 and boric acid H_3BO_3 are mixed in the stoichiometric ratio and sufficient aqueous solution of the desired dopant (Mn, Fe, Co, Mo, Ag, Cu, in form of chlorides, nitrates, or oxides) is added, to obtain an impurity content ranging from 0.03 to 0.5 wt%. After stirring and desiccation, the mixture is melted at 950°C in a silica or platinum crucible, then rapidly cooled to room temperature. The resultant glassy mass is reheated at 650°C for 0.5 h, which assures a complete crystallization, and then ground and sieved to obtain a 100 to 200 mesh crystalline powder.

- sintering method: an acetone or alcohol solution of the activator is added to raw $Li_2B_40_7$ powder, and the mixture is stirred and dried. The resulting powder is heated in air in a platinum container for 1 h at 910°C and then cooled to room temperature.

Good TL performances are reported for $Li_2B_4O_7$ doped with Cu, Ag, Mn and Cu+In impurities.

All these phosphors exhibit two glow peaks, the first one occúrring at 110 to 120°C (very low temperature for dosimetric purposes), and the second one in the range between 185 and 230°C, depending on the activator.

Linearity of the TL response in $Li_2B_4O_7$:Cu and $Li_2B_4O_7$:Cu,In is observed from $2x10^{-4}$ up to 10^3 Gy. The energy dependence of TL output in $Li_2B_4O_7$: Cu and $Li_2B_4O_7$:Cu,In for photons is almost flat from 30 keV to Co^{60} energy.

Fading is very quick for the loe temperature peak, but the dosimetric peak fades less than 10% after 3 months [7].

References

1. Schulman J.H., Kirk R.D. and West E.J., Proc. 1st Int. Conf. Lumin. Dos., Stanford (USA), 1967
2. Moreno y Moreno A., Archundia C. and Salsberg L., Proc. 3rd Int. Conf. Lumin. Dos., Riso (Denmark), 1971
3. Botter-Jensen L. and Christensen P., Acta Radiol., Suppl. **313** (1972) 247
4. Takenaga M., Yamamoto O. and Yamashita T., Proc. 5th Int. Conf. Lumin. Dos., San Paulo (Brazil), 1977
5. Takenaga M., Yamamoto O. and Yamashita T., Nucl. Instr. Meth. **175** (1980) 77
6. Takenaga M., Yamamoto O. and Yamashita T., Health Phys. **44** (1983) 387
7. Furetta C., Prokic M., Salamon R., Prokic V. and Kitis G., Nucl. Instr. Meth. **A456** (2001) 411

Lithium fluoride family (LiF)

Lithium fluoride is among the most widely used TL phosphors in dosimetric applications, because it provides a good compromise between the desired dosimetric properties. Its effective atomic number (8.14) is sufficiently close to that of the biological tissue (7.4) so as to provide a response which varies only slightly with photon energy. Thus it can be considered as tissue equivalent.

LiF:Mg,Ti

This phosphor is produced commercially by the Harshaw Chemical Co., USA. LiF:Mg,Ti dosemeters are known as TLD-100, TLD-600, and TLD-700, depending on their preparation from natural lithium or lithium enriched with ^{6}Li or ^{7}Li, respectively: ^{6}Li 95.6% and ^{7}Li 4.4% for TLD-600, ^{6}Li 0.01% and ^{7}Li 99.99% for TLD-700. Harshaw patent [1] describes two preparation methods for LiF:Mg,Ti TL phosphor powders: the solidification method and the single crystal method.

- in the solidification method, lithium fluoride (10^6 parts by weight), magnesium fluoride (400 parts by weight), lithium cryolite (200 parts by weight), and lithium titanium fluoride (55 parts by weight) are mixed in a graphite crucible. The mixture is homogeneously fused in vacuum and the product slowly cooled, then crushed and sieved between 60 and 200 μm.

- in the single crystal method, the above mixture is placed in a vacuumn or inert-atmosphere oven to grow a single crystal by the Czochralski method at a temperature sufficiently high to obtain a homogeneous fusion mixture. The mixture is then slowly moved to a lower temperature zone to allow progressive solidification (about 15 mm/h). Once the material is cooled, it is crushed and sieved between 60 and 200 μm.

In both cases the resulting TL phosphor powder is annealed at 400°C during some hours and then at 80°C during 48 h.

- the same patent also describes the preparation of extruded LiF dosemeters. To obtain them, the LiF powder mixture is placed in a neutral atmosphere and pressed at 3.5 x 108 Pa at a temperature of 700°C, pushing the mixture with a piston through a hole which acts as a die. The bar obtained is cut in sections to prepare pellets of uniform thickness and finally the faces of the pellets are polished. The extruded dosemeters have identical TL characteristics as the TL phosphor powder.

- another method [2] describes how to prepare sodium stabilized LiF dosemeters. In this method, 200 ppm of magnesium fluoride and 2 wt% of sodium fluoride are added to the LiF powder. The powder mixture is homogenized, put in an aluminum oxide crucible, and held at the crystallization temperature for about 3 h in a nitrogen flow oven. Then, the temperature is reduced to 60°C in 45 min and the sample taken out of the oven to be cooled quickly. The product is finely pulverized and the treatment repeated. Finally the product is repulverized and sieved between 60 and 200 μm. In order to favor the creation of traps, the product is annealed in an ordinary oven at 500°C over 72 h. The crystals are quenched by pouring them on a cold metal plate. To make pellets, the TL powder is finely sieved, compressed at about 5 x 10^8 Pa in the desired form, and submitted to a thermal treatment in a nitrogen oven at a temperature sightly lower than that of fusion. Before the use, the pellets must be annealed at 500°C.

Other methods have been developed to prepare LiF:Mg,Ti phosphor powder, LiF:Mg,Ti + PTFE (polytetrafluoroethylene) and LiF sintered pellets [3].

- the preparation of LiF : Mg, Ti phosphor powder is the following. A few ml of a solution 0.1 M of $MgCl_2$ are added to 40 ml of a LiCl solution (0.9 g/ml). Meanwhile, metallic titanium is dissolved in 50 ml of hydrofluoric acid (HF, 48 to 50%), then the first mixture is slowly added. Once LiF is precipitated, the sample is centrifuged and washed repeatedly. The precipitate is dried in a Pt crucible at a temperature of 30°C for 1 h. Then the material is cooled to room temperature adding a few ml of LiCl solution. This wet material is dried at 100°C for 1 h, placed in a Pt crucible, and then in an oven with nitrogen atmosphere at 300°C for 15 min. After that the temperature is raised up to 640°C and kept constant for 1 h. The sample is slowly moved to a lower temperature zone (400°C) to allow crystallization, and then taken out of the oven to be rapidly cooled to room temperature. Finally, the product is crushed and sieved to select powder with grain sizes between 80 and 200 μm.

- To obtain LiF : Mg, Ti + PTFE pellets, a mixture 2: 1 ofthe phosphor powder and PTFE resin powder is placed in a stainless steel die to be pressed, at room temperature, at about 1 GPa. Pellets thus obtained (5 mm diameter and 0.7 mm thickness), weighing approximately 30 mg, are thermally treated for a period longer than 5 h in a nitrogen oven at a temperature sightly lower than that of PTFE fusion.

- Sintered LiF : Mg, Ti pellets are obtained by pressing the TL powder into a stainless steel die at about 10 GPa. These compressed pellets undergo a thermal treatment in a nitrogen oven at a temperature slightly lower than that of LiF fusion to be sintered.

The TL glow curve of LiF:Mg,Ti, shows at least six peaks; it is quite complicated because ofits complex trap dynamics. The main peak (indicated as peak 5) normally used for dosimetric purposes, and then called the dosimetric peak, appears at a temperature of about 225°C corresponding to a very stable trap level. The low temperature peaks 1, 2, and 3 are relatively unstable and must be suppressed by a thermal treatment.

The linearity is maintained from 100 mGy up to about 6 Gy, beyond which superlinearity appears.

LiF containing ^{6}Li is sensitive to thermal neutrons. Peak 5 shows a response which deceases with increasing LET of ionizing particles (protons, α-particles, etc.). Peak 6 is particularly sensitive to α-particles. This difference in behavior is useful to measure thermal neutrons in a mixed radiation field.

LiF:Mg,Cu,P

LiF : Mg, Cu, P has been developed as a phosphor of low effective atomic number which exhibits a simple glow curve, low fading rate, and high sensitivity.

The preparations of this phosphor are the following:

- LiF of special grade in the market, used as starting material, is mixed in water with activators, CuF 2 (0.05 mol%) and MgCl2 (0.2 mol%), and added with ammonium phosphate. The wet mixture is heated in a Pt crucible at 1050°C for 30 min in nitrogen gas after being dried at about 80°C for 4 h. The melted LiF material is rapidly cooled to 400°C during 30 min and the polycrystalline mass is powdered and sieved. Powder of size between 80 and 150 mesh is used as LiF: Mg, Cu, P TL phosphor [4].

- another method [5] consists of obtaining first undoped LiF from the reaction LiCl + HF = LiF + HCl. Once LiF was precipitated, activators MgCI2, (NH4)2HPO4, and CuF 2 in aqueous solutions are incorporated until the required concentrations are reached. The material obtained in this way is dried (70 to 80°C for 4 h) and washed repeatedly. This dried material, placed in a Pt crucible, is oven heated in nitrogen atmosphere at 400°C during 15 min. After that the temperature is raised to 1150°C and kept constant for 15 min, then lowered to 400°C, and subsequently suddenly to room temperature. The resulting polycrystalline material is crushed and sieved selecting powder with grain sizes between 100 and 300 μm. The final product is the TL phosphor powder.

- pellets of LiF: Mg, Cu, P + PTFE are obtained in the same way as those of LiF: Mg,Ti + PTFE.

LiF: Mg, Cu, P obtained following the first reported preparation [4] shows linearity in the dose range between 5 x 10^{-5} and 10 Gy, beyond which the response becomes sublinear, a property quite different from superlinearity. The phosphor prepared following the second suggested procedure [5] gives linear response between 10^{-4} and 10^{2} Gy.

LiF:Cu^{2+}

The growth of single crystals is carried out by Kyropoulos method from Merck 99.6% powder. Doping with Cu^{2+} ions is obtained by adding to the melt various amounts of CuF_2 according to the required dopant concentrations. The glow curve of LiF:Cu^{2+} shows a very preminent and intense peak at 155°C (H.R.=3°C/s) and a minor peak at about 205°C overlapped, at high doses, by a third peak at around 230°C [6,7].

References

1. Patent Harshaw Chemical Co., USA
2. Portal G., Rep. CEA-R-4943 (1978)
3. Azorin J., Gutierrez A. and Gonzalez P., Tech. Rep. IA-89-07 ININ (Mexico) (1989)
4. Nakajima T., Morayama Y., Matsuzawa T. and Koyano A., Nucl. Instr. Meth. **157** (1978) 155
5. Azorin J., Tech. Rep. IA-89-08 ININ. Mexico (1989)
6. Furetta C., Mendozzi V., Sanipoli C., Scacco A., Leroy C., Marullo F. and Roy P., J. Phys. D: Appl. Phys. **28** (1995) 1488
7. Scacco A., Furetta C., Sanipoli C. and Vistoso G.F., Nucl. Instr. Meth. **B116** (1996) 545

Localized energy levels

Trapping levels within the material's forbidden energy gap.

Lower detection limit (D_{ldl})

The lower detection limit, D_{ldl}, is defined as three times the standard deviation of the zero dose reading:

$$3\sigma_{BKG}$$

Luminescence (general)

Luminescence [1-3] is the energy emitted by a material as light, after absorption of the energy from an exciting source which provokes the rise of an electron from its ground energy level to another corresponding to a bigger energy (excited level). The light emitted, when the electron comes back to its ground energy level, can be classified according a characterisctic time, τ, between the absorption of the exciting energy and the emission of light.

If this time is less than 10^{-8} sec, the luminescence is called fluorescence. The light is emitted with a wavelenght larger than the wavelenght of the absorbed light owing to dispersion of energy by the molecula. If the time between absoprtion and emission is larger than 10^{-4} sec, the luminescence is then called phosphorescence. The process of phosphorescence is explained with the presence of a metastable level, between the fondamental and the excited levels, which acts as a trap for the electron.

If the transition arrives at a temperature T and the energy difference E, between the excited and the metastable levels, is much larger than kT, the electron has an high probability to remain trapped for a very long time.

Assuming a maxwellian distribution of the energy, the probability of escaping by the trap is given by

$$p = s\exp\left(-\frac{E}{kT}\right)$$

As a consequence, the period of time between the excitation and the transition back to the ground state is delayied for the time the electron spends in the metastable state.

In the previous equation, the probability p is a function of the stimulation method, which can be thermal or optical and will assume a different form according to the type of stimulation.

References

1. McKeever S.W.S., Thermoluminescence of Solids, Cambridge University Press (1985)
2. Chen R. and McKeever S.W.S., Theory of Thermoluminescence and Related Phenomena, World Scientific (1997)
3. Chen R. and Kirsh Y., Analysis of Thermally Stimulated Processes, Pergamon Press (1981)

Luminescence (thermal stimulation)

Let us define N to be the concentration of the metastable states occupied by electrons. The intensity of luminescence, I, is proportional to the decrease, as a function of time, of the concentration of the metastable states in the system (i.e. the crystal structure) [1,2]:

$$I(t) = \frac{dN}{dt}$$

the quantity $\frac{dN}{dt}$ can be also expressed as

$$\frac{dN}{dt} = -p \cdot N$$

In the case of thermal stimulation, the probability p is expressed as following

$$p = \nu \cdot \kappa \exp\left(-\frac{F}{kT}\right)$$

where

ν is the vibrational frequency of phonons within the crystal structure,

κ is the transition probability,

F is the free energy of Helmholtz and

k is the Boltzmann's constant.

Because the free energy can be expressed as

$$F = E - T\Delta S$$

where ΔS is the entropy change during the transition and E is the thermal energy imparted to the electron, the probability can be then expressed as

$$p = \nu \cdot \kappa \cdot \exp\left(\frac{\Delta S}{k}\right) \cdot \exp\left(-\frac{E}{kT}\right)$$

from which

$$s = \nu \cdot \kappa \cdot \exp\left(\frac{\Delta S}{k}\right)$$

where s is called frequency factor (sec^{-1}); it is also called "attempt-to-escape frequency".

Because in this description $N = n$, where n is the concentration of the trapped electrons, the detrapping rate is given by

$$\frac{dn}{dt} = -p \cdot n$$

and then, the intensity of luminescence is

$$I = -c \cdot \frac{dn}{dt} = c \cdot p \cdot n$$

with c a constant.

References

1. Bube R.H., Photoconductivity of Solids, Wiley & Sons, N.Y. (1960)
2. Mahesh K., Weng P.S. and Furetta C., Thermoluminescence in Solids and Its Application, Nuclear Technology Publishing, England (U.K.) (1989)

Luminescence centres

The luminescent centres are atoms or group of atoms, called activators, positioned in the lattice of the host material and serve as discrete centres for localised absorption of excitation energy. In other words, a luminescent centre is a quantum state in the band gap of an insulator which acts as a centre of recombination of charge carriers when it captures a carrier and holds it for a period of time until another carrier of opposite sign is also trapped and both combine. The recombination causes the release of the energy in excess as photons or phonons.

Luminescence dosimetric techniques

The main luminescence dosimetric techniques are:

(i) radio-thermoluminescence (RTL) or thermoluminescence (TL) which consists in a transient emission of light from an irradiated solid when heated;
(ii) radio-photoluminescence (RPL) which consists in the emission of light from an irradiated solid by excitation with ultra-violet light;
(iii) radio-lyoluminescence (RLL) which consists in a transient emission of light from an irradiated solid upon dissolving it in water or some other solvent

Luminescence dosimetry

Luminescence dosimetry is an important part of solid state dosimetry and incorporates processes whereby energy absorbed from ionising radiation is later released as light.

Luminescence efficiency

The luminescence efficiency of a material, η, is defined as the ratio of the total energy emitted ($h\nu$) in the form of light to the energy observed ($h\nu_0$) by the material during the process of excitation:

$$\eta = \frac{h\nu}{h\nu_0} = \frac{\lambda}{\lambda_0} \tag{1}$$

The emission of luminescence following irradiation and the absorption of energy, depends on the relative probabilities of the radiative and non-radiative transitions. Eq. (1) can be then expressed in another form:

$$\eta = \frac{radiative \quad events}{total \quad events} = \frac{P_r}{P_r - P_{nr}} \tag{2}$$

where P_r is the probability of luminescent transitions and P_{nr} the probability of non-radiative transitions.

Experiments have shown that η is strongly temperature dependent: the efficiency remains quite constant up to a critical temperature beyond which it decreases rapidly.

Eq. (2) can be also written as following:

$$\eta = \frac{1}{1 + c\exp\left(-\frac{\Delta E}{kT}\right)} \tag{3}$$

because the radiative probability P_r is not affected by temperature, while the non-radiative probability P_{nr} is depending on temperature through the Boltzman factor.

In the above Eq. (3), the quantity ΔE represents the thermal energy absorbed by an electron, which is in an excited state at the minimum of energy, for rising to an higher excited state. From this higher state the electron can transfer to the ground state without emission of radiation. The decrease of luminescence efficiency as the temperature increases (**thermal quencing**) has been explained introducing the so called **killer centres**.

Luminescence phenomena

Luminescence is the emission of light from certain solids called phosphors.

This emission, which does not include black body radiation, is the release of energy stored within the solid through certain types of prior excitation of the electronic system of the solid. This ability to store is important in luminescence dosimetry and is generally associated with the presence of **activators**.

The following table lists the luminescence phenomena and the methods of excitation.

LUMINESCENCE PHENOMENA	METHODS OF EXCITATION
Bioluminescence	Biochemical reactions
Cathodoluminescence	Electron beam
Chemiluminescence	Chemical reactions
Electroluminescence	Electric field
Photoluminescence	U.V. and infrared light
Piezoluminescence	Pressure (10 tons m^{-2})
Triboluminescence	Friction
Radioluminescence	Ionising radiation
Sonoluminescence	Sound waves
Fluorescence Phosphorescence Thermoluminescence Lyoluminescence	Ionising radiation, U.V. and visible light

In particular, when some of the radiation energy is absorbed by a material, it can be re-emitted as light having a longer wavelength, according to the Stoke's law. Furthermore, the wavelength of the emitted light is caracteristic of the material.

M

Magnesium borate ($MgO \times nB_2O_3$)

This phosphor is a near tissue equivalent material with an effective atomic number for photoelectron absorption equal to 8.4.

The preparation of polycrystalline magnesium borate activated by dysprosium has been reported at first in 1974 [1]. A certain quantity of magnesium carbonate $MgCO_3$, boric acid H_3BO_3, and dysprosium nitrate $Dy(NO_3)_3$ is placed in a quartz cup and dried at a temperature ranging between 80 and 100°C. After that the material is annealed in a furnace, then cooled, ground, and screened.

The most sensitive material is obtained at the proportion of boric anhydride and magnesium oxide 2.2 to 2.4 and at the dysprosium concentration of about 1 mg-atom per g-mol of the base.

The glow curve of such a material shows a single peak located in the region from 190 to 200°C.

The sensitivity is reported to be 10 to 20 times larger than that of LiF. The energy responseat 40 keV is about 30% larger than that of LiF. The TL response Vs dose is linear from 10^{-5} to 10 Gy. Fading at room temperature is about 25% over a period of 40 days.

A development of the preparation method of magnesium borate activated by Dy and Tm and other unknown impurities added as co-activators, was presented in 1980 [2]. The sensitivity has been reported to be about seven times greater than that of LiF; other investigators reported a factor of four [3]. The glow curve of MgB_4O_7:Dy is composed by a single peak ; the TL response is linear from 10^{-5} to 102 Gy.

Further investigations [3,4] reported high variability of the TL features within a batch as well as among different batches. This suggested the necessity of improving the material preparation in order to use such a phosphor widely in personnel and environmental dosimetry without problems of individual detector calibration.

A new production of MgB_4O_7, activated by Dy + Na shows very good performances: reproducibility within 2% from 1 mGy to 0.25 Gy and a linear range from 6×10^{-8} Gy to 40 Gy [5].

References

1. Kazanskaya V.A., Kuzmin V.V., Minaeva E.E. and Sokolov A.D., Proc. 4th Int. Conf. Lumin. Dos., Krakow (Poland), 1974
2. Prokic M., Nucl. Instr. Meth. **175** (1980) 83
3. Barbina V., Contento G., Furetta C., Malisan M. and Padovani R., Rad. Eff. Letters **67** (1981) 55
4. Driscoll C.M.H., Mundy S.J. and Elliot J.M., Rad. Prot. Dos. **1** (1981) 135
5. Furetta C., Prokic M., Salamon R. and Kitis G., Nucl. Instr. Meth. **B160** (2000) 65

Magnesium fluoride (MgF_2)

A mixture of MgF_2 and individual dopant as Mn, Tb, Tm or Dy is heated at 1200°C during 1 hr in a nitrogen atmosphere. The molten mass is then cooled to room temperature. The atomic number of the obtained phosphor is about 10. The glow cuves of both pure or doped phosphors show 10 peaks from room temperature and 400°C. The dopants enhance the thermoluminescence emission. The highest sensitive phosphor is obtained with Mn. The TL response is linear up to about 40 R [1-3].

References

1. Paun J., Jipa S. and Ilie S., Radiochem. Radioanal. Lett. **40** (1979) 169
2. Braunlich P., Hanle W. and Scharmann A.Z., Z. Naturf. **16a** (1961) 869
3. Nagpal J.S., Kathuria V.K. and Bapat V.N., Int. J. Appl. Rad. Isot. **32** (1981) 147

Magnesium orthosilicate (Mg_2SiO_4)

Doping of Mg_2SiO_4 with terbium impurities produces a TL dosimetry phosphor, showing highest sensitivity and moderate photon energy dependence, particularly useful for dosimetry in high temperature areas. TL properties of this system, whose effective atomic number is about 11, are reported since 1970 [1-3] and are strongly dependent on the preparation procedure. Magnesium oxide MgO, freshly prepared by decomposition at 600°C of $Mg(N0_3)_2$ and silica gel are mixed in the molar ratio 2: 1 and added of Tb_4O_7 dopant. After thorough stirring in distilled water, the mixture is dried in an oven and then melted in a silica crucible by directly blowing a petroleum gas-oxygen flame (temperature of about 2750°C) over it.

The weight of dosemeter samples is typically 5 mg of powder. Solid discs are also available.

The TL glow curve of Mg_2SiO_4:Tb contains distinct peaks at 50, 90, 170, 300, 420°C (with an extra peak at 485°C for exposures greater than 12 KR), but 95% of the total intensity belongs to the 300°C peak.

The sensitivity of this material is 50 to 80 times higher than that of LiF TLD-100, depending on the sample quality. The exposure response is linear in the range from about 20 mR to 400 R.

Annealing at 500°C for 2 to 3 h is necessary for re-using the TL detector.

Mg_2SiO_4: Tb exhibits intense TL under irradiation with 254 nm UV light. This sensitivity to biologically active UV light (typical of germicidal lamps) can be very useful for UV dosimetry.

References

1. Hashizume T., Kato Y., Nakajima T., Toryu T., Sakamato H., Kotera N. and Eguchi S., Adv. Phys. Biol. Rad. Detec. IAEA, Vienna (1971)
2. Jun J.S. and Becker K., Health Phys. **28** (1975) 459
3. Bhasin B.D., Sasidharan R. and Sunta C.M., Health Phys. **30** (1976) 139

May–Partridge model (general order kinetics)

When the conditions of first or second order kinetics are not satisfied, one obtains the so called general order kinetics which deals with intermediate cases. May and Partridge (1964) wrote an empirical expression for taking into account experimental situations which indicated intermediate kinetics processes. They started with the assumption that the energy level of traps is single, as already assumed for the first and second orders.

Let's assume that the number n of charge carriers present in a single energy level is proportional to n^b. Then, the probability rate of escape is:

$$\frac{dn}{dt} = -s'' n^b \exp\left(-\frac{E}{kT}\right) \tag{1}$$

where s'' is the pre-exponential factor.

Equation (1) is the so called general order kinetics relation, and usually b is ranging in the interval between 1 and 2. The pre-exponential factor s'' is now expressed in $cm^{3(b-1)}sec^{-1}$. It has to be stressed that the dimensions of s'' change with the order b. Furthermore, s'' reduces to s' when b=2.

From Eq. (1) we can deduce the relation describing the TL emission.

Rearranging Eq. (1) we have:

$$\frac{dn}{n^b} = -s'' \exp\left(-\frac{E}{kT}\right) dt \tag{2}$$

$$\int_{n_0}^{n} n^{-b} dn = -\int_0^t s'' \exp\left(-\frac{E}{kT}\right) dt$$

$$n^{1-b} = n_0^{1-b}\left[1 + s'' n_0^{b-1}(b-1)t \exp\left(-\frac{E}{kT}\right)\right]$$

$$n = n_0\left[1 + s(b-1)t \exp\left(-\frac{E}{kT}\right)\right]^{\frac{1}{1-b}} \tag{3}$$

in which

$$s = s'' n_0^{b-1} \tag{4}$$

where s has units of sec^{-1}.

With this definition the difficulty with respect to the variation of dimensions has been bypassed. Any way, the frequency factor s is constant for a given dose and would vary when the dose is varied.

The intensity *I(t)* is then given by:

$$I(t) = -\frac{dn}{dt} = s'' n^b \exp\left(-\frac{E}{kT}\right) =$$

$$= sn_0 \exp\left(-\frac{E}{kT}\right)\left[1 + s(b-1)t \exp\left(-\frac{E}{kT}\right)\right]^{\frac{b}{1-b}} \tag{5}$$

Assuming a linear heating rate $dT=\beta dt$, we obtain from Eq. (2):

$$\frac{n^{1-b} - n_0^{1-b}}{1-b} = -\frac{s''}{\beta}\int_{T_0}^{T} \exp\left(-\frac{E}{kT'}\right) dT'$$

$$n^{1-b} = n_0^{1-b}\left[1 + \frac{s''(b-1)n_0^{b-1}}{\beta}\int_{T_0}^{T} \exp\left(-\frac{E}{kT'}\right) dT'\right]$$

Derivation of the root from both members and using expression (4) yields

$$n = n_0\left[1 + \frac{s(b-1)}{\beta}\int_{T_0}^{T} \exp\left(-\frac{E}{kT'}\right) dT'\right]^{\frac{1}{1-b}} \tag{6}$$

The intensity *I(T)* is now given by

$$I(T) = sn_0 \exp\left(-\frac{E}{kT}\right)\left[1 + \frac{s(b-1)}{\beta}\int_{T_0}^{T} \exp\left(-\frac{E}{kT'}\right) dT'\right]^{-\frac{b}{b-1}} \tag{7}$$

It must be observed that two factors contribute to *I(T)*:

- the exponential factor which constantly increases with *T*;
- the factor included in brackets, decreasing as *T* increases.

So we have again the explanation of the bell shape of the glow-curve as experimentally observed.

To conclude, Eq. (7) includes the **second order case** (b=2). Equation (7), which is not valid for the case b=1, reduces to the **first order equation** when $b \to 1$.

It must be stressed that Eq. (1) is entirely empirical, in the sense that no approximation can be found which is able to derive Eq. (1) from the set of differential equations governing the traffic of charge carriers and so, as a consequence, a physical model leading to general order kinetics does not exist.

Reference

May C.E. and Partridge J.A., J. Chem. Phys. **40** (1964) 1401

McKeever method

McKeveer gives a method for estimating both number and position of individual TL peaks within a complex glow curve. The method is based on the monitoring of the position of the peak temperature at the maximum, T_M, with increasing the heating temperature during a certain number of readings of a TL sample. The position of the peak temperature at the maximum is independent of the initial trapped charge concentration in a first order kinetics process; a slight shift in the peak position, on the contrary, can be observed in non-first order kinetics process when the initial trapped charge concentration is reduced.

This insensitivity of the peak position to small changes in the trap concentration has been used by McKeever to analyze TL curves consisting of several overlapping peaks.

An irradiated sample is heated, in the TL reader, at a linear heating rate, from room temperature to a temperature T_{STOP} corresponding to a position on the low temperature tail of the first glow peak in the glow curve. The sample is then cooled down rapidly and then re-heated to record the whole glow curve and the T_M value of the peak is noted. The process is repeated on new irradiated samples using different values of T_{STOP} each time, increasing T_{STOP} from 2 to 5 K each time and the T_M values are noted again. According to the structure of the glow curve, different plots of T_M as a function of T_{STOP} can be obtained.

1. A straight line of slope 0 corresponds to a simple first order peak.
2. A stair-case curve indicates several first order peaks; each flat region in the curve corresponds to individual first order peak.
3. A smooth line of slope about 1 indicates the presence of closely overlapping peaks, or a quasi-continuous distribution of peaks. There is no indication about the kinetics order.
4. In case of an individual peak following a second order kinetics, the plots is a curved line where T_M increases as T_{STOP} increases.
5. Similar to the second case, a smooth stair-case curve indicates a series of second order peaks.
6. In case of a quasi-continuous distribution of peaks, the plots is the same of case 3.

After that, the kinetics parameters can be obtained using a curve fitting procedure.

Reference

McKeever, S.W.S., Phys. Stat. Sol (a), **62** (1980) 331

Mean and half-life of a trap

The half-life ($t_{1/2}$), at a constant temperature, of a trap and, as a consequence, of the corresponding peak in the glow curve, is defined as the time for the number of trapped electrons to fall to half of its original value.

Starting from the **first order kinetics** equation

$$\frac{dn}{dt} = -ns\exp\left(-\frac{E}{kT}\right)$$

from which

$$-\int_{n_0}^{n_0/2} \frac{dn}{n} = s\exp\left(-\frac{E}{kT}\right)\int_0^{t_{1/2}} dt$$

and then

$$t_{1/2} = \frac{0.693}{s\exp\left(-\frac{E}{kT}\right)} \tag{1}$$

The temperature effect on the half life is showing in Figs. 1 and 2. Fig. 1 shows the variation of the half life as a function of the activation energy for given values of the frequency factor. Fig. 2 shows the same plot for given values of the activation energy.

The mean life of the decay process expressed by the equation:

$$n = n_0 \exp\left[-s \cdot t \cdot \exp\left(-\frac{E}{kT}\right)\right]$$

can be easily calculated substituting in the equation the n value with n_0/e and using τ instead of t. So, the mean life for the first order kinetics is then obtained as

$$\tau = \frac{1}{s\exp\left(-\frac{E}{kT}\right)} \tag{2}$$

From (1) and (2) result

$$t_{1/2} = \tau \ln 2 \tag{3}$$

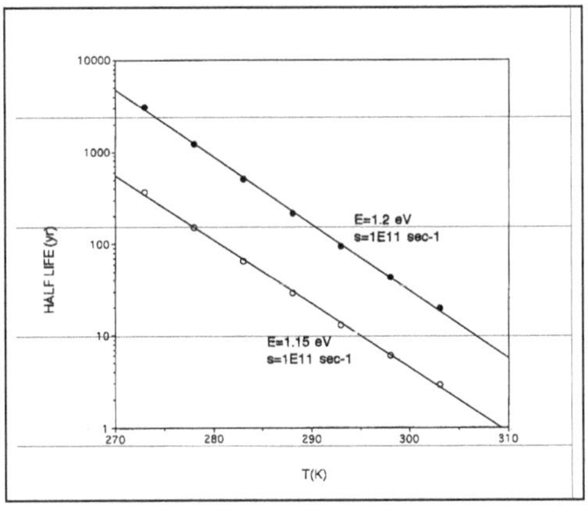

Fig. 1. Variation of the half life, Eq. (1), as a function of E for given values of s [1].

The mean life concept cannot be applied to a second or general order kinetics because the isothermal decay is not exponential any more. Furthermore, as it can be seen in the following calculations, in the hypothetical expression of the half life for any order different from the first one, it is always present the value of n_0 [2].

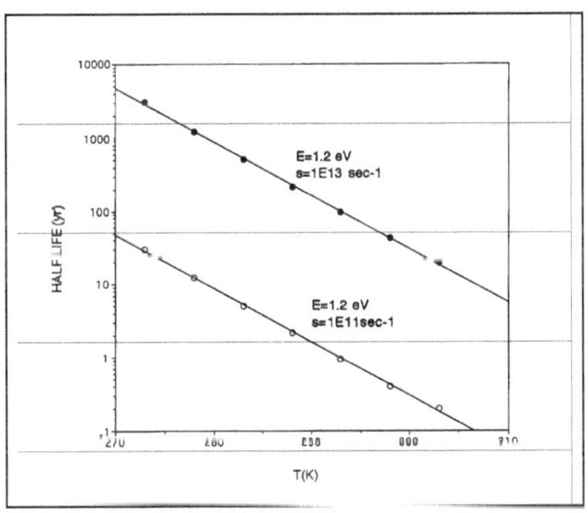

Fig. 2. Variation of the half life, Eq. (1) as a function of s for given values of E [1].

The half-life for the second order process could be calculated as follows

$$\int_{n_0}^{n_0/2} \frac{dn}{n^2} = -s' \exp\left(-\frac{E}{kT}\right) \int_0^{t_{1/2}} dt$$

its integration gives

$$t_{1/2} = \frac{1}{n_0 s' \exp\left(-\frac{E}{kT}\right)} = \frac{N}{n_0 s \exp\left(-\frac{E}{kT}\right)} \tag{4}$$

There is a substantial difference between the half life for a first order and the one for a second order. Indeed, the half life in the case of the first order kinetics is independent on the initial concentration of the trapped charges, which means to be independent on the dose. In the case of the second order kinetics, the situation is totally different because the half life is dose dependent (i.e., n_0): so, for an initial value of n_0, $t_{1/2}$ will have a given value; after a time from the initial one, n_0 changes to a value n'_0 ($n'_0 < n_0$) and the same does $t_{1/2}$ ($t'_{1/2} > t_{1/2}$). So that, as the period of time from the initial irradiation increases, the same does the half life. The same happens for a general order case.

For the general order, starting from the **general order eq**uation, one has

$$\int_{n_0}^{n_0/2} \frac{dn}{n^b} = -s'' \exp\left(-\frac{E}{kT}\right) \int_0^{t_{1/2}} dt$$

from which

$$t_{1/2} = \frac{n_0^{1-b}}{s''(1-b)} \left(1 - \frac{1}{2^{1-b}}\right) \exp\left(\frac{E}{kT}\right) \tag{5}$$

References

1. Furetta C. and Weng P.S., Operational Thermoluminescence Dosimetry, World Scientific (1998)
2. McKeever S.W.S., Thermoluminescence of Solids, Cambridge University Press (1985)

Metastable state

The metastable state is a level within the forbidden gap. This level is associated to a trapping level.

Method based on the temperature at the maximum (Randall-Wilkins)

The intensity I of a first order thermoluminescence peak is given by

$$I(T) = n_0 s \exp\left(-\frac{E}{kT}\right) \exp\left[-\frac{s}{\beta}\int_{T_0}^{T} \exp\left(-\frac{E}{kT'}\right) dT'\right] \tag{1}$$

At the beginning the intensity rises exponentially with temperature; the concentration of trapped electrons reduces and the intensity, after reaching a maximum at a temperature T_M, begins to fall and reaches to zero when the traps have emptied.

Randall and Wilkins did not solve Eq. (1) but they considered that at the maximum temperature the probability of electron escaping from a trap is equal to unity. So, they wrote

$$s \exp\left\{-\frac{E}{kT_M}[1 + f(s,\beta)]\right\} = 1 \tag{2}$$

from which

$$E = T_M\,[1 + f(s,\beta)] \cdot k \cdot \ln(s) \tag{3}$$

and where $f(s, b)$ is a function of the frequency factor and the heating rate. Assuming the average time t, during which the charge carrier remains in the trap, to be the reciprocal of the electron escape probability and plotting ln(t) against T_M one obtains the linear relation

$$\ln(t) = T_M \left\{\frac{[1 + f(s,\beta)]\ln(s)}{T}\right\} - \ln(s) \tag{4}$$

where T is the temperature at which the material is left to decay by phosphorescence. The value in double brackets corresponds to the slope of the straight line and -ln(s) to the intercept. They showed from Eq. (3), using the values of Bunger and Flechng for s and E in KCl:Tl phosphor, that the function f is small compared to unity when the heating rate is in the range from 0.5 to 2.5°C/s. Eq. (3) becomes, using $s = 2.9 \cdot 10^9$ s^{-1}:

$$E = 25kT_M \tag{5}$$

The E value determined in this way is very inaccurate because the value of s which changes from peak to peak and from a material to another.

Reference

Randall J.T. and Wilkins M.H.F., Proc. R. Soc. London, Ser. **A184** (1945) 366

Method based on the temperature at the maximum (Urbach)

Urbach gave the following relation using $s = 10^9 s^{-1}$:

$$E = \frac{T_M}{500}$$

The numerical factors in this equation dependents upon the *s* value and hence the value of E is only approximated because *s* may be different for each trap in the same substance as well as for different materials.

Reference

Urbach F., Winer Ber. IIa, **139** (1930) 363

Methods for checking the linearity

For checking the linearity of the experimental data, some methods are suggested in the following [1]:

graphical method.

The points of co-ordinates ($D_i, \overline{m_i}$) are reported on a log-log paper, each one with the respective error bar. An interpolation with a straight line having a slope equal to one. The best interpolation is obtained using the confidence interval, 2I($\overline{m_i}$), associated to each average $\overline{m_i}$, with

$$I(\overline{m_i}) = t_{n_j-1} \frac{(S.D.)_i}{\sqrt{n_j}} \tag{1}$$

with t_{n_j-1} is the value of the Student-t distribution for n_j-1 degree of freedom at the confidence level required (95%-99%).

UNI, IEC and IAEA methods.

Both UNI and IEC technical recommendations suggest to convert the average values m_i in evaluated kerma (K_{vi}) with the relative errors and compare these values to the conventional real kerma (K_{ci}). The maximum error between this two values for each group must not be larger than ±10%:

$$0.90 \le \frac{K_{vi} \pm t_{n_j-1} \frac{(S.D.)_i}{\sqrt{n_j}}}{K_{ci}} \le 1.10 \tag{2}$$

The IAEA method suggests to use three groups, here numbered 1, 2 and 3, of ten dosemeters each. All groups are processed as already said in point a) and irradiated at the specified doses, D_s, of 1, 10 and 100 mGy for groups 1, 2 and 3 respectively. The readings are then converted in evaluated doses (D_e). D_s and D_e are then substituted in the following expressions:

$$0.95 < \frac{\frac{D_e}{D_s}(group1)}{\frac{D_e}{D_s}(group2)} < 1.05$$

$$0.95 < \frac{\frac{D_e}{D_s}(group3)}{\frac{D_e}{D_s}(group2)} < 1.25 \tag{3}$$

regression analysis.

This method allows to adapt the experimental values obtained with the various TLDs groups to a regression straight line crossing the origin of the axis.

The starting point is to consider the equation of the type

$$y = ax^k$$

and its logarithmic transformation

$$\log y = \log a + k \log x$$

This equation, using the previous symbols, becomes

$$\log \overline{m} = \log(\frac{1}{F_c}) + k \log D \tag{4}$$

The straight line described in Eq. (4) has the property to have a slope k equal to 1 in case of proportionality between dose and TL emission. The slope can be calculated using the method of the least squares:

$$k = \frac{S_{xy}}{S_x} = \frac{\sum_{i=1}^{h}(x_i - \bar{x})}{\sum_{i=1}^{h}(x_i - \bar{x})^2} \tag{5}$$

where

$$S_{xy} = (S.D.)_{xy}$$
$$S_x = (S.D.)_x$$
$$x_i = \log D_i$$
$$y_i = \log \overline{m}_i$$
$$\overline{x} = \frac{1}{h}\sum_{i=1}^{h} x_i$$
$$\overline{y} = \frac{1}{h}\sum_{i=1}^{h} y_i \tag{6}$$

The standard deviation of k is then given by

$$S_k = \left(\frac{S_x S_y - (S_{xy})^2}{(h-2)(S_x)^2} \right)^{1/2} \tag{7}$$

where

$$S_y = \sum_{i=1}^{h} (y_i - \overline{y})^2 \tag{8}$$

One cannot reject the hypothesis of linearity and proportionality only if

$$|1-k| \le t_{h-2} S_k \tag{9}$$

in other words if k is not significatively different from 1.

analysis of variance.

Among the methods here outlined this is the more complex because it needs a numerical analysis not only for the h avarages but also for all the experimental N data.

Let us call y_{ij} the net and corrected readings for each jth dosemeter belonging to the ith group irradiated at the dose x_i. Let us indicate now n_i the number of dosemeters belonging to the ith group. We calculate now the following quantities:

$$D_1 = \sum_{i=1}^{h} n_i \left(\bar{y}_i - y'_i\right)^2 \tag{10}$$

with

$$\bar{y}_i = \frac{1}{n_i}\sum_{j=1}^{n_i} y_{ij} \qquad\qquad y'_i = x_i \frac{1}{N}\sum_{i=1}^{h}\frac{1}{x_i}\sum_{j=1}^{n_i} y_{ij} \tag{11}$$

and

$$D_2 = \sum_{i=1}^{h}\sum_{j=1}^{n_i}\left(y_{ij} - \bar{y}_i\right)^2 \tag{12}$$

Then calculate the Fischer's index, F:

$$F = \frac{D_1}{h-2} \times \frac{N-h}{D_2} \tag{13}$$

Let us say F_{tab} the value relative to [(h-2)(N-h)] degree of freedom at the desired confidence level (95 or 99%); if we get

$$F \leq F_{tab}$$

the hypothesis of linearity is accepted.

Reference

Scarpa G. in Corso sulla Termoluminescenza Applicata alla Dosimetria, 15-17 February 1994, Rome University La Sapienza (I)

Model of non-ideal heat transfer in TL measurements

An interesting model for heat transfer from the heating element to the sample and from the sample to the surrounding, assuming that all heat transmission is due to conduction (neglecting the convection from the sample to the surroundings) has been treated in [1,2] and this model is reported below.

The following assumptions are made:

- heat homogeneous distribution inside both heating element and TLD sample (temperature gradients are present if fast heating rate are used)

- surrounding temperature constant
- heat capacities at the interfaces (contact layers between heating planchet-sample and sample-surrounding gas) are zero
- heat capacities and thermal conductance of all the elements are temperature independent.

Let us indicate T_1 and T_2 the temperatures of the heating element and of the sample respectively, the rate transfer through the contact layer between the planchet and the sample is

$$\frac{dQ_c}{dt} = H_c(T_1 - T_2) \tag{1}$$

where Q_c (in J) is the energy transferred from the planchet to the sample and H_c is the thermal conductance of the contact layer (in J K^{-1} s^{-1}).

The change of the sample temperature is then given by

$$\frac{dT_2}{dt} = \frac{1}{c_S}\left(\frac{dQ_c}{dt} - \frac{dQ_d}{dt}\right) \tag{2}$$

where Q_d (in J) is the energy transferred from the sample to the surroundings and c_S is the heat capacity of the sample (JK^{-1}).

The rate of heat transfer from sample to gas is

$$\frac{dQ_d}{dt} = H_d(T_2 - T_g) \tag{3}$$

where Q_d (in J) is the energyb transferred from the sample to the gas, H_d is the thermal conductance of the sample-gas interface and T_g the gas temperature.

The quantities expressed by Eqs. (1) and (3) can be substituted in Eq. (2):

$$\frac{dT_2}{dt} = \frac{H_c}{c_s}(T_1 - T_2) - \frac{H_d}{c_s}(T_2 - T_g) \tag{4}$$

Considering now a linear heating rate β, $T_1 = T_0 + \beta \cdot t$, Eq. (4) becomes

$$\frac{dT_2}{dt} = \frac{H_c T_0 + H_d T_g + (H_c + H_d)\left(\frac{H_c \beta}{H_c + H_d} \cdot t - T_2\right)}{c_s} \tag{5}$$

where T_0 is the sample and planchet temperature at time $t=0$.

The solution of Eq. (5) is

$$T_2(t)=\frac{H_cT_0+H_dT_g}{H_c+H_d}+\frac{H_c\beta}{H_c+H_d}\cdot t-\left[1-\alpha\exp\left(-\frac{H_c+H_d}{c_s}\cdot t\right)\right]\times \times\left[\frac{H_cc_s\beta}{(H_c+H_d)^2}\right] \tag{6}$$

where α is a coefficient depending on the initial condition T_2 at time $t=0$.

A simulation of Eq. (6) shown that after a transit period (less than 10 seconds in the simulation) the factor in the first square brachets approaches to unity, so that Eq. (6) can be approximated by

$$T_2(t)=\frac{(H_c+H_d)(H_cT_0+H_dT_g)-H_cc_s\beta}{(H_c+H_d)^2}+\frac{H_c\beta}{H_c+H_d}\cdot t$$

or

$$T_2(t)=T_0'+\beta'\cdot t \tag{7}$$

Eq. (7) means that after a transitory period, the temperature profile of the sample is the same as that of the planchet but with the heating rate β replaced by an effective heating rate β' and the initial temperature T_0 replaced by an effective initial temperature T_0'.

The temperature lag, ΔT, between the sample temperature and the planchet temperature is then

$$\Delta T=T_1-T_2=\frac{c_s\beta'+H_d(T_1-T_g)}{H_c+H_d} \tag{8}$$

References

1. Piters T.M. and Bos A.J.J., J. Phys. D: Appl. Phys. **27** (1994) 1747
2. Piters T.M., A study into the mechanism of thermoluminescence in a LiF:Mg,Ti dosimetry material (Thesis, 1998), D.U.T.

Multi-hit or multi-stage reaction models

These models of thermoluminescence are based on the assumption that a trap may be subjected to a two or more stage reaction before its activation in the thermoluminescent process.

The multi-hit models were introduced to explain the supralinear growth of thermoluminescence, i.e., the thermoluminescence intensity, I, increases as a function of D^l, where D is the absorbed dose and l is not necessary one or an integer value.

Halperin and Chen [1] found a relation of the type $I \propto D^3$ for the supralinearity of semiconducting diamond, concluding that the growth of TL intensity as a function of the dose was ruled by a three stage reaction. This model needs two intermediate energy levels where the electrons are rised by two successive doses of irradiation. A third irradiation finally rises the electrons to the CB from where they are trapped.

In the works of Larson and Katz [2], Katz [3], Waligorski and Katz [4] a two-hit model was presented. In this model a trap is only produced after trapping first one and then a second electron. With this model Katz and colleagues were able to explain the supralinearity of certain peaks in LiF.

A similar model has been used in the works of Takeuchi et al. [5].

References

1. Halperin A. and Chen R., Phys. Rev. **148** (1966) 839
2. Larsson L. and Katz R., Nucl. Instr. Meth. **138** (1976) 631
3. Katz R., Nucl. Track Detect. **2** (1978) 1
4. Waligorski M.P.R. and Katz R., Necl. Instr. Meth. **172** (1980)
5. Takeuchi N., Inabe K., Kido H. and Yamashita J., J. Phys. C: Sol. St. Phys. **11** (1978) L147

N

Non linearity

The plot of the TL signal vs. Dose may present different zones. A hypothetical curve is shown in Fig. 1. As it can be seen from the figure, the TL emission is not linear in the low dose region and is not linear any more at high doses. To bypass some linguistic ambiguities concerning the terms superlinearity and supralinearity, two universal indices have been proposed by Chen and MacKeever [1] to mathematically describe all forms of nonlinearity.

The first of these indices is called "superlinearity index", g(D); it gives the indication of change in the slope of the dose response in all cases.

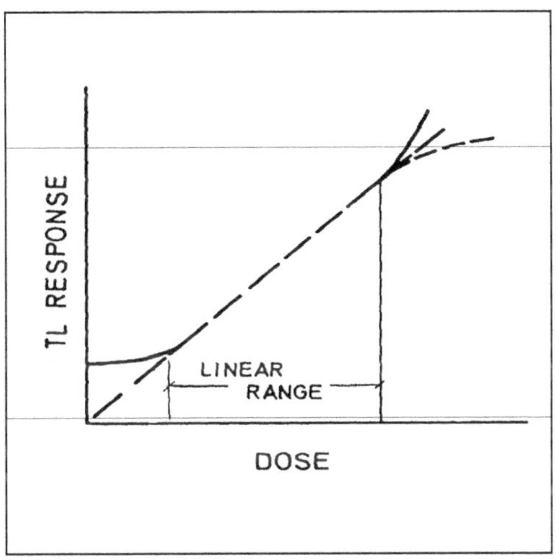

Fig. 1. The various zones which could be observed in a plot of TL as a function of dose.

The second one is the well known "supralinearity index", or dose response function, f(D), used to quantify the size of the correction required for extrapolation of the linear dose region.

As already discussed by Chen and Bowman [2], the term superlinearity is reserved to indicate an increase of the derivative of the M = M(D) function, where M indicates, as usual, the measured TL signal, both the peak high at the maximum or the peak area. Let us indicate by M' the first derivative of M at a point D and M" the second derivative. Then, if

M''(D) > 0 → M'(D) increases in D → M(D) increases and then is superlinear;
if
M''(D) < 0 → M'(D) decreases in D → M(D) decreases and then is sublinear;
if
M''(D) = 0 → M'(D) is constant in D → M(D) is linear.

To quantify the amount of superlinearity (or sublinearity) the authors have proposed the function

$$g(D) = \left[\frac{D \cdot M''(D)}{M'(D)}\right] + 1 \tag{1}$$

called the "superlinearity index".

The following cases are possible:

- g(D) > 1 indicates superlinearity
- g(D) = 1 signifies linearity
- g(D) < 1 means sublinearity

The second quantity, the f(D) index, concerns the supralinearity effect. The authors have suggested a slightly modified definition of the old dose response function. The old expression was

$$f(D) = \frac{\frac{M(D)}{D}}{\frac{M(D_l)}{D_l}} \tag{2}$$

where D_l is the normalisation dose in the linear region. The authors have proposed the following modified expression

$$f(D) = \frac{\frac{M(D)-M_0}{D}}{\frac{M(D_l)-M_0}{D_l}} \tag{3}$$

where M_0 is the intercept on the TL response axis.

The advantage of the new Eq. (3) lies in the possibility of applying it to cases in which the supralinear region precedes the linear region. In this case M_0 is negative but is still valid since it has no physical meaning.

M(D) values above the extrapolated linear region produce f(D) to be larger than 1, and the supralinearity appears in the TL response. M(D) values below the extrapolated linear region cause f(D) < 1 and underlinearity occurs [3]. When f(D) approaches to zero, saturation is occurring. Of course f(D) = 1 means linearity. As already stated, f(D) monitors the amount of deviation from linearity; that is the quantity needs for extrapolation to the linear region.

The main problem in the use of the previous indices concerns g(D) because it is not trivial problem to fit the experimental values of a TL response vs. dose with an analytical expression. Nevertheless, from a practical point of view the f(D) function is enough to characterise the TL vs. dose behaviour.

In the following some examples are given for a better understanding on the use of the new indices.

Figure 2 depicts a situation where the TL response at high doses is below the extrapolated linear range; on the contrary, at low doses the TL response is above the linearity. The experimental data are given in the following Table 1. The values in bold correspond to the linearity region. The third column corresponds to the TL net response. The dose dependence curve can be analytically expressed by the equation

$$M = 8,4539D^4 - 70.873D^3 + 170.74D^2 - 27.930D + 0.4909 \tag{4}$$

Dose (Gy)	TL (a.u.)	TL_{net} (a.u.)
0.000	13.932	0.000
0.001	13.932	0.000
0.100	13.932	0.000
0.120	13.990	0.058
0.250	17.182	3.250
0.500	34.553	20.621
0.750	**62.008**	**48.076**
1.000	**95.691**	**80.759**
1.500	**160.513**	**146.581**
2.000	209.951	196.019
2.500	234.355	220.423
3.000	238.495	224.563
3.500	238.154	224.222

Table 1. TL vs. dose. TL_{net} corresponds to the reading minus background.

The linear region is given by the equation

$$M = 131.38D - 50.527 \tag{5}$$

In both equations M is the net TL response. Some points of the curve can now be considered.

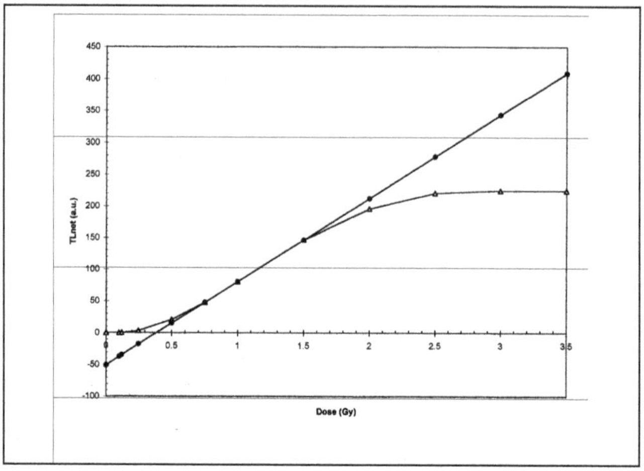

Fig. 2. Plot of TL vs dose showing under-response at high doses and over-response at low doses.

D = 2 Gy

One obtains:

M'(2) = 75.0788 > 0 which indicates an increase of M in D = 2.

M'' (2) = - 103.2088 < 0 which means that the M(D) function has the concavity facing the bottom in D = 2 and that M' is decreasing in the same point.

Then the values of the g(D) and f(D) functions are

g(2) = -1.7493 < 1

f(2) = 0.9390 < 1

The value of g(D) indicates sublinearity of the M(D) function in D = 2 and the value of f(D) depicts a situation of underlinearity or, in other words, it means that saturation starts to appear.

For the low dose region one can consider the value

D = 0.250 Gy

In this case one obtains:

M'(0.250) = 44.6797 > 1 which means that M is an increasing function in D = 0.250 Gy.

M'' (0.250) = 241.5109 > 1: M has the concavity facing the top in D = 0.250 Gy and, furthermore, M' is increasing. Then g(D) and f(D) are

g(0.250) = 2.3513 > 1

f(0.250) = 1.6385 > 1.

The two above values indicate superlinearity and supralinearity in the region preceding the linear part of the curve.

For a value of D = 1 Gy, i.e., a dose value situated in the linear range of the curve, both g(D) and f(D) give approximately 1.

A further example is the one given in Fig. 3. The plot has been obtained using the following equation [4]:

$$M = M_{sat}(1 - e^{-\alpha D}) - 13De^{-\alpha D} \tag{6}$$

where M_{sat} is the TL response at saturation level (=4844 a.u.) and $\alpha = 2.89\text{x}10^{-3}\text{Gy}^{-1}$. The data (calculated using the previous equation) are given in the following Table 2.

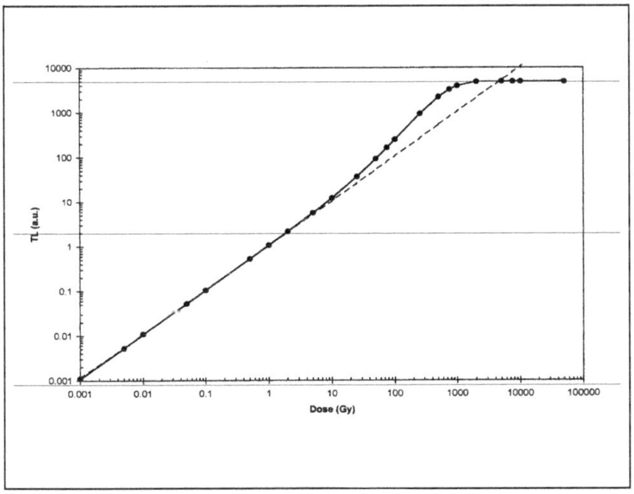

Fig. 3. Plot of TL vs dose according to Eq. (6).

The linear zone, numbers in bold in Table 2, is given by the following equation

$$M = 1.0472D + 9.4260 \cdot 10^{-5} \tag{7}$$

Some points of the plot can then be analysed:

<u>D = 50 Gy</u>:
M' > 0 → M is increasing
M'' > 0 → M' is increasing and the concavity is facing the top
g > 1 → M is superlinear
f > 1 → M is supralinear

Dose (Gy)	TL (a.u.)
0.001	**0.0011**
0.005	**0.0052**
0.010	**0.0110**
0.050	**0.0520**
0.100	**0.1050**
0.500	0.5280
1.000	1.0650
2.000	2.1640
5.000	5.6700
10.00	12.200
25.00	36.490
50.00	91.570
75.00	162.46
100.0	246.67
250.0	923.88
500.0	2183.0
750.0	3186.0
1000	3862.0
2000	4751.0
5000	4844.0
7500	4844.0
10000	4844.0
50000	4844.0

Table 2. Data calculated from Eq. (6).

D = 500 Gy:
$M' > 0 \rightarrow M$ is increasing
$M'' < 0 \rightarrow M'$ is decreasing and the concavity is facing the bottom
$g < 1 \rightarrow M$ is sublinear
$f > 1 \rightarrow M$ is supralinear

D = 10^4 Gy:
$M' > 0 \rightarrow M$ is increasing
$M'' < 0 \rightarrow M'$ is decreasing and M has the concavity facing the bottom
$g < 1 \rightarrow M$ is sublinear
$f < 1 \rightarrow M$ is underlinear and approaches saturation.

Table 3 gives a summary of the various configurations which can be find in case of nonlinearity TL response.

S''	S'	S	concavity of S	g	behaviour	supralinear	underlinear	
> 0	incr > 0	incr	↑	> 1	superlinear	f > 1		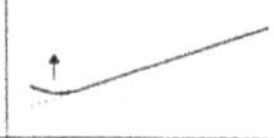
> 0	incr > 0	incr	↑	> 1	superlinear	f > 1		
< 0	decr > 0	incr	↓	< 1	sublinear		f < 1	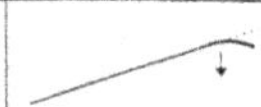
< 0	decr > 0	incr	↓	< 1	sublinear	f > 1		
< 0	decr > 0	incr	↓	< 1	sublinear		f< 1	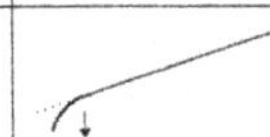
< 0	decr > 0	incr	↓	< 1	sublinear		f< 1	
> 0	incr > 0			1		f = 1 linearity		
> 0	incr > 0	incr	↓	1		f> 1		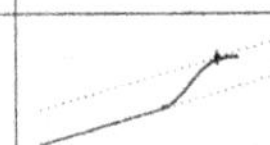 saturation starts
< 0	decr > 0	incr	↓	1			f < 1	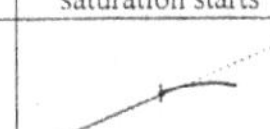 saturation starts

Table 3. Summary of the various configurations.

References

1. Chen R. and McKeever S.W.S., Rad. Meas. **23** (1994) 667
2. Chen R. and Bowman S.G.E., European PACT J. **2** (1978) 216
3. Furetta C. and Kitis G. (unpublished data)
4. Inabe K. and Takeuchi N., Jap. J. Appl. Phys. **19** (1980) 1165

Non-ideal heat transfer in TL measurements (generality)

There are various types of heating a thermoluminescent sample during read out. The most popular is the contact way realised using a planchet heating.

Because the temperature control is usually achieved by mean a thermocouple mounted on the back of the planchet, this method gives only a control of the planchet's temperature and not of the sample. The temperature lag between planchet and sample, as well as the temperature gradient across the TLD, can strongly influence the analysis of the glow curve, specially in the calculation of the kinetic parameters, where an accurate temperature determination is absolutely necessary.

The problem of non-ideal heat transfer has been studied by various authors and corrections have been also proposed [1-7]

References

1. Taylor G.C. and Lilley E., J. Phys. D: Appl. Phys. **15** (1982) 2053
2. Gotlib V.I., Kantorovitch L.N., Grebenshicov V.L., Bichev V.R. and Nemiro E.A., J. Phys. D: Appl. Phys. **17** (1984) 2097
3. Betts D.S., Couturier L., Khayrat A.H., Luff B.J and Townsend P.D., J. Phys. D: Appl. Phys. **26** (1993) 843
4. Betts D.S. and Townsend P.D., J. Phys. D: Appl. Phys. **26** (1993) 849
5. Piters T.M. and Bos A.J.J., J. Phys. D: Appl. Phys. **27** (1994) 1747
6. Facey R.A., Health Phys. **12** (1996) 720
7. Kitis G. and Tuyn J.W.N., J. Phys. D: Appl. Phys. **31** (1998) 2065

Numerical curve fitting method (Mohan-Chen: first order)

Mohan and Chen suggested the following method for first-order TL curves.

Haake has given an asymptotic series for evaluating the integral comparing in the expression of $I(T)$ for the first order:

$$\int_{T_0}^{T} \exp(-\frac{E}{kT})dT \cong T\exp(-\frac{E}{kT})\sum_{1}^{n}\left(\frac{kT}{E}\right)^{n}(-1)^{n-1}n! \tag{1}$$

Using only the first two terms of expression (1), one has

$$\int_{T_0}^{T} \exp(-\frac{E}{kT})dT \cong T\frac{kT}{E}\exp(-\frac{E}{kT}) - T_0\frac{kT_0}{E}\exp(-\frac{E}{kT_0}) \tag{2}$$

Since the first term on the right hand side is very strongly increasing function of T, it is conventional to neglect the second term in comparison to the first one. In this assumption the equation of the first order kinetics

$$I(T) = n_0 s\exp\left(-\frac{E}{kT}\right)\exp\left[-\frac{s}{\beta}\int_{T_0}^{T}\exp\left(-\frac{E}{kT'}\right)dT'\right] \tag{3}$$

becomes

$$I(T) \cong C \exp\left[-\frac{E}{kT} - \frac{sE}{\beta k}\left(\frac{kT}{E}\right)^2 \exp(-\frac{E}{kT})\right] \tag{4}$$

In Eq. (4) the term $sE/\beta k$ can be approximated by the following way: using $B = sE/\beta k$ and $x = E/kT$, Eq. (4) can be written as

$$I(T) \cong C \exp\left[-x - Bx^{-2} \exp(-x)\right]$$

Making the logarithm of the previous expression one gets

$$\ln I(T) = \ln C + \left[-x - Bx^{-2} \exp(-x)\right]$$

and then its derivative at the maximum, for $T=T_M$, is

$$\left(\frac{dI}{dT}\right)_{T=T_M} = -1 + 2Bx^{-3} \exp(-x) + Bx^{-2} \exp(-x) = 0$$

which gives

$$B = \frac{x^3 \exp(x)}{2 + x}$$

so that

$$\frac{sE}{\beta k} \approx \frac{\left(\frac{E}{kT_M}\right)^3}{\frac{E}{kT_M} + 2} \exp(\frac{E}{kT_M}) \tag{5}$$

The intensity is then given by

$$I(T) \cong C \exp\left[-\frac{E}{kT} - \frac{\left(\frac{E}{kT_M}\right)^3}{\frac{E}{kT_M} + 2} \cdot \left(\frac{kT}{E}\right)^2 \exp(\frac{E}{kT_M}) \exp(-\frac{E}{kT})\right] \tag{6}$$

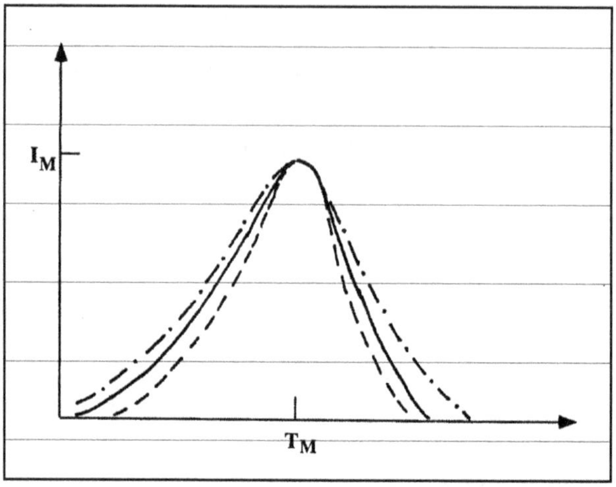

Fig. 1. Comparison between experimental and theoretical glow-peaks. ——— experiment, - - - - - - - - E is too high, - · - · - · - E is too small

Expression (6) leads to a convenient method of fitting because only one parameter, E, is free.

The procedure is now as follows: an experimental glow-curve is measured and an E value is estimated by using one of the experimental methods reported. Then a theoretical glow-curve is plotted using Eq. (6) and the constant is adjusted so that the intensity at maximum (I_M) of the experimental and theoretical curves coincide. The fitting of the remaining curve is then checked. If the chosen value of E is too small or too high the theoretical curve will lie above or below the experimental curve (except for the maximum) as shown in Fig. 1. In these case a new value of E is chosen and the procedure is repeated until the desired fit is obtained.

Reference

Mohan N.S. and Chen R., J. Phys. D: Appl. Phys. **3** (1970) 243

Numerical curve fitting methods (Mohan-Chen: second order)

In the case of a second-order kinetics, the Garlick and Gibson equation is used:

$$I(T) = -\frac{dn}{dt} = n^2 s' \exp\left(-\frac{E}{kT}\right) = \frac{n_0^2 s' \exp\left(-\frac{E}{kT}\right)}{\left[1 + \frac{s' n_0}{\beta}\int_{T_0}^{T} \exp\left(-\frac{E}{kT'}\right) dT'\right]^2} \tag{1}$$

From Eq. (1) the maximum intensity $I(T_M)$ is found; after that the intensity $I(T_i)$ corresponding to a certain number N of temperatures T_i is chosen and the normalised intensity is obtained by dividing each $I(T_i)$ by $I(T_M)$ as following

$$\frac{I(T_i)}{I(T_M)}=\frac{\exp(-\frac{E}{kT_i})\left[1+\left(\frac{s'n_0}{\beta}\right)\int_{T_0}^{T_M}\exp(-\frac{E}{kT})dT\right]^2}{\exp(-\frac{E}{kT_M})\left[1+\left(\frac{s'n_0}{\beta}\right)\int_{T_0}^{T_i}\exp(-\frac{E}{kT})dT\right]^2} \tag{2}$$

Using the condition for the maximum

$$\frac{\beta E}{2kT_M^2}\left[1+\frac{s'n_0}{\beta}\int_{T_0}^{T_M}\exp\left(-\frac{E}{kT'}\right)dT'\right]=s'n_0\exp\left(-\frac{E}{kT_M}\right) \tag{3}$$

and the integral approximation

$$\int_{T_0}^{T}\exp(-\frac{E}{kT})dT\cong T\frac{kT}{E}\exp(-\frac{E}{kT})-T_0\frac{kT_0}{E}\exp(-\frac{E}{kT_0}) \tag{4}$$

The procedure for the curve fitting is similar to the **numerical curve fitting** for the first-order case. However, a better fit may be expected if only points below the maximum temperature are taken, since the main difference between first- and second-order peaks is in the region above the maximum.

Reference

Mohan N.S. and Chen R., J. Phys. D: Appl. Phys. **3** (1970) 243

Numerical curve fitting method (Shenker-Chen: general order)

The numerical curve fitting procedure for the case of general-order has been carried out by Shenker and Chen.

The equation for the general-order case is the following

$$\frac{dn}{dt}=-s''n^b\exp\left(-\frac{E}{kT}\right) \tag{1}$$

where s'' is the pre-exponential factor, expressed in $cm^{3(b-1)}s^{-1}$ and b is the order of the kinetics, ranging from 1 to 2.

The solution of Eq. (1) is given by

$$I(T) = sn_0 \exp\left(-\frac{E}{kT}\right)\left[1+\frac{s(b-1)}{\beta}\int_{T_0}^{T}\exp\left(-\frac{E}{kT'}\right)dT'\right]^{-\frac{b}{b-1}} \quad (2)$$

where $s = s'' n_0^{b-1}$, expressed in s^{-1}.

Also in this case, since E/kT has values of 10 or more, the integral on the right-side of Eq. (2) can be resolved by using the asymptotic series. Eq. (2) can be normalised by dividing $I(T)$ by $I(T_M)$. The frequency factor s is found using the condition at the maximum and then some points $I(T_i)$ have to take from the experimental glow-curve and processed as for first and second cases (see **numerical curve fitting method for first- and second-order**).

Reference

Shenker D. and Chen R., J. Phys. D: Appl. Phys. **4** (1971) 287

O

Optical bleaching

Optical bleaching indicates the effect of light, of a specific wavelength, on irradiated TL samples, in the sense that charge carrier stimulation of a particular defect centre can be achieved via absorption of optical energy, resulting then in a photodepopulation of the centre. The charge carriers released may recombine with opposite sign carriers, emitting light during the illumination (bleaching light), or may be retrapped in other trapping centres. Observing then the changes occurring in the glow-curve resulting after the optical stimulation, relationships between thermoluminescence traps and optically activated centres can be obtained.

The term "beaching" is taken from the vocabolary of colour centres: a crystal is coloured by high dose of ionizing radiation and a subsequent illumination produces the colour fading, i.e., the sample is bleached.

Optical fading

The effect of light on an irradiated thermoluminescent sample consists in a reduction of the TL signal, depending on the light intensity, its wavelength and duration of exposure.

For practical applications (personel, environmental and clinical dosimetry), the sensitivity to the light of different TL materials can be avoided by wrapping the dosemeters in light-tight envelopes. If this procedure is not applied, fading correction factors have to be determined carrying out experiments in dark and light conditions.

Oven (quality control)

The oven used for annealing should be able to keep predetermined temperature oscillations within well specified margins. However, it must be noted that the reproducibility of the annealing procedure, concerning both heating up and cooling down processes, is much more important than the accuracy of the temperature setting.

Temperature overshoots due to the high thermal capacity of the oven walls can be minimised using ovens with circulating hot air. In this way the problem related to a non-ideal thermal conductivity of the annealing trays is also solved.

In some cases, when surface oxidation of chips is possible (i.e., in the case of carbon loaded chips), it would be advantageous to operate the annealing under inert gas atmosphere. This facility could also reduce any possible contamination.

It would be better to use different annealing ovens depending on the various needs: one of them should be suitable for high temperature annealing, another one for low temperature annealing and a third one for any pre-readout thermal cycles.

As far as are concerned the trays where the TLDs are located for the annealing procedure, the following suggestions may be useful:

- the tray should have between 50 to 100 recesses to accommodate the dosimeters,
- each position in the tray should be identified,
- the tray must be as thin as possible and with a flat bottom surface to get a very good thermal contact,
- the tray material can be ceramic (in particular porcelain), Pyrex and pure aluminium. Ceramic is preferable for its chemical inertia and good thermal conductivity. Good results have also been obtained using Ni-Cu and any light compound not oxidable,
- it should be possible to insert in the tray a thin thermocouple to monitor the actual temperature of the tray as well as that of the dosimeters during the annealing cycle.

The quality control program of the annealing procedure should include the following points:

- determination of the heating rate of the oven from the switch-on time to the steady condition,
- determination of the temperature accuracy and setup of a correction factor which is needed,
- check on the temperature stability,
- check on the temperature distribution inside the oven chamber,
- determination of the heating rate of the tray.

A quality control program concerning the ovens has been suggested by Scarpa and takes into account the various quantities which have to be checked, displayed graphically in next Fig. 1. The accuracy is related to the difference between the temperature set and the temperature monitored; the instability of the oven concerns the oscillations of the temperature monitored.

Fig. 2 shows an example concerning the heating up profile of a muffle oven. Because the heating time is a characteristic of each oven, it must be checked accurately. It is convenient to switch on the oven several hours before use.

Fig. 3 depicts the temperature oscillations during the heating up phase (temperature set at 240°C) and successive Fig. 4 shows a typical thermal conditioning for a ceramic tray, inserted in a preheated oven.

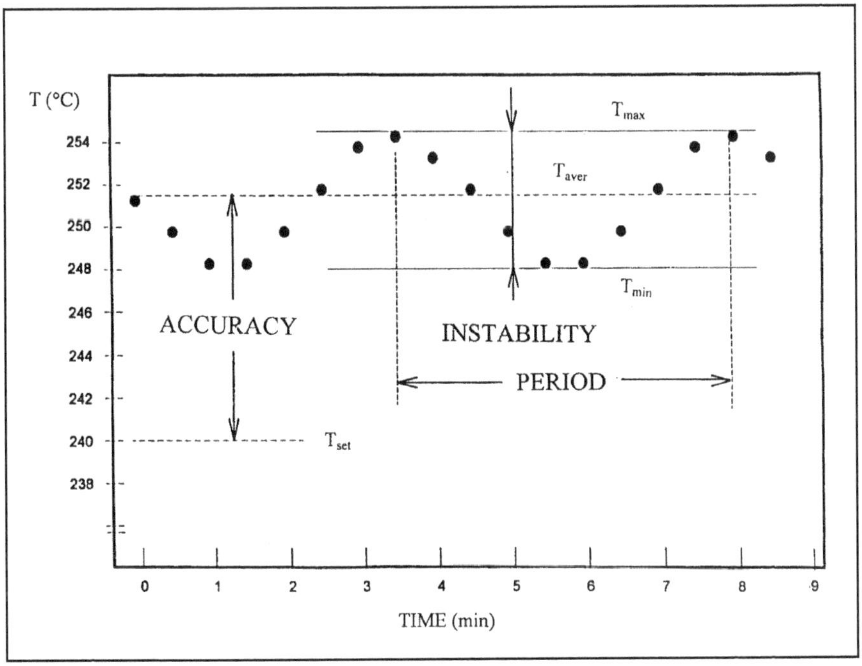

Fig. 1. Quantities to be checked for the quality control of the ovens.

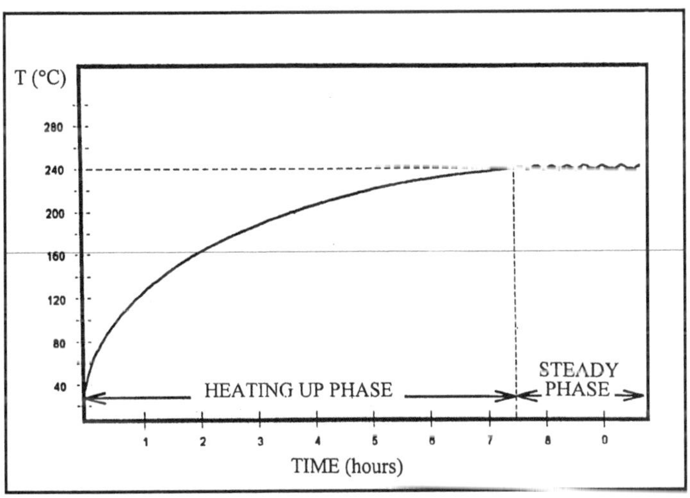

Fig. 2. Heating up phase of a muffle oven.

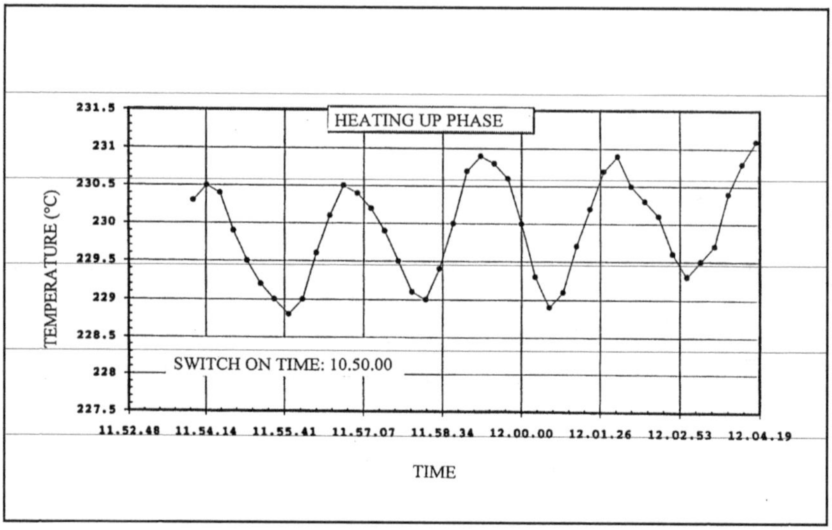

Fig. 3. Temperature oscillations during heating up phase.

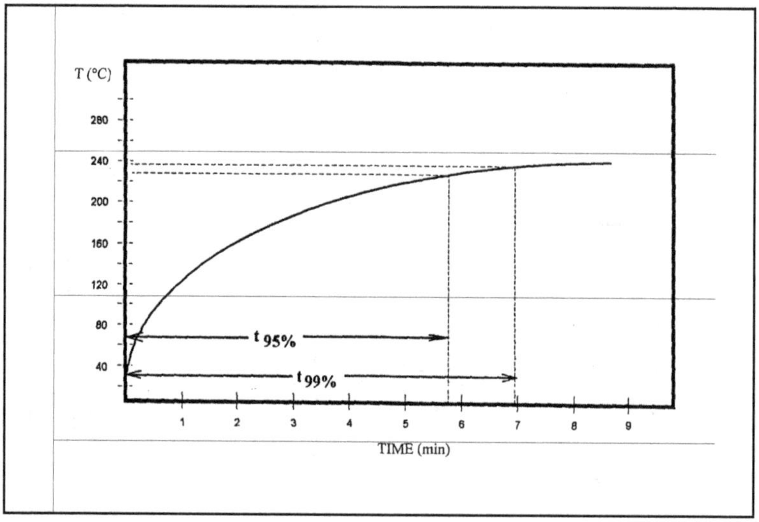

Fig. 4. Heating rate of a ceramic tray inserted in a preheated oven.

During the steady phase of the oven the temperature, normally, is not stable. The oscillations around the temperature set are depending on the quality of the oven. This parameter has to be reported in the list of the characteristics of any new oven. As an example, Fig. 5 depicts the temperature oscillations during the steady phase (temperature set at 240°C).

Another effect to be taken into account is that one which arises when the door of a preheated oven is opened to put the tray inside; the temperature drops to a lower value and then increases above the pre-set value. An example of this behaviour, measured for an oven without forced air circulation, set at a steady temperature of 400°C and an opening time of the door of 60 seconds, is shown in Fig. 6. After closing the door, the temperature rises to about 410°C and then, slowly, goes back to the pre-set value in about 30 minutes. Of course, it is not a good procedure to open the oven during the annealing treatment.

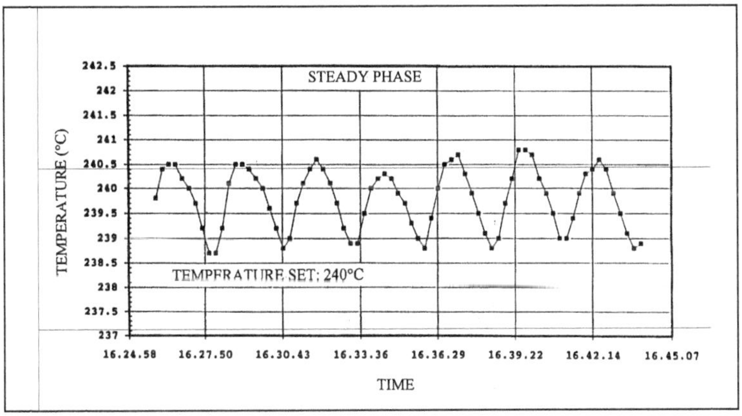

Fig. 5. Temperature oscillations in an oven during the steady phase.

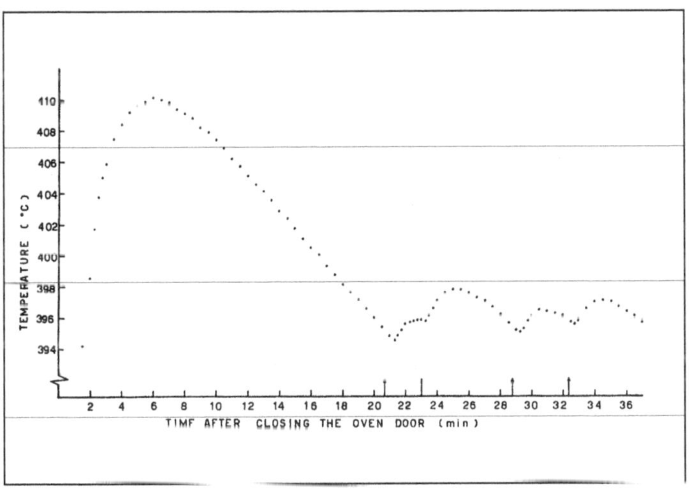

Fig. 6. Effect of "open door" on a preheated oven.

According to the previous effects, it is convenient to use at least two different ovens when the TL dosimeters need a complex annealing procedure, as in the case of LiF :Mg,Ti which needs an high temperature annealing followed by a low temperature treatment.

Fig. 7 shows the space distrution of temperatures inside an oven. Because the temperature gradients are always present inside an oven, the TLD tray must be positioned always at the same place.

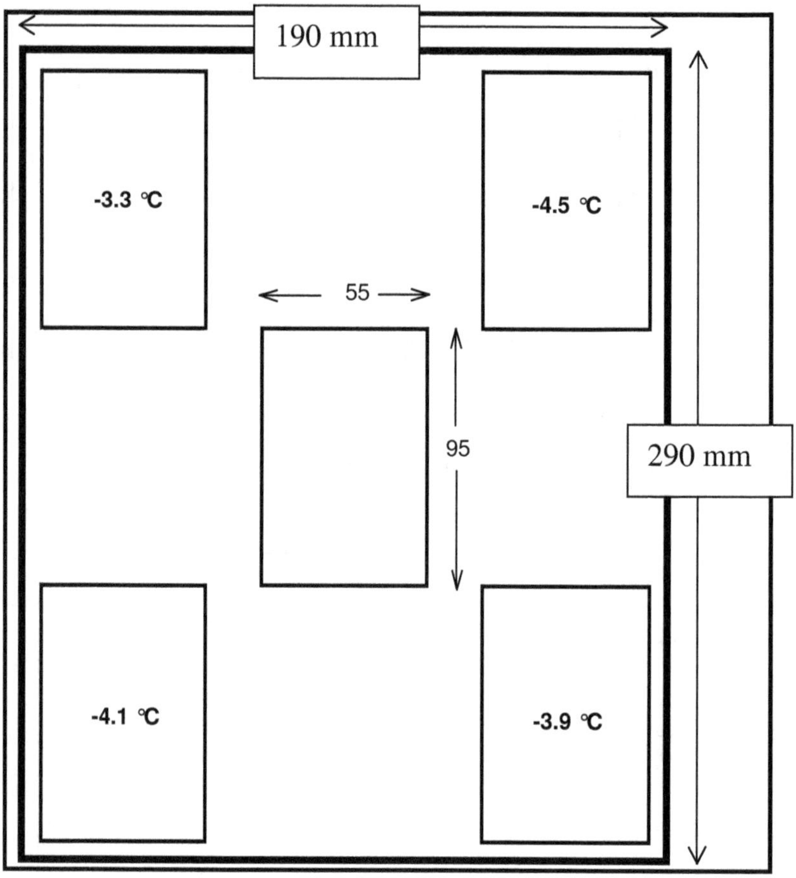

Fig. 7. Space distribution of temperatures inside an oven.

Reference

Scarpa G. in Corso sulla Termoluminescenza Applicata alla Dosimetria, Universita' di Roma "La Sapienza", 15-17 Febbraio 1994

Observations on the peak-shape methods based on the peak temperature at the maximum

The aim of this section is to give a critical analysis concerning some methods used for the activation energy determination in thermoluminescent (TL) experiments. Only the first-order kinetics is considered.

The methods are those that use only one temperature value of the TL glow curve: the temperature at the maximum, T_M. These methods are commonly called heuristic methods and are considered as a first approximation when compared with other methods based on two temperature values, which are well known as peak-shape methods.

The so-called heuristic methods are as follows:

Urbach [1], first suggested a very simple expression deduced from TL experiments on KCl

$$E = \frac{T_M}{500} = 23kT_M \tag{1}$$

which gives reasonable values if applied to other materials when the frequency factor, s, is $10^9\ s^{-1}$.

Randall and Wilkins [2], for alkaline earth and zinc sulphides, with $s = 10^8\ s^{-1}$, proposed the following expression

$$E = kT_M \ln s \tag{2}$$

Halperin and Braner [3] suggested, for NaCl,

$$E = 38kT_M \tag{3}$$

Finally, Dussel and Bube [4] gave a very similar expression for LiF

$$E = 39kT_M \tag{4}$$

The equations related to the various peak-shape methods are based on three characteristic temperatures of a single TL peak, namely T_1, on the low temperature side of the peak at its half intensity, T_2, as the temperature on the fall-off of the peak at half intensity, and T_M, the peak temperature at its maximum intensity. The other parameters are:

$\tau = T_M - T_1$, the half-width at the low-temperature side of the peak,

$\delta = T_2 - T_M$, the half-width towards the fall-off of the glow peak,

and

$\omega = T_2 - T_1$, the total half-width (FWHM).

The first-order kinetics equations for determining the activation energy according to the peak-shape methods are as follows:

Grossweiner [5]

$$E_{G,\tau} = 1.41\frac{T_1 T_M}{\tau} \tag{5}$$

The coefficient 1.41 in equation (5) has been computed by Dussel and Bube [4] for getting a better approximation.

Lushchik [6]

$$E_{L,\delta} = 0.976\frac{kT_M^2}{\delta} \tag{6}$$

where the coefficient has been computed by Chen [7].

Halperin and Braner [3]

$$E_{HB,\tau} = 1.72\frac{kT_M^2}{\tau}\left(1 - 1.58\Delta_M\right) \tag{7}$$

where $\Delta_M = \dfrac{2kT_M}{E}$. Equation (7) requires an iterative process owing to the presence of the quantity Δ_M. To avoid this difficulty, Chen modified the previous equation as following [7]:

$$E_{HB,\tau} = 1.52\frac{kT_M^2}{\tau} - 3.16kT_M \tag{8}$$

Balarin [8]

$$E_{B,\omega} = \frac{T_M^2}{4998 \cdot \omega} \tag{9}$$

All the previous expressions were summed up by Chen [7] who considered a general-order kinetics, ranging from 1 to 2, giving the possibility of non-integer values for the kinetics order:

$$E_\alpha = c_\alpha\left(\frac{kT_M^2}{\alpha}\right) - b_\alpha\left(2kT_M\right) \tag{10}$$

where α is τ, δ or ω. The values of c_α and b_α are summarized as

$$\begin{aligned} c_\tau &= 1.51 + 3.0(\mu - 0.42), \qquad & b_\tau &= 1.58 + 4.2(\mu - 0.42) \\ c_\delta &= 0.976 + 7.3(\mu - 0.42), \qquad & b_\delta &= 0 \\ c_\omega &= 2.52 + 10.2(\mu - 0.42), \qquad & b_\omega &= 1 \end{aligned} \tag{11}$$

where $\mu = \frac{\delta}{\omega} = \frac{T_2 - T_M}{T_2 - T_1}$

$\mu = 0.42$ for the first order kinetics.

All the previous equations have been modified to express then as a function of only the peak temperature at the maximum.
Indeed, asymmetrical property of the glow peaks following a first-order kinetics [2,9] allows the expression of the characteristic peak temperatures T_1 and T_2 as a function of T_M. For this, synthetic glow peaks for the first-order kinetics have been computed for a wide range of the activation energy values, from 0.7 to 2.1 eV. Taking into account the influence of the frequency factor s for each value of E, s varies from 10^8 to 10^{20}. This procedure allows us to determine the values of T_1, T_2, T_M, T_1/T_M, and T_2/T_M. After that, as a convenient first approximation one may take $0.95T_M$ for T_1 and $T_2 = 1.036T_M$. Other parameters used are: τ, δ and ω.

Table 1 lists the modified equations for the first-order kinetics. Table 2 lists the data obtained using the experimental methods given by the original equations as well as the approximated values obtained by the modified equations. The differences, in percentages, are also given for comparison.

It should be noted that there is quite a large discrepancy between the activation energy values obtained using the peak-shape methods and ones calculated using the modified equations where only peak temperature at the maximum is used.

To explain why the methods based on only one value of temperature, *i.e.* the temperature at the maximum, give only approximated values for the activation energy, first-order kinetics glow peaks have been simulated for various values of E and s, *i.e.* for E ranging from 0.5 to 2.5 eV and s from 109 to 1017 s. From the obtained glow peaks, only those with T_M between 250 and 750K were taken into account. From each simulated glow peak, the following parameters were extracted:

T_M, T_1, T_2, ω, δ, τ, T_1/T_M, T_2/T_M, T_M/ω, T_M/δ, T_M/τ

Original equations	Modified equations
Equation (5) – Grosweiner	$E\tau = 23.1\mathrm{E}-4 \cdot T_M$
Equation (6) – Lushchik	$E\delta = 23.4\mathrm{E}-4 \cdot T_M$
Equation (7) – Halperin and Braner	$E\tau = 23.5\mathrm{E}-4 \cdot T_M$
Equation (9) – Balarin	$E\omega = 23.3\mathrm{E}-4 \cdot T_M$
Equation (10) – Chen	$E\tau = 23.3\mathrm{E}-4 \cdot T_M$
Equation (10) – Chen	$E\delta = 23.4\mathrm{E}-4 \cdot T_M$
Equation (10) – Chen	$E\omega = 23.5\mathrm{E}-4 \cdot T_M$

Table 1

material	peak	b	T_M (K)	E (eV)	REF	E (eV)*	ΔEmax (%)
CaF2:Tm (TLD-300)	1	1	390	E_L=0.69 E_{HB} = 0.72	10	0.91 0.92	+32% +28%
CaF2:Tm (TLD-300)	3	1	437.8	E_L=1.20 E_{HB}=1.18 $E_{C\tau}$=1.10 $E_{C\delta}$=1.20 $E_{C\omega}$=1.10	10	1.02 1.03 1.02 1.02 1.03	-15% -13% -7% -15% -6%

Table 2

First case: general relationship between E and T_M

According to Chen, all the existing peak-shape methods can be expressed by the general formula (10). The second term of equation (10) is a correction factor, which is still a function of T_M. The correction factor will be ignored at the moment in the following, but we must keep in mind that it must be included in the final relations. Therefore, the main interest is in the first term of equation (10).

According to Kitis et al. [9] a relation for the activation energy determination as a function of only T_M, *i.e.* a relation of the form

$$E = const \cdot T_M \tag{12}$$

can be found.

The constant in equation (12) will be termed PreT_M factor. Taking into account only the first term of equation (11), the PreT_M factor will be of the form

$$\mathrm{Pre}T_M \text{ factor} = ckT_M/\alpha \tag{13}$$

Equation (13) means that an equation having the form of equation (12) exists when, and only when, for any glow peak, the ratio of T_M/α is a constant too.

Firstly, the simulation verifies that the activation energy is always a linear function of T_M, *i.e.* equation (12) always holds. However, as seen in figure 1, its slope depends of the frequency factor s. Therefore, the constant PreT_M in equation (13) is constant only for a given frequency factor because the ratio of T_M/α is nearly stable. An example of T_M/ω is given in figure 2.

Table 3 reports the explicit expressions related to figure 1.

As a consequence, the PreT_M quantity is constant only for a given frequency factor because the ratio T_M/α is nearly stable. As an example, figure 2 shows the behavior of T_M/ω *vs.* E for given values of s. Furthermore, the PreT_M factor, which is the slope of the functions reported in table 3, is a linear function of ln(s) as observed in figure 3.

The linear relationship is given by the following expression:

$$\mathrm{Pre}T_M = 3.963\cdot10^{-4}+8.35\cdot10^{-5}\cdot\ln(s) \tag{14}$$

then Eq. (12) becomes

$$E = \left[3.963\cdot10^{-4} + 8.35\cdot10^{-5}\cdot\ln(s)\right]\cdot T_M \tag{15}$$

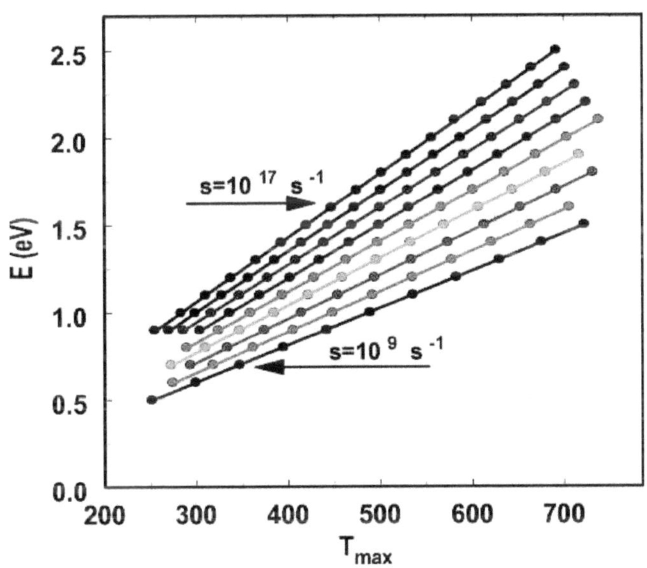

Fig. 1

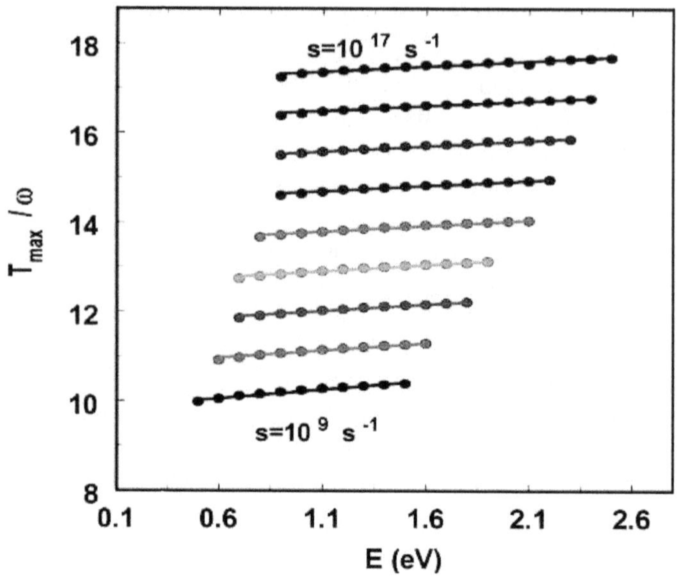
s=10 17 s -1
s=10 9 s -1
T max / ω
E (eV)
18
16
14
12
10
8
0.1
0.6
1.1
1.6
2.1
2.6

Fig. 2

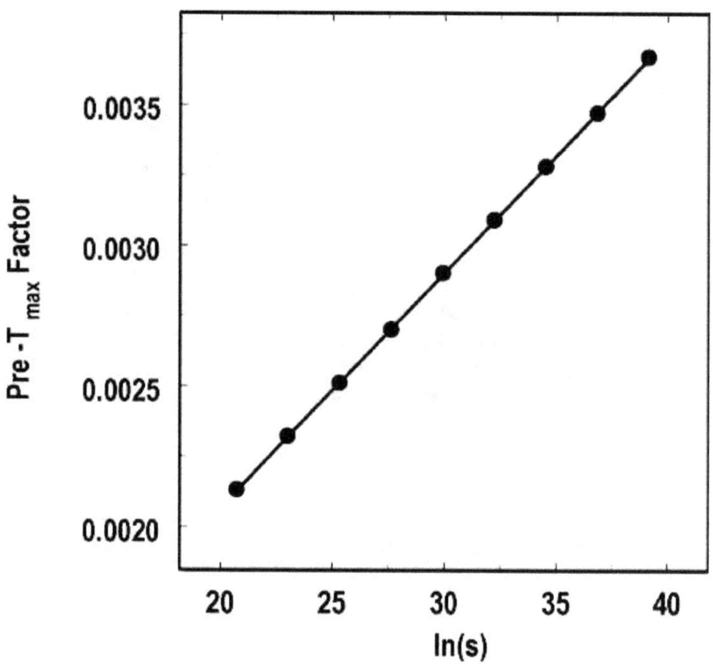
Pre -T max Factor
ln(s)
0.0035
0.0030
0.0025
0.0020
20
25
30
35
40

Fig. 3

$s(s^{-1})$	**Regression line**
10^9	$E = 0.0375 + 0.00213\ T_M$
10^{10}	$E = 0.0381 + 0.00232\ T_M$
10^{11}	$E = 0.0405 + 0.00251\ T_M$
10^{12}	$E = 0.0386 + 0.00279\ T_M$
10^{13}	$E = 0.0404 + 0.00290\ T_M$
10^{14}	$E = 0.0422 + 0.00309\ T_M$
10^{15}	$E = 0.0395 + 0.00328\ T_M$
10^{16}	$E = 0.0383 + 0.00347\ T_M$
10^{17}	$E = 0.0372 + 0.00367\ T_M$

Table 3

Second case: relations between E and T_M using T_1/T_M and T_2/T_M

More relationships between E and T_M are now obtained using the approximated ratios $T_1/T_M = \alpha$ and $T_2/T_M = \beta$. In this case, the peak shape expressions assume very simple forms

$$E_\omega = \frac{2.52kT_M^2}{\omega} \tag{16a}$$

$$E_\tau = \frac{1.51kT_M^2}{\tau} \tag{16b}$$

$$E_\delta = \frac{0.976kT_M^2}{\delta} \tag{16c}$$

Using now

$$\omega = T_2 - T_M\ ,\ \alpha = T_1/T_M \text{ and } \beta = T_2/T_M$$

Eqs. (16a-c) become

$$E_\omega = \frac{2.52kT_M}{\beta - \alpha} \tag{17a}$$

$$E_\tau = \frac{1.51kT_M}{1 - \alpha} \tag{17b}$$

$$E_\delta = \frac{0.976kT_M}{\beta - 1} \tag{17c}$$

The variation of both parameters α and β is very slow, <0.001% in an energy range between 0.5 and 2 eV. So one can assume that their values are constant for all E at given s.

On the other hand, the parameters $\beta - \alpha$, $1 - \alpha$, and $\beta - 1$ (figure 4) have great variation in the same range of energy.

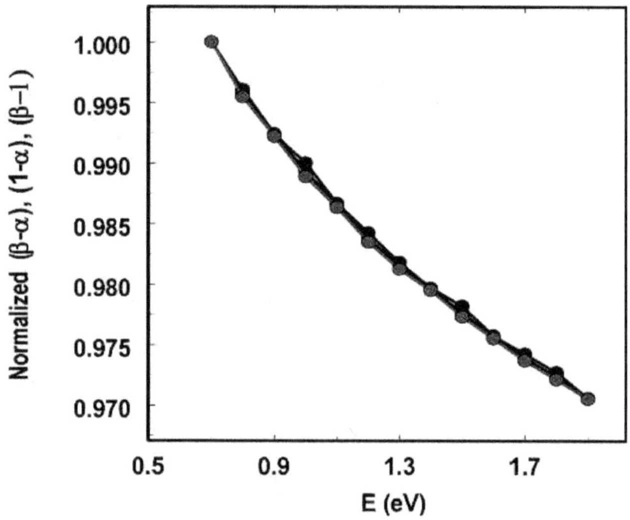

Fig. 4. Variation of $\beta - \alpha$, $1 - \alpha$, and $\beta - 1$ as a function of E.

In figure 4, it is observed that the variation of the parameters is ~3% in the energy region between 0.5 and 2 eV. As a consequence, the final stability for all the E values, having the same frequency factor s, is also of the order of 3%.

The following step is the analysis of the stability of the parameters as a function of the frequency factor. Figures 5 and 6 show the behaviour of α and β *vs.* $\ln(s)$ and of $\beta - \alpha$, $1 - \alpha$, and $\beta - 1$ *vs.* $\ln(s)$, respectively.

As shown in figure 5, the parameters α and β vary by only 2% for s ranging from 10^9 to 10^{17} s−1. This is an extreme stability for such a broad region of s values. However, as in the earlier cases, the influence of the parameters α and β passes through $(\beta - \alpha)$, $(1 - \alpha)$, and $(\beta - 1)$. Their variation as a function of s is shown in figure 6.

Here the situation is very different. The variation of the parameters of interest is almost 50% for s, ranging from 10^9 to 10^{17} s^{-1}. Therefore, the relatively low 2% stability of T_1/T_M, T_2/T_M becomes 50% through the expressions $\beta - \alpha$, $1 - \alpha$, and $\beta - 1$. This is the reason why the relations of the type $E = \text{const.}T_M$ give E values differing by 25% from those expected.

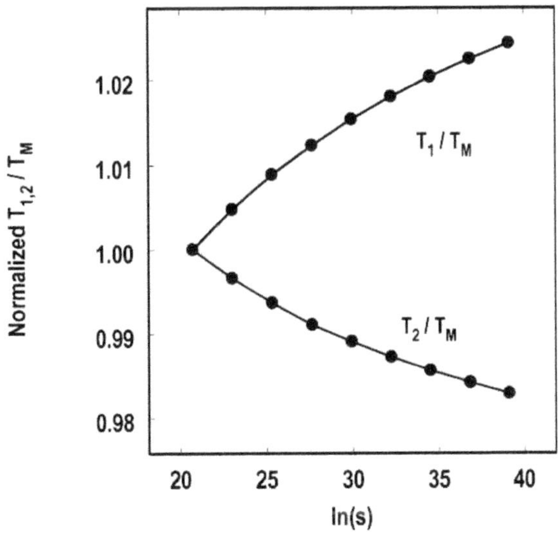
Normalized $T_{1,2}$ / T_M
1.02
1.01
1.00
0.99
0.98
T_1 / T_M
T_2 / T_M
20
25
30
35
40
ln(s)

Fig. 5

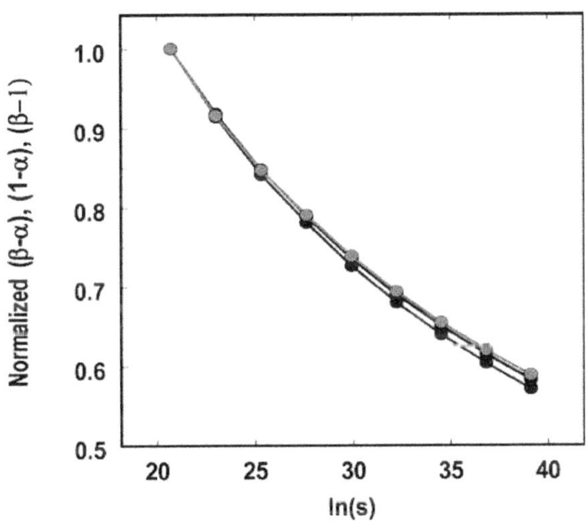
Normalized (β-α), (1-α), (β−1)
1.0
0.9
0.8
0.7
0.6
0.5
20
25
30
35
40
ln(s)

Fig. 6

In order to re-determine the expressions of E *vs.* T_M, the following relations, from the Eqs. (17a–c), must be fulfilled

$$\frac{2.52k}{\beta-\alpha} \approx \frac{1.51k}{1-\alpha} \approx \frac{0.976k}{\beta-1} \approx const \tag{18}$$

The results of the simulation give the following expressions listed in table 4. As seen from table 2, the PreT_M factor in equations (17a–c) has slightly different values for the cases of ω, τ and δ. The PreT_M factors of column 2 in table 4 are in very close agreement with those in table 3. These slight differences are due to the fact that in the Second Case, the constant 2.52 was used, whereas in the First Case, this constant is evaluated intrinsically, having values between 2.32 and 2.44. Therefore, as in the First Case, the values of Pre*T*M factor are unique only for all E values within the same s. They are not unique as a function of s.

The PreT_M factor has a close linear behaviour *vs.* ln(s), with the same slope as shown in figure 8.

The regression lines as obtained from the previous figure are, respectively,

$$\text{Pre } T_{M\omega} = 4.2 \text{ x } 10^{-4} + 8.33 \text{ x } 10^{-5} \ln(s)$$
$$\text{Pre } T_{M\delta} = 5.04 \text{ x } 10^{-4} + 8.68 \text{ x } 10^{-5} \ln(s) \tag{19}$$
$$\text{Pre } T_{M\tau} = 3.165 \text{ x } 10^{-4} + 8.25 \text{ x } 10^{-5} \ln(s)$$

The final set of equations is then

$$E_\omega = \text{Pre } T_{M\omega} \text{ x } T_M$$
$$E_\tau = \text{Pre } T_{M\tau} \text{ x } T_M \tag{20}$$
$$E_\delta = \text{Pre } T_{M\delta} \text{ x } T_M$$

References

1. Urbach F., Winer Ber **IIa 139** (1930) 363
2. Randall J.T. and Wilkins M.H.F., Proc. Roy. Soc. **A184** (1945) 366
3. Halperin A. and Braner A.A., Phys. Rev. **117** (1960) 408
4. Dussel G.A. and Bube R.H., Phys. Rev. **155** (1967) 764
5. Grossweiner L.I., J. Appl. Phys. **24** (1969) 1306
6. Lushihik L.I., Sov. Phys. **3** (1956) 390
7. Chen R., J. Appl. Phys. **40** (1969) 570
8. Kitis G., Cruz-Zaragoza E., Furetta C., REDS **161** (2006) 149

P-1

(from **Patridge-May model** to **Peak shape method: reliability expressions**)

Partridge-May model (zero-order kinetics)

Partridge and May have reported some observations concerning an apparent kinetic order less than the first one (b<1).

If a single process is involved, one has the following equation

$$I = -\frac{dn}{dt} = cn^b \tag{1}$$

where c is the rate constant.

On the contrary, if b is less than 1, they explained such behavior by means of two competing processes: a first radiative order and a zero order without radiative transitions.

For such a model one can write:

$$-\frac{dn}{dt} = c_1 n + c_2 \tag{2}$$

where c_2 is the zero-order rate constant.

The integration of Eq. (2) gets:

$$-\int \frac{dn}{c_1 n + c_2} = \int dt$$

$$-\frac{1}{c_1} \int \frac{d(c_1 n + c_2)}{c_1 n + c_2} = \int dt$$

from which

$$-\frac{1}{c_1} \ln(c_1 n + c_2) = t + k' \tag{3}$$

where k' is a constant of integration.

The intensity is given by

$$I = c_1 n \tag{4}$$

Combining Eq. (3) and Eq. (4) one obtains

$$-\ln(I + c_2) = c_1 t + k'' \tag{5}$$

Partridge and May have reported that in isothermal decay experiments some data fitted Eq. (5) better than an equation expressing order higher than the first one.

Reference

Partridge J.A. and May C.E., J. Chem. Phys. **42** (1965) 797

Peak-shape method (Balarin: first- and second-order kinetics)

Balarin [1,2] deduced some expressions for determining the activation energy, based on the quantity $\omega = T_2 - T_1$, where T_1 and T_2 correspond to the temperatures on either side of T_M, corresponding to half intensity. In the following the Ballarin's original symbolism is used.

He started from the following general rate equation

$$-\frac{dc}{d\tau} = \frac{c^{\gamma}}{c_0^{\gamma-1}} \cdot k_0 \cdot \exp\left(-\frac{E}{kT}\right) \tag{1}$$

with $c_0 = c(\tau_0)$ and where

c = concentration of some kind of reactant,
γ = kinetic order,
k_0 = frequency factor,
E = activation energy,
τ = time,
T = temperature,
k = Boltzmann constant.

The processes governed by Eq. (1) are enhanced when the temperature of the system is raised continuosly with a constant heating rate $\beta = \frac{dT}{d\tau}$. Eq. (1) can be then transformed as

$$-\frac{dc}{dT} = \frac{1}{\beta} \cdot \frac{c^{\gamma}}{c_0^{\gamma-1}} \cdot k_0 \cdot \exp\left(-\frac{E}{kT}\right) \tag{2}$$

The maximum condition is then obtained

$$\left.\frac{d^2c}{dT^2}\right|_{T=T_M} = -\frac{k}{\beta}\cdot\frac{1}{c_0^{\gamma-1}}\cdot\gamma\cdot c_M^{\gamma-1}\cdot\frac{E}{kT_M^2}\cdot\exp\left(-\frac{E}{kT_M}\right) = 0$$

$$\frac{k_0}{\beta} = \left(\frac{c_0}{c_M}\right)^{\gamma-1}\cdot\frac{1}{\gamma}\cdot\frac{E}{kT_M^2}\cdot\exp\left(\frac{E}{kT_M}\right) \tag{3}$$

Considering the following quantities:

$$C(T) = \frac{c(T)}{c_0}, \qquad y = \frac{kT}{E}, \qquad t = \frac{T}{T_M} = \frac{y}{y_M}$$

Eq. (2) becomes

$$-\gamma\cdot C_M^{\gamma-1}\cdot\frac{dC}{C^{\gamma}} = \frac{E}{kT_M^2}\cdot\exp\left[\frac{E}{kT_M}\left(1-\frac{T_M}{T}\right)\right]dT \tag{4}$$

Integration of Eq. (4), starting at $t=0, T=0, y=0$ and $C_0=1$ up to $C(T)$, gives, for various values of γ:

$$\left.\begin{array}{ll} \gamma\neq 1,\neq 0 & \dfrac{\gamma\cdot C_M^{\gamma-1}}{\gamma-1}\left(\dfrac{1}{C^{\gamma-1}}-1\right) \\ \gamma=1 & \ln\dfrac{1}{C} \\ \gamma=0 & 1-C \end{array}\right\rangle = t^2\exp\left(\frac{1-\frac{1}{t}}{y_M}\right)\cdot\eta(y) \tag{5}$$

where

$$\eta(y) = 1+\sum_{n=1}^{\frac{1}{y}-2}(-1)^n\cdot(n+1)!\,y^n \tag{6}$$

is a correction function which is always close to unity, $\eta(y)\leq 1$.

Using expression (6) in (5), we obtain

$$C_M = C(T_M) = \begin{cases} \left(1 - \dfrac{\gamma - 1}{\gamma}\eta_M\right)^{\frac{1}{\gamma-1}} & \gamma \neq 1, \neq 0 \\ \exp(-\eta_M) & \gamma = 1 \\ 1 - \eta_M & \gamma = 0 \end{cases} \tag{7}$$

and

$$\frac{C(T_M)}{C_M} = \begin{cases} \left[\dfrac{\gamma}{\gamma - (\gamma-1)\eta_M + (\gamma-1)\eta t^2 \exp\left(\dfrac{1-\frac{1}{t}}{y_M}\right)}\right]^{\frac{1}{\gamma-1}} \\ \exp\left[\eta_M - \eta t^2 \exp\left(\dfrac{1-\frac{1}{t}}{y_M}\right)\right] \\ \dfrac{\left[1 - \eta t^2 \exp\left(\dfrac{1-\frac{1}{t}}{y_M}\right)\right]}{1-\eta_M} \end{cases} \tag{8}$$

The temperature positions $T_i = T_1$ and T_2 are obtained when the intensity $J(T) = -\dfrac{dC}{dT}$ is half of the maximum intensity. By means of Eq. (1) or (4), we get

$$\frac{J(T_i)}{\frac{1}{2}J(T_M)} = 1 = 2\left(\frac{C_i^{(\gamma)}}{C_M^{(\gamma)}}\right)^{\gamma} \exp\left[\frac{E}{kT_M}\left(1 - \frac{T_M}{T_i}\right)\right] =$$

$$= 2\left(\frac{C_i^{(\gamma)}}{C_M^{(\gamma)}}\right)^{\gamma} \exp\left(\frac{1 - \frac{1}{t_i^{(\gamma)}}}{y_M}\right) \qquad (9)$$

where the supscript (γ) indicates the individula solution for every distinct kinetic order.

Inserting (8) in (9), expressions for $t_i^{(\gamma)}$ and $T_i^{(\gamma)}$ can be obtained. These expressions contain the quantities $\eta_M(y)$ and η_i which are polynomials in $y_M = \frac{kT_M}{E}$. Furthermore, the half-width $\omega = T_2 - T_1$ again depends on y_M. It is then deduced that the quantity $E \times \omega$, divided by T_M^2 is an invariant, different for every kinetic order, and the following relations can be then obtained

$$E = \frac{T_M^2}{4998 \cdot \omega} \qquad 1^{st} \text{ order kinetic} \qquad (10)$$

$$E = \frac{T_M^2}{3542 \cdot \omega} \qquad 2^{nd} \text{ order kinetic} \qquad (11)$$

According to Balarin, the previous expressions allow to determine the activation energy values accurately to within 0.5%. The experimental inaccuracy in the determination of ω becomes then dominant and limiting.

References

1. Balarin M., Phys. Stat. Sol. (a) **31** (1975) K111
2. Balarin M., J. Thermal Anal. **17** (1979) 319

Peak shape method (Chen: first- and second-order)

Chen [1] derived expressions for evaluating E using numerical approximations. The Chen's method is useful for a broad range of energies ranging between 0.1 eV and 2.0 eV and pre-exponential factors between 10^5 sec^{-1} and 10^{13} sec^{-1} and it does not make any use of iterative procedures. Furthermore, the method does not need any knowledge of the kinetics order which is directly found from the peak shape.

Its method is based on the shape of a TL peak, similarly to the **Lushchik** [2] and **Halberin-Braner methods** [3]. For semplicity, the parameters involved in a well resolved peak are here reported in Fig. 1.

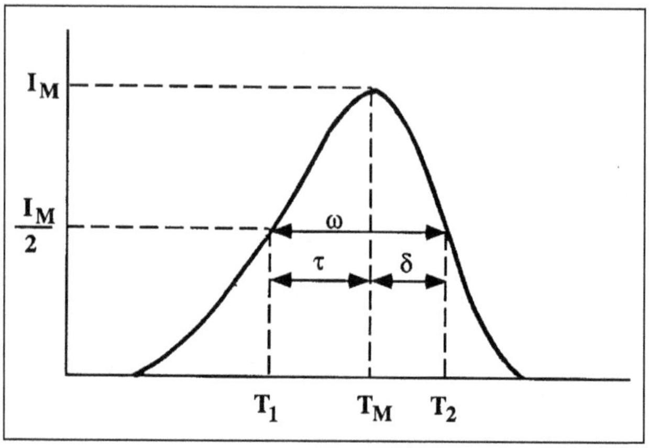

Fig. 1. Parameters characterising a single peak.

where

T_M, T_1, T_2: are respectively the peak temperature at the maximum and the temperatures on either side of the temperature at the maximum, corresponding to half intensity,

$\tau = T_M - T_1$: is the half-width at the low temperature side of the peak,

$\delta = T_2 - T_M$: is the half-width towards the fall-off of the glow peak,

$\omega = T \ - T_1$: is the total half-width,

$\mu_g = \dfrac{\delta}{\omega}$: is the symmetrical geometrical factor.

Total half-width peak method

- *First order kinetics*

Starting from the following **first order equation**, giving the TL intensity as a function of the temperature,

$$I(T) = n_0 s \exp\left(-\frac{E}{kT}\right) \exp\left[-\frac{s}{\beta}\int_{T_0}^{T} \exp\left(-\frac{E}{kT'}\right) dT'\right] \tag{1}$$

the equation for the maximum intensity, I_M , is given by

$$I_M = n_0 s \exp\left(-\frac{E}{kT_M}\right)\exp\left[-\frac{s}{\beta}\int_{T_0}^{T_M}\exp\left(-\frac{E}{kT'}\right)dT'\right] \tag{2}$$

The integral on the right-hand side can be approximated by means of an asymptotic expansion and a reasonable approximation is given by

$$\int_{T_0}^{T_M}\exp\left(-\frac{E}{kT}\right)dT \approx \left(\frac{kT_M^2}{E}\right)\exp\left(-\frac{E}{kT_M}\right)(1-\Delta_M) \tag{3}$$

with $\Delta_M = \dfrac{2kT_M}{E}$.

Then, Eq. (2) becomes

$$I_M = n_0 s \exp\left(-\frac{E}{kT_M}\right)\exp\left[\left(-\frac{s}{\beta}\right)\left(\frac{kT_M^2}{E}\right)\exp\left(-\frac{E}{kT_M}\right)(1-\Delta_M)\right] \tag{4}$$

Using now the **maximum condition for the first order kinetics**, Eq. (4) becomes

$$I_M = \left(\frac{n_0 E\beta}{ekT_M^2}\right)\exp(\Delta_M) \tag{5}$$

Using the assumption of **Lushchik** [2] that the area of the second half of the peak is equal to that of a triangle having the same height and half width, one can write

$$\frac{\delta \cdot I_M}{\beta \cdot n_M} = 1 \tag{6}$$

where n_M is the number of the trapped charges at T_M.

A similar assumption about the relation between the total glow peak and a triangle can be written as

$$\frac{\omega \cdot I_M}{\beta \cdot n_0} = 1 \tag{7}$$

Chen considers relation (7) as a constant different from 1 for obteining a result with higher precision:

$$\frac{\omega \cdot I_M}{\beta \cdot n_0} = C_\omega \tag{8}$$

Inserting now the quantity $\beta \cdot n_0$ from Eq. (8) into Eq. (5), one obtains

$$E\exp(\Delta_M) = \frac{eC_\omega kT_2^M}{\omega} \tag{9}$$

Since Δ_M is quite small, we get $\exp(\Delta_M) \approx 1 + \Delta_M$ and then Eq. (9) becomes

$$E_\omega = 2kT_M\left(\frac{eC_\omega T_M}{2\omega}\right) - 1 \tag{10}$$

Inserting Eq. (10) in the condition at the maximum, one obtains the frequency factor as following

$$s = \left(\frac{2\beta}{e^2 T_M}\right)\left(\frac{eC_\omega T_M}{2\omega} - 1\right)\exp\left(\frac{eC_\omega T_M}{\omega}\right) \tag{11}$$

Chen found $C_\omega = 0.92$, and so Eq. (10) becomes

$$E_\omega = 2kT_M\left(1.25\frac{T_M}{\omega} - 1\right) \tag{12}$$

Inserting Eq. (12) into the condition at the maximum

$$\frac{\beta E}{kT_M^2} = s\exp\left(-\frac{E}{kT_M}\right)$$

one has

$$s = \frac{2.29\beta}{\omega}\exp\left(\frac{2.29T_M}{\omega}\right)$$

which can be rewritten as

$$s = \frac{2.67\beta}{\omega}10^{T_M/\omega} \tag{13}$$

having taking into account that 2.29 is close to $ln 10$. Furthermore, 2.29 has been changed in 2.67 in order to compensate for addicional inaccuracies.

- *Second order kinetics*

According to the second order equation

$$I(T) = \frac{n_0^2 s' \exp\left(-\frac{E}{kT}\right)}{\left[1 + \frac{s' n_0}{\beta} \int_{T_0}^{T} \exp\left(-\frac{E}{kT'}\right) dT'\right]^2} \tag{13}$$

the intensity at the maximum is given by

$$I_M = n_0^2 s' \exp\left(-\frac{E}{kT_M}\right) \left[1 + \left(\frac{n_0 s'}{\beta}\right) \int_{T_0}^{T_M} \exp\left(-\frac{E}{kT'}\right) dT'\right]^{-2} \tag{14}$$

Inserting the **maximum condition for the second order** into Eq. (14), this becomes

$$I_M = n_0^2 s' \exp\left(-\frac{E}{kT_M}\right) \left[\frac{2kT_M^2 n_0 s'}{\beta E} \exp\left(-\frac{E}{kT_M}\right)\right]^{-2} \tag{15}$$

or better

$$I_M = \frac{1}{s'} \left(\frac{\beta E}{2kT_M^2}\right)^2 \exp\left(\frac{E}{kT_M}\right) \tag{16}$$

Using the integral approximation (3) in Eq. (13), one obtains

$$\left(\frac{n_0 s' kT_M^2}{\beta E}\right) \exp\left(-\frac{E}{kT_M}\right)(1 - \Delta_M) + 1 = \left(\frac{2kT_M^2 n_0 s'}{\beta E}\right) \exp\left(-\frac{E}{kT_M}\right)$$

and rearranging

$$\left(\frac{\beta E}{s' kT_M^2}\right) \exp\left(\frac{E}{kT_M}\right) = n_0 (1 + \Delta_M) \tag{17}$$

Inserting now Eqs. (8) and (17) into Eq. (16), one gets

$$E(1 + \Delta_M) = \frac{4C_\omega kT_M^2}{\omega}$$

or better

$$E = 2kT_M\left(\frac{2C_\omega T_M}{\omega} - 1\right) \tag{18}$$

In this case Chen calculated the coefficient C_ω equal to 0.878 and then, Eq. (18) becomes:

$$E_\omega = 2kT_M\left(1.756\cdot\frac{T_M}{\omega} - 1\right) \tag{19}$$

High temperature-side half peak

- *First-order kinetics*

The **method of Lushchik** [2] is slightly modified here in order to get more accurate values.

The **assumption of Lushchik** [2], given by the relation

$$n_M = \frac{1}{\beta}\int_{T_M}^{\infty} IdT \cong \frac{1}{\beta} AREA\Delta(ABC) = \frac{I_M 2\delta}{2\beta} = \frac{I_M\delta}{\beta} \tag{20}$$

has been changed by Chen as following

$$\frac{I_M\delta}{n_M\beta} = C_\delta \tag{21}$$

From equation

$$\frac{I_M}{n_M} = s\exp\left(-\frac{E}{kT_M}\right) \tag{22}$$

using the **maximum condition** for the first-order kinetics and Eq. (21), Eq. (22) yields

$$\frac{C_\delta\beta}{\delta} = \frac{E\beta}{kT_M^2}$$

and hence

$$E = \frac{C_\delta kT_M^2}{\delta} \tag{23}$$

Chen calculated C_δ to be 0.976, then Eq. (23) becomes

$$E = 0.976 \frac{kT_M^2}{\delta} \tag{24}$$

- *Second-order kinetics*

Using the solution for n and replacing n_M for n, one has

$$n_M = n_0 \left[1 + \frac{s' n_0}{\beta} \int_{T_0}^{T_M} \exp\left(-\frac{E}{kT'} \right) dT' \right]^{-1} \tag{25}$$

Using now the expression for the intensity at the maximum, I_M, and making the ratio between I_M and n_M, one has

$$\frac{I_M}{n_M} = \frac{s' n_0 \exp\left(-\dfrac{E}{kT_M} \right)}{\left[1 + \dfrac{s' n_0}{\beta} \displaystyle\int_{T_0}^{T_M} \exp\left(-\dfrac{E}{kT'} \right) dT' \right]} \tag{26}$$

The insertion of the maximum condition for the second-order kinetics in Eq. (26) yields

$$\frac{I_M}{n_M} = \frac{s' n_0 \exp\left(-\dfrac{E}{kT_M} \right)}{\dfrac{2kT_M^2 n_0 s'}{E\beta} \exp\left(-\dfrac{E}{kT_M} \right)}$$

and, rearranging

$$\frac{I_M}{n_M} = \frac{E\beta}{2kT_M^2} \tag{27}$$

Inserting in Eq. (27) the condition given by Eq. (21), the expression for E is found

$$E = C_\delta \left(\frac{2kT_M^2}{\delta} \right) \tag{28}$$

Chen found $C_\delta = 0.853$, so that Eq. (28) can be rewritten as

$$E = 0.853\left(\frac{2kT_M^2}{\delta}\right) \tag{29}$$

Low-temperature side half peak

Concerning the low-temperature side of the peak, Chen gave a more accurate expression for the activation energy respect to the expression of Halperin and Braner [3].

At first, Chen wrote that the Halperin-Braner expressions include two inaccuracies:

(i) the Lushichik's assumption $C_\delta = 1$

(ii) the approximation of $\mu_M = \dfrac{m_M}{n_0}$ by $\mu_M = \dfrac{\delta}{\omega}$

To give more accurate expressions, Chen introduced the quantity

$$\frac{I_M \cdot \tau}{\beta(n_0 - n_M)} = C_\tau \tag{30}$$

which means that the ratio between the first half of the peak and a triangle having the same height and half-width is a constant.

Eq. (30) can be rewritten again as

$$\frac{I_M \cdot \tau}{n_M \cdot \beta\left(\dfrac{n_0}{n_M} - 1\right)} = C_\tau$$

and rearranging, as

$$\frac{n_0}{n_M} - 1 = \left(\frac{\tau}{\beta \cdot C_\tau}\right) \cdot \left(\frac{I_M}{n_M}\right) \tag{31}$$

- *First-order kinetics*

The number of trapped charges at the maximum is given by

$$n_M = n_0 \exp\left[-\frac{s}{\beta}\int_{T_0}^{T_M} \exp\left(-\frac{E}{kT'}\right) dT'\right] \tag{32}$$

Using the **approximation for the integral** comparing in (32), one gets

$$\frac{n_0}{n_M} = \exp\left[\frac{s}{\beta}\cdot\frac{kT_M^2}{E}(1-\Delta_M)\exp\left(-\frac{E}{kT_M}\right)\right] \tag{33}$$

Inserting now in (33) the **condition at the maximum**, one obtains

$$\frac{n_0}{n_M} = \exp(1-\Delta_M) \approx (1-\Delta_M)\cdot e \tag{34}$$

Using now the following equation

$$\frac{I_M}{n_M} = s\exp\left(-\frac{E}{kT_M}\right)$$

and inserting in it the condition at the maximum, one has

$$\frac{I_M}{n_M} = \frac{E\beta}{kT_M^2} \tag{35}$$

Inserting now Eqs. (34) and (35) in Eq. (31), one obtains

$$(1-\Delta_M)\cdot e - 1 = \frac{\tau}{C_\tau\cdot\beta}\cdot\frac{E\cdot\beta}{k\cdot T_M^2}$$

which gives

$$E = C_\tau\cdot\frac{k\cdot T_M^2}{\tau}\left[e\cdot(1-\Delta_M)-1\right] \tag{36}$$

Chen determined a value equal to 0.885 for the constant C_τ, so that Eq. (36) becomes

$$E = 1.52\frac{kT_M^2}{\tau}(1-1.58\Delta_M) \tag{37}$$

which is the Halperin and Braner´s corrected expression.
Because this equation needs iterative calculations, Chen gave a new expression, using another approximation for $\frac{n_0}{n_M}$. In fact

$$\frac{n_0}{n_M} \approx \frac{e}{1+\Delta_M}$$

so that Eq. (32) becomes

$$\frac{e}{1+\Delta_M} - 1 = \frac{\tau}{C_\tau \cdot \beta} \cdot \frac{E \cdot \beta}{k \cdot T_M^2}$$

and then

$$E\left(\frac{1+\Delta_M}{1-0.58\Delta_M}\right) = 1.72\left(\frac{k \cdot T_M^2 \cdot C_\tau}{\tau}\right) \tag{38}$$

because $0.58\Delta_M$ is very small, one can write

$$\frac{1+\Delta_M}{1-0.58\Delta_M} \approx (1+\Delta_M)(1+0.58) \approx 1+1.58\Delta_M$$

by neglegting the second power of Δ_M. So, Eq. (38) becomes

$$E\left[1+1.58\left(\frac{2kT_M}{E}\right)\right] = 1.72\left(\frac{C_\tau \cdot k \cdot T_M^2}{\tau}\right)$$

and in final form

$$E_\tau = 1.52\left(\frac{kT_M^2}{\tau}\right) - 1.58(2kT_M) \tag{39}$$

where $C_\tau = 0.919$.

Eq. (39) is more useful than Eq. (37) because no iterative processes are necessary.

- *Second-order kinetics*

Remembering Eq. (25) and inserting in it the maximum condition, one has

$$n_M = n_0\left(\frac{2kT_M^2 n_0 s'}{E\beta}\right)\exp\left(-\frac{E}{kT_M}\right)$$

Using the approximation given by Eq. (17), the previous equation gives

$$\frac{n_0}{n_M} = \frac{2}{1+\Delta_M} \tag{40}$$

Inserting Eq. (40) into Eq. (31), one obtains

$$\frac{2}{1+\Delta_M} - 1 = \left(\frac{\tau}{C_\tau \cdot \beta}\right)\left(\frac{I_M}{n_M}\right) \tag{41}$$

Inserting now Eq. (27) in Eq. (41), one has

$$\frac{2}{1+\Delta_M} - 1 = \frac{\tau \cdot E}{2C_\tau k T_M^2} \tag{42}$$

which, using the following approximation

$$\frac{2}{1+\Delta_M} \approx 2(1-\Delta_M)$$

gives the Halperin and Braner's corrected expression

$$E = 1.813\left(\frac{kT_M^2}{\tau}\right)(1-2\Delta_M) \tag{43}$$

where $C_\tau = 0.906$.

As before, Chen found a new expression without any iterative calculation. Indeed, Eq. (42) can be rewritten as

$$\frac{E(1+\Delta_M)}{1-\Delta_M} = \frac{2C_\tau k T_M^2}{\tau} \tag{44}$$

which can be changed using the approximation

$$\frac{1+\Delta_M}{1-\Delta_M} \approx 1 + 2\Delta_M$$

so that Eq. (44) becomes

$$E_\tau = \left(\frac{2C_\tau k T_M^2}{\tau}\right) - 2(2kT_M)$$

from which, using $C_\tau = 0.906$, the more convenient expression is obtained

$$E_\tau = 1.813\frac{kT_M^2}{\tau} - 4kT_M \tag{45}$$

References

1. Chen R., J. Appl. Phys. **40** (1969) 570
2. Lushchik C.B., Soviet Phys. JEPT **3** (1956) 390
3. Halperin A. and Braner A.A., Phys. Rev. **117** (1960) 408

Peak shape method (Chen: general-order kinetics)

Chen gave a method of calculating the activation energy for cases whose order is not necessarily first or second but rather may be a non-integer value [1]. This method is again based on measuring the maximum and the half intensity temperatures. From the equation proposed by May and Partridge for the **general order**

$$\frac{dn}{dt} = -s''n^b \exp\left(-\frac{E}{kT}\right) \tag{1}$$

where b is the order of the kinetics and s'' is the pre-exponential factor expressed in $cm^{3(b-1)}sec^{-1}$.

The solution of Eq. (1) is given by

$$I(T) = sn_0 \exp\left(-\frac{E}{kT}\right)\left[1+\frac{s(b-1)}{\beta}\int_{T_0}^{T}\exp\left(-\frac{E}{kT'}\right)dT'\right]^{-\frac{b}{b-1}} \tag{2}$$

where $s = s''n_0^{b-1}$ has the units of sec^{-1}.

This equation can be rearranged using the **condition at the maximum** and then solved numerically by **approximating the integral** by a certain number of terms of the asymptotic series and using the iterative Newton-Raphson method.

The Chen's method consists of finding the temperature at the maximum, T_M, by computer calculations for given values of b, s, E, β. The used values were

$$0.7 \le b \le 2.5$$

$$10^5 s^{-1} \le s \le 10^{13} s^{-1}$$

$$0.1eV \le E \le 1.6eV$$

$$\beta = 0.5Ks^{-1}$$

Once the value of T_M is found, the intensity I_M can be found by inserting the temperature at the maximum in Eq. (2) and using again the asymptotic series. The values T_1, T_2, the low and high temperatures of half intensity, can be calculated by solving numerically the equation

$$I(T) = \frac{I_M}{2}$$

Using the asymptotic series for the integral approximation, Chen found

$$T_1 = 0.95 T_M$$

$$T_2 = 1.05 T_M$$

After that , the **geometrical parameters** of the peak, i.e. $\delta, \tau, \omega, \mu_g$, are found.

Interpolating and extrapolating the constants appearing in the equations for the first- and second-order, Chen gave a general expression which summarises all the previously given expressions.

The equations can be summed up as:

$$E_\alpha = c_\alpha \left(\frac{kT_M^2}{\alpha} \right) + b_\alpha \left(2kT_M \right) \tag{3}$$

where α is τ, δ or ω. The values of c_α and b_α are summarised as below

$$c_\tau = 1.51 + 3.0(\mu_g - 0.42) \qquad b_\tau = 1.58 + 4.2(\mu_g - 0.42)$$

$$c_\delta = 0.976 + 7.3(\mu_g - 0.42) \qquad b_\delta = 0$$

$$c_\omega = 2.52 + 10.2\left(\mu_g - 0.42\right) \qquad b_\omega = 1$$

with $\mu_g = 0.42$ for 1st order

and $\mu_g = 0.52$ for 2nd order

Chen [1] calculated a graph of μ_g, ranging from 0.36 to 0.55 for values of b between 0.7 and 2.5 which can be used for the evaluation of b from a measured μ_g (see Fig. 1).

Another graph has been proposed by Balarin [2] which gives the kinetics order as a function of $\gamma = \delta/\tau$ (Fig. 2).

Once the activation energy is obtained, one can find the frequency factor using the following equation

$$s=\left(\frac{\beta}{T_M^2}\right)\left(\frac{E}{k}\right)\frac{1}{1+(b-1)(\frac{2kT_M}{E})}\exp\left(\frac{E}{kT_M}\right)$$

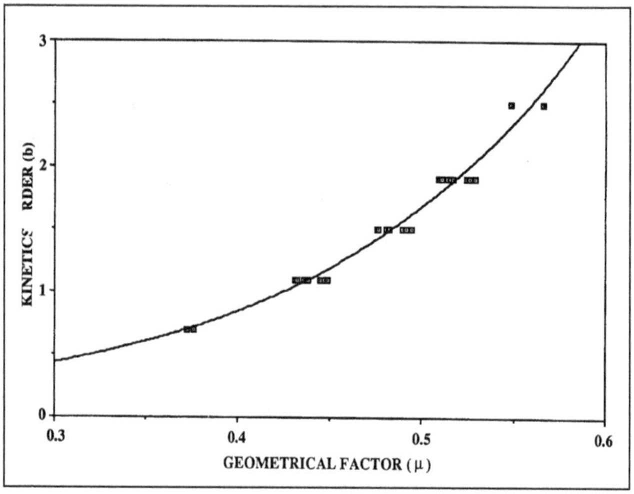

Fig. 1. Plot of the kinetics order b as a function of the geometrical factor $\mu_g = \delta/\omega$ [3].

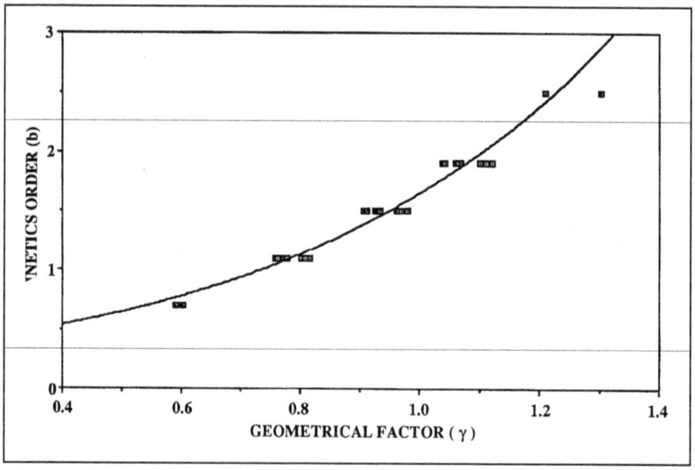

Fig. 2. Plot of the kinetics order b as a function of the geometrical factor $\gamma = \delta/\tau$ [3].

References

1. Chen R., J. Electrochem. Soc.: Solid State Science, **116** (1969) 1254
2. Balarin M., Phys. Stat. Sol. (a), **54** (1979) K137
3. Furetta C. and Weng P.S., Operational Thermoluminescence Dosimetry, World Scientific Pub. (1998)

Peak shape method (Christodoulides: first- and general-order)

Christodoulides [1] developped some expressions for the determination of the activation energy, *E*, of a first order peak, using the widths or half-widths of the peaks. These widths correspond to temperatures at which the signal level is 1/4, 1/2 or 3/4 of the peak height, on both sides of the peak temperature at the maximum, T_M. Fig. 1 shows the various temperatures previously defined.

The expressions are valid in the region of E/kT_M values between 10 and 100. Using the **first order kinetics equation** giving the variation of light intensity with temperature, and inserting in it the equation of the maximum for a constant heating rate, one gets the following expression in terms of the variable $\varepsilon = E/kT$ and its value at the peak maximum, $\varepsilon_M = E/kT_M$:

$$I = n_0 s \exp\left\{-\varepsilon - \varepsilon_M^2 \exp(\varepsilon_M) \int_{\varepsilon}^{\infty} \frac{\exp(-\varepsilon)}{\varepsilon^2} d\varepsilon\right\} \tag{1}$$

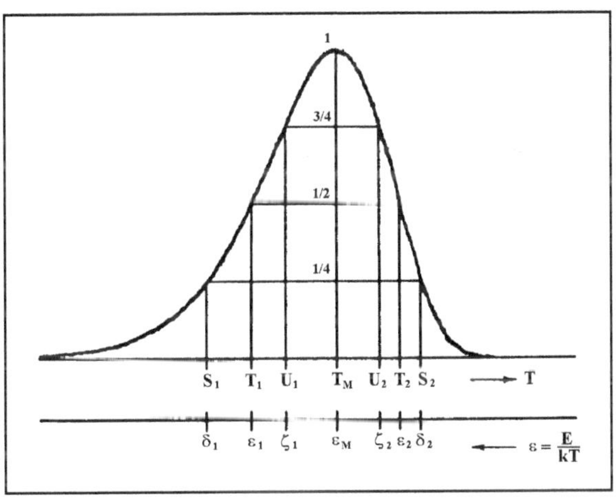

Fig. 1. A glow-peak and the various temperatures used in the method.

The integral in Eq. (1) can be expressed in terms of the second exponential integral

$$E_2(\varepsilon)=\int_1^{\infty}\frac{\exp(-\varepsilon t)}{t^2}dt=\varepsilon\int_{\varepsilon}^{\infty}\frac{\exp(-z)}{z^2}dz \tag{2}$$

So that Eq. (1) becomes

$$I=n_0 s\exp\left\{-\varepsilon-\varepsilon_M^2\exp(\varepsilon_M)\frac{E_2(\varepsilon)}{\varepsilon}\right\} \tag{3}$$

which has a maximum value, for $\varepsilon=\varepsilon_M$ equal to

$$I_M=n_0 s\exp\{-\varepsilon_M-\varepsilon_M\exp(\varepsilon_M)E_2(\varepsilon_M)\} \tag{4}$$

and finally

$$I=I_M\exp\left\{-(\varepsilon-\varepsilon_M)-\varepsilon_M^2\exp(\varepsilon_M)\left[\frac{E_2(\varepsilon)}{\varepsilon}-\frac{E_2(\varepsilon_M)}{\varepsilon_M}\right]\right\} \tag{5}$$

which gives the signal output I as a function of ε for a TL peak of given I_M and ε_M. For a given value of ε_M, the ratio I/I_M depends only on ε. An iteration procedure is needed to solve Eq. (5) because one must be sure about the convergence of the procedure itself. This is done using ε_M as a starting value of ε. Eq. (5) may be rewritten as

$$\begin{aligned}(\varepsilon-\varepsilon_M)+\left(\frac{\varepsilon_M}{\varepsilon}\right)^2\exp(\varepsilon_M-\varepsilon)[\varepsilon\exp(\varepsilon)E_2(\varepsilon)]=\\ =[\varepsilon_M\exp(\varepsilon_M)E_2(\varepsilon_M)]+\ln\left(\frac{I_M}{I}\right)\end{aligned} \tag{6}$$

Using tables of exponential integrals [2], one finds that for $5 \le \varepsilon \le \infty$ the quantity $\varepsilon\cdot\exp(\varepsilon)\cdot E_2(\varepsilon)$ is ranging between 0.7 and 1: therefore it is of order unity.

- for $T < T_M$, i.e. $\varepsilon > \varepsilon_M$, we have $\exp(\varepsilon_M - \varepsilon) < 1$ and the term $(\varepsilon - \varepsilon_M)$ dominates in Eq. (6):

$$\varepsilon=\ln\left(\frac{I_M}{I}\right)+\varepsilon_M\left[1+\exp(\varepsilon_M)E_2(\varepsilon_M)\right]$$
$$-\left(\frac{\varepsilon_M}{\varepsilon}\right)^2\exp(\varepsilon_M)\exp(-\varepsilon)\left[\varepsilon\exp(\varepsilon)E_2(\varepsilon)\right] \tag{7}$$

because $\varepsilon\cdot\exp(\varepsilon)\cdot E_2(\varepsilon)\cong 1$ and then $\exp(-\varepsilon)\cong\varepsilon\cdot E_2(\varepsilon)$:

$$\varepsilon=\ln\left(\frac{I_M}{I}\right)+\varepsilon_M\left[1+\exp(\varepsilon_M)E_2(\varepsilon_M)\right]-\left(\frac{\varepsilon_M}{\varepsilon}\right)^2\exp(\varepsilon_M)\varepsilon E_2(\varepsilon) \tag{8}$$

- for $T > T_M$, i.e., $\varepsilon < \varepsilon_M$, $\exp(\varepsilon - \varepsilon_M)$ dominates over $(\varepsilon - \varepsilon_M)$ in Eq. (6):

$$\varepsilon=\varepsilon_M-\ln\left\{\frac{\ln\left(\frac{I_M}{I}\right)+\varepsilon_M\left[1+\exp(\varepsilon_M)E_2(\varepsilon_M)\right]-\varepsilon}{\varepsilon_M^2\exp(\varepsilon)E_2(\varepsilon)/\varepsilon}\right\} \tag{9}$$

In expressions (8) and (9) a rational approximation may be used for the trascendental function $E_2(\varepsilon)$. Such an expression is [3]

$$\exp(\varepsilon)E_2(\varepsilon)=\frac{0.99997\varepsilon+3.03962}{\varepsilon^2+5.036373\varepsilon+4.19160}+\Delta(\varepsilon) \tag{10}$$

where $\left|\Delta(\varepsilon)\right|<10^{-7}$ for $\varepsilon \geq 10$.

The values of ε, corresponding to I/I_M = ¼, ½ and ¾, are defined in Table 1 along the corresponding temperatures.

I/I_M	ε	T(K)	peak zone
¼	δ_1	S_1	low temperature side of the peak
½	ε_1	T_1	
¾	ξ_1	U_1	
1	ε_M	T_M	maximum
¾	ξ_2	U_2	high temperature side of the peak
½	ε_2	T_2	
¼	δ_2	S_2	

Table 1. Definitions of the ε and corresponding temperatures.

Simple linear relations can be then searched for connecting pairs of the quantities $(\varepsilon_1, \varepsilon_M)$, $(\varepsilon_2, \varepsilon_M)$, $(\varepsilon_1, \varepsilon_2)$, $(\delta_1, \varepsilon_M)$, $(\delta_2, \varepsilon_M)$, , (δ_1, δ_2), (δ_1, ξ_1) and (δ_2, ξ_2). Similar expressions are also given which allow to know the width of a peak whose E and T_M are known.

References

1. Christodoulides C., J. Phys. D: Appl. Phys. **18** (1985) 1501
2. Abromowitz M. and Stegun I.A., Handbook of Mathematical Functions. Dover, N.Y. (1955)
3. Hastings C. Jr, Approximation for digital computers. Univ. Press. Princepton (1955)

Peak shape method (Gartia, Singh & Mazumdar: (b) general order)

These Authors presented a new set of expressions for general order [1]. The prior knowledge of the kinetics order is required. The method uses any points of a peak. The mathematical procedure is similar to the one already given in **Mazumdar, Singh & Gartia peak shape method (a)**. Using Eq. (1) for b = 1, given in **Peak shape method (Christodoulides: first- and general-order)** [2], and Eqs. (12) and (13) for b $\neq$ 1, given in **Mazumdar, Singh & Gartia peak shape method (a)** [3], and solving them by an iterative method, it is possible to write the following expression for the activation energy

$$E = \frac{CkT_M^2}{|T_x - T_y|} + DkT_M \tag{1}$$

where

$$|T_x - T_y| = \tau, \delta, or \omega$$

The coefficients C and D are found using the method of least squares for different order of kinetics b in the range from 0.7 to 2.5 and for x = 1/2, 2/3 and 4/3. For a particular value of x the coefficients result to be dependent on b and then can be expressed as a quadratic function of b itself. So that, the previous equation can be rewritten as

$$E = \frac{\left(C_0 + C_1 b + C_2 b^2\right) kT_M^2}{|T_x - T_y|} + (D_0 + D_1 b + D_2 b^2) kT_M \tag{2}$$

Table 1 gives the coefficients for different values of x.

The Authors claim the validity and the superiority of their method in comparison to those of Chen. Indeed, the E values obtained by using expression for x = 1/2 are more accurate than those of Chen. Furthermore, it is pointed out that E_τ, E_δ and E_ω are in excellent agreement among themselves, whereas the Chen's values for E_δ and E_ω yeld poor results.

ratio	parameter	C_0	C_1	C_2	D_0	D_1	D_2
1/2	τ	1.019	0.504	-0.066	-1.059	-1.217	0.109
	δ	0.105	0.926	-0.048	0.154	-0.205	-0.128
	ω	1.124	1.427	-0.113	-0.902	-0.346	-0.061
2/3	τ	0.684	0.426	-0.055	-0.720	-1.21	0.098
	δ	0.146	0.683	-0.048	0.184	-.0.432	-0.094
	ω	0.830	1.108	-0.103	-0.529	-0.607	-0.029
4/5	τ	0.449	0.342	-0.043	-0.480	-1.184	0.085
	δ	0.153	0.487	-0.041	0.180	-0.606	-0.062
	ω	0.602	0.829	-0.084	-0.293	-0.777	-0.006

Table 1. Numerical values of the coefficients comparing in Eq. (2).

References

1. Gartia R.K., Singh S.J. and Mazumdar P.S., Phys. Stat. Sol.(a) **114** (1989) 407
2. Christodoulides C., J. Phys. D: Appl. Phys. **18** (1985) 1501
3. Mazumdar P.S., Singh S.J. and Gartia R.K., J. Phys. D: Appl. Phys **21** (1988) 815

Peak shape method (Grossweiner: first order)

Grossweiner [1] was the first to use the shape of the glow-peak to calculate the trap depth and the frequency factor. His method is based on the temperature at maximum and on the low temperature at half intensity, T_1. Using the first order-kinetics one can write

$$I_M = I(T_M) = n_0 s \exp(-\frac{E}{kT_M}) \exp\left[-\frac{s}{\beta}\int_{T_0}^{T_M} \exp(-\frac{E}{kT})dT\right] \tag{1}$$

$$I_{\frac{M}{2}} = I(T_1) = n_0 s \exp(-\frac{E}{kT_1}) \exp\left[-\frac{s}{\beta}\int_{T_0}^{T_1} \exp(-\frac{E}{kT})dT\right] \tag{2}$$

their ratio is

$$\frac{1}{2} = \exp\left[-\frac{E}{k}\left(\frac{1}{T_1} - \frac{1}{T_M}\right)\right] \exp\left[\frac{s}{\beta}\int_{T_1}^{T_M} \exp\left(-\frac{E}{kT}\right)dT\right] \tag{3}$$

The integral in brackets can be resolved by asymptotic expansion as indicated before. By dropping terms after the first in the series, expression (3) changes in the following

$$\frac{1}{2}=\exp\left[-\frac{E}{k}\left(\frac{1}{T_1}-\frac{1}{T_M}\right)\right]\exp\left[\frac{s}{\beta}\frac{kT_M^2}{E}\exp(-\frac{E}{kT_M})-\frac{s}{\beta}\frac{kT_1^2}{E}\exp(-\frac{E}{kT_1})\right] \quad (4)$$

Doing the logarithm and rearranging using the maximum condition one obtains

$$\frac{E}{k}\left(\frac{1}{T_1}-\frac{1}{T_M}\right)-0.693=1-\left(\frac{T_1}{T_M}\right)^2\exp(-\frac{E}{kT_1}+\frac{E}{kT_M}) \quad (5)$$

For E/kT larger than 20, the exponential of the last expression becomes equal to 0.184. Furthermore, the term $(T_1/T_M)^2$ may be neglected because it affects E by less than 2% if s/β is larger than 10^7. These approximations get the final form

$$E=1.51k\frac{T_M T_1}{T_M-T_1} \quad (6)$$

This expression was empirically modified by Chen [2] with 1.41 replacing 1.51 to get a better accuracy in the calculation of E, i.e.

$$E=1.41k\frac{T_M T_1}{T_M-T_1} \quad (7)$$

The frequency factor can be directly obtained by the following expression

$$s=\frac{1.41\beta T_1}{\tau T_M}\exp\left(\frac{1.41T_1}{\tau}\right) \quad (8)$$

References

1. Grossweiner L.I., J. Appl. Phys. **24** (1969) 1306
2. Chen R., J. Appl. Phys. **24** (1969) 570

Peak shape method (Halperin-Braner)

Halperin and Braner [1] proposed a method, for determining the activation energy, based on the temperatures on either side of the temperature at the maximum, corresponding to the half maximum intensity of the peak. They considered the luminescence emission as mainly due to two different kinds of recombination processes. In the first one the electrons raise to an excited state within the forbidden gap below the conduction band and recombine with holes by tunnelling process (model A). In the second process, the recombination takes place via conduction band (model B). Figs. 1and 2 shows the two recombination processes treated in the text.

The kinetics equations are formuled as showed below:

Model A

$$-\frac{dm}{dt} = mn_e A_m \tag{1a}$$

$$-\frac{dn}{dt} = \gamma n - sn_e \tag{1b}$$

$$-\frac{dn_e}{dt} = \gamma n - n_e\left(mA_m + s\right) \tag{1c}$$

Model B

$$-\frac{dm}{dt} = mn_c A_m \tag{2a}$$

$$-\frac{dn}{dt} = \gamma n - n_c\left(N - n\right)A_n \tag{2b}$$

$$-\frac{dn_c}{dt} = \gamma n - n_c mA_m + \left(N - n\right)A_n \tag{2c}$$

where

n_e = concentration of electrons in the excited states N_e,

n_c = concentration of electrons in the conduction band (CB),

n = number of trapped electrons in the elctron traps N,

m = number of trapped holes in hole traps M,

A_m = probability of recombination,

A_n = probability of retrapping,

$\gamma = s\exp\left(-\frac{E}{kT}\right)$ = probability of thermal excitation,

s = frequency factor or probability per second of retrapping.

In Model B one assumes that s is temperature dependent, i.e. $s = s_0 T^2$.

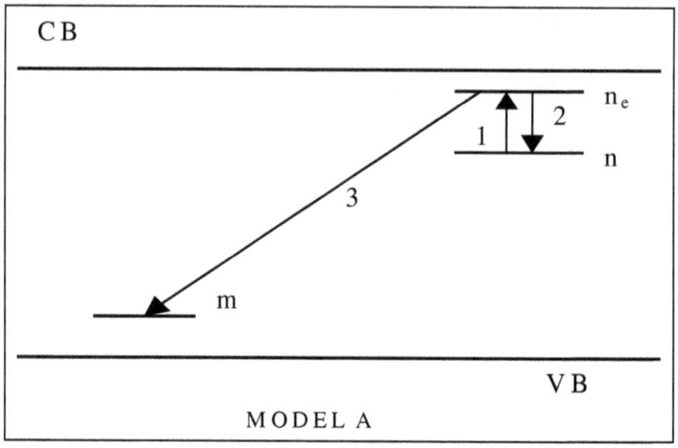

Fig. 1. Electrons raise from N to an excited state within the forbidden gap below the conduction band (N_e) and recombine with holes, in M, by tunnelling process.

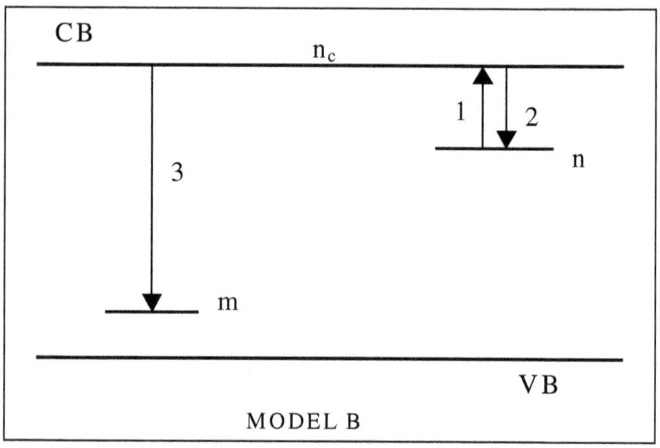

Fig. 2. Recombination process takes place via conduction band.

Assuming that transitions 2 or 3 are fast enough, one can put

$$\frac{dn_e}{dt} = 0 \qquad \text{for model A,} \tag{3}$$

and

$$\frac{dn_c}{dt} = 0 \qquad \text{for model B} \tag{4}$$

The neutrality condition is expressed by

$$n_0 - n = m_0 - m \tag{5}$$

where n_0 and m_0 being the concentration at time t_0 and n, m at time t.

<u>Model A</u>

In this case, from Eq. (1c), using the assumption of Eq. (3), one has

$$n_e = \frac{\gamma n}{mA_m + s} \tag{6}$$

Eq. (1a) becomes

$$-\frac{dm}{dt} = \frac{\gamma nmA_m}{mA_m + s} \tag{7}$$

To find an expression for the activation energy, E, one defines the ratio of the initial concentrations of trapped electrons to trapped holes:

$$\rho = \frac{n_0}{m_0} \succ 1 \tag{8}$$

Let us introduce now the following paprameters:

$$\mu = \frac{m}{m_0} \tag{9}$$

$$\chi = \frac{N}{n_0} \tag{10}$$

$$A = A_m \tag{11}$$

$$B = \frac{s}{m_0} \tag{12}$$

from the neutrality condition gives by Eq. (5) one obtains

$$n = n_0 + m - m_0$$

$$n = m_0(\rho + \mu - 1) \tag{13}$$

from Eq. (9) one obtains

$$\frac{dm}{dt} = m_0 \frac{d\mu}{dt}$$

and so Eq. (7) becomes

$$-m_0 \frac{d\mu}{dt} = \frac{\gamma n m A_m}{m A_m + s} \tag{14}$$

Rearranging and inserting Eqs. (9), (11), (12) and (13), one gets

$$-\frac{d\mu}{dt} = \gamma A \frac{m m_0(\rho + \mu - 1)}{m_0(mA + s)}$$

or better

$$-\frac{d\mu}{dt} = \frac{\mu A(\rho + \mu - 1)}{(\mu A + B)} \tag{15}$$

From Eq. (1a) one has now

$$I = -\frac{dm}{dt} = -m_0 \frac{d\mu}{dt} \tag{16}$$

Using a linear heating rate $\beta = dT/dt$, Eq. (16) becomes

$$I = -\beta m_0 \frac{d\mu}{dT} \tag{17}$$

and again, using Eq. (15), one obtains

$$\frac{I}{\beta m_0} = -\frac{d\mu}{dT} = \left(\frac{A\gamma\mu}{\beta}\right)\frac{\rho + \mu - 1}{\mu A + B} \tag{18}$$

Now, Halperin and Braner introduced the following parameters concerning an isolated TL glow peak.

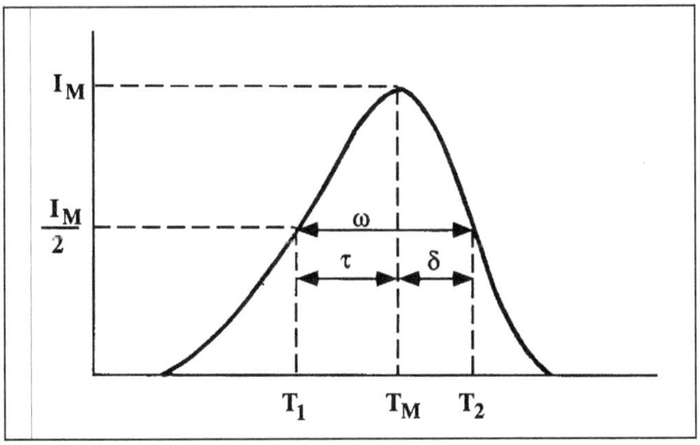

Fig. 3. The geometrical parameters characterezing an isolated peak.

According to the Fig. 3, the defined parameters are:

T_M, T_1, T_2: are respectively the peak temperature at the maximum and the temperatures on either side of the temperature at the maximum, corresponding to half intensity,

$\tau = T_M - T_1$: is the half-width at the low temperature side of the peak,

$\delta = T_2 - T_M$: is the half-width towards the fall-off of the glow peak,

$\omega = T \ - T_1$: is the total half-width,

$\mu_g = \dfrac{\delta}{\omega}$: is the symmetrical geometrical factor.

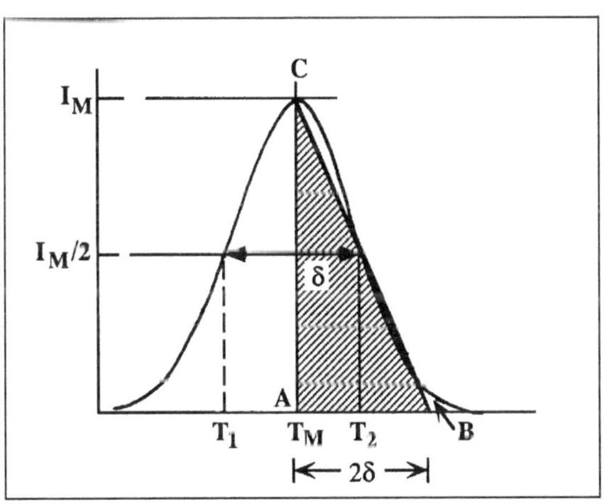

Fig. 4. A glow peak assimilated to a triangle.

Considering Fig. 4, where a glow peak may be regarded as a triangle, the concentration of the carriers at the maximum, n_M, can be calculated, with a good approximation, as

$$n_M = \frac{1}{\beta}\int_{T_M}^{\infty} I dT \approx AREA(ABC) = \frac{1}{2\beta} I_M 2\delta = \frac{I_M \delta}{\beta} \tag{19}$$

where I_M is the maximum intensity.

Then, from Eq. (18), calculated at the maximum of the glow peak, it follows

$$\mu_M = \frac{m_M}{m_0} \approx \frac{I_M \delta}{\beta m_0} \tag{20}$$

Hence, Eq. (18), calculated at the maximum, becomes

$$\frac{\mu_M}{\delta} = \frac{I_M}{\beta m_0} = \left(-\frac{d\mu}{dT}\right)_M = \frac{A\gamma_M \mu_M (\rho + \mu_M - 1)}{\rho(\mu_M A + B)} \tag{21}$$

Taking now its logarithm:

$$\ln\left(\frac{I}{\beta m_0}\right) = \ln\mu + \ln(\rho + \mu - 1) - \ln(A\mu + B) + \ln\left(\frac{As}{\beta}\right) - \frac{E}{kT}$$

and equating its temperature derivative at maximum to zero, one gets

$$\left(\frac{d\mu}{dT}\right)_{T_M} \cdot \left(\frac{1}{\mu_M} + \frac{1}{\rho + \mu_M - 1} - \frac{A}{A\mu_M + B}\right) + \frac{E}{kT_M^2} = 0 \tag{22}$$

Inserting now Eq. (21) into Eq. (22) and rearranging, one gets

$$\frac{E}{kT_M^2} = \frac{1}{\delta}\left(\frac{\mu_M}{\rho + \mu_M - 1} + \frac{A\mu_M + B - A\mu_M}{A\mu_M + B}\right)$$

from which

$$E = H\frac{kT_M^2}{\delta} \tag{23}$$

where

$$H = \frac{\mu_M}{\rho + \mu_M - 1} + \frac{B}{A\mu_M + B} \tag{24}$$

with the approximation

$$\mu_M = \frac{m_M}{m_0} = \frac{\delta}{\omega} \tag{25}$$

where ω is the half-intensity width of the peak.

Model B

For this model, using the condition expressed by Eq. (4), one obtains

$$n_c = \frac{\gamma n}{mA_n + (N-n)A_n} \tag{26}$$

and therefore

$$-\frac{dm}{dt} = mn_c A_m = \frac{\gamma mnA_m}{mA_m + (N-n)A_n} \tag{27}$$

The following expression are now introduced

$$A^* = A_m - A_n \tag{28}$$

$$B^* = A_n(\rho\chi - \rho + 1) \tag{29}$$

Also in this case it is considered $\rho \succ 1$.

The expression $N - n$ can be trasformed using Eqs. (9) and (13):

$$N - n = m\left(\frac{N}{m} - \frac{n}{m}\right) - m_0\mu\left(\frac{N}{m_0\mu} - \frac{m_0(\rho+\mu-1)}{m_0\mu}\right) - \\ = m_0(\rho\chi - \mu - \rho + 1) \tag{30}$$

By introducing Eq. (30) into Eq. (27) one gets

$$-m_0\left(\frac{d\mu}{dt}\right) = \gamma A_m \frac{m_0\mu(\rho+\mu-1)}{\mu A_m + A_n(\rho\chi - \mu - \rho + 1)}$$

and rearranging

$$-\frac{d\mu}{dt}=\frac{\gamma A_m\mu(\rho+\mu-1)}{A^*\mu+B^*} \tag{31}$$

Using a linear heating rate β, Eq. (31) becomes

$$-\frac{d\mu}{dT}=\left(\frac{\gamma A_m}{\beta}\right)\frac{\mu(\rho+\mu-1)}{A^*\mu+B^*} \tag{32}$$

Using Eq. (17) for the intensity, Eq. (32) gives

$$\frac{I}{\beta m_0}=-\frac{d\mu}{dT}=\left(\frac{\gamma A_m}{\beta}\right)\frac{\mu(\rho+\mu-1)}{A^*\mu+B^*} \tag{33}$$

Using the approximation expressed by Eq. (20), Eq. (33), calculated at the maximum, becomes

$$\frac{\mu_M}{\delta}=\frac{I_M}{\beta m_0}=\left(-\frac{d\mu}{dT}\right)_{T=T_M}=\frac{\gamma_M A_m\mu_M(\rho+\mu_M-1)}{\mu_M A^*+B^*} \tag{34}$$

The logarithm of Eq. (33) yields now

$$\ln\left(\frac{I}{\beta m_0}\right)=\ln\mu+\ln(\rho+\mu-1)-\ln\left(\mu A^*+B^*\right)+2\ln T-\frac{E}{kT}+\cos$$

and its derivative at maximum equating to zero is

$$\left(\frac{1}{\mu_M}+\frac{1}{\rho+\mu_M-1}-\frac{A^*}{\mu_M A^*+B^*}\right)\left(\frac{d\mu}{dT}\right)_{T_M}+$$
$$+\frac{E}{kT_M^2}\left(1+\frac{2kT_M^2}{E}\right)=0 \tag{35}$$

Inserting Eq. (34) into Eq. (35), one has

$$-\left(\frac{\rho+\mu_M-1+\mu_M}{\mu_M(\rho+\mu_M-1)}-\frac{A^*}{\mu_M A^*+B^*}\right)\left(\frac{\mu_M}{\delta}\right)+\frac{E}{kT_M^2}(1+\Delta_M)=0$$

where $\Delta_M = \dfrac{2kT_M}{E}$. Rearranging the last expression we get

$$\left(\frac{E}{kT_M^2}\right)(1+\Delta_M) = \left(\frac{1}{\delta}\right)\left(\frac{\mu_M}{\rho+\mu_M-1}+\frac{B^*}{\mu_M A^*+B^*}\right) \tag{36}$$

by using the parameter

$$H = \frac{\mu_M}{\rho+\mu_M-1}+\frac{B^*}{A^*\mu_M+B^*}$$

Eq. (36) becomes

$$\frac{E}{kT_M^2} = \frac{H}{\delta}\left(\frac{1}{1+\Delta_M}\right)kT_M^2 \tag{37}$$

Since $\Delta_M \prec\prec 1$, we can write

$$(1-\Delta_M)^{-1} \approx 1-\Delta_M$$

and Eq. (37) becomes

$$E = \left(\frac{HkT_M^2}{\delta}\right)(1-\Delta_M) \tag{38}$$

The values of H are different for the first and second order kinetics. Exactly:

$$H = 1.72\cdot\left(\frac{\mu_M}{1-\mu_M}\right)(1-1.58\cdot\Delta_M) \quad \text{first order} \tag{39}$$

$$H = 2\cdot\left(\frac{\mu_M}{1-\mu_M}\right)(1-2\cdot\Delta_M) \quad \text{second order} \tag{40}$$

Halperin and Braner gave also a very easy way to decide the type of kinetics involved in the process.
If

$$\mu_M \le \frac{1+\Lambda_M}{e} \tag{41}$$

the process is of the first order, while if

$$\mu_M \succ \frac{1+\Delta_M}{e} \tag{42}$$

the process is of the second order.

Eqs. (39) and (40) can be changed by introducing the half-width at the low temperature side of the peak

$$\tau = \omega - \delta \tag{43}$$

This is very useful because if it is easy to eliminate any interferring glow appearing at low temperature side, it is impossible to eliminate shouldering peaks at the high temperature side of the observed peak. Using Eq. (43) and $\mu_M = \delta/\omega$ in Eq. (39) and Eq. (40), Eq. (38) becomes

$$E = \frac{1.72 \cdot kT_M^2}{\tau}(1 - 2.58 \cdot \Delta_M) \quad \text{for the 1}^{\text{st}} \text{ order} \tag{44}$$

and

$$E = \frac{2 \cdot kT_M^2}{\tau}(1 - 3 \cdot \Delta_M) \quad \text{for the 2}^{\text{nd}} \text{ order} \tag{45}$$

The equations of Halperin and Braner require iterative process to find E owing the presence of Δ_M. To overcome this difficulty a new approximated method was proposed by Chen [2] without any iterative process (see **Peak shape method. Chen: first- and second-order**).

References

1. Halperin A. and Braner A.A., Phys. Rev. **117** (1960) 408
2. Chen R., J. Appl. Phys. **40** (1969) 570

Peak shape method (Lushchik: first and second order)

Lushchik [1] also proposed a method based on the glow-peak shape for both first- and second-order kinetics. Introducing the parameter $\delta = T_2 - T_M$, a glow-peak can be approximated to a triangle as shown in Fig. 1.

In this case, with a good approximation, one has

$$n_M = \frac{1}{\beta}\int_{T_M}^{\infty} I dT \cong \frac{1}{\beta} AREA\Delta(ABC) = \frac{I_M 2\delta}{2\beta} = \frac{I_M \delta}{\beta} \tag{1}$$

where n_M is the carriers concentration at the maximum.

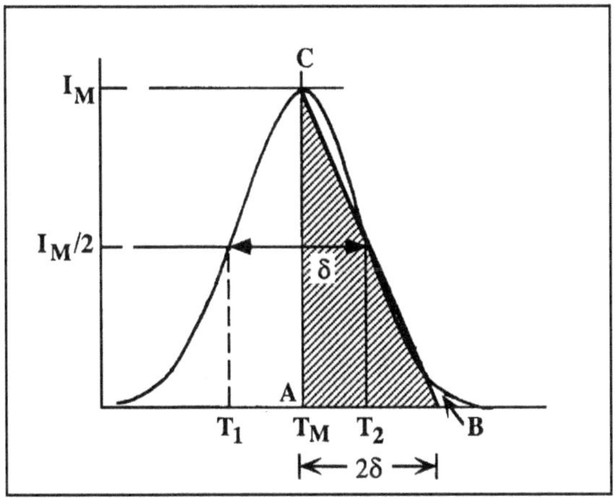

Fig. 1. Approximation of a glow-peak in a triangle.

For the first-order kinetics, the equation

$$I = cpn$$

at the maximum point becomes

$$\frac{I_M}{n_M} = s\exp\left(-\frac{E}{kT_M}\right) \tag{2}$$

Using the condition at the maximum

$$\frac{\beta E}{kT_M} = s\exp\left(-\frac{E}{kT_M}\right)$$

Eq. (2) gets

$$\frac{I_M}{n_M} = \frac{\beta E}{kT_M^2} \tag{3}$$

The substitution of expression (1) in (3) allows us to obtain the Lushchik's expression for the activation energy for the **first-order process**:

$$E = \frac{kT_M^2}{\delta} \tag{4}$$

The Lushchik formula for a second-order kinetics is obtained using the solution for n, valid for a second order kinetics, replacing n with n_M:

$$n_M = n_0 \left[1 + \frac{s' n_0}{\beta} \int_{T_0}^{T_M} \exp\left(-\frac{E}{kT'} \right) dT' \right]^{-1} \tag{5}$$

Using now the expression for the intensity at the maximum, I_M, and doing the ratio between I_M and n_M, one has

$$\frac{I_M}{n_M} = \frac{s' n_0 \exp\left(-\frac{E}{kT_M} \right)}{1 + \frac{s' n_0}{\beta} \int_{T_0}^{T_M} \exp\left(-\frac{E}{kT'} \right) dT'} \tag{6}$$

The insertion of the maximum condition for the second order in Eq. (6) yields

$$\frac{I_M}{n_M} = \frac{s' n_0 \exp\left(-\frac{E}{kT_M} \right)}{\frac{2kT_M^2 n_0 s'}{\beta E} \exp\left(-\frac{E}{kT_M} \right)} = \frac{\beta E}{2kT_M^2} \tag{7}$$

Using again Eq. (1) and rearranging, the expression of Lushchik for the **second order** is obtained:

$$E = \frac{2kT_M^2}{\delta} \tag{8}$$

Chen [2] modified the two previous equations for a better accuracy in the E value by multiplying by 0.978 Eq. (4) and by 0.853 Eq. (8), i.e.

$$E = 0.978 \frac{kT_M^2}{\delta}$$

$$E = 1.706 \frac{kT_M^2}{\delta}$$

The frequency factor for the first-order process is obtained by the following expression:

$$s = 0.976 \left(\frac{\beta}{\delta} \right) \exp\left(0.976 \frac{T_M}{\delta} \right) \tag{9}$$

References

1. Lushihik L.I., Sov. Phys. JEPT **3** (1956) 390
2. Chen R., J. Appl. Phys. **40** (1969) 570

Peak shape method (Mazumdar, Singh & Gartia: (a) general order)

A new set of expressions, to evaluate the thermal activation energy, E, of a thermoluminescent peak following a general order of kinetics, are given by Mazumadar et al. [1]. The work is an extension of the **peak shape method suggested by Christodoulides** [2]. The involved temperatures are now the ones at which the intensity of the peak is, respectively, 1/2, 2/3 and 4/5 of the maximum. The Authors claim that the selection of these points is based on the fact that the upper half of the peak, in general, is expected to be free from interference from satellite peaks.

Taking into consideration the intentensity at any temperature, T, for a peak obeying a general-order kinetics, given by

$$I(T) = sn_0 \exp\left(-\frac{E}{kT}\right)\left[1+\frac{s(b-1)}{\beta}\int_{T_0}^{T}\exp\left(-\frac{E}{kT'}\right)dT'\right]^{\frac{b}{1-b}} \tag{1}$$

and the condition for the maximum intensity given by

$$\frac{kT_M^2 bs}{\beta E}\exp\left(-\frac{E}{kT_M}\right) = 1+\frac{s(b-1)}{\beta}\int_{T_0}^{T_M}\exp\left(-\frac{E}{kT'}\right)dT' \tag{2}$$

Putting $T_0 = 0$ in it, as well as $\varepsilon = E/kT$ and $\varepsilon_M = E/kT_M$, we have

$$\frac{s}{\beta} = \frac{k}{E}\left\{\frac{\varepsilon_M^2}{b\exp(-\varepsilon) - (b-1)\varepsilon_M^2 J_M}\right\} \tag{3}$$

where

$$J_M = \int_{\varepsilon_M}^{\infty}\frac{\exp(-z)}{z^2}dz \tag{4}$$

Equation (1) becomes then

$$I = sn_0\exp(-\varepsilon)\left\{\frac{b\exp(-\varepsilon_M) + (b \quad 1)(J - J_M)\varepsilon_M^2}{b\exp(-\varepsilon_M) - (b-1)J_M\varepsilon_M^2}\right\}^{-\frac{b}{b-1}} \tag{5}$$

with

$$J = \int_{\varepsilon}^{\infty} \frac{\exp(-z)}{z^2} dz \tag{6}$$

The intensity at the maximum is then given by

$$I_M = sn_0 \exp(-\varepsilon_M) \left\{ \frac{b\exp(-\varepsilon_M)}{b\exp(-\varepsilon_M) - (b-1)J_M \varepsilon_M^2} \right\}^{-\frac{b}{b-1}} \tag{7}$$

In both expressions the integrals are expressed in terms of the second-exponential integral, i.e., $E_2(\varepsilon) = \varepsilon J$.

Finally, one can write

$$\frac{I}{I_M} = \exp(\varepsilon_M - \varepsilon) \left\{ 1 - \left[\frac{b-1}{b} \right] F(\varepsilon, \varepsilon_M) \right\}^{-\frac{b}{b-1}} \tag{8}$$

where

$$F(\varepsilon, \varepsilon_M) = \varepsilon_M^2 \exp(\varepsilon_M) \left[\frac{E_2(\varepsilon_M)}{\varepsilon_M} - \frac{E_2(\varepsilon)}{\varepsilon} \right] \tag{9}$$

Putting now $I/I_M = x$ and $\varepsilon = \varepsilon_x$, one gets

$$x = \exp(\varepsilon_M - \varepsilon_x) \left[1 - \frac{b}{b-1} F(\varepsilon_x, \varepsilon_M) \right]^{-\frac{b}{b-1}} \tag{10}$$

and then

$$\ln x = \varepsilon_M - \varepsilon_x - \left(\frac{b}{b-1} \right) \ln \left[1 - \frac{b}{b-1} F(\varepsilon_x, \varepsilon_M) \right] \tag{11}$$

The procedure is now very similar to the one already used in **Christodoulides' method**.

Indicating $\varepsilon_x = \varepsilon_x^-$ for $T < T_M$:

$$\varepsilon_x^- = \varepsilon_M - \ln x - \frac{b}{b-1} \ln \left[1 - \frac{b-1}{b} F(\varepsilon_x^-, \varepsilon_M) \right] \tag{12}$$

Having now $\varepsilon_x = \varepsilon_x^+$ for $T > T_M$, one gets:

$$\varepsilon_x^+ = \varepsilon_M - \ln\left[\frac{\left[\left(\frac{1}{x}\right)\exp(\varepsilon_M - \varepsilon_x^+)\right]^{\frac{b-1}{b}} - 1 + \frac{b-1}{b}\varepsilon_M \exp(\varepsilon_M) E_2(\varepsilon_M)}{\frac{b-1}{b}\varepsilon_M^2 \exp(\varepsilon_x^+)\frac{E_2(\varepsilon_x^+)}{\varepsilon_x^+}}\right] \tag{13}$$

It must be noted that the above equations are not valid for $b = 1$. For this case the previous equations given by **Christodoulides** have to be used.

For a given value of the ratio I/I_M the corresponding values of ε_x^- and ε_x^+ are then determinated. The iteration procedure is the same already used by **Christodoulides**, using ε_M as a starting value of ε.

Now, if i and j denote the intensity ratios, the expression for the activation energy can be written as

$$E(T_i, T_j) = \frac{T_i T_j}{C(T_i - T_j)} - \frac{T_j}{D} \tag{14}$$

for $T_i > T_j$ where T_i and T_j are the temperature at a given ratio at the falling and rising side respectively of the peak. The values of coefficients C and D are listed in the following Table 1.

Temperature relation	b = 1.0		b = 1.5		b = 2.0	
	C	D	C	D	C	D
$T_M, T_{1/2}^-$	7941	14978	7124	10430	6584	8126
$T_{1/2}^+, T_M$	11779	10001	8372	6351	6585	4577
$T_{1/2}^+, T_{1/2}^-$	4742	11967	3846	7659	3289	5543
$T_M, T_{2/3}^-$	10965	14025	9659	9717	8816	7539
$T_{2/3}^+, T_M$	14816	10351	10926	6687	8817	4890
$T_{2/3}^+, T_{2/3}^-$	6299	11819	5124	7726	4405	5698
$T_M, T_{4/5}^-$	15444	13362	13375	9211	12065	7116
$T_{4/5}^+, T_M$	19304	10690	14653	6997	12067	5176
$T_{4/5}^+, T_{4/5}^-$	8578	11765	6990	7796	6030	5822

Table 1. Numerical values of coefficients C and D comparing in Eq. (14).

References

1. Mazumdar P.S., Singh S.J. and Gartia R.K., J. Phys. D: Appl. Phys. **21** (1988) 815
2. Christodoulides C., J. Phys. D: Appl. Phys. **18** (1985) 1501

Peak shape method (parameters)

An isolated TL glow peak, obtained using a linear heating rate, can be characterized by some parameters as it can be seen in the figure below.

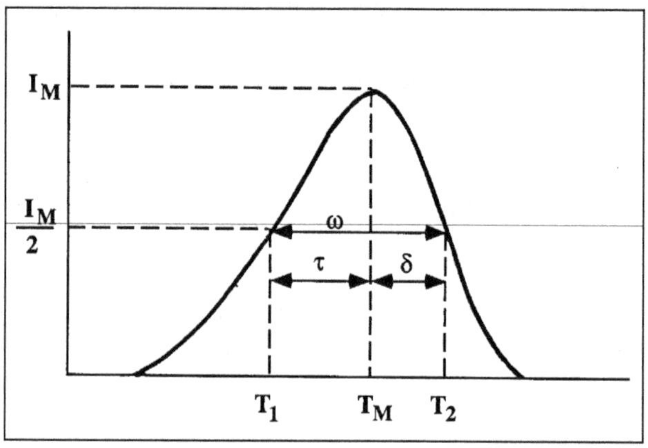

Fig. 1. The geometrical parameters characterizing an isolated peak.

As a first approach, it is possible to check the symmetry properties of the peak:

- a first-order peak has an asymmetrical shape.
- a second-order peak is characterised by a symmetrical shape.

According to the figure, the following parameters can be defined:

T_M, T_1, T_2: are respectively the peak temperature at the maximum and the temperatures on either side of the temperature at the maximum, corresponding to half intensity,

$\tau = T_M - T_1$: is the half-width at the low temperature side of the peak,

$\delta = T_2 - T_M$: is the half-width towards the fall-off of the glow peak,

$\omega = T_2 - T_1$: is the total half-width,

$\mu_g = \dfrac{\delta}{\omega}$: is the symmetrical geometrical factor.

It has to be noted that:

- According to the asymmetrical property of a first-order peak, τ is almost 50% bigger than δ,
- The geometrical factor μ_g is equal to 0.42 for a first-order kinetics, and 0.52 in the case of a second-order, hence, the following relation can be deduced

$$\left|0.52-\mu_g\right|-\left|0.42-\mu_g\right| \tag{1}$$

and two possibilities can be obtained. If relation (1) is less than zero, a second-order kinetics or a tendecy to has to be considered; if relation (1) results to be larger than one, a first-order or a tendency to is possible.

- μ_g is practically independent of *E*, in the range from 0.1 to 1.6 eV, and of s'', from 10^5 to 10^{13} s^{-1},
- μ_g is strongly dependent on the kinetic order, *b*, in the range $0.7 \leq b \leq 2.5$,
- Another factor, namely $\gamma = \dfrac{\delta}{\tau}$, is ranging from 0.7 to 0.8 for a first-order peak, and from 1.05 to 1.20 for a second-order.

References

1. Grosswiener L.I., J. Appl. Phys. **24** (1953) 1306
2. Lushchik C.B., Sov. Phys. JEPT **3** (1953) 390
3. Halperin A. and Braner A.A., J. Appl. Phys. **46** (1960) 408
4. Chen R., J. Appl. Phys. **46** (1969) 570
5. Chen R., J. Electrochem. Soc. **106** (1969) 1254
6. Balarin M., Phys. Stat. Sol. (a) **31** (1975) K111

Peak-shape methods based on the peak temperature at the maximum

The methods are those that use only one temperature value of the TL glow curve: the temperature at the maximum, T_M. These methods are commonly called heuristic methods and are considered as a first approximation when compared with other methods based on two temperature values, which are well known as peak-shape methods.

The so-called heuristic methods are as follows:

Urbach [1], first suggested a very simple expression deduced from TL experiments on KCl

$$E = \frac{T_M}{500} = 23kT_M \tag{1}$$

which gives reasonable values if applied to other materials when the frequency factor, s, is 10^9 s^{-1}.

Randall and Wilkins [2], for alkaline earth and zinc sulphides, with $s = 10^8$ s^{-1}, proposed the following expression

$$E = kT_M \ln s \tag{2}$$

Halperin and Braner [3] suggested, for NaCl,

$$E = 38kT_M \tag{3}$$

Finally, Dussel and Bube [4] gave a very similar expression for LiF

$$E = 39kT_M \tag{4}$$

The equations related to the various peak-shape methods are based on three characteristic temperatures of a single TL peak, namely T_1, on the low-temperature side of the peak at its half intensity, T_2, as the temperature on the fall-off of the peak at half intensity, and T_M, the peak temperature at its maximum intensity. The other parameters are:

$\tau = T_M - T_1$, the half-width at the low-temperature side of the peak,

$\delta = T_2 - T_M$, the half-width towards the fall-off of the glow peak,

and

$\omega = T_2 - T_1$, the total half-width (FWHM).

The first-order kinetics equations for determining the activation energy according to the peak-shape methods are as follows:

Grossweiner [5]

$$E_{G,\tau} = 1.41\frac{T_1 T_M}{\tau} \tag{5}$$

The coefficient 1.41 in equation (5) has been computed by Dussel and Bube [4] for getting a better approximation.

Lushchik [6]

$$E_{L,\delta} = 0.976\frac{kT_M^2}{\delta} \tag{6}$$

where the coefficient has been computed by Chen [7].

Halperin and Braner [3]

$$E_{HB,\tau} = 1.72\frac{kT_M^2}{\tau}(1 - 1.58\Delta_M) \tag{7}$$

where $\Delta_M = \frac{2kT_M}{E}$. Equation (7) requires an iterative process owing to the presence of the quantity Δ_M. To avoid this difficulty, Chen modified the previous equation as following [7]:

$$E_{HB,\tau} = 1.52\frac{kT_M^2}{\tau} - 3.16kT_M \tag{8}$$

Balarin [8]

$$E_{B,\omega} = \frac{T_M^2}{4998 \cdot \omega} \tag{9}$$

All the previous expressions were summed up by Chen [7] who considered a general-order kinetics, ranging from 1 to 2, giving the possibility of non-integer values for the kinetics order:

$$E_\alpha = c_\alpha\left(\frac{kT_M^2}{\alpha}\right) - b_\alpha(2kT_M) \tag{10}$$

where α is τ, δ or ω. The values of c_α and b_α are summarized as

$$\begin{aligned} c_\tau &= 1.51 + 3.0(\mu - 0.42), & b_\tau &= 1.58 + 4.2(\mu - 0.42) \\ c_\delta &= 0.976 + 7.3(\mu - 0.42), & b_\delta &= 0 \\ c_\omega &= 2.52 + 10.2(\mu - 0.42), & b_\omega &= 1 \end{aligned} \tag{11}$$

where $\mu = \frac{\delta}{\omega} = \frac{T_2 - T_M}{T_2 - T_1}$ μ = 0.42 for the first order kinetics.

References

1. Urbach F., Winer Ber **IIa 139** (1930) 363
2. Randall J.T. and Wilkins M.H.F., Proc. Roy. Soc. **A184** (1945) 366
3. Halperin A. and Braner A.A., Phys. Rev. **117** (1960) 408
4. Dussel G.A. and Bube R.H., Phys. Rev. **155** (1967) 764
5. Grossweiner L.I., J. Appl. Phys. **24** (1969) 1306
6. Lushihik L.I., Sov. Phys. **3** (1956) 390

7. Chen R., J. Appl. Phys. **40** (1969) 570
8. Kitis G., Cruz-Zaragoza E., Furetta C., REDS **161** (2006) 149

Peak shape method when $s = s(T)$. (Chen: first-, second- and general-order)

Total half-width of a peak

- *First-order kinetics*

Starting from the TL intensity expression, I=I(T), given in case of s=s(T), having used the integral approximation [1,2]:

$$I(T) = n_0 s_0 T^\alpha \exp(-\frac{E}{kT})\left[-\frac{s_0}{\beta}\int_{T_0}^{T} T^\alpha \exp(-\frac{E}{kT'})dT'\right] \tag{1}$$

the expression for the intensity at the maximum is given by

$$I_M = s_0 n_0 T^\alpha \exp\left(-\frac{E}{kT_M}\right)\exp\left[-\frac{s_0 k T_M^{\alpha+2}}{\beta \cdot E}\exp\left(-\frac{E}{kT_M}\right)\left[1-\left(1+\frac{1}{2}\alpha\right)\Delta_M\right]\right] \tag{2}$$

Inserting in Eq. (2) the condition at the maximum:

$$\frac{\beta}{T_M^{\alpha+2}} = \exp(-\frac{E}{kT_M})\left[\frac{ks_0}{E\left(1+\frac{\alpha}{2}\Delta_M\right)}\right] \tag{3}$$

Eq. (2) becomes:

$$I_M = \frac{n_0 \cdot \beta \cdot E}{k \cdot T_M^2}\left(1+\frac{\alpha}{2}\Delta_M\right)\exp\left(-1+\Delta_M+\frac{\alpha}{2}\Delta_M^2+\frac{\alpha^2}{4}\Delta_M^2\right)$$

Neglecting in the above equation the second-order Δ_M terms, one has

$$I_M = \frac{n_0 \cdot \beta \cdot E}{k \cdot T_M^2}\left(1+\frac{\alpha}{2}\Delta_M\right)\cdot\frac{\exp(\Delta_M)}{e} \tag{4}$$

Because

$$\exp(\Delta_M) \approx 1 + \Delta_M$$

Eq. (4) becomes

$$I_M = \frac{n_0 \cdot \beta \cdot E}{e \cdot k \cdot T_M^2}\left(1 + \frac{\alpha}{2}\Delta_M\right) \cdot (1 + \Delta_M)$$

which can be rearranged as

$$I_M = \frac{n_0 \cdot \beta \cdot E}{e \cdot k \cdot T_M^2}\left[1 + \left(1 + \frac{\alpha}{2}\right)\Delta_M\right] \tag{5}$$

Remembering the Chen's assumption between the total glow area and a triangle

$$\frac{\omega \cdot I_M}{\beta \cdot n_0} = C_\omega \tag{6}$$

Eq. (5) becomes

$$E\left[1 + \left(1 + \frac{\alpha}{2}\Delta_m\right)\right] = \frac{C_\omega \cdot e \cdot k \cdot T_M^2}{\omega}$$

from which the expression for the activation energy is obtained:

$$E_\omega = 2kT_M\left[1.26 \cdot \frac{T_M}{\omega} - \left(1 + \frac{\alpha}{2}\right)\right] \tag{7}$$

- *Second-order kinetics*

The condition at the maximum (**see dependence of**…..) is given by

$$\frac{2s_0' n_0 T_M^{\alpha+2} k \exp\left(-\frac{E}{kT_M}\right)}{E\beta\left(1 + \frac{\alpha}{2}\Delta_M\right)} = \left[1 + \frac{s_0' n_0}{\beta}\int_{T_0}^{T} T^\alpha \exp\left(-\frac{E}{kT}\right) dT\right] \tag{8}$$

which becomes, using the **integral approximation**:

$$\frac{2s_0' n_0 T_M^{\alpha+2} k}{E \cdot \beta\left(1+\frac{\alpha}{2}\Delta_M\right)} \exp\left(-\frac{E}{kT_M}\right) =$$
$$= \left\{1+\frac{s_0' n_0 k T_M^{\alpha+2}}{\beta \cdot E}\left[1-\left(1+\frac{\alpha}{2}\right)\Delta_M\right]\exp\left(-\frac{E}{kT_M}\right)\right\}$$

Rearranging the above expression and neglecting the second-order Δ_M terms, we obtain

$$\exp\left(\frac{E}{kT_M}\right) = \frac{s_0' n_0 k T_M^{\alpha+2}}{\beta \cdot E}\left(\frac{1+\Delta_M}{1+\frac{\alpha}{2}\Delta_M}\right) \tag{9}$$

Inserting Eq. (9) in the expression of the intensity

$$I_M = \left[\frac{E \cdot \beta \cdot \left(1+\frac{\alpha}{2}\Delta_M\right)}{2 \cdot T_M^{\alpha+2} \cdot k}\right]^2 \left(\frac{T_M^{\alpha}}{s_0'}\right)\exp\left(\frac{E}{kT_M}\right) \tag{10}$$

and using expression (6) and rearranging, we obtain

$$\frac{C_\omega}{\omega} = \frac{E\left(1+\frac{\alpha}{2}\Delta_M\right)(1+\Delta_M)}{4kT_M^2}$$

Neglecting again the second-order Δ_M terms and substituting Δ_M with $2kT_M/E$, we have

$$E_\omega = 2kT_M\left[\frac{2T_M C_\omega}{\omega} - \left(1+\frac{\alpha}{2}\right)\right] \tag{11}$$

High-temperature side half peak

- *First-order kinetics*

The maximum intensity can be expressed as follows

$$I_M = n_M s_0 T_M^{\alpha} \exp\left(-\frac{E}{kT_M}\right) \tag{12}$$

and using the condition at the maximum we can write

$$\frac{I_M}{n_M} = s_0 T_M^{\alpha}\left(\frac{\beta \cdot E}{k s_0 T_M^{\alpha+2}}\right)\left(1+\frac{\alpha}{2}\Delta_M\right)$$

which, substituting $\Delta_M = \dfrac{2kT_M}{E}$, yields

$$E = \left(\frac{I_M k T_M^2}{\beta \cdot n_M}\right) - \alpha k T_M \tag{13}$$

Using the **Lushchik assumption**, modified by Chen,

$$\frac{\delta \cdot I_M}{\beta \cdot n_M} = C_\delta \tag{14}$$

Eq. (13) becomes

$$E_\delta = \frac{C_\delta}{\delta} k T_M^2 - \alpha \cdot k \cdot T_M \tag{15}$$

- *Second-order kinetics*

The expression of n_M is given by

$$n_M = n_0\left[1+\left(\frac{s_0' n_0}{\beta}\right)\int_{T_0}^{T_M} T^{\alpha} \exp\left(-\frac{E}{kT'}\right) dT'\right]^{-1}$$

and using the expression of the intensity as a function of the temperature

$$I(T) = \frac{n_0^2 s_0' T^{\alpha} \exp(-\frac{E}{kT})}{\left[1+\frac{s_0' n_0}{\beta}\int_{T_0}^{T} T^{\alpha} \exp\left(-\frac{E}{kT}\right) dT\right]^2}$$

we obtain

$$\frac{I_M}{n_M} = \frac{n_0 s_0' T_M^{\alpha} \exp\left(-\frac{E}{kT_M}\right)}{\left[1 + \frac{s_0' n_0}{\beta} \int_{T_0}^{T_M} T^{\alpha} \exp\left(-\frac{E}{kT'}\right) dT'\right]}$$

Inserting in this expression the **condition at the maximum**, one has

$$\frac{I_M}{n_M} = \frac{E \cdot \beta \left(1 + \frac{\alpha}{2}\Delta_M\right)}{2kT_M^2}$$

Using the assumption (13), we finally obtain

$$E_\delta = \frac{2kT_M^2 C_\delta}{\delta} - \alpha k T_M \tag{15}$$

Low-temperature side half peak

- *First-order kinetics*

The number of trapped charges, at the temperature at the maximum, is given by

$$n_M = n_0 \exp\left[-\left(\frac{s_0}{\beta}\right) \int_{T_0}^{T_M} T^{\alpha} \exp\left(-\frac{E}{kT'}\right) dT'\right] \tag{16}$$

Using the **integral approximation**, we obtain

$$\frac{n_M}{n_0} = \exp\left[-\left(\frac{s_0}{\beta}\right)\left(\frac{kT_M^{\alpha+2}}{E}\right) \exp\left(-\frac{E}{kT_M}\right)\left[1 - \left(1 + \frac{\alpha}{2}\Delta_M\right)\right]\right]$$

Inserting in the above expression the **condition at the maximum** and rearranging, we have

$$\frac{n_M}{n_0} = \exp\left[-\frac{(k\alpha T_M + E)(1 - \Delta_M)}{E}\right]$$

or better

$$\frac{n_M}{n_0} = \exp\left[-\left(\frac{\alpha}{2}\Delta_M + 1\right)\left(1 - \Delta_M - \frac{\alpha}{2}\Delta_M\right)\right]$$

from which, neglecting the second-order Δ_M terms, one has

$$\frac{n_M}{n_0} = \exp\left[-\left(1-\Delta_M\right)\right] \approx \frac{1+\Delta_M}{e} \tag{17}$$

Remembering the relation given by Chen (**low-temperature side half peak method**):

$$\frac{n_0}{n_M} - 1 = \left(\frac{\tau}{\beta \cdot C_\tau}\right) \cdot \left(\frac{I_M}{n_M}\right)$$

the insertion in the last expression of Eqs. (12) and (17) gives

$$\frac{e}{1+\Delta_M} - 1 = \frac{\tau}{C_\tau} \cdot \frac{E}{kT_M^2} \cdot \left(1 + \frac{\alpha}{2}\Delta_M\right) \tag{18}$$

To obtain the Halperin and Braner's corrected formula, Chen used $\alpha = 2$, so that the previous equation becomes

$$E = \frac{C_\tau k T_M^2}{\tau}\left[\frac{e}{\left(1+\Delta_M\right)^2} - \frac{1}{1+\Delta_M}\right] \tag{19}$$

Eq. (18) can be semplified using the following approximations:

$$\frac{1}{\left(1+\Delta_M\right)^2} \approx 1 - 2\Delta_M$$

$$\frac{1}{1+\Delta_M} \approx 1 - \Delta_M$$

hence obtained

$$E_\tau = 1.515\left(\frac{kT_M^2}{\tau}\right)\left(1 - 2.58\Delta_M\right) \tag{20}$$

To obtain an expression without iterative calculations, one can starts from the previous Eq. (18):

$$\frac{1.718-\Delta_M}{1+\Delta_M}=\frac{\tau\cdot E}{C_\tau kT_M^2}\left(1+\frac{\alpha}{2}\Delta_M\right)$$

$$\frac{1.718(1-0.58\Delta_M)}{1+\Delta_M}=\frac{\tau\cdot E}{C_\tau kT_M^2}\left(1+\frac{\alpha}{2}\Delta_M\right)$$

$$\frac{1.718}{(1+\Delta_M)(1+0.58\Delta_M)}=\frac{\tau\cdot E}{C_\tau kT_M^2}\left(1+\frac{\alpha}{2}\Delta_M\right)$$

from which, neglecting the second-order Δ_M terms. One has

$$E_\tau=1.515\left(\frac{kT_M^2}{\tau}\right)-\left(\frac{\alpha}{2}+1.58\right)\cdot\left(2kT_M\right) \tag{21}$$

- *Second-order kinetics*

Inserting Eq. (8) into the equation which gives the intensity in the case of a second order kinetics with the frequency factor depending on the temperature:

$$I(T)=\frac{n_0^2 s_0' T^\alpha \exp(-\frac{E}{kT})}{\left[1+\frac{s_0' n_0}{\beta}\int_{T_0}^{T} T^\alpha \exp\left(-\frac{E}{kT}\right)dT\right]^2} \tag{22}$$

we obtain

$$I_M=s_0' n_0^2 T_M^\alpha \exp\left(-\frac{E}{kT_M}\right)\left[2\frac{s_0' n_0 kT_M^{\alpha+2}\exp\left(-\frac{E}{kT_M}\right)}{\beta\left(\alpha kT_M+E\right)}\right]^{-2} \tag{23}$$

from which

$$I_M=s_0' n_0^2 T_M^\alpha \exp\left(-\frac{E}{kT_M}\right)\left[\frac{s_0' kT_M^{\alpha+2}}{\beta E\exp\left(-\frac{E}{kT_M}\right)}\left(\frac{2n_0}{1+\frac{\alpha}{2}\Delta_M}\right)\right]^{-2} \tag{24}$$

Coming back to Eq. (8) and using the approximation for the integral comparing in that equation, we get

$$1=\frac{s_0' n_0 k T_M^{\alpha+2}\exp\left(-\frac{E}{kT_M}\right)}{\beta E}\cdot\left\{\frac{2}{1+\frac{\alpha}{2}\Delta_M}-\left[1-\left(1+\frac{\alpha}{2}\right)\Delta_M\right]\right\}$$

The previous expression can be modified considering the following approximations:

$$\frac{1}{1\pm\Delta_M}\approx 1\mp\Delta_M \tag{25}$$

$$\Delta_M \prec\prec 1 \tag{26}$$

It then becomes

$$1=\frac{s_0' n_0 k T_M^{\alpha+2}\exp\left(-\frac{E}{kT_M}\right)}{\beta E}\left[2\left(1-\frac{\alpha}{2}\Delta_M\right)-1+\left(1+\frac{\alpha}{2}\right)\Delta_M\right]$$

which, rearranged, yields

$$n_0\left[1+\left(1-\frac{\alpha}{2}\right)\Delta_M\right]=\frac{\beta E}{s_0' k T_M^{\alpha+2}}\exp\left(\frac{E}{kT_M}\right) \tag{27}$$

Inserting Eq. (27) into Eq. (24), allows to obtain, using the approximations (25) and (26):

$$I_M=\frac{n_0\beta E}{4kT_M^2}\left[1+\left(1+\frac{\alpha}{2}\right)\Delta_M\right]$$

To find the expression for the activation energy, we need the expression of n:

$$n=n_0\left[1+\frac{s_0' n_0}{\beta}\int_{T_0}^{T}T^{\alpha}\exp\left(-\frac{E}{kT}\right)dT\right]^{-1}$$

and insert it in Eq. (8) for obtaining

$$\frac{n_M}{n_0} = \left[\frac{\beta E\left(1+\frac{\alpha}{2}\Delta_M\right)}{2s_0' n_0 k T_M^{\alpha+2}} \right] \exp\left(\frac{E}{kT_M}\right)$$

This last expression can be now inserted in Eq. (27), getting

$$\frac{n_M}{n_0} = \frac{1+\Delta_M}{2}$$

or better

$$n_0 - n_M = \frac{n_0\left(1-\Delta_M\right)}{2} \tag{28}$$

Remembereing the **Chen's expression for the low-temperature side half peak**, i.e.

$$\frac{\tau I_M}{\beta\left(n_0 - n_M\right)} = C_\tau$$

and inserting in it Eq. (28):

$$\frac{\frac{\tau n_0 \beta E}{4kT_M^2}\left[1+\left(1+\frac{\alpha}{2}\right)\Delta_M\right]}{\beta n_0\left(\frac{1-\Delta_M}{2}\right)} = C_\tau \tag{29}$$

from which the activation energy is obtained

$$E = \left(\frac{2C_\tau kT_M^2}{\tau}\right)\left[1-\left(2+\frac{\alpha}{2}\right)\Delta_M\right] \tag{30}$$

Beacause this expression needs an iterative procedure, it can be expressed in another way. Rearranging Eq. (29) as following

$$C_\tau = \frac{\tau}{2}\left[\left(\frac{E}{kT_M^2}\right)+\left(2+\frac{\alpha}{2}\right)\frac{2}{T_M}\right]$$

from which we get

$$E_\tau = \frac{2C_\tau k T_M^2}{\tau} - \left(2 + \frac{\alpha}{2}\right) 2kT_M \tag{31}$$

Inserting in this equation the value $2C_\tau = 1.81$, an expression without resorting to iterative process is obtained.

Chen gave a general expression for the activation energy, i.e.

$$E_\gamma = c_\gamma \left(\frac{kT_M^2}{\alpha}\right) + b_\gamma \left(2kT_M\right) \tag{32}$$

where γ is τ, δ or ω. The values of c_γ and b_γ are summarised as below

$$c_\tau = 1.51 + 3.0(\mu_g - 0.42) \qquad b_\tau = 1.58 + 4.2(\mu_g - 0.42) + \frac{\alpha}{2}$$

$$c_\delta = 0.976 + 7.3(\mu_g - 0.42) \qquad b_\delta = \frac{\alpha}{2}$$

$$c_\omega = 2.52 + 10.2\left(\mu_g - 0.42\right) \qquad b_\omega = 1 + \frac{\alpha}{2}$$

with $\mu_g = 0.42$ for 1st order

and $\mu_g = 0.52$ for 2nd order

References

1. Chen R., J. Appl. Phys. **40** (1969) 570
2. Chen R., J. Electrochem. Soc. **116** (1969) 1254

Peak shape method: reliability expressions

An important and widely used method for investigating the trapping levels in crystals is based, among the various TL methods introduced during the years, on the geometrical characterization of a TL glow peak, the well known peak shape (PS) methods. In fact, for calculating the activation energy of the trapping level corresponding to a peak in the glow curve, one needs to measure three temperature values on the peak itself: the temperature at the maximum, T_M, and the first and second half temperatures, T_1 and T_2.

The formulas proposed [1,2] for finding the activation energy usually include the following factors:

$\tau = T_M - T_1$ the half width at the low temperature side of the peak,

$\delta = T_2 - T_M$ the halfwidth towards the falloff of the peak,

$\omega = T_2 - T_1$ the total halfwidth (FWHM).

In the following is given a list of the various expressions, for both first and second order of kinetics, allowing for the activation energy determination. All the expressions have been modified by Chen for getting a better accuracy in the E values.

Grosswiener (G)

$$1^{st} \quad ORDER \quad (E_G)_\tau = 1.41k\frac{T_1 T_M}{\tau} \tag{1}$$

$$2^{nd} \quad ORDER \quad (E_G)_\tau = 1.68k\frac{T_1 T_M}{\tau} \tag{2}$$

Lushchik (L)

$$1^{st} \quad ORDER \quad (E_L)_\delta = 0.976\frac{kT_M^2}{\delta} \tag{3}$$

$$2^{nd} \quad ORDER \quad (E_L)_\delta = 1.706\frac{kT_M^2}{\delta} \tag{4}$$

Halperin & Braner (HB)

$$1^{st} \quad ORDER \quad (E_{HB})_\tau = 1.72\frac{kT_M^2}{\tau}(1-2.58\Delta_M) \tag{5}$$

$$2^{nd} \quad ORDER \quad (E_{HB})_\tau = \frac{2kT_M^2}{\tau}(1-3\Delta_M) \tag{6}$$

where $\Delta_M = 2kT_M/E$

Chen also gave two more expressions based on the ω factor:

Chen's additional expressions (C_{aex})

$$1^{st} \quad ORDER \quad E_\omega = 2.29k\frac{T_M^2}{\omega} \tag{7}$$

$$2^{nd}\ ORDER \quad E_\omega = 2kT_M\left(1.756\frac{T_M}{\omega}-1\right) \tag{8}$$

Chen's expressions (general) (C_{gex})

The previous methods were summed up by Chen, who considered general order kinetics, l, ranging from 1 to 2, then giving the possibility of non-integer value of the kinetics order. The general expression is

$$E_\alpha = c_\alpha\left(\frac{kT_M^2}{\alpha}\right) - b_\alpha(2kT_M)$$

where α is τ, δ or ω. The values of c_α and b_α are summarized as below

$$c_\tau = 1.51 + 3.0(\mu - 0.42) \qquad b_\tau = 1.58 + 4.2(\mu - 0.42)$$

$$c_\delta = 0.976 + 7.3(\mu - 0.42) \qquad b_\delta = 0$$

$$c_\omega = 2.52 + 10.2(\mu - 0.42) \qquad b_\omega = 1$$

with

$$\mu = \frac{\delta}{\omega} = \frac{T_2 - T_M}{T_2 - T_1}$$

being $\mu = 0.42$ for a first order kinetics and $\mu = 0.52$ for a second order.

The previous general expression, developed just for a 1st and a 2nd order, gives:

1st *ORDER*

$$(E_C)_\tau = 1.51\frac{kT_M^2}{\tau} - 3.04kT_M \tag{9}$$

$$(E_C)_\delta = 0.976\frac{kT_M^2}{\delta} \tag{10}$$

$$(E_C)_\omega = 2.52\frac{kT_M^2}{\omega} - 2kT_M \tag{11}$$

2nd *ORDER*

$$(E_C)_\tau = 1.81\frac{kT_M^2}{\tau} - 4kT_M \tag{12}$$

$$(E_C)_\delta = 0.706\frac{kT_M^2}{\delta} \tag{13}$$

$$(E_C)_\omega = 3.54\frac{kT_M^2}{\omega} - 2kT_M \tag{14}$$

Furthermore, the following parameter, introduced by Balarin, is also used:

$$\gamma = \frac{\delta}{\tau} = \frac{T_2 - T_M}{T_M - T_1}$$

Using the previous parameters, some relations among them can be obtained as following:

1st order-kinetics:

$$\mu = 0.42 \qquad \gamma = 0.72 \qquad \delta = 0.72\tau \qquad \delta = 0.42\omega$$

2nd order-kinetics:

$$\mu = 0.52 \qquad \gamma = 1.09 \qquad \delta = 1.09\tau \qquad \delta = 0.52\omega$$

As a first approximation, the following relations among the peak's temperatures can be also used:

$$T_1 = 0.95T_M \quad \text{and} \quad T_2 = 1.05T_M$$

The expressions so far given have been handled for getting a criteria of reliability of the E values obtained using the PS methods. In most of the cases the Chen's expressions have been used as reference because they have a more general meaning respect to the others and also give more accurate values of E.

1st ORDER

- $$\left(\frac{E_L}{E_C}\right)_\delta = \frac{0.978K\frac{T_M^2}{\delta}}{0.976K\frac{T_M^2}{\delta}} = 1.002$$

- $$\left(\frac{E_G}{E_C}\right)_\tau = \frac{1.41K\frac{T_M T_1}{\tau}}{1.51K\frac{T_M^2}{\tau} - 3.16KT_M} = \frac{T_1}{1.071(T_M - 2.09\tau)}$$

$$= \frac{T_1}{1.07(2.09T_1 - 1.09T_M)} \approx 0.9915$$

- $$\left(\frac{E_{HB}}{E_C}\right)_\tau = \frac{\frac{1.72kT_M^2}{\tau}(1-2.58\Delta_M)}{\frac{1.51kT_M^2}{\tau} - 3.16kT_M} = 1.139\frac{1-2.58\Delta_M}{1-2.093\frac{T_M - T_1}{T_M}}$$

Limits:

$$\Delta_M = 0 \rightarrow \frac{1.139}{1-2.039\frac{T_M - T_1}{T_M}} = \frac{1.042}{1.915\frac{T_1}{T_M} - 1}$$

$$\Delta_M = 0.1 \rightarrow \frac{0.742}{1-2.093\frac{T_M - T_1}{T_M}} = \frac{0.679}{1.915\frac{T_1}{T_M} - 1}$$

$$\left(\frac{0.679}{1.915\frac{T_1}{T_M} - 1}\right)_{\Delta_M = 0.1} \leq \left(\frac{E_{HB}}{E_C}\right)_\tau \leq \left(\frac{1.042}{1.915\frac{T_1}{T_M} - 1}\right)_{\Delta_M = 0}$$

- $$\left(\frac{E_G}{E_{HB}}\right)_\tau = \frac{1.41k\frac{T_1 T_M}{\tau}}{1.72\frac{kT_M^2}{\tau}(1-2.58\Delta_M)} = \frac{0.8198T_1}{T_M(1-2.58\Delta_M)}$$

limits:

$$\Delta_M = 0 \rightarrow 0.8198\frac{T_1}{T_M}$$

$$\Delta_M = 0.1 \rightarrow 1.1048\frac{T_1}{T_M}$$

$$\left(0.8198\frac{T_1}{T_M}\right)_{\Delta_M = 0} \leq \left(\frac{E_G}{E_{HB}}\right)_\tau \leq \left(1.1048\frac{T_1}{T_M}\right)_{\Delta_M = 0.1}$$

2nd ORDER

- $$\left(\frac{E_L}{E_C}\right)_\delta = \frac{1.706\dfrac{kT_M^2}{\delta}}{1.71\dfrac{kT_M^2}{\delta}} = 0.998$$

- $$\left(\frac{E_G}{E_C}\right)_\tau = \frac{1.68k\dfrac{T_M T_1}{\tau}}{1.81k\dfrac{T_M^2}{\tau} - 4kT_M} = \frac{0.7671T_1}{1.8265T_1 - T_M} \approx 0.9913$$

- $$\left(\frac{E_{HB}}{E_C}\right)_\tau = \frac{\dfrac{2kT_M^2}{\tau}(1-3\Delta_M)}{1.81\dfrac{kT_M^2}{\tau} - 4kT_M} = 0.917\frac{1-3\Delta_M}{1.83\dfrac{T_1}{T_M} - 1}$$

Limits:

$$\Delta_M = 0 \rightarrow \frac{0.917T_M}{1.83T_1 - T_M}$$

$$\Delta_M = 0.1 \rightarrow \frac{0.7T_M}{1.83T_1 - T_M}$$

so that

$$\left(\frac{0.7T_M}{1.83T_1 - T_M}\right)_{\Delta_M = 0.1} \leq \left(\frac{E_{HB}}{E_C}\right)_\tau \leq \left(\frac{0.917T_M}{1.83T_1 - T_M}\right)_{\Delta_M = 0}$$

- $$\left(\frac{E_G}{E_{HB}}\right)_\tau = \frac{1.68k\dfrac{T_M T_1}{\tau}}{\dfrac{2kT_M^2}{\tau}(1-3\Delta_M)} = \frac{0.84T_1}{T_M(1-3\Delta_M)}$$

limits:

$$\Delta_M = 0 \rightarrow 0.84\frac{T_1}{T_M} \qquad \Delta_M = 0.1 \rightarrow 1.2\frac{T_1}{T_M}$$

so that

$$\left(0.84\frac{T_1}{T_M}\right)_{\Delta_M=0} \leq \left(\frac{E_G}{E_{HB}}\right)_\tau \leq \left(1.2\frac{T_1}{T_M}\right)_{\Delta_M=0.1}$$

Some more expressions derived by the original ones, using the geometrical factors μ and γ.

<u>Grosswiener expressions given as a function of δ and ω.</u>

$$(E_G)_\delta = 1.0152k\frac{T_1T_M}{\delta} \qquad 1st \quad order$$

$$(E_G)_\delta = 1.8313k\frac{T_1T_M}{\delta} \qquad 2nd \quad order$$

$$(E_G)_\omega = 2.4171k\frac{T_1T_M}{\omega} \qquad 1st \quad order$$

$$(E_G)_\omega = 3.5217k\frac{T_1T_M}{\omega} \qquad 2nd \quad order$$

<u>Lushchik expressions given as a function of τ and ω.</u>

$$(E_L)_\tau = 1.3556\frac{kT_M^2}{\tau} \qquad 1st \quad order$$

$$(E_L)_\tau = 1.5651\frac{kT_M^2}{\tau} \qquad 2nd \quad order$$

$$(E_L)_\omega = 2.3238\frac{kT_M^2}{\omega} \qquad 1st \quad order$$

$$(E_L)_\omega = 3.2808\frac{kT_M^2}{\omega} \qquad 2nd \quad order$$

<u>Halperin-Braner expressions given as a function of δ and ω.</u>

$$(E_{HB})_\delta = 1.2384\frac{kT_M^2}{\delta}(1-2.58\Delta_M) \qquad 1st \quad order$$

$$(E_{HB})_\delta = 2.1801\frac{kT_M^2}{\delta}(1-3\Delta_M) \qquad 2nd \quad order$$

$$(E_{HB})_{\omega} = 2.9487 \frac{kT_M^2}{\omega}(1 - 2.58\Delta_M) \qquad 1st \quad order$$

$$(E_{HB})_{\omega} = 4.1929 \frac{kT_M^2}{\omega}(1 - 3\Delta_M) \qquad 2nd \quad order$$

Comparison of the previous derived expressions to the corresponding Chen's expressions

Grosswiener's modified expressions related to Chen's expressions

$$\left(\frac{E_G}{E_C}\right)_{\delta} = \frac{1.0152k\dfrac{T_1 T_M}{\delta}}{0.976k\dfrac{T_M^2}{\delta}} = 1.0402\frac{T_1}{T_M} \qquad 1^{st} \text{ order}$$

$$\left(\frac{E_G}{E_C}\right)_{\delta} = \frac{1.8313k\dfrac{T_1 T_M}{\delta}}{0.706k\dfrac{T_M^2}{\delta}} = 2.5939\frac{T_1}{T_M} \qquad 2^{nd} \text{ order}$$

$$\left(\frac{E_G}{E_C}\right)_{\omega} = \frac{2.4171k\dfrac{T_1 T_M}{\omega}}{2.52k\dfrac{T_M^2}{\omega} - 2kT_M} = \frac{1.2086T_1}{1.26T_M - T_2 + T_1} \qquad 1^{st} \text{ order}$$

$$\left(\frac{E_G}{E_C}\right)_{\omega} = \frac{3.5217k\dfrac{T_1 T_M}{\omega}}{3.54k\dfrac{T_M^2}{\omega} - 2kT_M} = \frac{1.7609T_1}{1.77T_M - T_2 + T_1} \qquad 2^{nd} \text{ order}$$

Error analysis

According to the error propagation rules, having a function of various independent variables, i.e.

$$\overline{\Phi} = f(\overline{x}_1, \overline{x}_2, \ldots, \overline{x}_n)$$

the error is given by

$$\Delta\Phi = \pm\left[\left(\frac{\partial f}{\partial x_1}\delta x_1\right)^2 + \left(\frac{\partial f}{\partial x_2}\delta x_2\right)^2 + \cdots + \left(\frac{\partial f}{\partial x_n}\delta x_n\right)^2\right]^{1/2}$$

The previous expression can be applied, for instance, to the Chen's equation

$$E = \frac{2.29 \cdot k \cdot T_M^2}{\omega}$$

According to the error propagation one gets

$$\sigma_E = \pm \left[\left(\frac{\partial E}{\partial T_M} \Delta T_M \right)^2 + \left(\frac{\partial E}{\partial \omega} \Delta \omega \right)^2 \right]^{1/2}$$

The errors associated to the various expressions can been calculated in the same way.

References

1. Kitis G., private communication
2. Furetta C., Sanipoli C. and Kitis G., J. Phys. D: Appl. Phys. **34** (2001) 857

P-2

(from **Peak shift** to **Properties of the maximum conditions**)

Peak shift

The TL intensity for first (**Randal and Wilkins model**) and second (**Garlick and Gibson model**) order of kinetics are respectively

$$I(T) = n \cdot s \cdot \exp\left(-\frac{E}{kT}\right) \quad (1^{st}\text{ order}) \tag{1}$$

and

$$I(T) = n^2 \cdot s' \cdot \exp\left(-\frac{E}{kT}\right) \quad (2^{nd}\text{ order}) \tag{2}$$

where n is the trapped carrier concentration, s is the frequency factor, $s' = {}^{s}/_{N}$ is the so-called pre-exponential factor, with N the available trap concentration.

Eq. (2) can be rewritten as

$$I(T) = s'' \cdot n \cdot \exp\left(-\frac{E}{kT}\right) \tag{3}$$

with $s'' = s'n$, which is equivalent to s in the first order case.

Considering Eqs. (1) to (3), it can be seen that the peak temperature at the maximum, T_M, depends on E and s, s' or s''; then, if n changes, increase or decreases, a first-order peak remains in the same position, but a second-order peak shifts, i.e., to higher temperatures as n decreases because the variation in s'.

Figs. (1) and (2) show the different behaviours expected for glow peaks following a first-order kinetics or a second-order.

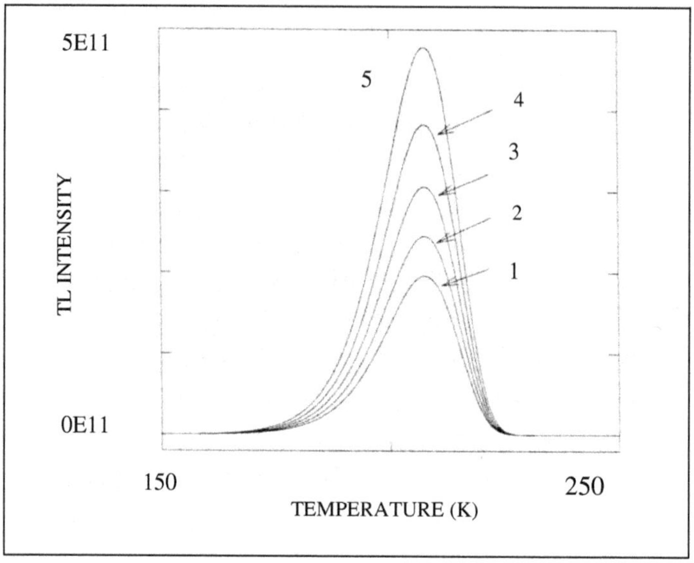

Fig. 1. Glow curves of first-order kinetics as a function of the given dose; n_0 increases from (1) to (5).

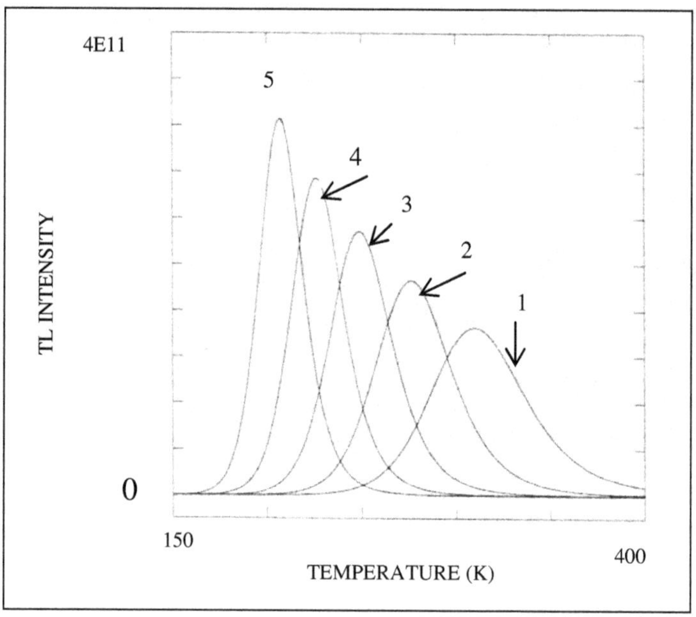

Fig. 2. Behaviour of the second-order glow curves as a function of the given dose; n_0 increases from (1) to (5).

Peak shift as a function of dose

- Second order kinetics

The variation in the dose levels, i.e. the initial concentration of the trapped charges, n_0, changes the shape of the glow curve. It is noted that the peak temperature at the maximum, T_M, decreases as n_0 increases. The temperature shift [1] can be approximated by

$$T_1 - T_2 \approx T_1 T_2 \frac{k}{E} \ln f \tag{1}$$

where T_1 is the temperature of maximum TL intensity at a certain dose and T_2 the temperature of maximum intensity at an f times higher dose.
It can be also noted, from the previous equation, that to smaller values of E correspond bigger peak shifts.

- General order kinetics

The temperature shift can be estimated using the following equation

$$T_1 - T_2 \approx T_1 T_2 \frac{k(b-1)}{E} \ln f \tag{2}$$

where b is the kinetics order.

A more general vision of the peak shift has been reported in [2].

- General order kinetics

Starting from the general order equation

$$I(T) = s n_0 \exp\left(-\frac{E}{kT}\right)\left[1 + \frac{s(b-1)}{\beta}\int_{T_0}^{T} \exp\left(-\frac{E}{kT'}\right) dT'\right]^{-b/(b-1)} \tag{3}$$

with $s = s'' n_0^{b-1}$.
The maximum TL emission occurs at the temperature T_M for which

$$\frac{d(\ln I)}{dT}\Big|_{T=T_M} = 0$$

The logarithm of Eq. (3) is

$$\ln[I(T)] = \ln(sn_0) - \frac{E}{kT} - \frac{b}{b-1}\ln\left[1 + \frac{s(b-1)}{\beta}\int_{T_0}^{T}\exp\left(-\frac{E}{kT'}\right)dT'\right]$$

and then

$$\frac{d(\ln I)}{dT}\Big|_{T=T_M} = \frac{E}{kT_M^2} - \frac{b}{b-1}\left[1 + \frac{s(b-1)}{\beta}\int_{T_0}^{T_M}\exp\left(-\frac{E}{kT'}\right)dT'\right] \times$$
$$\times \frac{s(b-1)}{\beta}\exp\left(-\frac{E}{kT_M}\right) = 0$$

from which we get

$$\frac{kT_M^2 bs}{\beta E}\exp\left(-\frac{E}{kT_M}\right) = 1 + \frac{s(b-1)}{\beta}\int_{T_0}^{T_M}\exp\left(-\frac{E}{kT'}\right)dT' \tag{4}$$

From Eq. (4) we can obtain the expression for the pre-exponential factor:

$$s = \left[\frac{kT_M^2 b\exp\left(-\frac{E}{kT_m}\right)}{\beta E} - \frac{(b-1)\int_{T_0}^{T_M}\exp\left(-\frac{E}{kT'}\right)dT'}{\beta}\right]^{-1} \tag{5}$$

Using now the integral approximation, we obtain

$$s = \left[\frac{kT_M^2\exp\left(-\frac{E}{kT_M}\right)}{\beta E}\left(1 + \frac{2kT_M(b-1)}{E}\right)\right]^{-1} \tag{6}$$

which is expressed in s^{-1}.
The expression of s" is then

$$s'' = \frac{1}{n_0^{b-1}}\left[\frac{kT_M^2\exp\left(-\frac{E}{kT_M}\right)}{\beta E}\left(1 + \frac{2kT_M(b-1)}{E}\right)\right]^{-1} \tag{7}$$

expressed in $cm^{3(b-1)}s^{-1}$.

- <u>Second order kinetics</u>

For the second order kinetics, Eq. (7) becomes, just using b=2,

$$s' = \frac{\beta E \exp\left(\frac{E}{kT_M}\right)}{kn_0 T_M^2}\left[1+\frac{2kT_M}{E}\right]^{-1} \tag{8}$$

expressed now in s^{-1}.

To determine the peak shift as a function of dose, n_0, Eq. (7) can be rearranged and written for two different doses, i.e. n_1 and n_2, and temperatures at the maximum T_1 and T_2

$$\frac{\beta E}{kT_1^2} = n_1^{b-1} s'' \exp\left(-\frac{E}{kT_1}\right)\left[1+(b-1)\frac{2kT_1}{E}\right] \tag{9}$$

$$\frac{\beta E}{kT_2^2} = n_2^{b-1} s'' \exp\left(-\frac{E}{kT_2}\right)\left[1+(b-1)\frac{2kT_2}{E}\right] \tag{10}$$

If $n_2 > n_1$, then $T_1 > T_2$ according to the property of a general order peak.

The ratio of the two previous equations gives

$$\frac{T_2^2\left[E+2(b-1)kT_2\right]n_2^{b-1}}{T_1^2\left[E+2(b-1)kT_1\right]n_1^{b-1}} = \exp\left(\frac{E}{kT_2}-\frac{E}{kT_1}\right) \tag{11}$$

The logarithm of Eq. (11) gives

$$2\ln\left(\frac{T_2}{T_1}\right)+(b-1)\ln\left(\frac{n_2}{n_1}\right)+\ln\left(\frac{E+2(b-1)kT_2}{E+2(b-1)kT_1}\right) - = \frac{E}{k}\left(\frac{1}{T_2}-\frac{1}{T_1}\right) \tag{12}$$

The peak shift ΔT is then

$$\Delta T = T_1 - T_2 = T_1 T_2\left\{\frac{k}{E}(b-1)\ln\left(\frac{n_2}{n_1}\right)+2\frac{k}{E}\ln\left(\frac{T_1}{T_2}\right)+ +\frac{k}{E}\ln\left[\frac{E+2(b-1)kT_2}{E+2(b-1)kT_1}\right]\right\} \tag{13}$$

Because the low value of the Boltzmann's constant, the last term is of the order of 10^{-6} and then it can be dropped and Eq. (13) is simplified in the following expression:

$$\Delta T = T_1 - T_2 = T_1 T_2 \left\{ \frac{k}{E}(b-1)\ln\left(\frac{n_2}{n_1}\right) - 2\frac{K}{E}\ln\left(\frac{T_1}{T_2}\right) \right\} \tag{14}$$

From Eq. (14) it is not possible to derive an explicit expression for T_2. Anyway, because it is valid the following relation:

$$\ln\left(\frac{n_2}{n_1}\right) \succ\succ \ln\left(\frac{T_1}{T_2}\right)$$

Eq. (14) becomes

$$\Delta T = T_1 - T_2 = T_1 T_2 \left\{ \frac{k}{E}(b-1)\ln\left(\frac{n_2}{n_1}\right) \right\} \tag{15}$$

from which an explicit but approximated expression for T_2 is obtained

$$T_2 = \left[\frac{k}{E}(b-1)\ln\left(\frac{n_2}{n_1}\right) + \frac{1}{T_1} \right]^{-1} \tag{16}$$

It is possible to refine the solution by substituting Eq. (16) into Eq. (14) and solving it in steps by iteration as described below.

Let us indicate by i the step index of the solution $T_2^{(i)}$, we compute the approximate solution:

$$T_2^{(0)} = \left[\frac{k}{E}(b-1)\ln\left(\frac{n_2}{n_1}\right) + \frac{1}{T_1} \right]^{-1} \tag{17}$$

and then we use Eq. (14) in an iterative way, i.e.

$$T_2^{(i+1)} = T_1 - T_1 T_2^{(i)} \left\{ \frac{k}{E}(b-1)\ln\left(\frac{n_2}{n_1}\right) + \frac{2k}{E}\ln\left(\frac{T_2^{(i)}}{T_1}\right) \right\} \tag{18}$$

where $i = 0, 1, 2, \ldots$

Hence, in each step, we refine the solution until

$$\left| T_2^{(i+1)} - T_2^{(i)} \right| \prec \varepsilon \tag{19}$$

where ε has been fixed at 10^{-4}.

- Second order

For a second order peak, the peak shift is given by a simplified expression analogous to Eq. (16), i.e.

$$T_2 = \left[\frac{k}{E} \ln\left(\frac{n_2}{n_1} \right) + \frac{1}{T_1} \right]^{-1} \tag{20}$$

Also in this case the iteration algorithm follows the same scheme as described before, but using the following iteration expression:

$$T_2^{(i+1)} = T_1 - T_1 T_2^{(i)} \left\{ \frac{k}{E} \ln\left(\frac{n_2}{n_1} \right) + \frac{2k}{E} \ln\left(\frac{T_2^{(i)}}{T_1} \right) \right\} \tag{21}$$

where $T_2^{(0)}$ is the approximate solution obtained by Eq. (18) and $i = 0, 1, 2, \ldots$..

Eqs. (18) and (21) have been processed for different values of the activation energy E, from 0.75 up to 2 eV, in steps of 0.25 eV, and for kinetics order b equal to 1.5 and 2. The following figures show some of the obtained results.

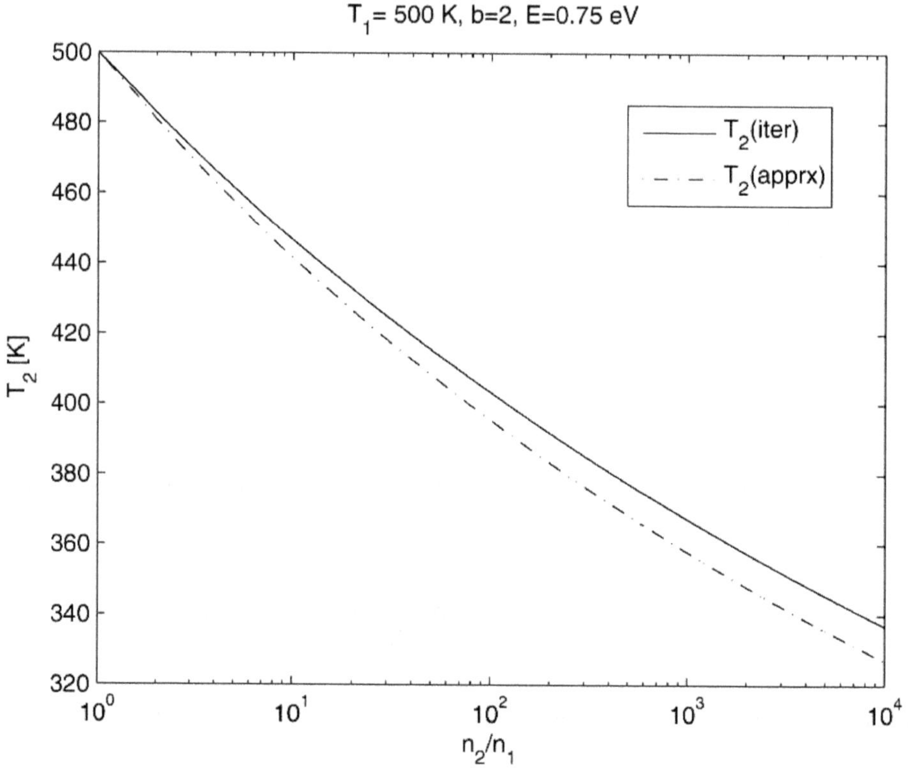

Fig. 1. T_2 Vs n_2/n_1 (T_1=500K, b=2, E=0.75 eV).

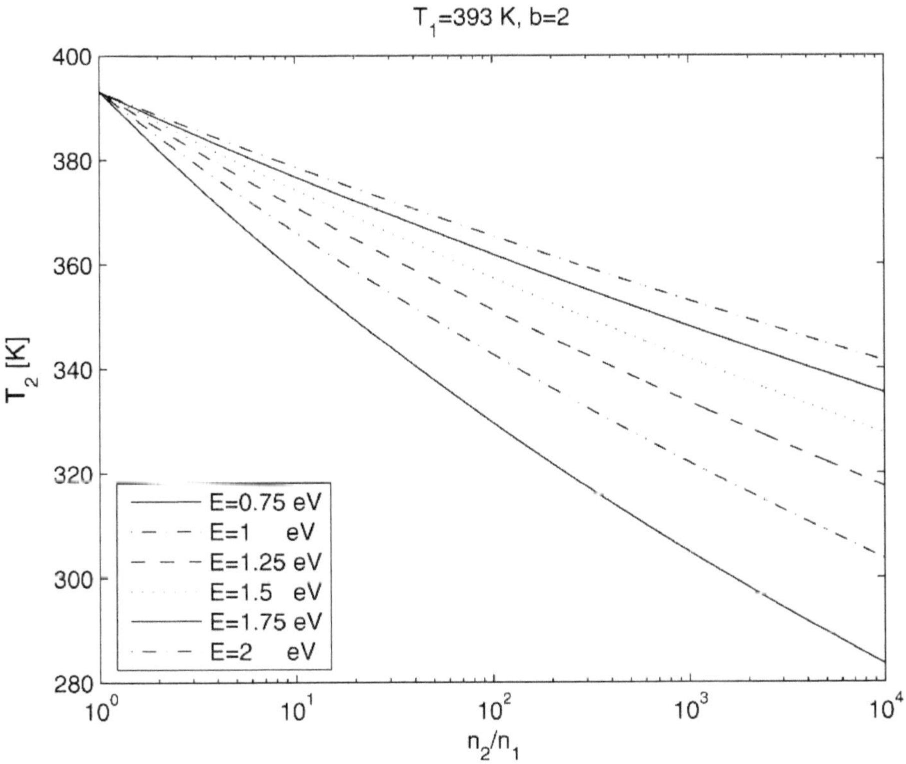

Fig. 2. T_2 Vs n_2/n_1 for various E values (T_1=393 K, b=2).

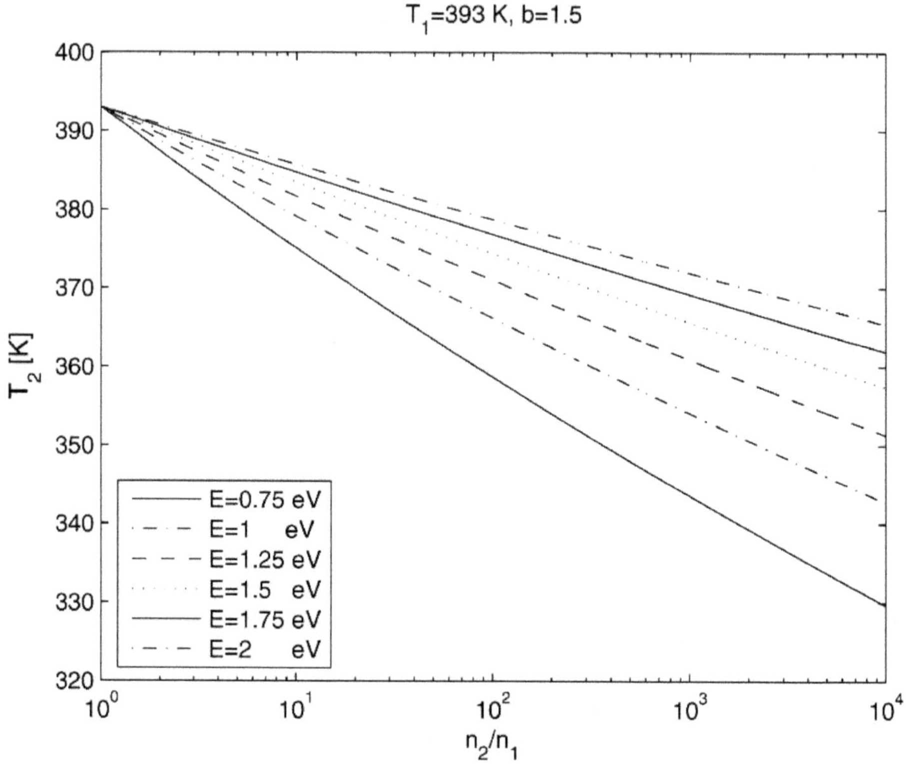

Fig. 3. T_2 Vs n_2/n1 for various values of E (T_1=393 K, b=1.5).

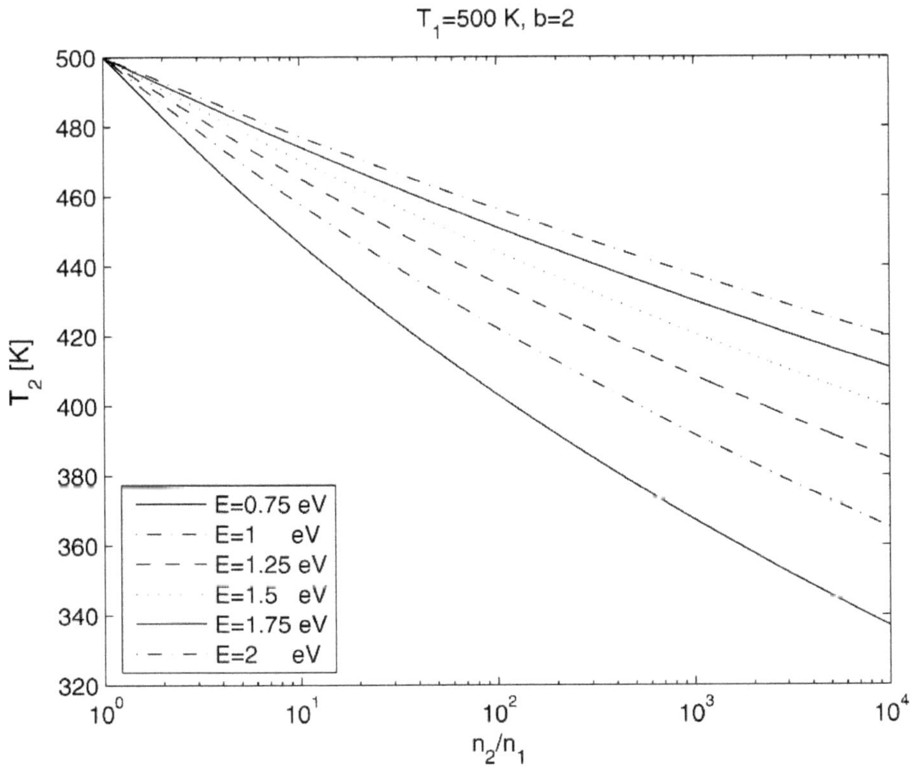

Fig. 4. T_2 Vs n_2/n_1 for various values of E (T_1=500 K, b = 2).

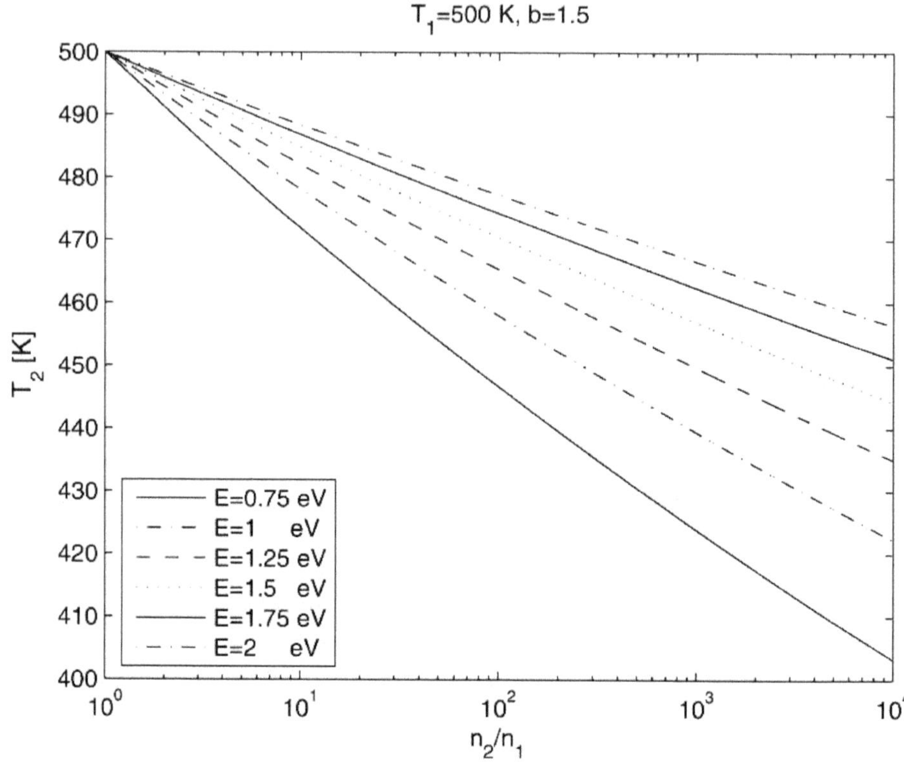

Fig. 5. T_2 Vs n_2/n_1 for various values of E (T_1=500 K, b = 1.5).

References

1. Boss A.J.J., "Theory of thermoluminescence", Rad. Meas. **41**, Suppl.1 (2006)
2. Favalli A., Furetta C. and Cruz-Zaragoza E. Rad. Eff. Def. Sol., **161**, N.o 5 (2006) 279

Peak singularity

M.S. Rasheedy suggested a method for cheeking the isolation and singularity of a thermoluminescence glow peak [1]. The method depends on obtaining the order of the kinetics b at several position of the glow peak. If all values of b are identical with each other, the glow peak is considered isolated.

Making reference to Fig. 1 and to the general order kinetics equation modified by Rasheedy, the following expressions can be written:

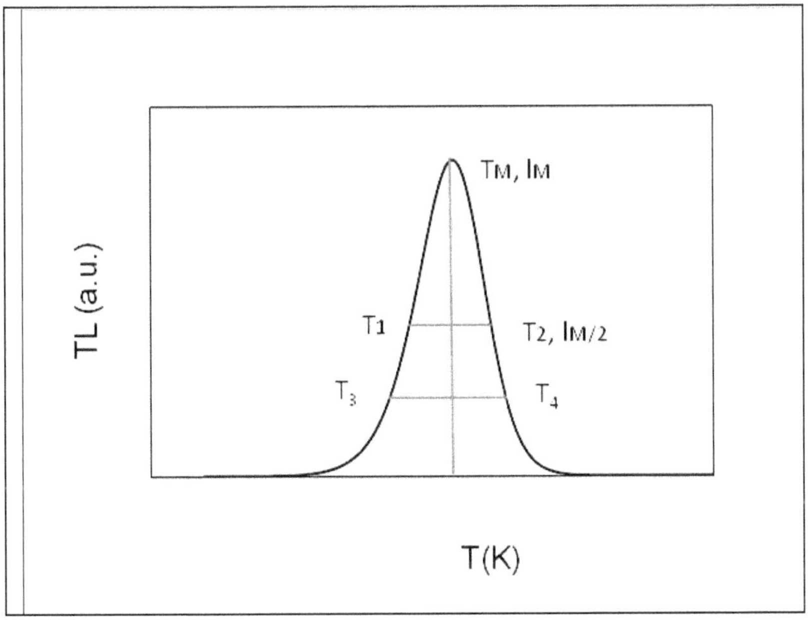

Fig. 1.

$$I_M(T) = \frac{A_M^b}{N^{b-1}} s \exp\left(-\frac{E}{kT_M}\right) \tag{1}$$

$$I_1(T) = \frac{I_M}{2} = \frac{A_1^b}{N^{b-1}} s \exp\left(-\frac{E}{kT_1}\right) \tag{2}$$

$$I_2(T) = \frac{A_2^b}{N^{b-1}} s \exp\left(-\frac{E}{kT_2}\right) \tag{3}$$

$$I_3(T) = \frac{A_3^b}{N^{b-1}} s \exp\left(-\frac{E}{kT_3}\right) \tag{4}$$

$$I_4(T) = \frac{A_4^b}{N^{b-1}} s \exp\left(-\frac{E}{kT_4}\right) \tag{5}$$

Where A_{M1}, A_1, A_2, A_3 and A_4 ibdicate the peak area from T_M to T_f, T_1 to T_f, T_2 to T_f, T_3 to T_f and T_4 to T_f respectively. From Eqs. (1) to (5) the following expressions can be obtained

$$b = \frac{T_M (T_1 - T_2) \cdot \ln 2}{T_1 (T_M - T_2) \ln \frac{A_M}{A_1} - T_2 (T_M - T_1) \ln \frac{A_M}{A_2}} \tag{6}$$

$$b = \frac{T_2 (T_M - T_4) \cdot \ln 2 - T_4 (T_M - T_2) \cdot \ln 4}{T_2 (T_M - T_4) \ln \frac{A_M}{A_2} - T_4 (T_M - T_2) \ln \frac{A_M}{A_4}} \tag{7}$$

$$b = \frac{T_M (T_3 - T_4) \cdot \ln 4}{T_3 (T_M - T_4) \ln \frac{A_M}{A_3} - T_4 (T_M - T_3) \ln \frac{A_M}{A_4}} \tag{8}$$

The previous expressions of b are covering different portions of the peak; any way, more expressions of b may be obtained by selecting different portions on the peak in order to get a better statistics.

Reference

1. Rasheedy, M.S., Physica B **228** (1996) 342

Perovskite's family (ABX_3)

Perovskite compounds corresponding to the general formula ABX_3 (where A is an alkali metal, B is an alkaline earth metal, and X is a halogen, usually fluorine) constitute a class of TL phosphors with good performances, especially when doped with proper activators.

Considerable experimental work has been carriedput on these TL materials, pure or doped with rare earth or transition metal impurities [1-4].

Preparation of these materials in crystalline form is achieved by growing polycrystals or single crystals from a melt, obtained by mixing fluorides of the desired alkali and alkaline earth metals in the stoichiometric ratio. The dopant is usually added to the starting powder before the growth, which can be performed with various techniques (Czochralski, Bridgman, slow cooling).

TL signals of undoped compounds are in generalless intense than those obtained from doped samples. Rare earth impurities show high efficiency as activators in perovskites.

For dosimetry purposes, $KMgF_3$:Eu and $KMgF_3$:Ce can be considered a very interesting phosphor. Its sensitivity is higher (about two to four times) than that of LiF, the response to the radiation dose is linear up to 1 Gy, the most prominent peak at 340°C shows no fading effect in a time of 15 h. Since its effective atomic number (about 13) is higher than that of the biological tissue, a good application would be in the environmental dosimetry.

References

1. Altshuler N.S., Kazakov B.N., Korableva S.L., Livanova L.D. and Stolov A.L., Soviet Phys.- Optics and Spectroscopy **33** (1972) 207
2. Alcala R., Koumvakalis N. and Sibley W.A., Phys. Stat. Sol. (a) **30** (1975) 449
3. Kantha Reddy B., Somaiah K and Hari Babu V., Cryst. Res. Technol. **18** (1983) 1443
4. Furetta C, Bacci C., Rispoli B., Sanipoli C. and Scacco A., Rad. Prot. Dos. **33** (1990) 107
5. Scacco A., Furetta C., Bacci C., Ramogida G. and Sanipoli C., Nucl. Instr. Meth. Phys. Res. **B91** (1994) 223
6. J. Phys. Chem. Solids, 55(11) (1994) 1337
7. Kitis G., Furetta C., Sanipoli C. and Scacco A., Rad. Prot. Dos. **65**(1-4) (1996) 545
8. Kitis G., Furetta C., Sanipoli C. and Scacco A., Rad. Prot. Dos. **82**(2) (1999) 151
9. Furetta C., Sanipoli C. and Kitis G., J. Phys. D: Appl. Phys. **34** (2001) 857
10. Furetta C., Santopietro F, Sanipoli C. and Kitis G., Appl. Rad. Isot. **55** (2001) 533
11. Le Masson N.J.M., Bos A.J.J., Van Eijk C.W.E., Furetta C. and Chaminade J.P., to be published on Rad. Prot. Dos.

Phosphorescence

Phosphorescence takes place for a time longer than 10^{-8} s and it is also observable after removal of exciting source. The decay time of phosphorescence is dependent on the temperature. Referring to Fig. 1, one can observe that this situation arises when an electron is excited (e.g. by ionizing radiation) from a ground state E_o to a metastable state E_m (electron trap), from which it does not return to the ground level with emission of a photon (e.g. the transition from E_m to E_o), because it is completely or partially forbidden by the selection rules.

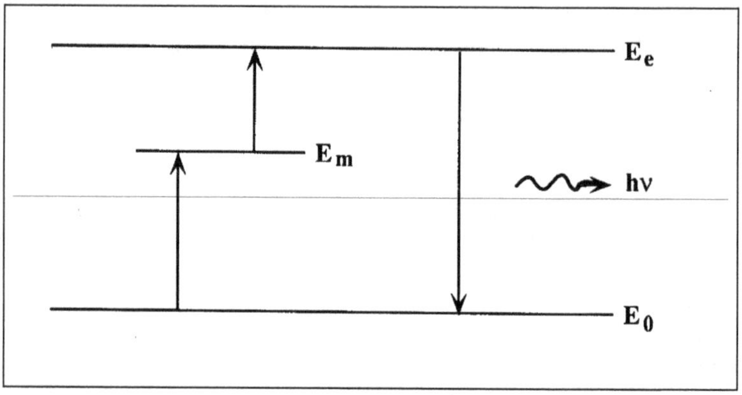

Fig. 1. Phosphorescence phenomenon.

If one supposes that a higher excited level, E_e, exists to which the system can be raised by absorption of the energy E_e - E_m and that the radiative transition E_e - E_m is allowed, then one can provide the energy E_e - E_m by thermal means at room temperature. After that a continuing luminescence emission (phosphorescence) can be observed even after the excitation source is removed. This emission will continue with diminishing intensity until there are no longer any charges in the metastable state. For a short delay time, let us say less than 10^{-4} s, it is difficult to distinguish between fluorescence and phosphorescence. The only way is then to check if the phenomenon is temperature dependent or not. If the system is raised to a higher temperature, the transition from E_m to E_e will occur at an increased rate; consequently the phosphorescence will be brighter and the decay time will be shorter due to the faster depopulation of the metastable state. The phosphorescence is then called thermoluminescence. The delay between excitation and light emission is now ranging from minutes to about 10^{10} years.

The delay observed in phosphorescence corresponds, then, to the time the trapped charge (i.e. an electron) spends in the electron trap. The mean time spent by the electron in the trap, at a given temperature T , is expressed by

$$\tau = s^{-1} \exp\left(\frac{E}{kT}\right) \tag{1}$$

where s is called frequency factor (sec^{-1}), E is the energy difference between E_e and E_m, called trap depth (eV) and k is the Boltzmann's constant ($8.62 \cdot 10^{-4}$ eV K^{-1}). Once the electron is in the elctron trap, it needs an energy E, provided by thermal stimulation, for rising to E_e from E_m and then to fall back to E_0 emitting a photon.

Randall and Wilkins in 1945 [1], presented the first mathematical treatment of phosphorescence, which is also the fondation of the thermoluminescence theory, making the assumption that once the electron has done the transition $E_m \rightarrow E_e$, the probability of retrapping in E_m is much less than the probability to reach E_0. According to their formalism, the emission intensity of phosphorescence at any instant, $I(t)$, is proportional to the rate of recombination (i.e. rate of the transictions $E_e \rightarrow E_0$); because these transitions are depending on the $E_m \rightarrow E_e$ transictions, the intensity of phosphorescence is proportional to the rate of release of electrons, ${}^{dn}\!/_{dt}$, from E_m:

$$I(t) = -c\frac{dn}{dt} \tag{2}$$

where c is a constant (which can be assumed equal to 1).

Eq. (2) can be rewritten as

$$I(t) - c\frac{n}{\tau} \tag{3}$$

where τ^{-1} is the probability per second, p, concerning the thermally stimulated process and n is the concentration of the trapped electrons.

Using Eqs. (2) and (3) one gets, by integration

$$n = n_0 \exp\left(-\frac{t}{\tau}\right) \tag{4}$$

where n_0 is the initial concentration of the trapped electrons.

Eq. (4), together Eq. (2), gives

$$I(t) = I_0 \exp\left(-\frac{t}{\tau}\right) \tag{5}$$

where I_0 is the intensity at time $t = 0$.

Eq. (5) is the equation of the phosphorescence decay at a given constant temperature, which is an exponential decay, also termed first-order decay.

Randall and Wilkins, in their theory, also postulated the probability that the decay of phosphorescence is not-exponential. In fact, the electron released from the trap may return to the trap (**retrapping**) or recombine at E_0. In this case, the recombination rate is proportional to both the concentration of the trapped electrons in E_m and to the concentration of recombination sites in E_0. Assuming that the concentrations are equal (n in E_0 = n in E_m), the intensity is now given by

$$I(t) = -c\frac{dn}{dt} = \alpha \cdot n^2 \tag{6}$$

where α ($cm^3\ sec^{-1}$) is a constant.

The solution of Eq. (6) is

$$I(t) = \frac{I_0}{\left(1 + \alpha \cdot n_0 \cdot t\right)^2} \tag{7}$$

which is related to a hyperbolic decay of phosphorescence, termed second-order decay. The physical process is called bimolecular.

E.I. Adirovitch, in 1956 [2], used a set of three differential equations to explain the decay of phosphorescence in a more general case.

References

1. Randall J.T. and Wilkins M.H.F., Proc. Roy. Soc. **A184** (1945) 366
2. Adirovitch E.I., J. Phys. Rad. **17** (1956) 705

Phosphors (definition)

The term phosphors is used to designate all solid or liquid luminescent materials. This term is also used, in particular, for thermoluminescent materials (i.e., TL phosphors).

Photon energy response (calculation)

In any dosimetric applications in the field of photon radiation, the energy response is one of the main characteristics that must be known. The energy response, or energy dependence, is the measure of the energy absorbed in the thermoluminescent material in comparison to the energy absorbed in a reference material (i.e. air or human tissue), when irradiated at the same dose [1,2,3]. Let us indicate with S(E) the energy response; then, according to the definition so far given, S(E) is given by

$$S(E) = \frac{\left(\frac{\mu_{en}}{\rho}\right)_{TLD}}{\left(\frac{\mu_{en}}{\rho}\right)_{air}} \tag{1}$$

where $\left(\frac{\mu_{en}}{\rho}\right)_{TLD}$ and $\left(\frac{\mu_{en}}{\rho}\right)_{air}$

are the mass energy absorption coefficients for the TLD and for air respectively.

Because the ^{60}Co (1.25 MeV) is normally considered as the reference photon source, it is convenient to introduce the relative energy response, RER, of the TLD material, at the photon energy E, normalized to the ^{60}Co energy:

$$RER = \frac{S(E)}{S(E)_{^{60}Co}} \tag{2}$$

Since TLDs are complex media, the law of mixture can be applied:

$$\left(\frac{\mu_{en}}{\rho}\right)_{TLD} = \sum_i \left(\frac{\mu_{en}}{\rho}\right)_i \cdot W_i \tag{3}$$

where $\left(\frac{\mu_{en}}{\rho}\right)_i$ is the mass energy absorption coefficient of the i-th element and W_i is its fraction by weight.

As an example, the RER has been calculated for Ge-doped optical fibre [4]. Table 1 shows the fiber composition detectected by Scan Electron Microscope (SEM).

element	W_i (%)
Si	46.12
O	53.64
Ge	0.233

Table 1. Ge-doped optical fiber composition.

Table 2 shows the mass energy absorption coefficients for each element and for each energy [5].

Energy (MeV)	μ_{en}/ρ (cm^2/g)			
	Si	O	Ge	Air
0.015	9.794	1.545	62.56	1.334
0.03	1.164	1.729x10^{-1}	11.26	1.537x10^{-1}
0.05	2.43x10^{-1}	4.414x10^{-2}	2.759	4.098x10^{-2}
0.1	4.513x10^{-2}	2.355x10^{-2}	3.803x10^{-1}	2.325x10^{-2}
1.25	2.652x10^{-2}	2.669x10^{-2}	2.353x10^{-2}	2.666x10^{-2}
6	1.827x10^{-2}	1.668x10^{-2}	2.027x10^{-2}	1.647x10^{-2}
10	1.753x10^{-2}	1.483x10^{-2}	2.208x10^{-2}	1.45x10^{-2}
20	1.757x10^{-2}	1.36x10^{-2}	2.452x10^{-2}	1.311x10^{-2}

Table 2. Mass energy absorption coefficients for the elements of Ge-doped optical fiber and for air.

Table 3 shows, at each energy, both the energy dependence, Eq. (1), as well as the experimental and theoretical RER, Eq. (2), for Ge-doped optical fiber.

Energy (MeV)	Energy Dependence	Relative Energy Response	
		Theoretical	Experimental
0.015	4.116	4.126	---
0.03	4.26	4.269	---
0.05	3.468	3.475	3.92
0.1	1.44	1.443	1.497
1.25	0.997	1	1
6	1.05	1.052	1.02
10	1.109	1.11	0.93
20	1.17	1.172	1.11

Table 3. Energy dependence, Eq. (1), and relative energy response (RER), Eq. (2), for Ge-doped optical fiber.

Fig. 1 shows the energy dependence according Eq. (1) and Table 3. Fig. 2 shows the relative energy response (RER), both theoretical and experimental results, according to Table 3.

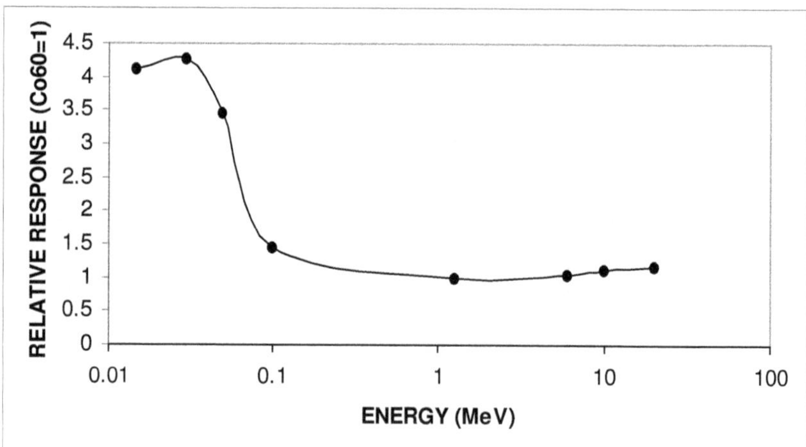

Fig. 1. Energy dependence according to the data given in Table 3.

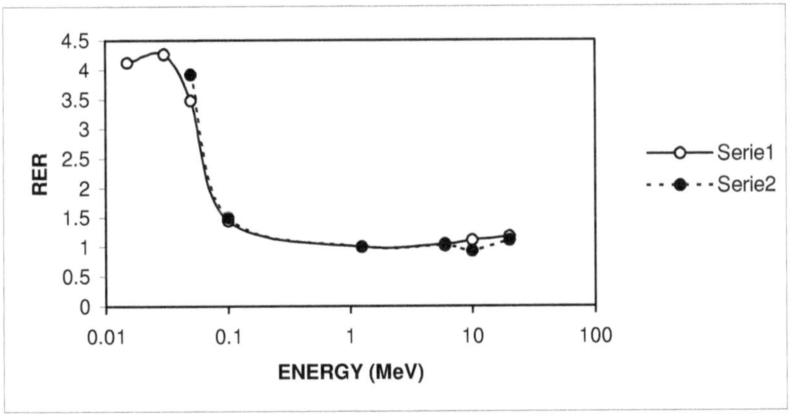

Fig. 2. Theoretical and experimental relative energy response (RER) according to the data in Table 3.

References

1. Busuoli G. in Applied Thermoluminescence Dosimetry, ISPRA Courses, Edited by M. Oberhofer and A. Scharmann (Adam Hilger Ltd, Bristol, 1981)
2. Furetta C. and Weng P.S., Operational Thermoluminescence Dosimetry (World Scientific, 1998)
3. S.W.S. McKeever, Thermoluminescence of Solids (Cambridge University Press, 1985)
4. Youssef Abdulla (private communication)
5. Hubbell J.H. and Selzer S.M., Int. J. Appl. Radiat. Isot. **33** (1982) 1269

Photon energy response (definition)

The energy response is a measure of the energy absorbed in the TL material used in comparison to the energy absorbed in a material taken as reference (i.e., air or tissue), when irradiated at the same exposure dose.

The energy response is a characteristic of each thermoluminescent material and its direct measurement is only possible when the TL sample is irradiated under electronic equilibrium conditions.

The following Table 1 lists the energy response at 30 keV, normalised to the response to ^{60}Co and to ^{137}Cs, for many different kind of phosphors.

phosphors	30 keV/^{60}Co	30 keV/^{137}Cs
LiF:Mg,Ti LiF:Mg,Ti,Na LiF:Mg,Cu,P	1.3	1.3
$Li_2B_4O_7$:Mn $Li_2B_4O_7$:Cu Li2B4O7:Cu,Ag	0.9 – 0.98 0.8 – 0.98 0.98	0.9 0.8 - 0.9
MgB_4O_7:Dy MgB_4O_7:Tm	1.5	1.3 - 2.4
Mg_2SiO_4:Tb	4 - 5	~ 4.5
$CaSO_4$:Dy $CaSO_4$:Tm $CaSO_4$:Mn	11.5 11.5 11.2	~ 13
CaF_2 (natural) CaF_2:Dy CaF_2:Mn	14.5 15.6 15.4	~ 15
BeO	0.87 - 1.4	0.9 - 1
Al_2O_3	3.5	~ 4.5

Table 1. Normalised energy response.

Phototransferred thermoluminescence (PTTL) (general)

Phototransfered TL technique is based on the phenomenon of re-excited TL by UV illumination after annealing or read-out of a thermoluminescent sample. UV irradiation induces transfer of electrons from deeper traps (not involved in the annealing or read-out procedures) into shallower traps. The efficiency of the phenomenon is temperature dependent. The phototransfer effect was first observed by Stoddard (1960).

A practical use of phototransfer is in TL dosimetry and TL dating: i.e., measurement of carriers accumulated in very deep traps as a measure of the absorbed dose.

The phototransfer technique consists in giving to the sample a certain amount of UV light which allows the transfer of carriers from a deep trap to a shallower one. The TL intensity of the transfered peak is proportional to the original concentration of the carriers in the deep trap.

Reference

Stoddard A.E., Phys. Rev. **120** (1960) 114

Phototransferred thermoluminescence (PTTL): model

The most simple model for phototransferred thermoluminescence (PTTL) is the one which considers one shallow trap, one deep trap and one reconbination centre [1-3].

Let us indicate:

n_a = concentration of electrons in the shallow traps (acceptor),
n_d = concentration of electrons in the deep traps (donor),
m = concentration of holes in the recombination centres,
n_c = electrons in the conduction band (CB),
N_d = total number of deep traps (donor traps),
N_a = total number of shallow traps (acceptor traps),
M = recombination centres,
$\tau = (n_c A_m)^{-1}$ is a lifetime,
A_d, A_a = retrapping probability for free electrons into empty traps,
A_m = recombination probability,
f = rate of loss of electrons from deep traps (donor) owing to light excitation,
γ_a = thermal excitation from the shallow (acceptor) traps.

The initial conditions, at the end of the ionizing radiation and before the light illumination ($t = 0$), are

$n_{a0} = 0$

$n_{d0} = m_0$

Considering now that the illumination excites electrons from the deep traps (donor traps) to the shallow traps (acceptor traps), one can write the following rate equations, valid during the illumination period (0 – t*):

$$\frac{dn_c}{dt} = -\frac{dn_a}{dt} - \frac{dn_d}{dt} + \frac{dm}{dt} \tag{1}$$

$$\frac{dn_a}{dt} = n_c A_a (N_a - n_a) \tag{2}$$

$$\frac{dn_d}{dt} = n_c A_d (N_d - n_d) - n_d f \tag{3}$$

$$\frac{dm}{dt} = A_m m n_c = \frac{m}{\tau} \tag{4}$$

Considering the equilibrium condition

$$\left|\frac{dn_c}{dt}\right| \prec\prec \left|\frac{dn_a}{dt}\right|, \left|\frac{dm}{dt}\right| \tag{5}$$

and the condition of no retrapping into the donor traps:

$$n_d f \succ\succ n_c A_d (N_d - n_d) \tag{6}$$

Integration of Eq. (3), taking into account the condition (6)

$$\int_{n_{d0}}^{n_d} \frac{dn_d}{dt} = -\int_0^{t^*} f dt$$

gives

$$n_d^* = n_{d0} \exp(-ft^*) \tag{7}$$

From Eq. (2), with the initial condition $n_{a0} = 0$, we get

$$\int_{N_a}^{N_a - n_a} \frac{dn_a}{(N_a - n_a)} = \int_0^{t^*} n_c A_a dt$$

we get

$$n_a^* = N_a \left[1 - \exp\left(-n_c A_a t^* \right) \right] \tag{8}$$

Finally, from integration of Eq. (4)

$$\int_m^m \frac{dm}{dt} = -\frac{1}{\tau} \int_0^{t^*} dt$$

we obtain

$$m^* = m_0 \exp\left(-\frac{t}{\tau} \right) \tag{9}$$

where $n_c A_a$ and τ are approximately constant if $dn_c/dt \approx 0$.

At the end of illumination, according to the Eqs. (7), (8) and (9), a certain concentration of charges will be then in traps and centers.

The heating phase of the sample follows the illumination phase. The heating phase is similar to the situation of competition during heating, so that the mathematical treatment is very similar.

The set of the new equations is now:

$$\frac{dn_c}{dt} = -\frac{dn_a}{dt} - \frac{dn_d}{dt} + \frac{dm}{dt} \tag{10}$$

$$\frac{dn_a}{dt} = A_a n_c (N_a - n_a) - \gamma_a n_a \tag{11}$$

$$\frac{dn_d}{dt} = A_d n_c (N_d - n_d) \tag{12}$$

$$I = -\frac{dm}{dt} = A_m m n_c \tag{13}$$

Eqs. (12) and (13) can be rewritten as

$$n_c = -\frac{1}{A_d}\frac{d}{dt}\left[\ln\left(N_d - n_d\right)\right] \tag{14}$$

$$n_c = -\frac{1}{A_m}\frac{d}{dt}\left(\ln m\right) \tag{15}$$

which can be integrated taking into account that the initial values of m and n_d (at the end of illumination) are, respectively, m^* and n_d^*.
Then, the integration yields to

$$N_d - n_d = \left(N_d - n_d^*\right)\left(\frac{m}{m^*}\right)^{A_d/A_m} \tag{16}$$

Considering now the quasi-equilibrium condition (5), Eq. (10) can be written as

$$\frac{dm}{dt} \approx \frac{dn_a}{dt} + \frac{dn_d}{dt} \tag{17}$$

By substitution of Eqs. (11), (12) and (13) in (17) we obtain:

$$-A_m m n_c \approx A_a n_c \left(N_a - n_a\right) - \gamma_a n_a + A_d n_c \left(N_d - n_d\right)$$

from which we get an explicit expression for n_c:

$$n_c = \frac{\gamma_a n_a}{A_d\left(N_d - n_d\right) + A_a\left(N_a - n_a\right) + A_m m} \tag{18}$$

Then, the intensity is given by

$$I=-\frac{dm}{dt}=A_m m n_c=\frac{A_m m\gamma_a n_a}{A_d\left(N_d-n_d\right)+A_a\left(N_a-n_a\right)+A_m m} \tag{19}$$

where the first term in the denominator is the probability of retrapping in donor level, the second concerns the retrapping in the acceptor level and the third is the recombination probability.
The integration of Eq. (17) yields

$$m-m^*=\left(n_a-n^*\right)+\left(n_d-n^*\right) \tag{20}$$

Substituting (16) and (20) in Eq. (19), we obtain

$$I(t)=-\frac{dm}{dt}=\gamma_a A_m m F(m) \tag{21}$$

where

$$F(m)=\frac{\left(n_a^*+n_d^*-m^*-N_d\right)+m+\left(N_d-n_d^*\right)\left(\frac{m}{m^*}\right)^{A_d/A_m}}{A_a\left(N_a+N_d-n_a^*-n_d^*+m^*\right)+\left(A_m-A_a\right)\left(N_d-n_d^*\right)\left(\frac{m}{m^*}\right)^{A_d/A_m}}$$

Assuming now that trapping in donor level is larger than both retrapping in acceptor level and in recombination centre, this means

$$A_d\left(N_d-n_d\right)\succ\succ A_a\left(N_a-n_a\right)+A_m m \tag{22}$$

so Eq. (19) becomes

$$I(t)=-\frac{dm}{dt}=\frac{A_m\gamma_a m n_a}{A_d\left(N_d-n_d\right)} \tag{23}$$

Furthermore, assuming that retrapping in the shallow traps is very little compared to the rate of release of trapped electrons, i.e.

$$\gamma_a n_a \succ\succ A_a n_c\left(N_a-n_a\right)$$

Eq. (11) becomes

$$\frac{dn_a}{dt} = -\gamma_a n_a \tag{24}$$

which gives the following solution

$$n_a = n_a^* \exp\left(-\int_0^t \gamma_a d\tau\right)$$

Substituting now Eqs. (16) and (24) into Eq. (23), we get

$$I(t) = -\frac{dm}{dt} = \frac{A_m \gamma_a n_a^* m \exp\left(-\int_0^t \gamma_a d\tau\right)}{A_d\left(N_d - n_d^*\right)\left(\dfrac{m}{m^*}\right)^{A/A_m}}$$

which can be integrated as following

$$\int_{m^*}^{m} -\frac{1}{m}\left(\frac{m}{m^*}\right)^{A_d/A_m} \frac{A_d\left(N_d - n_d^*\right)}{A_m n_a^*} dm = \int_0^t \gamma \exp\left(-\int_0^t \gamma \cdot d\tau\right) dt$$

Since the untegral on the right side of the previous expression is equal to unity, we get

$$m = m^* \left[1 - \frac{n_a^*}{N_d - n_d^*}\right]^{A_m/A_d}$$

which, using the approximation $n_a^* \prec\prec N_d - n_d^*$, becomes

$$m \approx m^* \left[1 - \frac{A_m}{A_d}\left(\frac{n_a^*}{N_d - n_d^*}\right)\right]$$

The area S under the glow curve is then given by

$$S = \int_0^t I(t) dt = \int_0^t -\frac{dm}{dt} dt = m^* - m$$

so that

$$S \approx m^* \left[\frac{A_m n_a^*}{A_d \left(N_d - n_d^* \right)} \right]$$

which can be transformed using Eqs. (7), (8) and (9):

$$S = \frac{m_0 \exp\left(-\frac{t}{\tau} \right) N_a A_m \left[1 - \exp\left(-n_c A_a t^* \right) \right]}{A_d \left[N_d - n_{d0} \exp\left(-f t^* \right) \right]}$$

with the condition $n_{d0} = m_0$ at the end of irradiation and immediately before illumination, the glow curve area becomes

$$S = C \frac{\exp\left(-\frac{t}{\tau} \right) N_a A_m \left[1 - \exp\left(-n_c A_a t^* \right) \right]}{A_d \left[\frac{N_d}{n_{d0}} - \exp\left(-f t^* \right) \right]}$$

which describes the PTTL peak produced by the shallow trap as a function of the illumination period 0 ----t*.

References

1. McKeveer S.W.S. and Chen R., Rad. Meas. **27** (1997) 625
2. Chen R. and McKeever S.W.S., Theory of Thermoluminescence and Related Phenomena (World Scientific, 1997)
3. Alexander C.S. and McKeever S.W.S., J. Phys. D: Appl. **31** (1998) 2908

Post-irradiation annealing

The post-irradiation annealing is the thermal procedure having the aim to erase all the low temperature peaks which could be errors in the dose estimation because their high fading rate (see **Annealing: general considerations**)

Post-readout annealing

The post-readout annealing is another way to indicate the annealing procedure, i.e. the standard annealing, to be used before using again the thermoluminescent dosemeters.

Precision and accuracy (general considerations)

Before of the identification of the sources of error in thermoluminescent dosimetry, to classify them and finally to give suggestions on the procedures to be used to optimize the experimental results, some general considerations should be given [1,2].

The results obtained by a dosimetric evaluation, based on thermoluminescence phenomenon, present a large dispersion and then high uncertainty. To identify all the sources of uncertainty it is necessary to write the general relationship between the dose D and the correlated TL emission signal. Several factors may be present in the dose determination

$$D = M_{net} \cdot S_i \cdot F_c \cdot F_{st} \cdot F_{en} \cdot F_{lin} \cdot F_{fad} \tag{1}$$

where

- D is the absorbed dose in the phosphor,
- M_{net} is the net TL signal (i.e., the TL signal corrected for the intrinsic background signal M_0: $M_{net} = M - M_0$),
- S_i is the relative intrinsic sensitivity factor or individual correction factor concerning the ith dosimeter,
- F_c is the individual calibration factor of the detector, relative to the beam quality, c, used for calibration purposes,
- F_{st} is the factor which takes into account the possible variations of F_c due to variations of the whole dosimetric system and of the experimental conditions (electronic instabilities of the reader, changes in the planchet reflectivity, changes in the light transmission efficiency of the filters interposed between the planchet and the PM tube, temperature instabilities of the annealing ovens, variation of the environmental conditions in the laboratory, changes in the dose rate of the calibration source, etc.),
- F_{en} is the factor which allows for a correction for the beam quality, q, if the radiation beam used is different from the one used for the detector calibration,
- F_{lin} is the factor which takes into account for the non-linearity of the TL signal as a function of the dose,
- F_{fad} is the correction factor for fading which is a function of the temperature and the period of time between the end of irradiation and readout.

All the conversion and correction factors, let us say to be in the number of m, can be indicated by using the general symbol α_j. In this sense, the relation between dose and TL reading, Eq. (1), can be rewritten as

$$D = (M_i - M_0)\prod_{j=1}^{m} \alpha_j \tag{2}$$

Before going into a deep discussion, we have to say that a measurement, which is the "reading" in the present case of thermoluminescence, can be affected by two types of errors: the random and the systematic errors.

The random errors are variable in both magnitude and sign. For random uncertainties a statistical procedure can be applied since their probability distribution is known.

On the contrary, a source of systematic errors has a constant relative magnitude and is always of the same sign. A statistical procedure cannot then be applied because the distribution is not known.

Furthermore, two terms are very important to discriminate between errors. These two terms are "precision" and "accuracy ".

Precision is a term related to the reproducibility of a system and concerns statistical methods applied to a number of repeated measurements. Low precision means that random uncertainties are very high.

Accuracy concerns the closeness with which a measurement is expected to approach the true value and includes both types of uncertainties. The value of a quantity is considered "true" either by theoretical considerations or by comparison with fundamental measurements. The true value is also called "actual value". The measured value is called "indicated value" or " reading". High accuracy means that the measured value and the true value are nearly the same.

Random uncertainties

Repeated measurements follow a normal distribution, which is characterized by the standard deviation σ of the group of results. From a statistical point of view, for an infinite number of results 95% of them fall within 2σ of the mean. This is the commonly applied criterion to specify the reproducibility or precision of a system.

Let us denote with X the measured value of any quantity. For a normal distribution, the probability of X to have a value between X and $X+dX$ is given by

$$P(X)dX = \frac{1}{\sigma\sqrt{2\pi}}\exp\left[-\frac{(X-\mu)^2}{2\sigma^2}\right]dX \tag{3}$$

where

μ is a constant equal to the value of X at the maximum of the distribution curve;

σ, the standard deviation, is a measure of the dispersion or width of the curve (FWHM). The quantity σ^2 is the variance of the distribution.

Performing N measurements of the same quantity X, the best estimate of μ is given by the mean value of the N measurements:

$$\overline{X} = \frac{1}{N}\sum_{i=1}^{N} X_i \tag{4}$$

The best estimate of σ^2 is the variance given by

$$S^2 = \frac{1}{N-1}\sum_{i=1}^{N}\left(X_i - \overline{X}\right)^2 \tag{5}$$

In the practical situations the $\overline{X}$ comes from a limited number of measurements. In this case, one can perform repeated determinations of the average, let us say *M*. It must be noted that if *M* is large, the average value will have a distribution very close to the normal one whatever the distribution of *X* is.

It is now possible to define the standard deviation of the distribution of the average, called standard error:

$$\overline{S}^2(\overline{X}) = \frac{1}{N(M-1)}\sum_{i=1}^{M}(X_i - \overline{X})^2 = \frac{S^2(X)}{M} \tag{6}$$

In many cases, as in the one of Eq. (2), measurements involve several quantities.

This means that the value *X* of a physical quantity, e.g. the dose *D*, is a function of other physical quantities, e.g. the parameters α_j. Each of the separate quantities has a proper variance, i.e.:

$$S_1^2(\alpha_1), \qquad S_2^2(\alpha_2), \qquad S_3^2(\alpha_3), \qquad \ldots \tag{7}$$

The variance of *X* is then given by

$$S^2(X) = \left(\frac{\partial X}{\partial \alpha_1}\right)^2 S_1^2(\alpha_1) + \left(\frac{\partial X}{\partial \alpha_2}\right)^2 S_2^2 + \ldots \tag{8}$$

A similar expression holds for the average.

Systematic uncertainties

Let us again consider a physical quantity X which depends on the independent measurements of separate physical quantitities α_j. Because the distribution functions for each of the quantitities α_j are not known in the case of systematic errors, the methods for combining the individual systematic uncertainties are less well defined than for the random uncertainties. Several methods can be used in practice to combine the different conmponents in order to give the overall systematic uncertainty ΔX.

The first method considers a simple arithmetic addition

$$\Delta X = (\Delta X)_{\alpha_1} + (\Delta X)_{\alpha_2} + ... = \left|\frac{\partial X}{\partial \alpha_1}\right| \Delta\alpha_1 + \left|\frac{\partial X}{\partial \alpha_2}\right| \Delta\alpha_2 + ... \quad (9)$$

The second method is to combine them in quadrature

$$\Delta X^2 = (\Delta X)^2_{\alpha_1} + (\Delta X)^2_{\alpha_2} ... = \left(\frac{\partial X}{\partial \alpha_1}\right)^2 \Delta^2\alpha_1 + \left(\frac{\partial X}{\partial \alpha_2}\right)^2 \Delta^2\alpha_2 + ... \quad (10)$$

Because the first method overestimate the total systematic uncertainty while the second tends to underestimate it, it has been suggested to multiply by 1.13 the result of systematic errors. The factor 1.13 is necessary to ensure a minimum confidence level of 95%.

References

1. Busuoli G. in Applied Thermoluminescent Dosimetry, ISPRA Courses, Edited by M. Oberhofer and A. Scharmann. Adam Hilger publisher (1981)
2. Marshall T.O. in Proceeding of the Hospital Physicists' Association. Meeting on Practical Aspects of TLD. Edited by A.P. Hufton, University of Manchester, 29th March, 1984

Precision concerning a group of TLDs of the same type submitted to one irradiation

One group of a same type of TLDs are annealed, irradiated and then readout. The variations in the precision are mainly due to the following causes:

- variation in the mass among the TLDs group
- variation in the opticval transmission from sample to sample
- instability of the TL reader during the period of the measurement

The precision is expressed by the following equation:

$$\sigma_{TOT} = \sqrt{\left(\frac{\sigma_S}{100} D\right)^2 + \sigma^2_{BKG}}$$

where

$\frac{\sigma_S}{100}$ is the percentage standard deviation of the dosemeter group irradiated at the dose D

σ_{BKG} is the standard deviation of the background readings of the unirradiated dosemeters.

Reference

Busuoli G. in Applied Thermoluminescence Dosimetry, ISPRA Courses, Edited by M. Oberhofer and A. Sharmann, Adam Hilger Ltd., Bristol (1981)

Precision concerning only one TLD undergoing repeated cycles of measurements (same dose)

A single TLD is repeatedly annealed, irradiated at exactly the same dose and read-out. All the experimental parameters must be kept constant. Variations in TL readings are then observed. The sources of the reading variations are mainly due to:

- dosimeter's background signal, or zero dose reading, and its variations;
- electronic instability of the reading system

The precision may be expressed by the following expression giving the total standard deviation in a series of repeated measurements, at any dose D, carried out on only one single dosimeter

$$\sigma_{TOT} = \sqrt{\left(\frac{\sigma_S}{100} D\right)^2 + \sigma_{BKG}^2}$$

where

- $\sigma_S/100$ is the percentage standard deviation of the repeated measurements when the background is negligible,
- D is the absorbed dose,
- σ_{BKG}^2 is the variance of the readings of the unirradiated dosimeter expressed in equivalent absorbed dose.

Reference

Busuoli G. in Applied Thermoluminescence Dosimetry, ISPRA Courses, Edited by M. Oberhofer and A. Sharmann, Adam Hilger Ltd., Bristol (1981)

Precision concerning several identical dosimeters submitted to different doses

The equation given for the **precision a group of TLDs of the same type submitted to one irradiation**

$$\frac{\sigma_T}{D} = \sqrt{(\sigma_S)^2 + \left(\frac{\sigma_B}{D}\right)^2} \qquad (1)$$

can now be used to test the reproducibility of a TL system using a batch of TLDs (let us say 10 TLDs of the same type) which are irradiated consecutively to doses starting with the lowest detectable dose of the system, LDD, up to 1000LDD. This is the approach used by Burkhardt and Piesh [1] and further developed by Zarand and Polgar [2, 3]. The relative standard deviations, obtained at different dose levels, are then compared to the theoretical two parameters Eq. (1). The lowest detectable dose, LDD, according to [4], is defined as three times the standard deviation of the zero dose reading of the unirradiated batch. The characteristic shape of $(\sigma_T/D)\%$ as a function of dose D is shown in Fig. 1.

The behaviour of the plot can be interpreted considering the effect of two components:

- σ_B, which is the main parameter affecting the reproducibility in the low dose region (<100LDD), and takes into account the intrinsic variability of the TL system (batch quality and readout technique), which is evaluated by the readings of the unirradiated TLDs;

- σ_S takes into account the variation of the TL system as a function of the irradiation (batch calibration, annealing and reader quality).

To test the validity of Eq. (1), a set of 4 LiF:Mg,Ti (TLD-700) has been irradiated at several doses from 0.3 mGy to 1 Gy. Each reading was then corrected by the individual background and the individual correction factor. After that the average value and the relative standard deviation were calculated. Table 1 shows the experimental results.

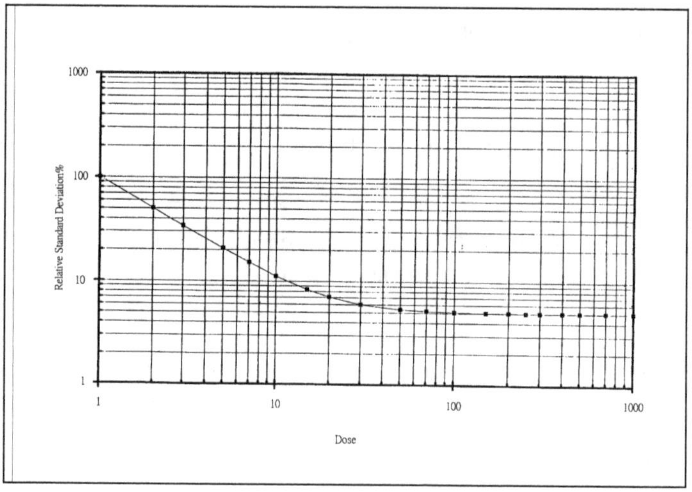

Fig. 1. Theoretical behaviour of the two parameters function given by Eq. (1).

D (mGy)	M	$\overline{M}$	σ% (exp)	σ% (th)	D (mGy)	M	$\overline{M}$	σ% (exp)	σ% (th)
0.3	1.194				20	69.637			
	1.092					67.857			
	0.991					68.637			
	0.887	1.041	12.7	11.4		67.647	68.440	1.5	1.8
0.55	2.005				30	101.437			
	1.742					104.437			
	2.017					103.137			
	1.787	1.888	7.6	6.4		106.617	103.82	2.1	1.8
1	3.493				50	170.737			
	3.737					169.937			
	3.547					165.737			
	3.357	3.530	4.5	3.8		164.737	168.00	1.8	1.8
2	6.863				100	332.078			
	6.812					339.518			
	7.017					331.420			
	6.547	6.800	2.8	2.5		341.010	333.01	1.5	1.8
5	16.737				300	1027.01			
	17.207					994.031			
	16.437					985.071			
	17.137	16.880	2.2	1.9		999.075	1001.3	1.8	1.8
10	33.837				1000	3371.51			
	33.450					3371.02			
	33.900					3500.15			
	35.437	34.406	2.2	1.9		3401.73	3411.1	1.8	1.8

Table 1. Comparison of experimental and theoretical data for σ_T.

Figure 2 shows the plot of the relative standard deviation, in percentage, as a function of the dose.

The following data were used:

$$\sigma_B = 3.4\ 10^{-2}\ \text{mGy and } \sigma_S = 1.7810^{-2}\ \text{reader units}$$

so that

$$\frac{\sigma_T}{D}\% = 100\times\sqrt{(\sigma_S)^2 + (\frac{\sigma_B}{D})^2} =$$

$$100\sqrt{3.17\times10^{-4} + \frac{1.144\times10^{-3}}{D^2}}$$

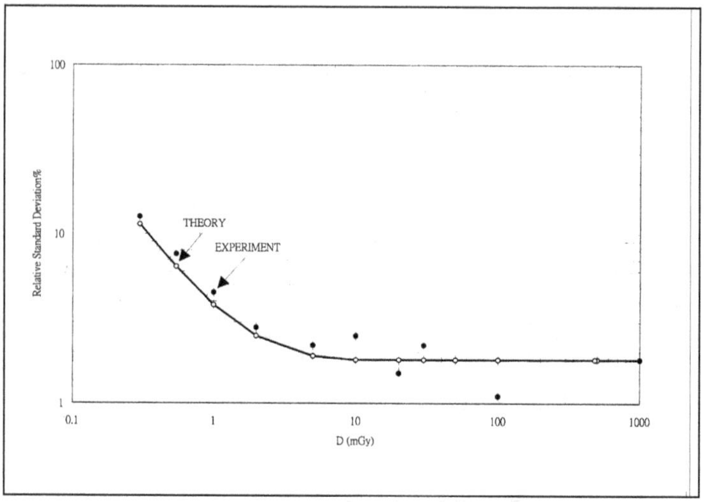

Fig. 2. Comparison between experimental and theoretical values of Eq. (1) as a function of dose.

References

1. Burkhardt B. and Piesh E., Nucl. Instr. Meth. **175** (1980) 159
2. Zarand P. and Polgar I., Nucl. Instr. Meth. **205** (1983) 525
3. Zarand P. and Polgar I., Nucl. Instr. Meth. **222** (1984) 567
4. Busuoli G. in Applied Thermoluminescent Dosimetry, ISPRA Courses, Edited by M. Oberhofer and A. Scharmann, Adam Hilger publisher, Bristol (1981)

Precision concerning several identical dosimeters undergoing repeated and equal irradiation (procedures)

First procedure

An accurate way for studying the reproducibility of a TLD system (detector + reader + annealing + irradiation) can be performed using several dosimeters (i.e., 10 - 20) of the same type [1]. All the dosimeters are annealed, irradiated and read out and the same procedure is repeated several times (10 - 20 times) as it is shown in Fig. 1. The analysis of the the coefficients of variation allows us to determine the different sources of variation: i.e., the variation of the system, the reader and the TL elements..

The following Table 1 shows, as an example, the matrix of the results obtained from a sample of 10 TLDs (LiF:Mg,Cu,P).

From the data it is possible to calculate the following quantities:

- $\overline{\%CV}$ which is the mean value of the percent standard deviations of each detector; it would give an indication of the reproducibility of the whole system. It is called here "system variability index" , SVI.

- $\%CV$ which is the percent standard deviation of the mean values of each cycle of readings; it would give an indication on the long term reader reproducibility. It is indicated here as "reader variability index", RVI.
- from the previous quantities it is now possible to define an index of variability only concerning the TL detectors. This index, called "detector variability index", DVI, is calculated as the square root of the difference between the system reproducibility index and reader reproducibility index, both squared:

$$DVI = \sqrt{(SVI)^2 - (RVI)^2} \tag{1}$$

Using the data in the example, we get

$$SVI = \%\overline{CV} = 1.16\%$$
$$RVI = \%CV = 0.62\%$$
$$DVI = 0.98\%$$

Second procedure

Another type of test allows a much more sophisticated statistical analysis concerning a TLD system [2]. This kind of analysis allows us to evidenziate if:

- there are differences in sensitivity between the TL dosimeters,
- there are differences between consecutive irradiations, at the same dose, produced by sensitivity variations of the TL elements, instability of the reader system and differences between irradiations.

These differences can be considered as systematic uncertainties if they are always of the same sign (i.e., a dosimeters which systematically presents a higher response) or as random if they are variable in both magnitude and sign. Let us call these kind of variability as "adjunctive variabilities". The following analysis is carried out in two steps: at first the analysis is based on the χ^2 analysis, which allows to recognize if there is or not an adjunctive variability. In a second step, using the F distribution, it is possible to know the cause of the adjunctive variability.

The analysis has been done using 10 selected and individually calibrated dosimeters LiF:Mg,Cu,P (GR-200A). The test dose was 130 mGy obtained from a calibrated ^{90}Sr-^{90}Y beta source. The TLDs were processed 10 times according to the procedure: annealing, cooling, irradiation, prompt readout. The matrix of the results is shown in Table 2.

The data can be affected by both systematic and random errors caused by the TL elements, the TL reader unit, the irradiation system, and the thermal treatment.

Let us define the following quantities:

m = number of TL elements = number of columns = 10

n = number of readings (cycles) = numbers of rows = 10

x_{ij} = jth reading of the ith dosimeter

i = index of column = 1,…, m
j = index of row = 1,…, n

$$\bar{X} = \text{average of the n×m data (=100)} = \frac{1}{nm}\sum_{i=1}^{m}\sum_{j=1}^{n} X_{ij} \tag{2}$$

The variance of the *n*×*m* data is given by

$$s^2 = \frac{1}{nm-1}\sum_i\sum_j (X_{ij} - \bar{X})^2 \tag{3}$$

A statistical estimate of variance can be performed when both systematic individual sensitivity differences among samples and the systematic changes of sensitivity in repeated experiments are eliminated. This means that the possibility to determine an interval, including the true value of the standard deviation, is with a probability of 90%

The following quantities can then be defined:

$$S = \sum_i\sum_j (X_{ij} - \bar{X})^2 \tag{4}$$

which takes into account all the variations within the values;

$$S_A = m\sum_{j=1}^{n} (\bar{X}_{rj} - \bar{X})^2 \tag{5}$$

which takes into account variations between repeated irradiations (cycles) and $\bar{X}_{rj}$ is the average value of the readings of all the dosimeters in reference to the jth cycle.

$$S_B = n\sum_{i=1}^{m} (\bar{X}_{ci} - \bar{X})^2 \tag{6}$$

which takes into account the differences in sensitivity between dosimeters and $\bar{X}_{ci}$ is the average of the repeated readings of the same dosimeter.

$$S_0 = S - S_A - S_B \tag{7}$$

which takes into account the residual variations excluding the previous effects due to differences among successive irradiations and among dosimeters. It takes into account the variations due to the random effects only.

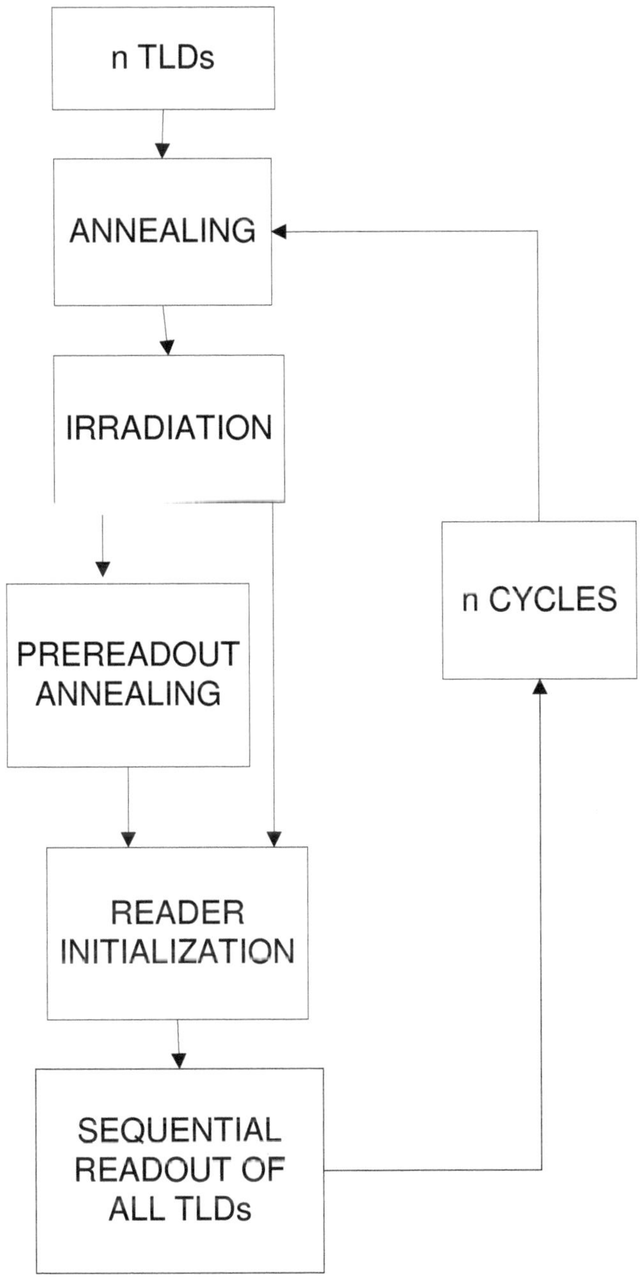

Fig. 1. Operational flow-chart for reproducibility measurements.

Table 1. Matrix of the results obtained in a reproducibility experiment (first procedure).

	dosimeter numbers												
cycle	1	2	3	4	5	6	7	8	9	10	$\overline{X}_r$	s_r	$\%CV_r$
1	8361	8126	8275	8348	8383	8643	8123	8157	8459	-	8323	173	2.07
2	8280	8177	8133	8114	8542	8591	8073	8440	8417	8365	8313	185	2.23
3	8459	8209	8242	8273	8346	8661	8093	8192	8442	8363	8328	164	1.96
4	8391	8194	8235	7909	8458	8564	7986	8235	8482	8282	8274	211	2.55
5	1094	8032	8086	8207	8491	8276	7966	8168	8459	8178	8196	172	2.10
6	8330	8729	8122	8201	8564	8611	8020	8237	8462	8329	8361	227	2.72
7	8386	8143	8255	8226	8620	8568	8079	8409	8491	8398	8358	179	2.14
8	8290	8060	8096	8160	8563	8504	7922	8303	8442	8330	8267	206	2.49
9	8327	8184	8126	8166	8508	8538	8021	8327	8429	8249	8278	155	1.88
10	8392	8127	8167	8142	8470	8415	7889	8268	8419	8241	8253	179	2.17
$\overline{X}_c$	8331	8198	8174	8178	8495	8527	8017	8274	8450	8304			
s_c	99	195	71	122	84	120	77	97	25	71			
$\%CV_c$	1.19	2.38	0.87	1.49	0.99	1.41	0.96	1.17	0.30	0.86			

Table 2. Matrix of the results obtained in a reproducibility experiment (second procedure).

	dosimeter numbers												
cycles	1	2	3	4	5	6	7	8	9	10	$\overline{X}_{rj}$	σ_j	σ_j%
1	14460	14469	14564	14587	14686	14395	14473	14715	14659	14247	14526	145	1.0
2	14842	14362	14681	14527	14886	14296	14779	14367	14680	14498	14592	212	1.5
3	14799	14392	14703	14373	14520	14261	14529	14092	14583	14489	14475	207	1.4
4	14670	14326	14493	14444	14685	14517	14654	14628	14453	14532	14540	117	0.8
5	14760	14634	14460	14454	14650	14428	14527	14444	14299	14425	14508	136	0.9
6	14723	14800	14597	14570	14659	14687	14487	14578	14411	14584	14610	113	0.8
7	14558	14915	14492	14714	14636	14774	14582	14702	14415	14683	14647	144	1.0
8	14453	14594	14703	14514	14735	14421	14430	14421	14778	14679	14573	142	1.0
9	14420	14963	14922	14595	14595	14557	14549	14802	14615	14611	14663	174	1.2
10	14850	14836	14832	14766	14932	14853	14638	14710	14882	14705	14796	91	0.6
$\overline{X}_{ci}$	14649	14629	14645	14554	14698	14519	14565	14546	14578	14545			
σ_i	162	238	153	121	125	199	103	215	182	140			
σ_i%	1.1	1.6	1.0	0.8	0.9	1.4	0.7	1.5	1.2	1.0			

For each of the above quantities it is possible to calculate the degree of freedom, n, as the number of the elements which contributes to the quantity minus a reference value which is the average one. Furthermore, the quantities

$$\frac{S}{\nu}, \quad \frac{S_A}{\nu_A}, \quad \frac{S_B}{\nu_B}, \quad \frac{S_0}{\nu_0} \tag{8}$$

are the estimates of the corresponding variances with the following means:

$$\sigma_0^2 = \frac{S_0}{\nu_0} \tag{9}$$

which concerns with the intrinsic variation that one should have if only one dosimeter is irradiated several times;

$$\sigma_A^2 = \frac{S_A}{\nu_A} \tag{10}$$

takes into account two different components: the first is the intrinsic variation and the second component is due to possible variations from one cycle to another (row effect).

The differences between the dosimeters (column effect) are not considered because the average values $\overline{X}_{rj}$ are used.

$$\sigma_B^2 = \frac{S_B}{\nu_B} \tag{11}$$

considers also two components: one is the intrinsic variation and the other concerns the differences between detectors (column effect). The row effect turns out to be uninportant because the average values $\overline{X}_{ci}$ are used.

Because S_0 deals with the random uncertainties only, the quantity S_0/ν_0^2 is a statistical variable which is χ^2 distributed with n_0 degrees of freedom. It is then possible to determine a $\chi^2_{95\%}(\nu),\left[\chi^2_{5\%}(\nu)\right]$, value such that the probability of finding a value smaller, [larger], than it, in a single evaluation of S_0/ν_0^2 is 95%,[5%]. This is expressed by the following relation:

$$P\left[\chi^2_{5\%}(\nu) < \frac{S_0}{\sigma_0^2} < \chi^2_{95\%}(\nu)\right] = 0.90 \tag{12}$$

where P stands for probability. Rearranging, one finds for the standard deviation

$$P\left[\sqrt{\frac{S_0}{\chi^2_{95\%}(\nu)}} < \sigma_0 < \sqrt{\frac{S_0}{\chi^2_{5\%}}}\right] = 0.90 \tag{13}$$

The experimental data are shown in Table 3.

Table 3. Values calculated using the experimental data in Table 2.

quantity	degree of freedom	variance
S=2883029.00	n=nm-1=99	s^2=S/n=29121.51
S_A=782366.25	n_A=n-1=9	$s_A^2=S_A/n_A$=86929.58
S_B=307833.06	n_B=m-1=9	$s_B^2=S_B/n_B$=34203.67
S_0=1792829.60		

Considering now that the χ^2 values for n_0 = 81 are respectively

$$\chi^2_{5\%} - 61.262 \quad \text{and} \quad \chi^2_{95\%} = 103.009$$

the previous expression for the probability P gives

$$P[131.926 < \sigma_0 < 171.071]$$

or in percentage

$$P[0.90\% < \sigma_0\% < 1.17\%]$$

Considering now that the average of the nm data is

$$\bar{X} = 14592.85$$

with an experimental standard deviation (%) equal to 1.16%, and comparing this value to the percent standard deviation interval, it is easily observed that the influence of the systematic uncertainty is negligible.

We can now invoke the F distribution which allows us to recognize if the variation in the observations depends on the dosimeters and/or on the cycles. The F function takes into account the ratio between two experimental variances. Using as usual a confidential limit of 5%, the following two quantities can be calculated

$$F_A = \frac{\sigma_A^2}{\sigma_0^2} = 3.93 \qquad F_B = \frac{\sigma_B^2}{\sigma_0^2} = 1.55 \tag{14}$$

F_A and F_B account for the differences among columns and rows respectively.

Having considered a confidential limit of 5%, from statistical tables a so called critical value F_{cr} is obtained as a function of the degrees of freedom of the variances. In the present case $F_{cr} = 1.93$. Since $F_A > F_{cr}$, there are statistically significant variations from cycle to cycle: annealing, irradiations and/or reader. On the other hand, because F_B is very close to F_{cr}, there is a slight tendency towards a different sensitivity among the dosemeters yet without statistically significant evidence at the 95% confidence level.

The same procedure can be applied to the residual (second reading) in all cases when an oven annealing cannot be carried out. It is the case of plastic cards which cannot be annealed at high temperature.

References

1. Scarpa G., Corso sulla termoluminescenza applicata alla dosimetria. University of Rome "La Sapienza", Italy, 15-17 February 1994
2. Furetta C., Leroy C. and Lamarche F., Med. Phys. **21** (1994) 1605

Precision in TL measurements (definition)

The precision concerning with TL measurements is related to the random uncertainties associated with the measurement itself. The standard deviation of a set of measurements may be used to quantify the precision.

Pre-exponential factor

Introduction

It is well known that the calculation of the so-called pre-exponential factor, comparing in the expressions for second [**1**] and general order kinetics [**2**] describing the thermoluminescent processes, is mainly complicated by two factors, i.e., (i) its dependence on the given dose, (ii) the presence of the integral

$$F(T,E) = \int_{T_0}^{T} \exp\left(-\frac{E}{kT'}\right) dT' \tag{1}$$

which cannot be solved analytically.

The aim of this paragraph has two folders. The first one is to compare the solutions of (1) obtained by the approximation method and by its numerical solution. The second aspect is to investigate the influence of the two different ways of calculation on the values of the pre-exponential factor s" [3]

In the following, a short review of the expressions of the pre-exponential factor is also given.

Integral approximation

A method which is usually followed for evaluating the value of the integral is by integration in parts, when the lower limit of integration is *0* instead of T_0. So, a good approximation is provided by the asymptotic series

$$F(T,E)=\int_0^T \exp\left(-\frac{E}{kT'}\right)dT'=T\exp\left(-\frac{E}{kT}\right)\sum_{n=1}\left(\frac{kT}{E}\right)^n(-1)^{n-1}n!\ldots\ldots \tag{2}$$

The value of (1) is then given by

$$\int_{T_0}^T \exp\left(-\frac{E}{kT'}\right)dT'=F(T,E)-F(T_0,E)$$

Since $F(T,E)$ is a very strong increasing function of T, $F(T_0,E)$ is negligible compared to $F(T,E)$, the right hand side of Eq. (2) can be considered to represent the integral value from T_0 as well.

In the most practical cases, a good approximation of the integral is given by the second order approximation, I2, of Eq. (2) [**4**]:

$$F(T,E)=\frac{kT^2}{E}\exp\left(-\frac{E}{kT}\right)\left(1-\frac{2kT}{E}\right) \tag{3}$$

Eq. (3) has been calculated as a function of E and T. The range of the activation energy E was from 0.5 eV to 2.25 eV and the temperature was ranging from 300 to 500 K. Table 1 gives the data obtained. I2 indicates the second order approximation.

T [K]	I2 (T,E=0.5)	I2 (T,E=0.75)	I2 (T,E=1)	I2 (T,E=1.25)	I2 (T,E=1.5)	I2 (T,E=1.75)	I2 (T,E=2)	I2 (T,E=2.25)
300	5.54E-08	2.42E-12	1.17E-16	5.96E-21	3.16E-25	1.72E-29	9.53E-34	5.36E-38
310	1.10E-07	6.58E-12	4.34E-16	3.02E-20	2.19E-24	1.63E-28	1.23E-32	9.48E-37
320	2.10E-07	1.68E-11	1.49E-15	1.39E-19	1.35E-23	1.34E-27	1.36E-31	1.40E-35
330	3.85E-07	4.07E-11	4.73E-15	5.83E-19	7.44E-23	9.75E-27	1.30E-30	1.77E-34
340	6.82E-07	9.35E-11	1.41E-14	2.25E-18	3.72E-22	6.32E-26	1.09E-29	1.92E-33
350	1.17E-06	2.05E-10	3.96E-14	8.05E-18	1.70E-21	3.69E-25	8.14E-29	1.82E-32
360	1.96E-06	4.32E-10	1.05E-13	2.69E-17	7.15E-21	1.95E-24	5.43E-28	1.53E-31
370	3.18E-06	8.76E-10	2.64E-13	8.43E-17	2.79E-20	9.46E-24	3.27E-27	1.15E-30
380	5.06E-06	1.71E-09	6.35E-13	2.49E-16	1.01E-19	4.23E-23	1.80E-26	7.75E-30
390	7.85E-06	3.23E-09	1.46E-12	6.98E-16	3.45E-19	1.75E-22	9.05E-26	4.75E-29
400	1.19E-05	5.93E-09	3.23E-12	1.86E-15	1.11E-18	6.76E-22	4.21E-25	2.66E-28
410	1.78E-05	1.06E-08	6.87E-12	4.72E-15	3.36E-18	2.45E-21	1.82E-24	1.37E-27
420	2.60E-05	1.83E-08	1.41E-11	1.15E-14	9.67E-18	8.35E-21	7.35E-24	6.56E-27
430	3.75E-05	3.10E-08	2.81E-11	2.68E-14	2.65E-17	2.69E-20	2.78E-23	2.92E-26
440	5.31E-05	5.13E-08	5.42E-11	6.04E-14	6.96E-17	8.23E-20	9.92E-23	1.21E-25
450	7.41E-05	8.31E-08	1.02E-10	1.31E-13	1.75E-16	2.40E-19	3.35E-22	4.74E-25
460	1.02E-04	1.32E-07	1.86E-10	2.76E-13	4.24E-16	6.68E-19	1.07E-21	1.75E-24
470	1.39E-04	2.05E-07	3.31E-10	5.62E-13	9.89E-16	1.78E-18	3.27E-21	6.10E-24
480	1.86E-04	3.14E-07	5.77E-10	1.11E-12	2.23E-15	4.57E-18	9.54E-21	2.02E-23
490	2.48E-04	4.73E-07	9.82E-10	2.15E-12	4.86E-15	1.13E-17	2.66E-20	6.39E-23
500	3.25E-04	7.01E-07	1.64E-09	4.04E-12	1.03E-14	2.69E-17	7.15E-20	1.93E-22

Table 1. Second order approximation, I2, of the integral given by the equation (3).

Numerical solution of the integral

The integral (1) has been also calculated, as a function of E and T, by means of the Matlab program. The numerical evaluation has been carried out using the adaptive Lobatto-Gauss quadrature embedded in Matlab^{TM} 7.0 [5]. Special attention was paid at the low temperature region where the exponential assumes initially very low values and then changes very fast as the temperature increases (Fig. 1).

The lower limit of the integration interval is the smallest floating point number, i.e. realmin in Matlab, while the upper limit value changes from 300 K to 700 K.

The integration limit has been divided in two subintervals: the first for $T \leq 400$ K and the second for $T > 400$ K in order to define the first region where the integration function assumes very low values and increases strongly with the temperature.

The function to be integrated is multiplied by a numerical coefficient to avoid to work with too low values.

In the following the numerical solution of the integral (1) is indicated by Q(T,E). Table 2 lists the values. The range values of E and T are the same as before.

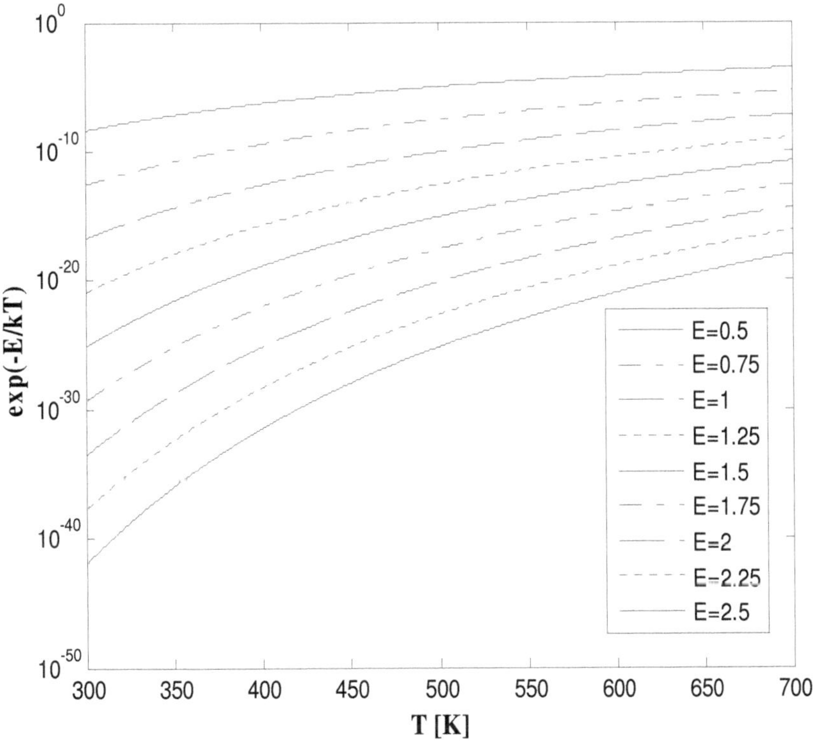

Fig. 1. Exponential function as a function of absolute temperature for several activation energies ranging from E=0.5 to 2.5 eV.

T [K]	Q(T,E=0.5)	Q(T,E=0.75)	Q(T,E=1)	Q(T,E=1.25)	Q(T,E=1.5)	Q(T,E=1.75)	Q(T,E=2)	Q(T,E=2.25)
300	5.62E-08	2.44E-12	1.17E-16	5.98E-21	3.16E-25	1.72E-29	9.54E-34	5.37E-38
310	1.12E-07	6.62E-12	4.35E-16	3.03E-20	2.19E-24	1.63E-28	1.23E-32	9.48E-37
320	2.13E-07	1.69E-11	1.49E-15	1.39E-19	1.35E-23	1.34E-27	1.36E-31	1.40E-35
330	3.92E-07	4.10E-11	4.75E-15	5.84E-19	7.46E-23	9.77E-27	1.30E-30	1.77E-34
340	6.95E-07	9.43E-11	1.42E-14	2.26E-18	3.73E-22	6.33E-26	1.09E-29	1.92E-33
350	1.20E-06	2.07E-10	3.98E-14	8.08E-18	1.70E-21	3.69E-25	8.15E-29	1.83E-32
360	2.00E-06	4.37E-10	1.05E-13	2.70E-17	7.17E-21	1.96E-24	5.44E-28	1.53E-31
370	3.26E-06	8.85E-10	2.66E-13	8.46E-17	2.80E-20	9.48E-24	3.28E-27	1.15E-30
380	5.18E-06	1.73E-09	6.39E-13	2.50E-16	1.02E-19	4.24E-23	1.80E-26	7.76E-30
390	8.04E-06	3.27E-09	1.47E-12	7.00E-16	3.46E-19	1.75E-22	9.07E-26	4.76E-29
400	1.22E-05	6.00E-09	3.25E-12	1.86E-15	1.11E-18	6.78E-22	4.22E-25	2.67E-28
410	1.83E-05	1.07E-08	6.92E-12	4.74E-15	3.37E-18	2.45E-21	1.82E-24	1.38E-27
420	2.68E-05	1.86E-08	1.42E-11	1.15E-14	9.70E-18	8.37E-21	7.36E-24	6.57E-27
430	3.86E-05	3.15E-08	2.83E-11	2.70E-14	2.66E-17	2.70E-20	2.79E-23	2.92E-26
440	5.48E-05	5.21E-08	5.46E-11	6.07E-14	6.99E-17	8.26E-20	9.94E-23	1.21E-25
450	7.66E-05	8.44E-08	1.03E-10	1.32E-13	1.76E-16	2.41E-19	3.35E-22	4.75E-25
460	1.06E-04	1.34E-07	1.87E-10	2.77E-13	4.26E-16	6.70E-19	1.07E-21	1.75E-24
470	1.44E-04	2.09E-07	3.34E-10	5.66E-13	9.93E-16	1.79E-18	3.28E-21	6.11E-24
480	1.93E-04	3.20E-07	5.82E-10	1.12E-12	2.24E-15	4.58E-18	9.56E-21	2.03E-23
490	2.57E-04	4.81E-07	9.92E-10	2.16E-12	4.88E-15	1.13E-17	2.67E-20	6.40E-23
500	3.39E-04	7.14E-07	1.66E-09	4.06E-12	1.03E-14	2.70E-17	7.17E-20	1.93E-22

Table 2. Calculated values of Q(T,E) for given values of E and T.

Condition at the maximum

The frequency factor expressions are here extracted by the TL equations.

The empirical equation for general order kinetics [**6**] is

$$I = -\frac{dn}{dt} = s''n^b \exp\left(-\frac{E}{kT}\right) \tag{4}$$

and its solution is

$$I(T) = s''n_0^b \exp\left(-\frac{E}{kT}\right)\left[1+\frac{s''n_0^{b-1}(b-1)}{\beta}\int_{T_0}^{T}\exp\left(-\frac{E}{kT'}\right)dT'\right]^{-\frac{b}{b-1}} \tag{5}$$

where s" has units of $cm^{3(b-1)}sec^{-1}$.

The condition at the maximum is obtained by Eq. (5) as following.

The logarithm of $I(T)$ is:

$$\ln\left[I(T)\right] = \ln(s''n_0^b) - \frac{E}{kT} - \frac{b}{b-1}\ln\left[1+\frac{s''n_0^{b-1}(b-1)}{\beta}\int_{T_0}^{T}\exp\left(-\frac{E}{kT'}\right)dT'\right]$$

then, the condition at the maximum is obtained quoting its derivative to zero:

$$\left.\frac{d(\ln I)}{dT}\right|_{T=T_M}$$

$$= \frac{E}{kT_M^2} - \frac{b}{b-1}\left[1+\frac{s''n_0^{b-1}(b-1)}{\beta}\int_{T_0}^{T_M}\exp\left(-\frac{E}{kT'}\right)dT'\right]^{-1} \times$$

$$\times\frac{s''n_0^{b-1}(b-1)}{\beta}\exp\left(-\frac{E}{kT_M}\right) = 0$$

from which we obtain

$$\frac{kT_M^2 bs''n_0^{b-1}}{\beta E}\exp\left(-\frac{E}{kT_M}\right) = 1+\frac{s''n_0^{b-1}(b-1)}{\beta}\int_{T_0}^{T_M}\exp\left(-\frac{E}{kT'}\right)dT' \tag{6}$$

From the last equation it is possible to obtain the expression for the pre-exponential factor. Rearranging Eq. (6), we obtain:

$$s'' = \frac{1}{n_0^{b-1}} \cdot \left[\frac{kT_M^2 b \exp\left(-\frac{E}{kT_M}\right)}{\beta E} - \frac{(b-1)\int_{T_0}^{T_M} \exp\left(-\frac{E}{kT'}\right) dT'}{\beta} \right]^{-1} \tag{7}$$

or, considering $s = s'' n_0^{b-1}$

$$s = \left[\frac{kT_M^2 b \exp\left(-\frac{E}{kT_M}\right)}{\beta E} - \frac{(b-1)\int_{T_0}^{T_M} \exp\left(-\frac{E}{kT'}\right) dT'}{\beta} \right]^{-1} \tag{8}$$

Using the second order approximation, we get

$$s'' = \frac{1}{n_0^{b-1}} \cdot \left[\frac{kT_M^2 \exp\left(-\frac{E}{kT_M}\right)}{\beta E} \left(1 + \frac{2kT_M(b-1)}{E}\right) \right]^{-1} \tag{9}$$

and finally

$$s'' = \frac{\beta E \exp\left(\frac{E}{kT_M}\right)}{n_o^{b-1} kT_M^2} \left[1 + \frac{2kT_M(b-1)}{E}\right]^{-1} \tag{10}$$

in $sec^{-1}cm^{3(b-1)}$, or, considering $s = s'' n_0^{b-1}$

$$s = \frac{\beta E \exp\left(\frac{E}{kT_M}\right)}{kT_M^2} \left[1 + \frac{2kT_M(b-1)}{E}\right]^{-1} \quad \text{expressed in sec}^{-1} \tag{11}$$

The second order expression can be obtained from Eqs. (7), (8), (9), (10) or (11) for $b = 2$.

Observing the previous equations, it is clear that both the heating rate, β, and the initial dose, n_0, are multiplication factors: for this reason we compare directly the expressions (8) and (11) to find the possible differences in the pre-exponential factor values when it is obtained by integral approximation, Eq. (11), or using the numerical solution of the integral in Eq. (8). In both cases only one parameter has been varied in each simulation, i.e. the kinetics order b, the heating rate β the temperature at the maximum T_M or the initial delivered dose n_0.

Higher order approximations

In this paragraph third and fourth order approximations are presented and results are compared with the second order approximation so far discussed.

In general by means of the asymptotic series, the expression of s becomes:

$$s = \frac{\beta E}{kT_M^2} e^{\frac{E}{kT_M}} \left\{ 1 + (b-1) \sum_{n=1}^{+\infty} (-1)^{n+1} (n+1)! \left(\frac{kT_M}{E} \right)^n \right\}^{-1} \tag{12}$$

where the second order approximation is obtained for n=1; the third order approximation for n=3, and so on.

In particular, the third order expression is

$$s3 = \frac{\beta E}{kT_M^2} e^{\frac{E}{kT_M}} \left\{ 1 + (b-1) \left[\frac{2kT_M}{E} - 6 \left(\frac{kT_M}{E} \right)^2 \right] \right\}^{-1} \tag{13}$$

and the fourth order approximation is given by

$$s4 = \frac{\beta E}{kT_M^2} e^{\frac{E}{kT_M}} \left\{ 1 + (b-1) \left[\frac{2kT_M}{E} - 6 \left(\frac{kT_M}{E} \right)^2 + 24 \left(\frac{kT_M}{E} \right)^3 \right] \right\}^{-1} \tag{14}$$

The equation (8) shows the general expression of the pre-exponential factor and the equation (11) shows the pre-exponential factor expression using a second order approximation.

As discussed previously, T_o in equation (8) has been fixed to 0.

Figures 2 to 5 present the behavior of s as a function of the kinetic order b with the activation energy ranging from 0.5 to 2eV. The temperature of the maximum T_M is fixed at 500K and the heating rate at 10K/s. As expected, the largest difference is found for low values of the activation energy E. As shown by the equation (8), for high values of kinetics order b the difference increases: as the kinetics order b increases, the weight of the exponential integral increases the value of s. Table 3-6 give the values of frequency s as a function of the kinetic order calculated directly by means of equation (8) and by means of the equation (11). A ratio of the two values is also added.

Figures 6-7 present the behavior of s as a function of the temperature of the maximum T_M. It has been considered a kinetics order b equal to 2, which is the worst case.

Table 7-8 summaries the results.

Figure 8 and table 9 compare the second order approximation to high orders.

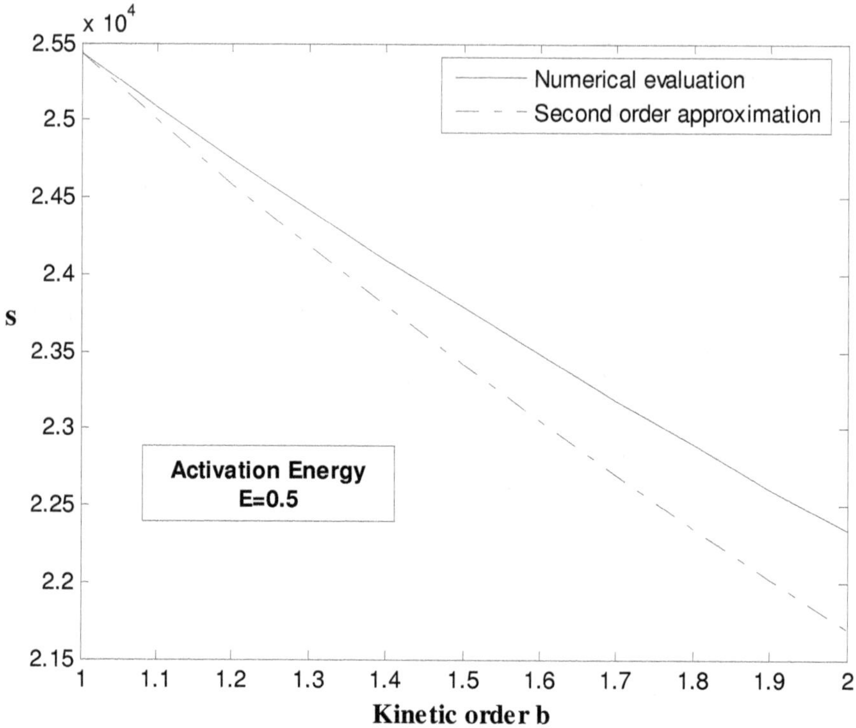
x 10^4
2.55
2.5
2.45
2.4
2.35
2.3
2.25
2.2
2.15
s
Numerical evaluation
Second order approximation
Activation Energy
E=0.5
1
1.1
1.2
1.3
1.4
1.5
1.6
1.7
1.8
1.9
2
Kinetic order b

Fig. 2

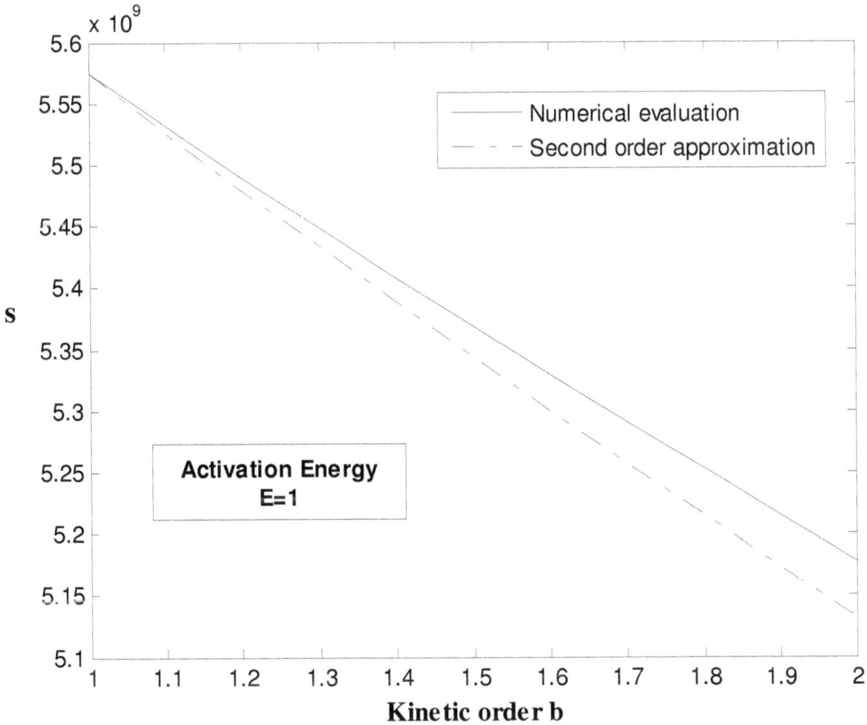
x 10^9
5.6
5.55
5.5
5.45
5.4
5.35
5.3
5.25
5.2
5.15
5.1
S
Numerical evaluation
Second order approximation
Activation Energy
E=1
1
1.1
1.2
1.3
1.4
1.5
1.6
1.7
1.8
1.9
2
Kinetic order b

Fig. 3

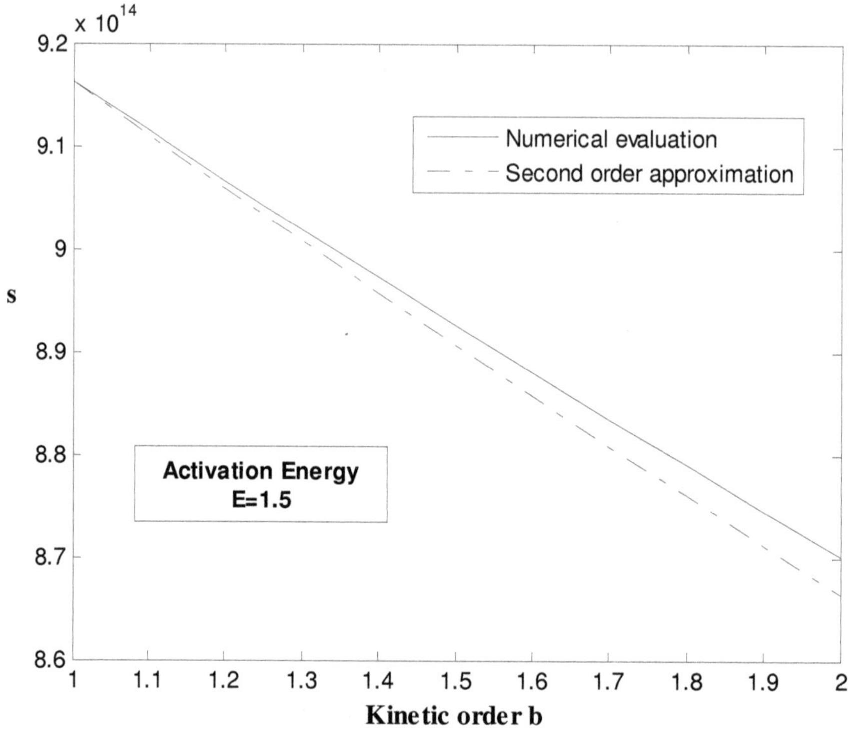
x 10^14
9.2
9.1
9
8.9
8.8
8.7
8.6
s
1
1.1
1.2
1.3
1.4
1.5
1.6
1.7
1.8
1.9
2
Kinetic order b
Numerical evaluation
Second order approximation
Activation Energy
E=1.5

Fig. 4

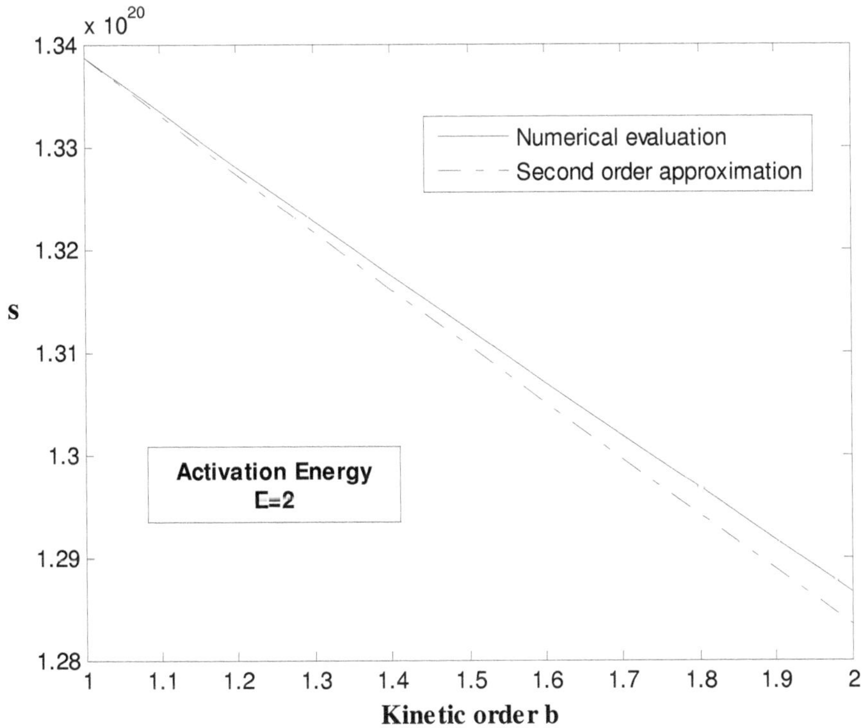
x 10^20
1.34
1.33
1.32
1.31
1.3
1.29
1.28
s
Numerical evaluation
Second order approximation
Activation Energy
E=2
1
1.1
1.2
1.3
1.4
1.5
1.6
1.7
1.8
1.9
2
Kinetic order b

Fig. 5

Activation energy: E=0.5 eV			
b (kinetics order)	s	s2	s2/s
1	2.543E+04	2.543E+04	1.000
1.2	2.475E+04	2.459E+04	0.993
1.4	2.410E+04	2.379E+04	0.987
1.6	2.349E+04	2.305E+04	0.981
1.8	2.290E+04	2.235E+04	0.976
2	2.235E+04	2.169E+04	0.971

Table 3

Activation energy E=1 eV			
b (kinetics order)	s	s2	s2/s
1	5.57E+09	5.57E+09	1.00
1.2	5.49E+09	5.48E+09	1.00
1.4	5.41E+09	5.39E+09	1.00
1.6	5.33E+09	5.30E+09	0.99
1.8	5.25E+09	5.21E+09	0.99
2	5.18E+09	5.13E+09	0.99

Table 4

Activation Energy E=1.5 eV			
b (kinetics order)	s	s2	s2/s
1	9.163E+14	9.163E+14	1.000
1.2	9.067E+14	9.059E+14	0.999
1.4	8.972E+14	8.957E+14	0.998
1.6	8.880E+14	8.857E+14	0.997
1.8	8.790E+14	8.760E+14	0.997
2	8.701E+14	8.665E+14	0.996

Table 5

Activation Energy E=2 eV			
b (kinetics order)	s	s2	s2/s
1	1.339E+20	1.339E+20	1.000
1.2	1.328E+20	1.327E+20	0.999
1.4	1.317E+20	1.316E+20	0.999
1.6	1.307E+20	1.305E+20	0.998
1.8	1.297E+20	1.294E+20	0.998
2	1.287E+20	1.283E+20	0.998

Table 6

Table 3-6. Frequency factor as a function of kinetic order b. s has been evaluated by means of numerical integration, s2 by means of the second order approximation; s2/s is the ratio of the two values. T_M and heating rate β are respectively fixed to 500 K and 10 K/s.

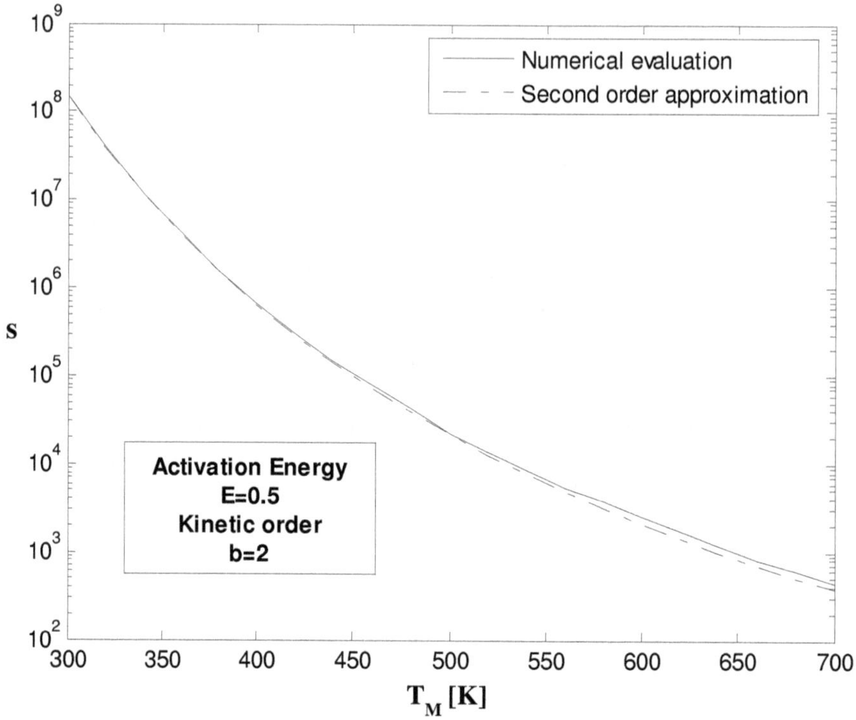
Numerical evaluation
Second order approximation
Activation Energy
E=0.5
Kinetic order
b=2
s
T_M [K]
10^9
10^8
10^7
10^6
10^5
10^4
10^3
10^2
300
350
400
450
500
550
600
650
700

Fig. 6

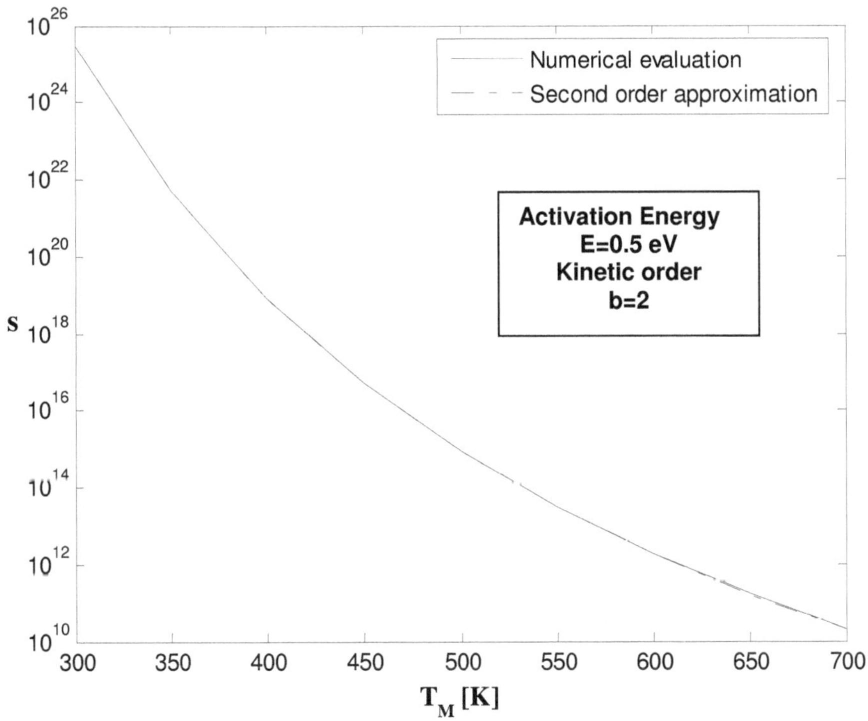

Numerical evaluation
Second order approximation
Activation Energy
E=0.5 eV
Kinetic order
b=2
S
10^26
10^24
10^22
10^20
10^18
10^16
10^14
10^12
10^10
300
350
400
450
500
550
600
650
700
T_M [K]

Fig. 7

E=0.5 eV, b=2			
TM	s	s2	s2/s
300	1.484E+08	1.466E+08	0.988
350	6.801E+06	6.693E+06	0.984
400	6.483E+05	6.354E+05	0.980
450	1.014E+05	9.869E+04	0.974
500	2.235E+04	2.169E+04	0.971
550	6.644E+03	6.153E+03	0.926
600	2.504E+03	2.116E+03	0.845
650	9.873E+02	8.447E+02	0.856
700	4.395E+02	3.796E+02	0.864

Table 7

E=1.5 eV, b=2			
TM	s	s2	s2/s
300	2.959E+25	2.955E+25	1.00
350	5.436E+21	5.425E+21	1.00
400	8.267E+18	8.244E+18	1.00
450	5.165E+16	5.147E+16	1.00
500	8.701E+14	8.665E+14	1.00
550	3.022E+13	3.007E+13	0.99
600	1.809E+12	1.798E+12	0.99
650	1.649E+11	1.636E+11	0.99
700	2.113E+10	2.072E+10	0.98

Table 8

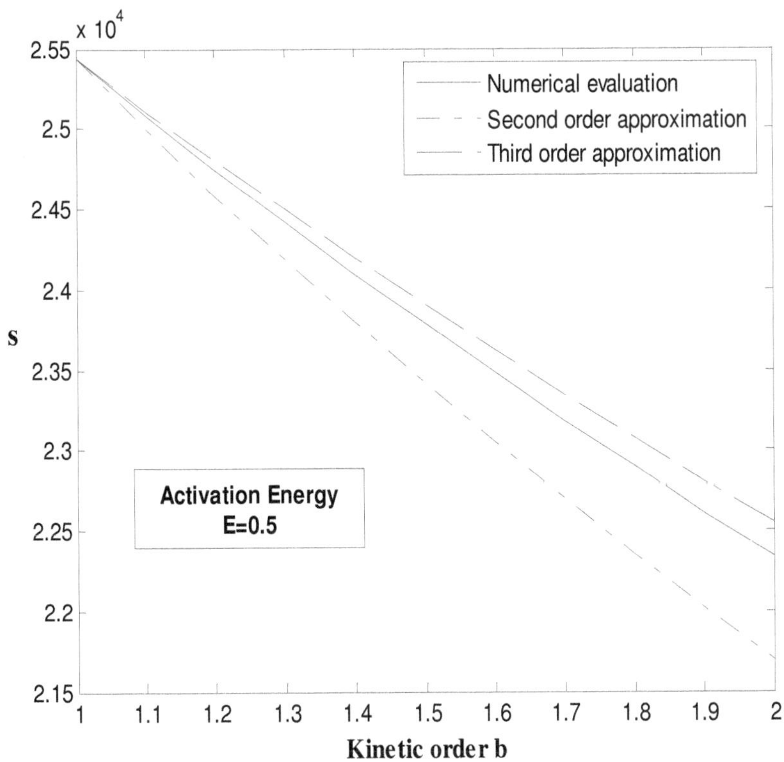
x 10^4
2.55
2.5
2.45
2.4
2.35
2.3
2.25
2.2
2.15
S
Numerical evaluation
Second order approximation
Third order approximation
Activation Energy
E=0.5
1
1.1
1.2
1.3
1.4
1.5
1.6
1.7
1.8
1.9
2
Kinetic order b

Fig. 8

Activation Energy E=0.5 eV							
b (kinetics order)	**s**	**s2**	**s2/s**	**s3**	**s3/s**	**s4**	**s4/s**
1	2.543E+04	2.543E+04	1.000	2.543E+04	1.00	2.543E+04	1.00
1.2	2.475E+04	2.459E+04	0.993	2.480E+04	1.00	2.473E+04	1.00
1.4	2.410E+04	2.379E+04	0.987	2.420E+04	1.00	2.406E+04	1.00
1.6	2.349E+04	2.305E+04	0.981	2.362E+04	1.01	2.342E+04	1.00
1.8	2.290E+04	2.235E+04	0.976	2.307E+04	1.01	2.282E+04	1.00
2	2.235E+04	2.169E+04	0.971	2.255E+04	1.01	2.225E+04	1.00

Table 9. s numerical evaluation; s2 second order approximation; s3 third order approximation; s4 fourth order approximation; s2/s , s3/s s4/s are respectively the ratios between second order approximation s2 and s, third order approximation s3 and s and finally fourth order approximation s4 and s.

References

1. J.T. Randall and M.H.F. Wilkins, Proc. Roy. Soc. **A184** (1945) 366
2. G.F.J. Garlick and A.F. Gibson, Proc. Phys. Soc. **60** (1948) 574
3. A. Favalli, private communication
4. C. Furetta, Handbook of thermoluminescence, World Scientific, Singapore (2003)
5. http://www.mathworks.com/
6. C.E. May and J.A. Partridge, J. Chem. Phys. **40** (1964) 1401

Pre-exponential factor (definition) [1]

The equation of the TL intensity, for the general order kinetics, has been given by May and Partridge [2]. It is

$$I = -\frac{dn}{dt} = s'n^b \exp\left(-\frac{E}{kT}\right) \tag{1}$$

In this equation s' is a constant quantity called pre-exponential factor.
Eq. (1) reduces to the first order kinetics equation of Randall-Wilkins [3] when b=1. In this case s' is equal to the frequency factor s, expressed in units of s^{-1}. When b=2, Eq. (1) reduces to the second order equation of Garlick and Gibson [4] and $s' = s/N$ having dimensions $s^{-1}m^3$; N is the total concentration of the traps.

According to the one trap and one recombination centre model, and using the quasi-equilibrium condition, the TL intensity is given by [5]

$$I = -\frac{dm}{dt} = \frac{A_m mns \exp\left(-\frac{E}{kT}\right)}{(N-n)A_n + mA_m} \tag{2}$$

where

m = concentration of recombination centres (m^{-3})
N = total concentration of traps (m^{-3})
A_n = capture coefficient of the traps ($m^3\ s^{-1}$)
A_m = capture coefficient of the recombination centres ($m^3\ s^{-1}$)
n = concentration of trapped charges (m^{-3})

If retrapping exceeds recombination, i.e. $(N-n)A_n \gg mA_m$, and $n \ll N$, $m=n$, Eq. (2) reduces to

$$I = \frac{A_m n^2 s \exp\left(-\frac{E}{kT}\right)}{NA_n} \tag{3}$$

Or better

$$I = s' n^2 \exp\left(-\frac{E}{kT}\right) \tag{4}$$

where

$$s' = \frac{sA_m}{NA_n} \tag{5}$$

Eq. (4) is the well known second order kinetics equation. If, instead of m = n, one uses the condition m = n + C, with C >> n, Eq. (2) becomes

$$\begin{aligned} I &= \frac{A_m mns \exp\left(-\frac{E}{kT}\right)}{NA_n} \\ &= \frac{A_m (n+C) ns \exp\left(-\frac{E}{kT}\right)}{NA_n} \\ &- \frac{A_m Cns \exp\left(-\frac{E}{kT}\right)}{NA_n} \\ &= s' n \exp\left(-\frac{E}{kT}\right) \end{aligned}$$

which is the first order kinetics equation, with

$$s' = \frac{sA_m C}{NA_n}$$

where C is the high concentration of recombination centres arising for the presence of thermally disconnected deep traps [5,6].

When the kinetics order b and the factor s' are both independent on the initial trap occupancy $\frac{n_0}{N}$, s' is defined as

$$s' = \frac{s}{N^{b-1}}$$

This definition has been given in [7].

The previous definition applies to the case of a two-levels scheme in which $m = n$.

Integration of Eq. (1), using the previous definition of s', gives

$$I = n_0 s\left(\frac{n}{N}\right)^{b-1} \exp\left(-\frac{E}{kT}\right)\left[(b-1)\frac{s}{\beta}\left(\frac{n_0}{N}\right)^{b-1} \times \right.$$
$$\left. \times \int_{T_0}^{T} \exp\left(-\frac{E}{kT'}\right) dT' + 1\right]^{-\frac{b}{b-1}}$$

This equation reduces to the Garlick-Gibson equation for $b = 2$ and to that of Randall-Wilkins for $b \to 1$.

The Authors computed glow curves for various values of b (= 1.01, 1.25, 1.5, 1.75, 2, 2.5) for three cases of trap occupancy n_0/N (= 1, 10^{-2}, 10^{-4}). The results of the calculations are the following:

- The glow peak shifts to higher temperature values as b increases and n_0/N decreases: the retrapping increases;
- The glow peak remains at nearly the same position: (i) irrespective of changes in b when $n_0/N = 1$, (ii) irrespective of changes in n_0/N when $b \to 1$;
- The shape factor A of the glow peak (defined as the ratio of the fall side half area to the full area of the peak) is mainly influenced by changes in b.

References

1. Sunta C.M., Feria Ayta W.E., Kulkarni R.N., Chen R. and Watanabe S., Pre-exponential factor in general order kinetics of thermoluminescence and its influence on glow curves, Rad. Prot. Dos. 71, No.2, 1997, 93-97
2. May C. and Partridge J.A., Thermoluminescence kinetics of alpha-irradiated alkali halides, J. Chem. Phys. 40, 1401-1409, 1964

3. Randall J.T. and Wilkins M.H.F., Phosphorescence and electron traps:I. The study of trap distribution, Proc. R. Soc. A184,366-389, 1945
4. Garlick G.F.J. and Gibson A.F., The electron trap mechanism of luminescence in sulphide and silicate phosphors, Proc. Phys. Soc. 60, 574-590, 1948
5. Chen R. and Kirsh Y., Analysis of thermally stimulated process. Oxford, Pergamon Press, 1981
6. Lewandowski A.C. and McKeever S.W.S., Generalised description of thermally stimulated process without the quasi equilibrium approximation, Phys. Rev. B43, 8163-8178, 1991
7. Rasheedy M., On the general order kinetics of thermoluminescence glow peak, J. Phys.: Cond. Mat. 5, 633-636, 1993

Pre-irradiation annealing

The purpose of the pre-irradiation annealing is to re-estabilish the thermodynamic defect equilibrium which existed in the material before irradiation and readout.

Pre-readout annealing

This is another way to indicate the post-irradiation annealing procedure.

Properties of the maximum conditions

An interesting feature results from the equations giving the maximum conditions for the first-, second- and general-orders respectively:

$$\frac{\beta E}{kT_M^2} = s\exp\left(-\frac{E}{kT_M}\right) \tag{1}$$

$$\frac{\beta E}{2kT_M^2}\left[1+\frac{s'n_0}{\beta}\int_{T_0}^{T_M}\exp\left(-\frac{E}{kT'}\right)dT'\right] = s'n_0\exp\left(-\frac{E}{kT_M}\right) \tag{2}$$

or

$$\frac{\beta E}{2kT_M^2}\left[1+\frac{s}{\beta}\int_{T_0}^{T_M}\exp\left(-\frac{E}{kT'}\right)dT'\right] = s\exp\left(-\frac{E}{kT_M}\right) \tag{2'}$$

with $s = s'n_0$, and

$$\frac{kT_M^2 bs''n_0^{b-1}}{\beta E}\exp\left(-\frac{E}{kT_M}\right) = 1+\frac{s''n_0^{b-1}(b-1)}{\beta}\int_{T_0}^{T_M}\exp\left(-\frac{E}{kT'}\right)dT' \tag{3}$$

or

$$\frac{kT_M^2 bs}{\beta E}\exp\left(-\frac{E}{kT_M}\right)=1+\frac{s(b-1)}{\beta}\int_{T_0}^{T_M}\exp\left(-\frac{E}{kT'}\right)dT' \tag{3'}$$

- Equation (1) does not include the initial concentration n_o, therefore the first order peak is not expected to shift as a function of the irradiation doses;
- on the contrary, owing to the dependence of s on n_o for $b \neq 1$, and through it, on the excitation dose, one should expect T_M - from Eqs. (2') and (3') - to be dose dependent.

Q

Quasiequilibrium condition

The quasiequilibrium assumption [1-4] is expressed by the following relation:

$$\left|\frac{dn_c}{dt}\right| \prec\prec \left|\frac{dn}{dt}\right|, \left|\frac{dm}{dt}\right| \qquad (1)$$

where

n_c = free electron concentration in the conduction band (CB),

n = trapped electron concentration,

m = hole concentration in the recombination centres.

The assumption (1) means that the number of free electrons in the conduction band is quasistationary. Furthermore, if the initial concentration of the free electrons is assumed to be to be very small, (1) means that the free charges do not accumulate in the conduction band. The quasiequilibrium assumption allows an analytical solution of the differential equations describing the charge carrier transitions between the energy levels during thermal excitation.

References

1. McKeever S.W.S., Thermoluminescence of Solids, Cambridge University Press, 1985
2. Chen R. and Kirsh Y., Analysis of Thermally Stimulated Processes, Pergamon Press, 1981
3. Chen R. and McKeever S.W.S., Theory of Thermoluminescence and Related Phenomena, World Scientific, 1997
4. McKeever S.W.S., Markey B.G. and Lewandowski A.C., Nucl. Tracks Radiat. Meas. **21** (1993) 57

R

Radiation-induced-defects

The radiation-induced-defects are localized electronic states occupied by non-equilibrium concentration of electrons.

Randall-Wilkins model (first-order kinetics)

In 1945, Randall and Wilkins extensively used a mathematical representation for each peak in a glow curve, starting from studies on phosphorescence. Their mathematical treatment was based on the energy band model and yelds the well-known first order expression.

The following figure shows the simple model used for the theoretical treatment. Between the delocalized bands, conduction band (CB) and valence band (VB), two localized levels (metastable states) are considered, one acting as a trap, T, and the other acting as a recombination center (R). The distance between the trap T and the bottom of the CB is called activation energy or trap depth: E. This energy is the energy required to liberate a charge, i.e., an electron, which is trapped in T.

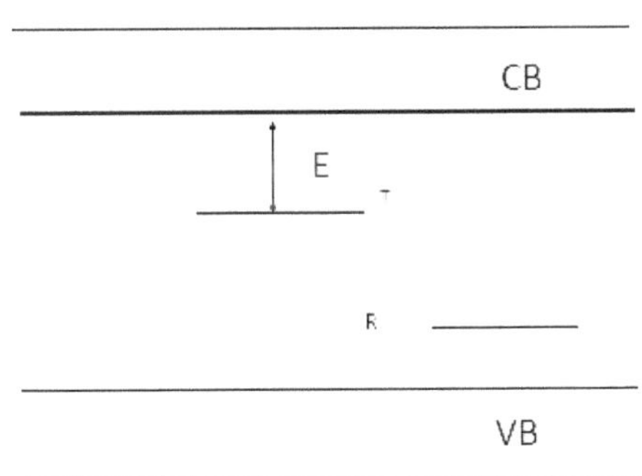

The probability p, per unit of time, that a trapped electron will escape from the trap, or the probability rate of escape per second, is given by the Arrhenius equation, having considered that the electrons in the trap have a Maxwellian distribution of thermal energies

$$p = s \cdot \exp\left(-\frac{E}{kT}\right) \tag{1}$$

where

E is the trap depth (eV), k the Boltzmann's constant, T the absolute temperature (K), s the frequency factor (sec^{-1}), depending on the frequency of the number of hits of an electron in the trap, seen as a potential well.

The life time, τ, of the charge carrier in the metastable state at temperature T, is given by

$$\tau = p^{-1} \tag{2}$$

If n is the number of trapped electrons in T, and if the temperature is kept constant, then n decreases with time t according to the following expression:

$$\frac{dn}{dt} = -pn \tag{3}$$

Integrating this equation

$$\int_{n_0}^{n} \frac{dn}{n} = -\int_{t_0}^{t} p \cdot dt \tag{4}$$

one obtains

$$n = n_0 \exp\left[-s \exp\left(-\frac{E}{kT}\right) \cdot t\right] \tag{5}$$

where n_0 is the number of trapped electrons at the initial time $t_0 = 0$.

Assuming now the following assumptions:

- irradiation of the thermoluminescent material at a low enough temperature so that no electrons are released from the trap,
- the life time of the electrons in the conduction band is short,
- all the released charges from trap recombine in luminescent center,
- the luminescence efficiency of the recombination centers is temperature independent,
- the concentrations of traps and recombination centers are temperature independent,
- no electrons released from the trap is retrapped

According to the previous assumptions, the TL intensity, I, at a constant temperature, is directly proportional to the detrapping rate, *dn/dt*:

$$I = -c\left(\frac{dn}{dt}\right) = cpn \tag{6}$$

where c is a constant which can be set to unity. Eq. (6) represents an exponential decay of phosphorescence.

Remembering Eq. (5), we obtain:

$$I(t) = n_0 s \exp\left(-\frac{E}{kT}\right) \exp\left[-st\exp\left(-\frac{E}{kT}\right)\right] \tag{7}$$

Heating now the material at a constant rate of temperature, $\beta = dT/dt$, from Eq. (4) we have:

$$\int_{n_0}^{n} \frac{dn}{n} = -\left(\frac{s}{\beta}\right) \int_{T_0}^{T} \exp\left(-\frac{E}{kT'}\right) dT'$$

$$\ln(n) - \ln(n_0) = -\left(\frac{s}{\beta}\right) \int_{T_0}^{T} \exp\left(-\frac{E}{kT'}\right) dT'$$

and again

$$n = n_0 \exp\left[-\frac{s}{\beta} \int_{T_0}^{T} \exp\left(-\frac{E}{kT'}\right) dT'\right] \tag{8}$$

Then, using Eq. (6)

$$I(T) = n_0 s \exp\left(-\frac{E}{kT}\right) \exp\left[-\frac{s}{\beta} \int_{T_0}^{T} \exp\left(-\frac{E}{kT'}\right) dT'\right] \tag{9}$$

This expression can be evaluated by mean of numerical integration, and it yields a bell shaped curve, as in Fig. 1, with a maximum intensity at a characteristic temperature T_M.

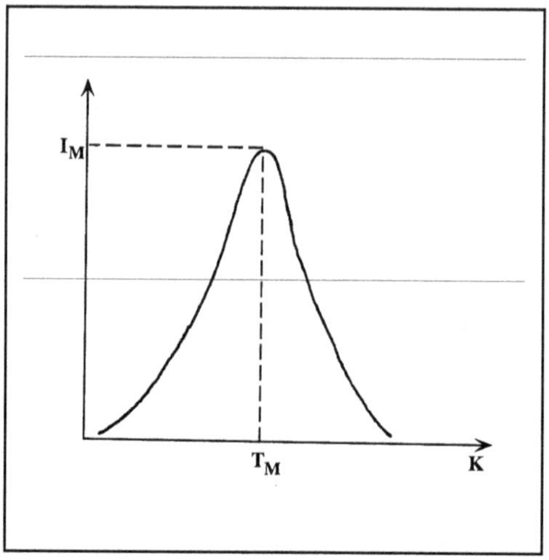

Fig. 1. Solution of Eq. (9). T_M is independent of the initial concentration of trapped electrons, n_0.

Some observation can be done on Eq. (9):

- $I(T)$ depends on three parameters E, s and b,
- E has values around $20kT$ in the range of occurance of TL peaks,
- $\exp\left(-\frac{E}{kT}\right)$ is of the order of 10^{-7},
- when T is slightly greater than of T_0 , the argument of the second exponential function is about equal to unity and decreases with increasing temperature. $I(T)$ is then dominated by the first exponential and increases very fast as the temperature increases. At a certain temperature, T_M, the behaviour of the two exponential functions cancel: at this temperature the maximum temperature occurs,
- Above T_M, the decrease of the second exponential is much more rapid than the increase of the first exponential and $I(T)$ decreases until the traps are totally emptied.

Reference

Randall J.T. and Wilkins M.H.F., Proc. Roy. Soc. **A184** (1945) 366

Rasheedy modified kinetics equations

M.S. Rasheedy [1] proposed a different way to write the general-order kinetics equation to overcome the difficulty related to the dimension of the pre-exponential factor which has an unacceptable meaning.

The original equations for first, second and general orders are the following:

- *first order kinetics*

The rate of release of electrons per unit time at a temperature T(K), for the first order case [2], is given by

$$-\frac{dn}{dt} = ns\exp\left(-\frac{E}{kT}\right) \tag{1}$$

where n is the trapped electrons concentration (cm^{-3}). Heating the sample using a linear rate $\beta = \frac{dT}{dt}$, Eq. (1) becomes

$$\frac{dn}{n} = -\frac{s}{\beta}\exp\left(-\frac{E}{kT}\right)dT \tag{2}$$

and finally, the equation related to the TL emission for a first-order kinetics, is given by

$$I(T) = n_0 s\exp\left(-\frac{E}{kT}\right)\exp\left[-\frac{s}{\beta}\int_{T_0}^{T}\exp\left(-\frac{E}{kT'}\right)dT'\right] \tag{3}$$

- ***second order kinetics***

The second order kinetics is related to the case of equal retrapping and recombination probabilities [3]. The rate of release of electrons per unit time is given by

$$-\frac{dn}{dt} = \frac{n^2 s}{N}\exp\left(-\frac{E}{kT}\right) \tag{4}$$

which transforms in the following using a linear heating rate:

$$-\frac{dn}{n^2} = \frac{s}{N\beta}\exp\left(-\frac{E}{kT}\right)dT \tag{5}$$

The TL intensity for the second order kinetics is then given by

$$I(T)=\frac{n_0^2 s\exp\left(-\frac{E}{kT}\right)}{N}\left[1+\frac{n_0 s}{N\beta}\int_{T_0}^{T}\exp\left(-\frac{E}{kT'}\right)dT'\right]^{-2} \tag{6}$$

which can be rewritten as

$$I(T)=n_0^2 s'\exp\left(-\frac{E}{kT}\right)\left[1+\frac{n_0 s'}{\beta}\int_{T_0}^{T}\exp\left(-\frac{E}{kT'}\right)dT'\right]^{-2} \tag{7}$$

where $s'=\frac{s}{N}$ is the pre-exponential factor having units cm^{-3} s^{-1}.

- ***general order kinetics***

The general order has been introduced to explain situations different from the first or the second order cases [4]. The rate of release of electrons per unit time is given by

$$-\frac{dn}{dt}=n^b s''\exp\left(-\frac{E}{kT}\right) \tag{8}$$

where s" is the pre-exponential factor and it is expressed in cm$^{3(b-1)}$s^{-1}.

The solution of Eq. (8), using a linear heating rate and $s=s''n_0^{b-1}$ having units of s^{-1}, is given by

$$I(T)=sn_0\exp\left(-\frac{E}{kT}\right)\left[1+\frac{(b-1)s}{\beta}\int_{T_0}^{T}\exp\left(-\frac{E}{kT'}\right)dT'\right]^{-\left(\frac{b}{b-1}\right)} \tag{9}$$

In order to overcome the difficulties introduced using the quantity $s=s''n_0^{b-1}$, which would vary as the dose is varied, Rasheedy suggested the following new equations. The first, second nd general order equations can be written as:

- ***first order kinetics***

$$-\frac{dn}{dt}=\frac{n^1}{N^0}s\exp\left(-\frac{E}{kT}\right) \tag{10}$$

- ***second order kinetics***

$$-\frac{dn}{dt}=\frac{n^2}{N^1}s\exp\left(-\frac{E}{kT}\right) \tag{11}$$

- ***general order kinetics***

$$-\frac{dn}{dt}=\frac{n^b}{N^{b-1}}s\exp\left(-\frac{E}{kT}\right) \tag{12}$$

The quantity s appearing in Eq. (12) has now units of s^{-1}, similar to the frequency factor in the first and second order kinetics; the difficulty related to the dimensions is now disappeared and, furthermore, it is independent of the change in dose.

The integration of Eq. (12) gives

$$n=n_0\left[1+\frac{s(b-1)\left(\frac{n_0}{N}\right)^{b-1}}{\beta}\int_{T_0}^{T}\exp\left(-\frac{E}{kT'}\right)dT'\right]^{-\frac{1}{b-1}} \tag{13}$$

and

$$I(T)=\frac{n_0^b s\exp\left(-\frac{E}{kT}\right)}{N^{b-1}}\left[1+\frac{s(b-1)\left(\frac{n_0}{N}\right)^{b-1}}{\beta}\int_{T_0}^{T}\exp\left(-\frac{E}{kT'}\right)dT'\right]^{-\frac{b}{b-1}} \tag{14}$$

References

1. Rasheedy M.S., J. Phys.: Condens. Matter **5** (1993) 633
2. Randall J.T. and Wilkins M.H.F., Proc. R. Soc. **A184** (1945) 366
3. Garlick G.F.J. and Gibson A.F., Proc. Phys. Soc. **60** (1948) 574
4. May C.E. and Partridge J.A., J. Chem. Phys. **40** (1964) 1401

Reading stability

To check the stability of the thermoluminescence response over a period of use, the following procedure is suggested.

Use N thermoluminescent dosimeters and a reader system for which the intrinsic sensitivity factors, S_i (see "Individual correction factor" section), and the calibration factor, F_C (see "Calibration factor" section), are known.

Anneal the dosimeters and irradiate them at about 10 mGy. After irradiation store the dosimeters in a place having a low background level (< 50 nGy/h). After a standard storage time after irradiation, i.e. 12 hours (any way, the storage time depends on the specific application: personal, clinical or environmental dosimetry), read out just few, n, of the irradiated dosimeters and calculate the average value of the readings:

$$\overline{M} = \frac{1}{n}\sum_{i=1}^{n}\frac{M_i}{S_i} \tag{1}$$

After 12 more hours of storage, read out other n dosimeters and calculate their average value, let say $\overline{M'}$. Repeat the procedure after 12 more hours and so on until the irradiated batch of dosimeters is finished.
The reading stability is achieved if

$$0.95 \le \frac{\overline{M'}}{\overline{M}} \le 1.05$$

In case the previous relation is not fulfilled, the calibration factor F_C has to be checked.

Recombination centre

A recombination centre is defined as the one in which the probability of recombination with an opposite sign charge carrier is greater than that of thermal excitation of the trapped carrier.

Recombination processes

The recombination processes between electrons and holes govern all luminescence phenomena. The following figure shows the possible electronic transitions in an insulator, as a thermoluminescent material is, involving both delocalized bands and localized levels.

The band-to-band recombinations are termed "direct" and the recombinations involving localized levels are termed "indirect".

For getting luminescence, recombinations must be accompanied by emission of light, which means "radiative" transitions. A "non-radiative" transition is accompanied only by phonon emission.

Reference and field dosimeters (definitions)

The main difference between the so called reference dosimeters and the field dosimeters is caused by their uses.

The sole function of the reference dosimeters is to provide a mean response to which the response of the field dosimeters is normalised in order to produce the **individual correction factors**. The reference dosimeters can be defined as a sub-batch of dosimeters which has a relative standard deviation smaller than 2-3%: this means that their responses are very close to the average value as defined in the homogeneity test.

The field dosimeters are used to monitor the radiation in all dosimetric applications and to calibrate the TLD readers.

The group of reference dosimeters, in a number of N_r depending on the size of the batch, is chosen from the previous batch itself; i.e. 10 dosimeters over a batch of 100 seems to be a proper sample. Their net TL signal must be much closer to the average value, calculated after an irradiation test, than those of all the samples. They are representative of the whole batch and will never be used for field applications (personnel, environmental or clinical dosimetry).

Only in the case of a very limited batch of dosimeters all of them can be used as reference dosimeters, in the sense that reference and field dosimeters are the same.

After annealing, irradiation and readout, the average value of the response of the N_r reference dosimeters is calculated as follows

$$\overline{M_r} = \frac{\sum_{i=1}^{N_R} (M_i - M_{oi})}{N_r} \tag{1}$$

The average is associated with the %CV, calculated as

$$\frac{\sigma}{\overline{\overline{M_r}}}\% \tag{2}$$

Relative intrinsic sensitivity factor or individual correction factor S_i (definition)

The general definition of S_i, where the index i stands for the ith dosimeter belonging to a given batch of N dosimeters, is the following

$$S_i = \frac{\overline{M}}{M_i - M_{0i}} \tag{1}$$

where

- $M_i - M_{0i} = M_{i,net}$;
- M_i is the reading of the ith dosimeter annealed and irradiated at a well defined dose D;
- M_{0i} is the background reading of the same dosimeter after annealing and not irradiated;
- $\overline{M}$ is the average of the net readings of the N dosimeters, annealed and irradiated at the dose D;

Using the previous definition of S_i, it becomes a multiplying factor of the actual net reading. However, in many scientific reports the S_i factor is defined as the inverse of the one defined by Eq. (1), so that it becomes a dividing factor of the reading.

It must be noted that the $M_i - M_{i0}$ values should be distributed around the average value $\overline{M}$ of the all readings, so that we should have

$$S_k \leq 1 \leq S_h \tag{2}$$

where S_k and S_h denote the individual correction factors for the kth and hth dosimeters respectively.

The S_i factor is associated with the proper dosimeter and used as a multiplying factor (according to its definition) of the net reading, to correct the dosimeter response at any delivered dose:

$$M_{i,net(cor)} = M_{i,net} \cdot S_i \tag{3}$$

The S_i factor is a correction factor which is needed to avoid any reading variations owing to the individual sensitivity of each dosimeter which, generally can vary from one to another dosimeter even belonging to the same batch.

During the use of the dosimeters, the S_i factors could vary owing to the irradiation levels (specially if high doses are used which can provoke some damage in the crystal lattice of the TL material), the thermal history and environmental factors, i.e., humidity and storage temperature. Because small variations in the sensitivity factors can produce large errors in the dose determination, it is imperative to check the S_i values for a given batch from time to time.

Tables 1 and 2 list the S_i factors determined for a batch of 28 TLD-100 and tested over a period of more than seven years. The factors have been calculated according to Eq. (1)

	01/07 1988 233.2 mR		28/11 1991 (?)		28/01 1994 523.3 mR		16/11 1995 380.1 mR		22/12 1995 535.1 mR	
N.	$r_{i_{net}}$	S_i	$r_{i_{net}}$	S_i	$r_{i_{net}}$	S_i	$r_{i_{net}}$	S_i	$r_{i_{net}}$	S_i
1	1.011	1.035	2.114	1.016	2.244	1.046	1.659	1.027	2.314	1.037
2	0.918	1.139	1.905	1.128	2.070	1.134	1.494	1.140	2.076	1.156
3	0.792	1.321	1.676	1.282	1.807	1.299	1.316	1.294	1.847	1.299
4	0.796	1.314	1.608	1.386	1.785	1.315	1.291	1.319	1.807	1.328
5	0.967	1.082	2.004	1.072	2.174	1.080	1.598	1.066	2.206	1.088
6	0.996	1.050	1.938	1.108	2.230	1.052	1.593	1.069	2.233	1.075
7	0.941	1.112	1.860	1.155	2.077	1.130	1.503	1.133	2.096	1.145
8	1.040	1.006	2.093	1.026	2.264	1.037	1.668	1.021	2.352	1.020
9	1.086	0.963	2.168	0.991	2.378	0.987	1.739	0.979	2.408	0.997
10	1.070	0.978	2.096	1.025	2.324	1.010	1.703	1.000	2.341	1.025
11	1.000	1.046	1.976	1.087	2.233	1.051	1.590	1.071	2.237	1.073
12	0.933	1.121	1.863	1.153	2.086	1.125	1.511	1.127	2.087	1.150
13	0.859	1.218	1.708	1.258	1.955	1.201	1.383	1.231	2.020	1.188
14	0.937	1.116	1.886	1.139	2.110	1.112	1.538	1.107	2.135	1.124
15	1.158	0.903	2.383	0.901	2.579	0.910	1.854	0.919	2.646	0.907
16	1.236	0.846	2.564	0.838	2.782	0.844	2.027	0.840	2.879	0.834
17	1.207	0.867	2.469	0.870	2.659	0.883	1.937	0.879	2.749	0.873
18	0.909	1.151	1.869	1.149	2.068	1.135	1.478	1.152	2.108	1.139
19	0.836	1.251	1.716	1.252	1.889	1.242	1.381	1.233	1.947	1.233
20	1.259	0.831	2.572	0.835	2.890	0.812	2.009	0.848	2.861	0.839
21	1.118	0.936	2.291	0.938	2.537	0.925	1.810	0.941	2.589	0.927
22	1.230	0.850	2.491	0.862	2.728	0.860	1.976	0.862	2.831	0.848
23	1.184	0.883	2.414	0.890	2.730	0.860	1.936	0.880	2.752	0.872
24	1.075	0.973	2.208	0.973	2.470	0.950	1.787	0.953	2.568	0.935
25	1.043	1.003	2.196	0.978	2.485	0.944	1.834	0.929	2.579	0.931
26	1.321	0.792	2.725	0.788	2.968	0.791	2.219	0.767	3.139	0.765
27	1.268	0.825	2.671	0.804	2.851	0.823	2.132	0.799	2.987	0.803
28	1.087	0.962	2.671	0.804	2.344	1.001	1.725	0.987	2.409	0.996
$\bar{R}$	1.046		2.148		2.347		1.703		2.400	

Table 1. Individual correction factors.

N.o	$\overline{S}_i \pm \sigma$	σ%
1	1.032±0.011	1.1
2	1.139±0.010	0.9
3	1.299±0.014	1.1
4	1.322±0.031	2.3
5	1.078±0.009	0.8
6	1.071±0.023	2.1
7	1.135±0.016	1.4
8	1.022±0.011	1.1
9	0.983±0.013	1.3
10	1.008±0.020	2.0
11	1.066±0.017	1.6
12	1.135±0.015	1.3
13	1.219±0.027	2.2
14	1.120±0.013	1.2
15	0.908±0.007	0.8
16	0.840±0.005	0.6
17	0.874±0.007	0.8
18	1.145±0.008	0.7
19	1.242±0.009	0.7
20	0.833±0.013	1.6
21	0.933±0.007	0.8
22	0.856±0.007	0.8
23	0.877±0.012	1.4
24	0.957±0.016	1.7
25	0.957±0.032	3.3
26	0.781±0.013	1.7
27	0.811±0.012	1.5
28	0.950±0.083	8.7

Table 2. Average values of the correction factors given in the previous Table 1.

Reference

Data provided by Dr. V. Klammert of the Nuclear Engineering Department of CESNEF, Milan, Italy

Relative intrinsic sensitivity factor or individual correction factor S_i (procedures)

General procedures for the determination of the individual correction factors are given in the following.

1st procedure

- annealing of all the samples according to the standard anneal procedure suitable for the material in use.
- readout of the samples, using the appropriate readout cycle, for determining the intrinsic background value of each dosimeter, M_{0i}.
- irradiation of the samples to a known dose, chosen in the region of the linear response and at a level which is supposed to be used for the dosimeters in the applications.
- readout of the irradiated samples, in only one session, using the same readout cycle used in the second point and determine the values M_i.
- calculate for each sample the quantity

$$M_{i,net} = M_i - M_{0i} \tag{1}$$

and calculate the mean response of the batch as

$$\overline{M_1} = \frac{1}{N}\sum_{i=1}^{N}(M_i - M_{0i}) \tag{2}$$

- perform a new annealing of the samples and re-irradiate at the same dose already delivered in the third point. Read all the samples and calculate a new average value

$$\overline{M_2}$$

- repeat the procedure 5 times.
- calculate the quantity

$$\overline{\overline{M}} = \sum_{j=1}^{5}\frac{\overline{M}_j}{5} \tag{3}$$

where j stands for the number of irradiations performed for the samples.

- calculate the average response for each sample of the batch according to the expression

$$\overline{M_i} = \sum_{j=1}^{5}\frac{M_{ij,net}}{5} \tag{4}$$

where i indicates the i-th dosimeter.

- calculate the relative intrinsic sensitivity for each single dosimeter as

$$S_i = \frac{\overline{\overline{M}}}{\overline{M}_i} \qquad (5)$$

This factor is quite stable during time so that it only needs to be checked no more than two times per year.

2nd *procedure*

The procedure just given above is the best but it is not easy to adopt with a large number of dosemeters, as it can be the case of a personnel dosimetry service.

When the batch of dosemeters is quite big, the correction factors can be calculated making use of a sub-group of dosemeters, the reference dosemeters, chosen from the same batch in use, and then normalise all the dosemeters to the response of the reference dosemeters.

Tables 1 (a,b) list the data concerning a batch of 78 dosemeters. From the batch, five dosemeters have been chosen as reference dosemeters, labelled with (*).

The total average on the 78 dosimeters is

$$\overline{M}_{net} = 7.556 \pm 0.434 \qquad (5.7\%)$$

The average of the reference dosimeters is

$$\overline{M}_{net}^{ref} = 7.595 \pm 0.040 \qquad (0.5\%)$$

The sensitivity factor for each dosemeter of the batch is then calculated as

$$S_i = \frac{\overline{M}_{net}^{ref}}{M_i - M_{0i}}$$

The effect of the correction factors is well proved by observing that the new average value of the remaining 73 dosemeters is now

$$\overline{M}_{net,cor} = 7.596 \pm 0.002$$

which means a %CV of 0.03%.

Dos. No.	$M_{i,net}$	S_i	$M_{i,net(cor)}$	Dos. No.	$M_{i,net}$	S_i	$M_{i,net(cor)}$
1	8.468	0.897	7.596	40	7.836	0.969	7.593
2	8.076	0.940	7.591	41	7.167	1.060	7.597
3	7.808	0.973	7.597	42	7.912	0.960	7.596
4	7.085	1.072	7.595	43	7.946	0.956	7.596
5	7.231	1.050	7.593	44	6.765	1.123	7.597
6	7.601	0.999	7.593	45	6.771	1.122	7.597
7	7.587	1.001	7.595	46	7.531	1.009	7.599
8	7.346	1.034	7.596	47	7.657	0.992	7.596
9*	7.634	-	-	48	7.045	1.078	7.596
10	6.916	1.098	7.594	49	7.434	1.023	7.605
11	7.394	1.027	7.594	50	7.476	1.016	7.596
12	7.491	1.014	7.596	51	7.239	1.049	7.594
13	8.094	0.938	7.592	52	7.480	1.015	7.592
14*	7.600	-	-	53	6.704	1.133	7.596
15	7.854	0.967	7.595	54	7.656	0.992	7.595
16	7.509	1.012	7.599	55	7.118	1.067	7.595
17	7.428	1.023	7.599	56	7.167	1.060	7.597
18	7.329	1.036	7.593	57	6.699	1.134	7.597
19	7.963	0.954	7.597	58	8.047	0.944	7.596
20	7.290	1.042	7.596	59	7.395	1.027	7.595

Table 1(a). Effect of the use of reference dosimeters.

Dos. No.	$M_{i,net}$	S_i	$M_{i,net(cor)}$	Dos. No.	$M_{i,net}$	S_i	$M_{i,net(cor)}$
21	7.676	0.989	7.592	60	7.984	0.951	7.593
22	7.294	1.041	7.593	61	8.057	0.943	7.598
23	8.387	0.906	7.599	62	8.014	0.948	7.597
24	7.677	0.989	7.593	63	7.555	1.005	7.593
25	8.232	0.923	7.598	64	6.968	1.090	7.595
26	8.143	0.933	7.597	65	6.720	1.130	7.594
27	7.839	0.969	7.596	66	8.320	0.913	7.596
28	8.111	0.936	7.592	67	7.778	0.977	7.599
29	7.464	1.018	7.598	68	7.487	1.014	7.592
30	7.374	1.030	7.595	69	6.786	1.119	7.594
31*	7.539	-	-	70	8.433	0.901	7.598
32	6.739	1.127	7.595	71	7.424	1.023	7.595
33	7.411	1.025	7.596	72	7.812	0.972	7.593
34*	7.574	-	-	73	7.402	1.026	7.594
35*	7.633	-	-	74	7.620	0.997	7.597
36	7.880	0.964	7.596	75	7.025	1.081	7.594
37	7.783	0.976	7.596	76	7.934	0.957	7.593
38	7.872	0.965	7.596	77	7.334	1.036	7.598
39	7.568	1.004	7.598	78	7.580	1.002	7.595

Table 1(b). Effect of the use of reference dosimeters.

3rd procedure

On the use of the reference dosemeters it is very useful to follow the procedure suggested by P. Plato and J. Miklos of the School of Public Health of the Michigan University. This procedure is well indicated when a large number of TLDs, larger than 10000, is used for dosimetric purposes.

As claimed by the Authors, their procedure should ensure that the individual correction factors take into account only variations among the TL samples of a given batch and not variations caused by the instability of the TLD reader.

The Authors suggest to divide a new batch into two batches:

- the reference dosemeters,
- the field dosemeters.

As stated before, the aim of the reference dosemeters is to provide a mean response to which the response of the field dosemeters is normalised to obtain the S_i factors. In this way the response of each field dosemeter will be the same as the mean response of the reference dosemeters. The number of the reference dosemeters should be about 2~5% of the whole batch, according to its size.

A problem can rise if some reference dosemeters are lost or they present changes in response owing to the age. To by-pass this potential problem, the procedure suggests the use of subsets of reference dosemeters rather than the whole reference group.

It must also be noted that the correction factors could be affected by irradiation, if this is not done uniformly, due to room scatter or if the beam is not isotropic. So, the irradiation geometry must be carefully checked for obtaining that all the dosemeters are irradiated uniformly.

The procedure consists of several steps and it is shortly reported here.

- annealed, irradiated and read the reference dosimeters,
- the same procedure is repeated at least three times,
- calculate the mean values for each irradiation and the coefficient of variation (CV) associated with

$$CV\% = 100\frac{\sigma}{\overline{M}_{net}}$$

It must be noted that the mean values obtained are calculated without the correction factors being applied because these factors do not exist at this level of the procedure.

- the individual correction factors for the each reference dosemeter are now calculated for each irradiation:

$$S_{ij} = \frac{M_{ij}}{\overline{M}_j}$$

with

$$\overline{M}_j = \sum_{i=1}^{N} \frac{M_{ij}}{N}$$

where
i=1,2,...N is the number of the reference dosemeters,
j=1,2,3,... is the number of the repeated irradiations,
S_{ij} is the individual correction factor for the ith dosemeter obtained after the jth irradiation,
M_{ij} is the response of the ith dosemeter after the jth irradiation,
$\overline{M}_j$ is the mean response for all the reference dosemeters after the j-th irradiation.
It must be stressed that during the whole procedure involving the three irradiations, the calibration of the TLD reader could change significantly from one readout session to another. However, the calibration factors are unaffected since they are based on the mean of a given irradiation.

- the average values of the correction factors are then calculated for each sample along the three successive irradiations

$$S_i = \sum_{j-1}^{3} \frac{S_{ij}}{3}$$

and the CV% is obtained as well.

- once the averages of S_i have been obtained, it is important to examine their distribution as well as the distribution of the associated CV%. If one or both of these quantities are abnormally large, it is better to reject the defective samples. It should be advisable to identify and eliminate all the elements having an S_i that is not within 20% of unity (the acceptable range is then from 0.80 to 1.20) and the elements which have a CV% greater than 5%.

The limits given for S_i and CV% can be dependent on several factors, i.e., the level of the delivered dose, the light emission from the phosphors, the light detection efficiency of the TLD reader. The CV is strongly dependent on the dose; one can expect to have a large value of the CV% at low doses and a little one for high doses. However, the suggested limits can be changed according to the specific use of the dosimeters.

When a large number of field dosimeters has to be used, it is better, as suggested by the authors, to divide the field dosemeters in sub-batches and to do the same for the reference dosemeters. As a consequence, each sub-group of field dosemeters will be related to a proper sub-group of reference dosemeters. This procedure is necessary to ensure that the TLD reader response will remain stable during the readout which, using a small quantity of TLDs, can be carried out in only one session.

When sub-groups of reference and field dosemeters are used, the S_i factors for field dosemeters are calculated using the TL response of the reference dosemeters corrected by the S_i factors already existing (see above); in this way, the mean response of the sub-set of reference dosemeters is the same as the mean response of the of the whole reference group.

The response of each sample of the sub-group of field dosemeters is corrected by the appropriate S_i calculated according to the following expression

$$S_i = \frac{M_i}{\overline{M}_{ref}}$$

where

M_i is the response of the ith field dosemeter and

$\overline{M}_{ref}$ is the mean response of the sub-group of reference dosemeters. Remember that that this value is coming from a set of values already corrected by the appropriate S_i.

Reference

Plato P. and Miklos J., Health Phys. **49**(5) (1985) 873

Residual TL signal

It is so called the TL signal obtained after the annealing procedure or after a second readout cycle of the same sample. The observation of a residual TL signal means that the annealing procedure or the second readout cycle have not obtained the effect to erase all the phosphor traps.

Any unerased TL signal may interfere with further TL measurements using the same sample.

The lower detection limit as well as the reproducibility are strongly affected by the residual signal. The residual signal depends on the phosphor type as well as its irradiation history.

Rubidium halide

RbCl and RbBr can be growth as single crystals from the melt by the Kiroupoulos method. Doping was achieved by adding suitable amounts of KOH to the melt. The suggested annealing, for getting high sensitive material, is 600°C for 30 minutes followed by quick quenching to room temperature. The TL sensitivity of RbCl:OH^- is decreasing as the dopant concentration increases. At the lowest dopant concentration, i.e. 0.13 mol %, the glow curve exibits a single peak at about 100°C. RbBr:OH- reveals a glow curve consisting in two peaks: one at 70°C and another, less intense, at about 175°C. After irradiation at more than 20 Gy, a third peak appears at 230°C. Any way, both materials are affected by high fading.

References

Furetta C., Laudadio M.T., Sanipoli C., Scacco A., Gomez-Ros J.M. and Correcher V., J. Phys. Chem. of Solids **60** (1999) 957

S

Second-order kinetics when $s' = s'(T)$

The detrapping rate in this case is given by [1-3]

$$\frac{dn}{dt} = -n^2 s_0' T^{\alpha} \exp(-\frac{E}{kT}) \tag{1}$$

Using a linear heating rate, Eq. (1) becomes

$$\frac{dn}{dT} = -\frac{n^2 s_0'}{\beta} T^{\alpha} \exp\left(-\frac{E}{kT}\right) \tag{2}$$

and the solution is then

$$n = n_0 \left[1 + \frac{s_0' n_0}{\beta} \int_{T_0}^{T} T^{\alpha} \exp\left(-\frac{E}{kT}\right) dT\right]^{-1} \tag{3}$$

while the intensity is given by

$$I(T) = \frac{n_0^2 s_0' T^{\alpha} \exp(-\frac{E}{kT})}{\left[1 + \frac{s_0' n_0}{\beta} \int_{T_0}^{T} T^{\alpha} \exp\left(-\frac{E}{kT}\right) dT\right]^2} \tag{4}$$

References

1. Chen R. and Kirsh Y., Analysis of Thermally Stimulated Processes (Pergamon Press, 1981)
2. McKeever S.W.S., Thermoluminescence of Solids (Cambridge University Press, 1985)
3. Furetta C. and Weng P.S., Operational Thermoluminescence Dosimetry (World Scientific Publ., 1998)

Self-dose in competition to fading (procedure)

The self-dose arises from the radioactive content of the thermoluminescent materials. With fading, the self-dose is an important factor in environmental radiation monitoring and it is strongly dependent on the packing materials constituting the dosimeter. Self-dose and fading are two effects in competition between them.

A precise estimation of self-dose effect needs an accurate experiment. A method of estimating accurately both self-dose and fading under conditions similar to the ones encountered in environmental monitoring applications consists in leaving a batch of TLDs in a sufficiently thick lead shield of about 5 cm to stop most of the external irradiations (only hard components of the cosmic rays will contribute to the radiation field inside the lead shield).

The experimental procedure is the same of the previous one and, as before, we have again three sub-groups of TLDs, group A, B and C.

After having done all the initial procedures, group A is irradiated at the test dose D_t and stored inside the lead shield together with the annealed one groups B and C but not irradiated.

At the end of the storage period, let us say 1 month, group B is irradiated at the same dose D_t and all three groups of TLDs are read.

Now we have three quantities, the averaged readings, which are linked by different equations:

$$\overline{M}_A = \frac{\dot{B}}{\lambda} + (\overline{M}_B - \frac{\dot{B}}{\lambda})\exp(-\lambda t_a) \tag{1}$$

$$\overline{M}_C = \frac{\dot{B}}{\lambda}[1 - \exp(-\lambda t_a)] \tag{2}$$

$$\overline{M}_A - \overline{M}_C = \overline{M}_B \exp(-\lambda t_a) \tag{3}$$

where t_a is the storage period of time and $\dot{B} = \dot{B}_f + \dot{B}_s$, with the first component being the field dose rate inside the shield and the second the self-dose rate.

The decay constant is now given by the following expression

$$\lambda = -\frac{1}{t_a}\ln\frac{\overline{M}_A - \overline{M}_C}{\overline{M}_B} \tag{4}$$

which substitutes in Eq. (2) gives:

$$\dot{B} = \frac{\lambda \overline{M}_C}{1 - \exp(-\lambda t_a)} \tag{5}$$

The component $\dot{B}_f$, the field dose rate inside the shield, can be measured by a high pressure ionisation chamber; after that the self-dose rate can easily be evaluated:

$$\dot{B}_s = \frac{\lambda \cdot \overline{M}_c}{1-\exp(-\lambda \cdot t_a)} - \dot{B}_f \tag{6}$$

Sensitisation (definition)

Sensitisation is a term used for indicating an increase of **sensitivity** in a TL sample due to a high dose of irradiation, usually followed by a heating treatment. This effect has been firstly found by Cameron in LiF [1-3].

References

1. Cameron J.R., Suntharalingam N. and Kenney G.N., Thermoluminescence Dosimetry, University of Wisconsin Press, Madison (1968)
2. McKeever S.W.S., Thermoluminescence in Solids, Cambridge University Press (1985)
3. Chen R. and Kirsh Y., Analysis of Thermally Stimulated Processes, Pergamon Press (1981)

Sensitivity (definition)

The sensitivity of a TLD, *S*, may be expressed , in general, as the TL response, in reader units, per unit of dose and unit of mass of the sample:

$$S = \frac{TL}{D \cdot m}$$

Variations in sensitivity, among dosemeters belonging to the same batch, can be encounterd in practice. The variations in sensitivity are mainly due to the following reasons:

- variation in the mass of the detectors,
- variation in the optical density from sample to sample,
- variation due to dirt contamination of the sample surface.

Sensitivity variations with optical transmission

If variations in $L_u^*(\lambda), k_2, D_0, G_{eff}$ are insignificant, Eq. (3), in "Sensitivity variations with thickness of TLD", becomes

$$L \propto \frac{1}{a}\left[1-\exp(-aT)\right]\cdot\left[1+r\exp(-aT)\right]$$

Inserting in this expression the light transmission coefficient

$$t_{eff} = \exp(-aT)$$

and

$$T = cw$$

one obtains

$$L \propto \frac{1}{a}\left(1-t_{eff}\right)\left(1+r\cdot t_{eff}\right)$$

And then

$$L \propto -\frac{w}{\ln t_{eff}}\left(1-t_{eff}\right)\left(1+r\cdot t_{eff}\right)$$

Sensitivity variations with thickness of TLD

From Eq. (1) and Eq. (2), in "Theory of thermoluminescent dose measurement" section, the light output from an uniformly irradiated TLD sample of thickness T is given by:

$$L = \frac{L_u(\lambda)k_2 D_0 G_{eff}}{aT}\left[1-\exp(-aT)\right]\cdot\left[1+r\exp(-aT)\right] \tag{1}$$

where a is related to the dosemeter coefficient of attenuation, $A(\lambda)$, of its internally generated light; and $L_u(\lambda)$ is the light generated in the dosemeter per unit of absorbed dose, D_0.
$L_u(\lambda)$ is linearly proportional to the thickness T, therefore

$$L_u(\lambda) = T\cdot L_u^*(\lambda) \tag{2}$$

where $L_u^*(\lambda)$ is the light generated per unit of absorbed dose and per unit of thickness, and it is independent of T.
Hence

$$L = \frac{L_u^*(\lambda)k_2 D_0 G_{eff}}{a}\left[1-\exp(-aT)\right]\cdot\left[1+r\exp(-aT)\right] \tag{3}$$

If variations in $L_u^*(\lambda), k_2, D_0, G_{eff}, a$ are insignificant for TLDs of slightly varying thickness and irradiated and readout under identical conditions, Eq. (3) becomes

$$L \propto \left[1-\exp(-aT)\right]\cdot\left[1+r\exp(-aT)\right] \tag{4}$$

If diameter and density are constant for a given batch of dosemeters, then T is proportional to the dosemeter weight w, i.e.

$$T = cw \tag{5}$$

and Eq. (4) can be rewritten as

$$L \propto \left[1-\exp(-acw)\right]\cdot\left[1+r\exp(-acw)\right] \tag{6}$$

Set up of a thermoluminescent dosimetric system (general requirements)

For the setting up of the system, one needs to fill up some requirements for initialising, characterising and calibrating the TL material according to the use for. These operations consist of several tests and measurements:

- initialisation procedure,
- determination of the batch homogeneity,
- to choose the reference dosimeters,
- to determine the relative intrinsic sensitivity of each dosimeter,
- the measurement of the threshold dose,
- to determine the linearity range of the system and its calibration factor,
- to carry out the reproducibility tests: of the reader, of the background, of the calibration factor, of the dosimeters,
- to study the appropriate annealing procedure in case using a new TL material for which no information in the scientific literature is available.
- to carry out a quality control of the instruments for the thermal treatments of the dosimeters (rise temperature curve, temperature stability, temperature distribution inside oven, etc.).

The following points are very important prerequisites before starting the experimental procedures listed above.

- select dosimeter elements having approximately equal mass.
- reject elements which are imperfect, discoloured or dashed .
- do not handle elements directly; use tweezers (preferably vacuum tweezers) or spatulas for TL powder. Avoid scratching the surfaces of the dosimeters.
- do not leave the dosimeters uncovered in the laboratory. It is better to store the dosimeters in opaque bags or containers.
- some dosimeters are sensitive to sunlight, UV light or develop background effects when exposed to UV light. It is advisable to use tungsten, filtered fluorescent lighting or red lamps and to keep the dosimeters away from direct sunlight.
- keep the dosimeters away from heat and radiation sources during storage. It could be better to store the dosimeters in lead boxes to avoid any background irradiations.

Simultaneous determination of dose and time elapsed since irradiation

Information about the time which has elapsed since an external radiation exposure is useful in determining the time of occurrence of an abnormal exposure in personal and accident dosimetry.

This information can be obtained from certain irradiated thermoluminescent dosimeters and the method consists essentially of a glow-curve behaviour study. The glow-curve is a finger print of the radiation effect in a particular TL phosphor. The glow-curve may consist of several peaks each having its maximum intensity at different temperatures. Those peaks which occur in the temperature region above 150°C are generally thermally stable, and the integrated area or the eight of such peaks is used to assess the absorbed dose from radiation exposure. However, in the glow-curve region below 150°C there are also less thermal stable peaks.

From a practical point of view, if one peak has faded and another has scarcely faded, the peak area or the peak height ratios will be function of the time after exposure and then the time can be estimated throughout the ratios. By using the decay rates of suitable peaks in the glow-curve and, in turn, the corresponding mean trap life times, τ, the time elapsed since a single exposure may be determined.

The feasibility of the method has been theoretically investigated simulating a glow-curve as shown in Fig. 1. Furthermore, the simulation considers the combined effects of a single exposure superposed to a continuous background exposure [1]. A further work [2] gave a theoretical model of the method as well as a comparison with experimental results. The same subject has been investigated in [3-4]

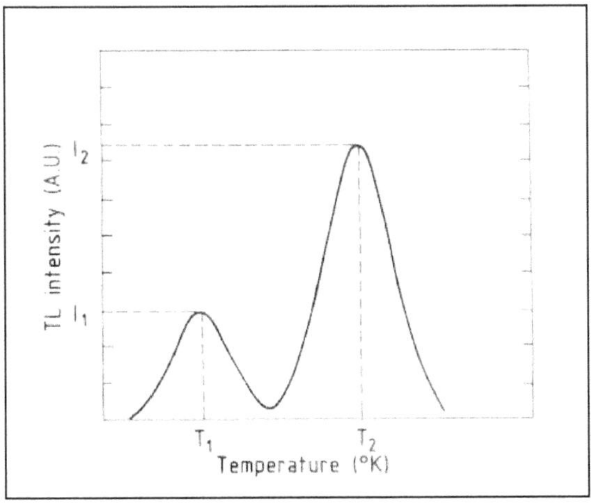

Fig. 1. A glow curve showing two well defined peaks.

- Theoretical model

Expressions for a single accidental exposure

The mathematical treatment starts from the first- and second-order equations:

$$\frac{dn}{dt} = -sn\exp\left(-\frac{E}{kT}\right) \qquad 1^{st} \text{ order} \tag{1}$$

$$\frac{dn}{dt} = -s'n^2\exp\left(-\frac{E}{kT}\right) \qquad 2^{nd} \text{ order} \tag{2}$$

Integration of the previous equations gives, respectively, the following solutions

$$n = n_0 \exp\left[-st\exp\left(-\frac{E}{kT}\right)\right] \tag{3}$$

$$n - n_0\left[1 + s'n_0 t\exp\left(-\frac{E}{kT}\right)\right]^{-1} \tag{4}$$

The TL intensity is given by

$$I(t) \propto -\frac{dn}{dt} \tag{5}$$

and then Eqs. (3) and (4) can be rewritten in the following way, respectively for the 1st and the 2nd order of kinetics:

$$I(t) = n_0 s \exp\left(-\frac{E}{kT}\right) \exp\left[-st \exp\left(-\frac{E}{kT}\right)\right] \tag{6}$$

$$I(t) = \frac{n_0^2 s' \exp\left(-\frac{E}{kT}\right)}{\left[1 + s' n_0 t \exp\left(-\frac{E}{kT}\right)\right]^2} \tag{7}$$

Taking into account the total TL light

$$\Phi_t = \int_0^t I(t)dt \tag{8}$$

and using Eq. (6), it turns out that Φ coincides numerically with n. Then, Eqs. (3) and (4) can be written as following

$$\Phi(t) = \Phi_0 \exp\left[-st \exp\left(-\frac{E}{kT}\right)\right] \tag{9}$$

$$\Phi(t) = \Phi_0 \left[1 + s' \Phi_0 t \exp\left(-\frac{E}{kT}\right)\right]^{-1} \tag{10}$$

Introducing the mean trap lifetime for both 1st and 2nd order kinetics respectively

$$\tau = s^{-1} \exp\left(\frac{E}{kT}\right) \tag{11}$$

$$\tau^* = \left(s' n_0\right)^{-1} \exp\left(\frac{E}{kT}\right) \tag{12}$$

Eqs. (9) and (10) can be written in the following way

$$\Phi = \Phi_0 \exp\left(-\frac{t}{\tau}\right) \qquad \text{1st order} \tag{13}$$

$$\Phi = \Phi_0 \left(1 + \frac{t}{\tau *}\right)^{-1} \qquad \text{2nd order} \tag{14}$$

Expressions including a continuous irradiation

The second contribution to the final equations is the signal due to a continuous irradiation, i.e. environmental irradiation background. The equations have to take into consideration a progressive extinction of the initially stored information, i.e. the accidental irradiation signal, whereas the environmental contribution leads to an increase of the TL signal. Under this condition, the previous Eqs. (1) and (2) assume the following forms

$$\frac{dn}{dt} = \alpha \cdot \dot{D} - \frac{n}{\tau} \qquad 1^{st} \text{ order} \tag{15}$$

$$\frac{dn}{dt} = \alpha \cdot \dot{D} - \frac{n}{\tau^*} \qquad 2^{nd} \text{ order} \tag{16}$$

where $\dot{D}$ is the environmental exposure rate and α is a constant, typical for each thermoluminescent material and giving its sensitivity (TL per unit of dose and mass).

Integration of Eq. (15) gives

$$n = C\exp\left(-\frac{t}{\tau}\right) + \alpha \cdot \dot{D} \cdot \tau$$

which, using the initial condition *n(0)=0* , becomes

$$n = \alpha \cdot \dot{D} \cdot t\left[1 - \exp\left(-\frac{t}{\tau}\right)\right] \tag{17}$$

As the elapsed time becomes very large, *n* gets more and more similar to the asymptotic value

$$n_\infty = \alpha \cdot \dot{D} \cdot t \tag{18}$$

Eq. (14) is explained assuming that, at infinity, a dynamical equilibrium is attained, providing the trapped charges to compensate at each instant those escaping owing to the fading phenomenon.

Eq. (17) can be changed using the total light

$$\Phi = \alpha \cdot \dot{D} \cdot \tau\left[1 - \exp\left(-\frac{t}{\tau}\right)\right] \tag{19}$$

Integration of Eq. (16) yields to the final 2nd order expression

$$\Phi = \left(\alpha \cdot \dot{D} \cdot \Phi_0 \cdot \tau^*\right)^{\frac{1}{2}} \frac{1 - \exp\left[-2\left(\dfrac{\alpha \cdot \dot{D}}{\Phi_0 \cdot \tau^*}\right)^{\frac{1}{2}} \cdot t\right]}{1 + \exp\left[-2\left(\dfrac{\alpha \cdot \dot{D}}{\Phi_0 \cdot \tau^*}\right)^{\frac{1}{2}} \cdot t\right]} \tag{20}$$

Eq. (20) can be rewritten in a more simply way as

$$\Phi = \left(\alpha \cdot \dot{D} \cdot \tau^* \cdot \Phi_0\right)^{\frac{1}{2}} \tanh\left[\left(\frac{\alpha \cdot \dot{D}}{\tau^* \cdot \Phi_0}\right)^{\frac{1}{2}} \cdot t\right] \tag{21}$$

Final expressions

The equations related to a single accidental exposure and to a continuous irradiation have now to be combined. The accidental exposure can be thought of as overwhelming; then a characteristic time t_i has to be introduced as the time interval elapsed from the zero instant to the time of the accidental exposure. Fig. 2 shows the superposition of the accidental exposure on the background irradiation for a 1st order. The accidental exposure has been assumed to occur in the middle of the observation period, i.e. $t_i = 15$ days over a period of 30 days. Until the 15th day only background is present. At the 15th day, as a

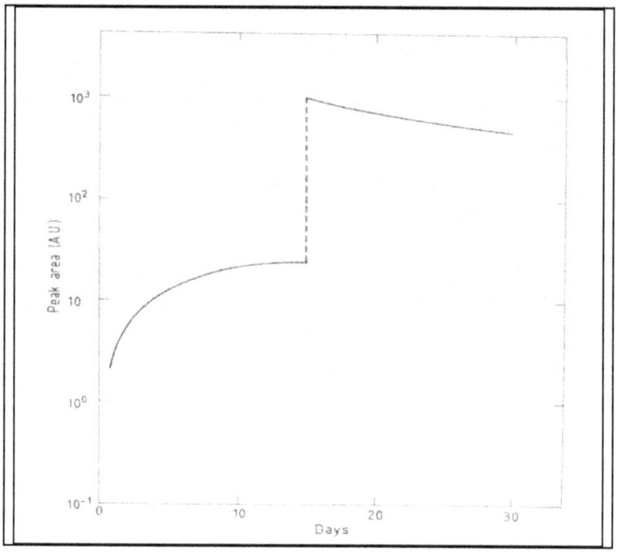

Fig. 2. Effect of an irradiation superposed to the background irradiation.

consequence of the external accidental irradiation, a sharp discontinuity occurs which is assumed to be as large as 1 Gy.

The overall equations can be written by combining together Eqs. (13) and (19) for the first order kinetics, and Eqs. (14) and (21) for the second order:

$$\Phi = \Phi_0 \exp\left(-\frac{t-t_i}{\tau}\right) + \alpha \cdot \dot{D} \cdot \tau \left[1 - \exp\left(-\frac{t}{\tau}\right)\right] \tag{22}$$

$$\Phi = \Phi_0 \left(1 + \frac{t-t_i}{\tau^*}\right)^{-1} + \left(\alpha \cdot \dot{D} \cdot \tau^* \cdot \Phi_0\right)^{1/2} \tanh\left[\left(\frac{\alpha \cdot \dot{D}}{\tau^* \cdot \Phi_0}\right)^{1/2} \cdot t\right] \tag{23}$$

Assuming now a glow-curve having two peaks, it is necessary to define the area ratio, R, between them. For the 1st order kinetics we get:

$$R = \frac{\Phi_2}{\Phi_1} = \frac{\Phi_{02} \exp\left(-\frac{t-t_i}{\tau_2}\right) + \alpha_2 \cdot \dot{D} \cdot \tau_2 \left[1 - \exp\left(-\frac{t}{\tau_2}\right)\right]}{\Phi_{01} \exp\left(-\frac{t-t_i}{\tau_1}\right) + \alpha_1 \cdot \dot{D} \cdot \tau_1 \left[1 - \exp\left(-\frac{t}{\tau_1}\right)\right]} \tag{24}$$

and

$$R = \frac{\Phi_2}{\Phi_1} = \frac{\Phi_{02} \left(1 + \frac{t-t_i}{\tau_2^*}\right)^{-1} + \left(\alpha_2 \cdot \dot{D} \cdot \tau_2^* \cdot \Phi_{02}\right)^{1/2} \tanh\left[\left(\frac{\alpha \cdot \dot{D}}{\tau_2^* \cdot \Phi_{02}}\right)^{1/2} \cdot t\right]}{\Phi_{01} \left(1 + \frac{t-t_i}{\tau_1^*}\right)^{-1} + \left(\alpha_1 \cdot \dot{D} \cdot \tau_1^* \cdot \Phi_{01}\right)^{1/2} \tanh\left[\left(\frac{\alpha \cdot \dot{D}}{\tau_1^* \cdot \Phi_{01}}\right)^{1/2} \cdot t\right]} \tag{25}$$

for the second order kinetics.

The indexes "1" and "2" refer to the first peak, i.e. high fading, and to the second peak , no fading, respectively.

The Eqs. (24) and (25) have been computed for some values of the mean trap klifetimes. Fig. 3 shows the trend of the peak area ratios for the parameters given in Table 1 and for different mean trap lifetimes.

Φ_{01}	Φ_{02}	α_1	α_2	$\dot{D}$	t	t-t$_i$	$\tau_2 = \tau_2^*$
90	100	90	100	3x10^{-3} mGy/d	30 days	40 days	400 days

Table 1. Parameters used for computing Eqs. (24) and (25).

From Fig. 3 it is evident that for practical application one needs a steep line: only in this case an accidental exposure can be accurately backdated. If the lines are too flat, the error in time determination will be very high even when two elapsed times are very different.

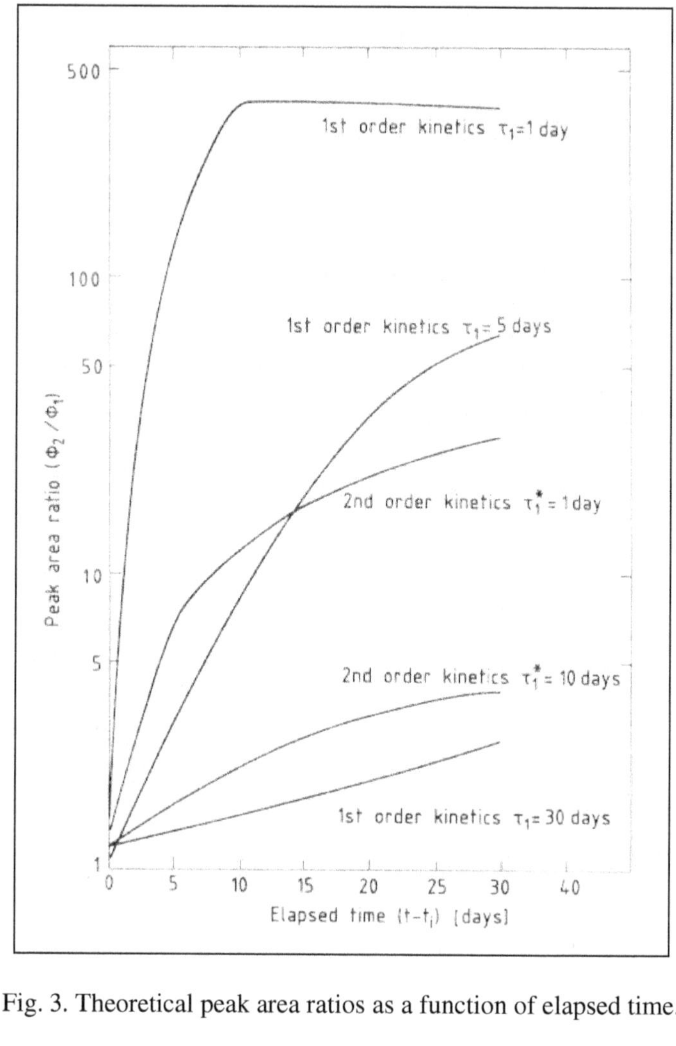

Fig. 3. Theoretical peak area ratios as a function of elapsed time.

Observing Fig. 4 it appears eveident that the peak-area ratio is more useful than the peak-height ratio because, in the former case, the elapsed time after irradiation can be estimated with a smaller error.

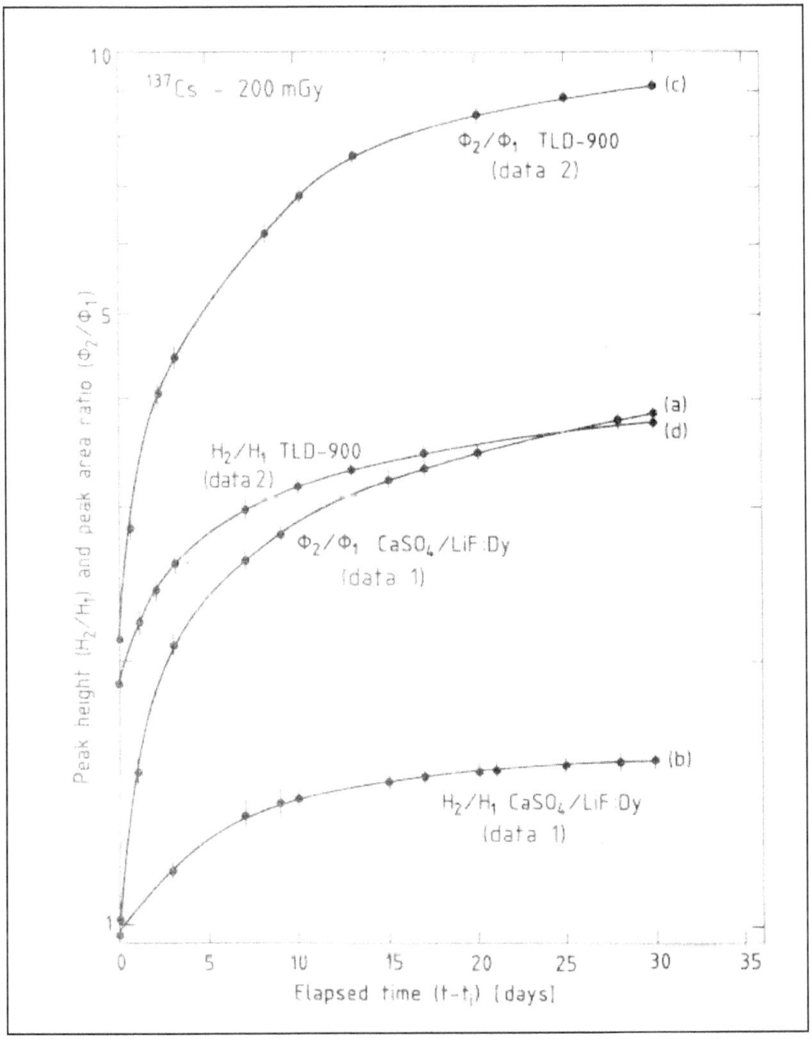

Fig. 4. Expermental data. Peak ratios Vs elapsed time.

Looking at the peak-area ratio in Fig. 4, the maximum error in time determination can be about ± 2 days when the accidental irradiation occurs in the range of 0 – 15 days. If the elapsed time between irradiation and readout is larger than 15 days, the uncertainty becomes larger. In any case, the peak-area ratio gives better figures than the peak-height ratio.

A mathematical approximation has been done for fitting the experimental data. For this purpose, a polynomial approximation, using the Tchebychev's norma, has been carried out. For example, the plots of Φ_2/Φ_1, labelled Data 1 and Data 2, have been fitted by the following 4th degree polynomials:

$$t_\Phi = -20.2 + 38R_\Phi - 23R_\Phi^2 + 5.7R_\Phi^3 - 0.37R_\Phi^4 \qquad \text{Data 1}$$

$$t_\Phi = 33.9 - 33.7R_\Phi + 11.3R_\Phi^2 - 1.5R_\Phi^3 + 0.07R_\Phi^4 \qquad \text{Data 2}$$

References

1. Furetta C., Pani V, Pellegrini R. and Driscoll V, Rad. Eff. **88** (1986) 59
2. Furetta C., Tuyn J.W.N., Louis F., Azorin-Nieto J., Gutierrez A. and Driscoll C.M.H., Appl. Radiat. Isot. **39** (1988) 59
3. Furetta C. and Azorin J., Nucl. Instr. Meth. **A280** (1989) 318
4. Budzanowski M., Saez-Vergara J.C., Gomez-Ros J.M., Romero-Gutierrez A.M. and Ryba E., Rad. Meas. **29** (1998) 361

Sodium pyrophosphate ($Na_4P_2O_7$)

This material, whose effective atomic number is about 11, is suitable when doped with dysprosium for obtaining TL dosemeters useful in accident monitoring.

The phosphor preparation consists of a mixing of commercially available sodium pyrophosphate and dysprosium oxide Dy_2O_3 in the ratio of 1000: 1 by weight. The mixture is heated at 100°C for one day under vacuum to remove all moisture, melted at 880°C at a pressure of 1.33 x 10^{-2} Pa, and then cooled down slowly. The poly-crystalline mass is then grounded into particles ranging from 60 to 100 mesh in size. The recommended annealing procedure is at 400°C for 1 h before exposure.

The glow curves of $Na_4P_2O_7$: Dy show three glow peaks at 90, 181, and 228°C. The 90°C peak fades away within a few hours after exposure, and the 228°C peak has a negligible intensity.

Linearity ofthe response to γ-rays is observed in the range between 1.6 KR and 13 KR. The sensitivity is comparable with that of LiF TLD-100. Thermal neutrons can be also detected.

The photon energy response is found to be not as good as that of LiF TLD-100.

Reference

Kundu H.K., Massand O.P., Marathe P.K. and Venkataraman G., Nucl. Instr. Meth. **175** (1980) 363

Solid state dosimeters

Common solid state dosimeters include:

(i) the photographic emulsion, which darkens upon exposure to radiation;

(ii) the silicon diode, which suffers radiation-induced changes in electrical resistance under fast neutron irradiation;

(iii) certain crystals which change colour upon irradiation

(iv) crystals which present luminescence phenomena (see luminescence dosimetry);

(v) irradiated crystals which present, upon heating, a transient increase in electrical conductivity (thermally stimulated conductivity, TSC);

(vi) irradiated crystals which present, upon heating, a transient emission of electrons from their surface (thermally stimulated exoelectron emission, TSEE).

Solid state dosimetry

Solid state dosimetry deals with the measurement of ionizing radiation by means of radiation-induced changes in the properties of certain materials (see solid state dosimeters)

Spurious thermoluminescence: chemiluminescence

Chemiluminescence is another spurious TL emission which can alter the radiation induced TL response, specially in the range of very low doses.

Chemiluminescence effect has origin from impurities which can contaminate the surface layer of the dosemeter. During readout of the TLD sample, the excitation of the impurities provokes a non-radiative signal which is superposed to the radiation induced signal.

The chemiluminescence effect is mainly produced by the oxidation of the surface of the TL phosphors.

Spurious thermoluminescence: surface-related phenomena

The TL light emitted during readout of a sample may be contaminated by **non-radiation-induced signals** (**spurious thermoluminescence**) which restrict the lower limit of detection.

Spurious thermoluminescence: triboluminescence

Triboluminescence indicates an emission of luminescence stimulated by mechanical stress, during readout of TL samples, and it is a **spurious signal** to be avoided otherwise it increases the detection threshold as well as the errors in the dose determination.

This phenomenon is much more evident in TL phosphors used in powder form than in solid chips. Furthermore it depends on the dose given to the dosemeter; in the range of high doses the phenomenon is less important.

Schulman and colleagues [1,2] carried out specific experiments to study the effect and showed that the triboluminescence signal can be eliminated just heating the TL sample in a oxygen-free atmosphere. The best results are obtained performing the TL readout in an atmosphere of inert gas, i.e. argon or nitrogen.

The effect of oxygen as well as of the inert gases on triboluminescence is not understood and no theoretical explanation has been given until now.

References

1. Schulman J.H., Attix F.H., West E.J., Ginther R.J. – Rev. Sc. Instrum. **32** (1960) 1263
2. Nash A.E., Attix F.H., Schulman J.H. – Proc. Int. Conf. Lumin. Dos. (Stanford), 244 (1965)

Stability factor F_{st} (definition)

This parameter is useful to check any possible variation in the stability of the reader and/or in the irradiation facility. Also in this case the procedures can be different from one another according to the various laboratories. In the following is reported the most usual procedure and some suggestions are given for its implementation.

Stability factor F_{st} (procedure)

The stability check of the reader and/or of the irradiation facility is carried out using a group of reference dosemeters and the procedure is based on the control of $F_{c,r}$ at any new session of readings. Making reference to the equation (see **Calibration factor F_c – 1st procedure**)

$$F_{c,r} = \frac{D_c}{\frac{1}{m}\sum_{r=1}^{m} M_{r,net} \cdot S_r} \qquad (1)$$

which is supposed to be determined at the beginning of the first use of a new batch of TLDs, one can use the same equation in all the period of use of the batch, and calculate the value of $F_{c,r}$ at the beginning of each session. It must be noted that in this way the period of time between a control and the subsequent readout could be of several weeks. Therefore, before starting a new session of readings, a new calibration factor is determined. Let us indicate it as

$$F'_{c,r} = \frac{D_c}{\frac{1}{m}\sum M'_{r,net} \cdot S_r} \tag{2}$$

Note that the dose in Eq. (2) is the same as that in Eq. (1); also the same are the S_r values. Only the TL response can be changed if variations occurred in the reader or/and in the irradiation facility.

The stability factor is then determined as a ratio between the two factors $F_{c,r}$ and $F'_{c,r}$ as following

$$F_{st} = \frac{F_{c,r}}{F'_{c,r}} \tag{3}$$

If all the experimental conditions are remained constant between the first determination of F and any other subsequent determinations, the F_{st} value will be more or less equal to 1; differences within 1-2% among the F values confirm a very good stability for both the reader and the irradiation system.

In some procedures it is suggested to determine $F_{c,r}$, the former calibration factor, as an average of several factors obtained by irradiating the TLDs of the control group several times, at least 5. On the contrary, the subsequent factors, F', are obtained after one irradiation only.

The previous procedure cannot be considered the best one because, as already mentioned, the time between a control and the subsequent readout can be long enough, and sometimes it does not allow this kind of procedure. On the contrary, the dose can be estimated using the actual calibration factor, F', without any references to the previously F's determinations. On the other hand, a stability check is very important if it is done during the readout session itself. This check is usually done at the beginning of the readout session and at the end of it. Dividing the reference dosemeters into two sub-groups, the first group is used as the reference at the beginning of the session and the second group is read at the end of the session. The two factors are then compared with the same procedure just mentioned above. If the number of dosemeters used for the field application is larger than 100, three or more sub-groups of reference dosemeters can be used: one at the beginning of the session, one at the end and the others during the session.

It must be noted that in this case the possible variations in the F values can be attributed to the reader only, because all the reference dosemeters have been irradiated all together at the beginning and so there are no uncertainties due to the irradiation facility introduced in the whole procedure.

An example concerning the stability of the calibration factor is given here. This test has been carried out over a period of 5 weeks. Five TLDs have been selected, prepared and irradiated at a dose of 12.44 mGy. The 5 TLDs have been read immediately after irradiation. After one week the procedure was repeated and so on over the whole test period of 5 weeks. Table 1 shows the results obtained with this test. For each dosemeter the individual background was determinated after annealing. After irradiation, the readings were corrected by background subtraction and by the individual correction factor. Note that the S_i factors are here dividing factors.

Considering the first calibration factor, $F_{c,1}$, determined at the beginning, as a normalisation factor, one gets the F_{st} values given in the Table. The average value of the F_{ci} factors, over the five weeks, is 0.211 (0.5%) and the average value of F_{st} is 1.002 (0.5%).

N.	M	M_0	M_{net}	S_i	$M_{net(c)}$	$\overline{M}_{net(c)}$	CV (%)	F_c	F_{st}
1	59.7	0.155	59.55	1.005	59.25				
2	58.6	0.119	58.48	0.999	58.54				
3	59.1	0.137	58.96	0.991	59.50				
4	60.1	9.144	59.96	1.016	59.02				
5	57.1	0.102	57.00	0.979	58.22	58.91	0.9	0.211	1.000
1	59.4	0.108	59.29	1.005	59.00				
2	58.2	0.111	58.09	0.999	58.15				
3	58.9	0.103	58.80	0.991	59.33				
4	59.7	0.133	59.57	1.016	58.63				
5	58.9	0.102	58.80	0.979	60.06	59.03	1.2	0.211	1.000
1	58.8	0.108	58.69	1.005	58.40				
2	58.9	0.114	58.79	0.999	58.85				
3	57.9	0.101	57.80	0.991	58.32				
4	58.6	0.092	58.51	1.016	57.59				
5	58.4	0.123	58.28	0.979	59.53	58.54	1.2	0.213	1.009
1	60.4	0.227	60.17	1.005	59.87				
2	59.1	0.157	58.94	0.999	59.00				
3	58.0	0.241	57.76	0.991	58.28				
4	60.2	0.122	60.08	1.016	59.13				
5	58.3	0.278	58.02	0.979	59.26	59.11	1.0	0.210	0.995
1	58.0	0.100	57.90	1.005	57.61				
2	60.1	0.171	59.93	0.999	59.99				
3	57.6	0.112	57.49	0.991	58.01				
4	59.3	0.204	59.10	1.016	58.17				
5	58.8	0.193	58.61	0.979	59.87	58.73	1.9	0.212	1.005

Table 1. Behaviour of the stability factor over 5 weeks.

Stability of the reading system background

The stability of the reding system depends on:

- environmental conditions (i.e. temperature, humidity)
- variations of the calibration light source placed inside the instrument,
- how long the instrument has been swithed one before use,
- variations of the electronic stability during the use.

Stability of the reading system background (procedure)

- $n(n \geq 5)$ consecutive readings without dosimeter
- calculate the average value $\overline{M}_S$
- repeat any time before using the TL reader $(\overline{M}_{Si})$
- verify $0.80 \leq \frac{\overline{M}_{Si}}{\overline{M}_S} \leq 1.20$

Stability of the TL response

The term stability referred to the TL response of a phophor means stability of the physicochemical properties of the phosphor. In other words, the repeated use of a phosphor, i.e. annealing – irradiation – readout cycles, should not change the phosphor's sensitivity and its glow curve. The stability can be checked on a group of TLDs, chosen randomly from a batch. The following Fig. (1) shows the stability plot obtained with 27 successive re-use cycles on LiF:Mg,Cu,P (GR-200A). The readings were obtained using a linear heting rate of 9°C/s. Fig. (2) shows the readings of 10 successive re-use cycles. In this case the readouts were undertaken by a plateau heating time at 230°C for 20 s.

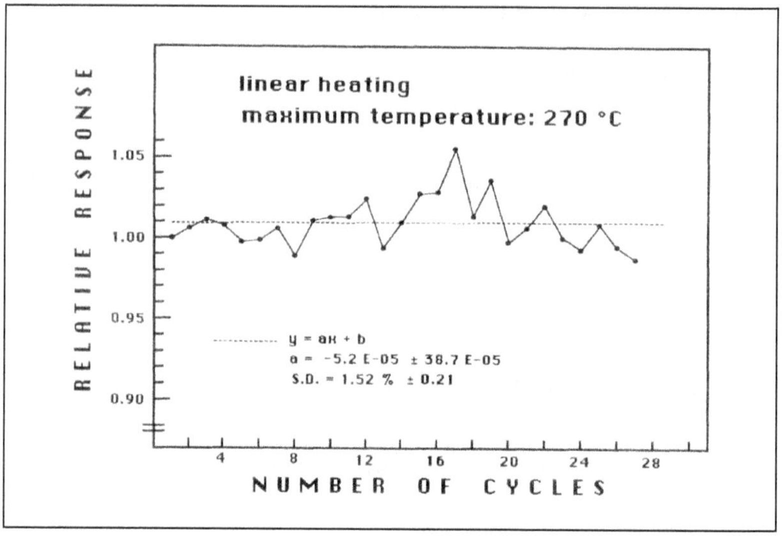

Fig. 1. Stability of GR-200A using a linear heating rate.

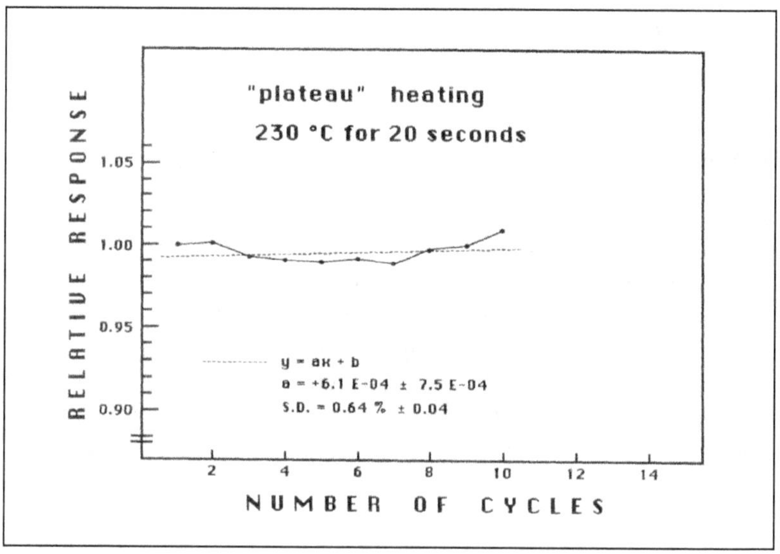

Fig. 2. Stability of GR-200A obtained with plateau readout.

Standard annealing

The standard annealing is the normal thermal procedure used for re-use of thermoluminescent phosphors (see **annealing general conditions** and **procedures**)

Stokes' law

G.G. Stokes formulated in 1852 the law of luminescence. The law states that the wavelength of the emitted light is greater than that of the exciting radiation.

Sulphate phosphors

The sulphate phosphors family is composed by many different compounds. A short review is given below [1-10].

Calcium sulphate ($CaSO_4$)

Two different kind of preparations can be used.

In the first, analytical-reagent-grade $CaSO_4x2H_2O$ and reagent-grade impurities (oxides of rare earths) are mixed in a proper ratio and dissolved in concentrated sulphuric acid to form a saturated solution of $CaSO_4$. The solution is then heated at about 300°C to allow the evaporation of the acid. Single crystals of doped calcium sulphate appear during evaporation. After cooling, the crystals are ground to powder and sieved to obtain grains ranging from 100 to 200 μ in size.

Another method consists in the dissolution of $Ca(NO_3)_2$ in 225 cm^3 of concentrated H_2SO_4. The dopants, in the required concentration, are added and the reagents thoroughly stirred in a flask, which is connected to a sealed condenser system with constant air flow as carrier for the acid vapour. A beaker containing a NaOH solutions captures and neutralizes the condensed acid. A hot plate provides the heat required to drive the reaction. An evaporation period of about 12 hs allows to obtain single crystals of $CaSO_4$. The crystals are repeatedly washed to remove any remaining acid, then they are placed in a Pt crucible and thermally treated for 1 h. After that the crystals are ground and sieved. The particle size ranges between 80 and 200 μ. Pellets of calcium sulphate with PTFE may also be obtained.

Calcium sulphate (Z_{eff} = 15.6) doped with Mn shows high sensitivity but a very high fading rate because it presents only one peak at about 90°C. $CaSO_4$:Dy and $CaSO_4$:Tm show similar glow curves with three peaks at about 80, 120, 220°C and a shoulder at 250°C. The third peak, to most prominent, is the dosimetric peak. Fading rate varies according different authors and preparation technique: from 7% to 30% in 6 months. The lower detection limit is about 1 μGy and the TL response is linear up to 3 Gy for Tm doped material and up to 100 Gy for Dy actvated calcium sulphate.

Strotium and barium sulphates Dy activated ($SrSO_4$:Dy, $BaSO_4$:Dy)

Analytical reagent grade $SrSO_4$ (Z_{eff} = 23) and $BaSO_4$ (Z_{eff} = 35) are dissolved in sulphuric acid together with dysprosium oxide Dy_2O_3. Crystals are formed after evaporation of the acid at 300°C. The crystals are then dried at 400°C during several hours, ground and sieved. The powder is annealed at 400°C for 5 hs; a second annealing at 400°C increases the sensitivity of about 40%. Both materials show a very intense peak in the temperature region 130-140°C. Their relative sensitivities, at the ^{60}Co energy, compared with that of LiF TLD-100, are 11 for $SrSO_4$ and 3 for $BaSO_4$.

Mixed sulphates ($K_2Ca_2(SO_4)_3$, $K_2Cd_2(SO_4)_3$)

For preparing $K_2Ca_2(SO_4)_3$, having an effective atomic number equal to about 14, K_2SO_4 and $CaSO_4$ powders in the molar ratio 1:2 are mixed and heated in a quartz tube at 1000°C for 24 hrs. The compound is formed by a process of solid state diffusion. The molten mass is slowely cooled and then crushed and sieved to obtain particles having a size of about 210μ. The glow curve shows four peaks in the region between 80°C and 500°C. The dosimetric peak, very intense, at 447°C does not show any fading.

$K_2Cd_2(SO_4)_3$ is prepared using the solid state diffusion technique. K_2SO_4 and $CdSO_4$ powders are mixed in the appropriate proportions and kept for 6 days at 600°C. The obtained mass is powdered and then melted at 770°C. Aftter cooling the powder is obtained as usual. This material has been also doped with Sm with an increase in sensitivity by a factor 40 respect to the undoped material. The undoped material presents a glow curve with two resolved peaks at 77°C and 200°C respectively. The doped Sm material presents only one prominent peak at 157°C.

References

1. Watanabe S. and Okuno E., Riso Rep. 249 (2) Danish AEC (1971) 864
2. Yamashita T., Nada N., Onish H. and Kitamura S., Proc. 2nd Intern. Conf. Luminescence Dosimetry, Gatlinburg (USA) (1968)
3. Yamashita T., Nada N., Onish H. and Kitamura S., Health Phys. **21** (1971) 295
4. Yamashita T., Proc. 4th Intern. Conf. Luminescence Dosimetry, Krakow (Poland) (1974)
5. Azorin J., Salvi R. and Moreno A., Nucl. Instr. Meth. **175** (1980) 81
6. Azorin J., Gonzalez G., Gutierrez A. and Salvi R., Health Phys. **46** (1984) 269
7. Azorin J. and Gutierrez A., Health Phys. **56** (1989) 551
8. Dixon R.L. and Ekstrand K.E., Phys. Med. Biol. **19** (1974) 196
9. Sahare P.D., Moharil S.V. and Bhasin B.D., J. Phys. D **22** (1989) 971
10. Deshmukh B.T., Bodade S.V. and Moharil S.V., Phys. Stat. Sol. (a) **98** (1986) 239

T

Temperature gradient in a TL sample

In case high heating rates are used during readout, a temperature difference between the bottom and the top of a sample can be observed and the glow peak becomes broader [1,2].

In case a temperature gradient across a TL sample is ignored, the TL intensity is given by

$$I(T) = ns\exp\left(-\frac{E}{kT}\right) \tag{1}$$

On the other hand, assuming a constant temperature gradient across a TLD sample, the emission can be written as

$$I(T_2) = \int_{(-1/2)\Delta T}^{(+1/2)\Delta T} \frac{n(T_2+\tau)}{\Delta T} s\exp\left(-\frac{E}{k(T_2+\tau)}\right) d\tau \tag{2}$$

where

T_2 = temperature of the sample

$n(T_2+\tau)$ = density of trapped charge carriers at temperature $T_2+\tau$ at a given position within the sample.

Assuming a linear time dependence of temperature at each position in the sample and neglecting the energy dissipation to the surroundings, we can write, for each position in the sample:

$$T_2 + \tau = T_0' + \beta'(t+t') \tag{3}$$

where T_0' and β' are, respectively, the effective starting temperature and the effective heating rate in the sample, and $\beta' t' = \tau$.

Indicating with dn the difference of n between two positions, corresponding to a difference in temperature of dT at a certain time, we can write

$$\frac{dn}{dT} = \frac{1}{\beta'}\frac{dn}{dt} \tag{4}$$

so that, the quantity $n(T_2+\tau)$ can be approximated by

$$n(T+\tau) \approx n(T) + \frac{dn}{dT}\cdot\tau \approx n(T) + \frac{\tau}{\beta'}\frac{dn}{dt}$$

$$\approx n(T)\left[1 - \frac{s\tau}{\beta'}\exp\left(-\frac{E}{kT}\right)\right] \tag{5}$$

Using last Eq. (5), Eq. (2) becomes

$$I(T_2) \approx n(T_2)\frac{s}{\Delta T}\int_{(-1/2)\Delta T}^{(+1/2)\Delta T}\left[1 - \frac{s\tau}{\beta'}\exp\left(-\frac{E}{kT_2}\right)\right]\exp\left(-\frac{E}{k(T_2+\tau)}\right)d\tau$$

which can be approximated as

$$I(T_2) = n(T_2)\frac{s}{\Delta T}\exp\left(-\frac{E}{kT_2}\right)\int_{(-1/2)\Delta T}^{(+1/2)\Delta T}\left[1 - \frac{s\tau}{\beta'}\exp\left(-\frac{E}{kT_2}\right)\right]\cdot\left(1 + \frac{E\tau}{kT_2^2}\right)d\tau \tag{6}$$

and again, solving the integral:

$$I(T_2) = n(T_2)s\exp\left(-\frac{E}{kT_2}\right)\exp\left[-\frac{s(\Delta T)^2}{12kT_2^2\beta'}\exp\left(-\frac{E}{kT_2}\right)\right] \tag{7}$$

The last exponential on the right of Eq. (7) can be developed into powers of $\left(\frac{1}{T_2} - \frac{1}{T_M}\right)$ around $\frac{1}{T_M}$:

$$\frac{1}{T_2^2}\exp\left(-\frac{E}{kT_2}\right) \approx \frac{1}{T_M^2}\exp\left(-\frac{E}{kT_M}\right) +$$
$$+\left[\frac{2}{T_M}\exp\left(-\frac{E}{kT_M}\right) - \frac{E}{kT_M^2}\exp\left(-\frac{E}{kT_M}\right)\right]\cdot\left(\frac{1}{T_2} - \frac{1}{T_M}\right)$$

and then Eq. (7) becomes

$$I(T_2) = n(T_2)s\exp\left(-\frac{E}{kT_2}\right)$$
$$\times\exp\left\{-\left(\frac{Es(\Delta T^2)}{12kT_M\beta'}\right)\cdot\left[\left(\frac{E}{kT_M^2}\right) + \left(\frac{E}{kT_M}\right)\cdot\frac{1}{T_2}\right]\exp\left(-\frac{E}{kT_2}\right)\right\} \tag{8}$$

From the comparison between Eq. (1) and Eq. (8) it is easily seen that if the temperature gradient across the TL sample is ignored, both activation energy and frequency factor are underestimated by the quantities

$$\Delta E \approx -\left[\frac{Es(\Delta T)^2}{12T_M\beta'}\right]\left(\frac{E}{kT_M}-2\right)\exp\left(-\frac{E}{kT_M}\right)$$

$$\Delta\ln(s) \approx -\left[\frac{Es(\Delta T)^2}{12T_M\beta'}\right]\left(\frac{E}{kT_M}-1\right)\exp\left(-\frac{E}{kT_M}\right)$$

References

1. Piters T.M. and Bos A.J.J., J. Phys. D: Appl. Phys. **27** (1994) 1747
2. Piters T.M., A study into the mechanism of thermoluminescence in a LiF:Mg,Ti dosimetry material (Thesis, 1998), D.U.T.

Temperature lag: Kitis' expressions for correction (procedure)

Kitis suggested the following procedure for corrected the temperature values when a temperature lag is suspected to be in TL measurements [1] (see **Temperature lag: Kitis' expressions for correction (theory)**)

- make a few measurements at very low heating rates, i.e. 1 and 2°C/s in order to evaluate the constant c from the relation

$$c = \frac{T_{M2} - T_{M1}}{\ln 2} \tag{1}$$

- using the following equation

$$T_{(M,x)corr} = T_{M1} - c\ln\left(\frac{\beta_1}{\beta_x}\right) \tag{2}$$

  evaluate the real temperature, $T_{(M,x)corr}$ at the maximum for the used heating rate

- evaluate the temperature lag at the heating rate β_x

$$\Delta T = T_{M,x} - T_{(M,x)corr}$$

  where $T_{M,x}$ is the peak maximum temperature of the glow peak with a temperature lag

- using the following equation

$$\beta_{eff} = \frac{T_{M,x} - T_0 - \Delta T}{T_{M,x} - T_0} \cdot \beta \tag{3}$$

calculate the effective heating rate. T_0 is the order of the room temperature (about 293 K)

<u>*Example*</u>

A glow peak shows the temperature at the maximum at 481.3 K when a heating rate of 1°C/s is used, and at 488 K with an heating rate of 2°C/s. Supposing that at those heating rates no temperature lag exists, we can calculate the constant *c* from Eq. (1):

$$c = \frac{488 - 481.3}{\ln 2} = 9.67$$

The same glow peak shows, using a heating rate of 40°C/s, a temperature at the maximum of 518 K. The correct value is then, using Eq. (2)

$$T_{M,corr}(40°C/s) = T_{M1} - 9.67 \cdot \ln\left(\frac{1}{40}\right) = 517K$$

The temperature lag at the heating rate of 40°C/s, is

$$\Delta T = 518 - 517 = 1K$$

and the effective heating rate of the sample is, using Eq. (3):

$$\beta_{eff} = \frac{518 - 293 - 1}{518 - 293} \cdot 40 = 39.8°C/s$$

It must be stressed that:

- Eq. (13) is valid in the range from 1°C/s to 50°C/s. It is a general equation, holding for every point of the glow peak. Each point of the glow peak shifts as a function of the heating rate, with its own constant *c*.

- The reference measurements at low heating rates need special attention to avoid any temperature lag. This can be achieved: (i) using silicon oil of high thermal conductivity when solid TL samples are used, (ii) using loose powder.

- The temperature lag is a linear function of the heating rate. This is in agreement with the theoretical prediction [2-4].

References

1. Kitis G. and Tuyn J.W.N., J. Phys. D: Appl. Phys. **31** (1998) 2065
2. Gotlib V.I., Kantorovitch L.N., Grebenshicov V.L., Bichev V.R. and Nemiro E.A., J. Phys. D: Appl. Phys. **17** (1984) 2097
3. Betts D.S. and Townsend P.D., J. Phys. D: Appl. Phys. **26** (1993) 849
4. Piters T.M. and Bos A.J.J., J. Phys. D: Appl. Phys. **27** (1994) 1747

Temperature lag: Kitis' expressions for correction (theory)

Kitis has provided a simple method to correct the effect of temperature lag in TL measurements. He derived expressions for temperature lag correction for both first- and general-order kinetics.

First-order kinetics

The equation describing the first-order kinetics is the following:

$$I(T) = n_0 s \exp\left(-\frac{E}{kT}\right)\exp\left[-\frac{s}{\beta}\int_{T_0}^{T}\exp\left(-\frac{E}{kT'}\right)dT'\right] \tag{1}$$

which can be rewritten in the following way

$$\ln I(T) = \ln(n_0 s) - \frac{E}{kT} - \frac{s}{\beta}\int_{T_0}^{T}\exp\left(-\frac{E}{kT'}\right)dT' \tag{2}$$

with the usual meaning of the various symbols.

Using two differen heating rates, $\beta_1 < \beta_2$, the glow curve obtained with the faster heating rate is shifted towards the higher temperature keeping its integral, any way, constant. Considering now the intensities of the two peaks at the same fraction of their maximum intensity, I_M, the following condition is verified:

$$\frac{d\ln(I_1)}{dT} = \frac{d\ln(I_2)}{dT} = \alpha \tag{3}$$

where $\alpha = 0$ at the peak maximum temperature.

From Eq. (2), according to Eq. (3), one then obtains

$$\frac{d\ln\left[\ln(I_1)\right]}{dT} = \frac{E}{kT_1^2} - \frac{s}{\beta_1}\exp\left(-\frac{E}{kT_1}\right) = \alpha \tag{4}$$

$$\frac{d\ln\left[\ln(I_2)\right]}{dT} = \frac{E}{kT_2^2} - \frac{s}{\beta_2}\exp\left(-\frac{E}{kT_2}\right) = \alpha \tag{5}$$

from which

$$\frac{E}{kT_1^2} - \alpha = \frac{s}{\beta_1}\exp\left(-\frac{E}{kT_1}\right) \tag{6}$$

$$\frac{E}{kT_2^2} - \alpha = \frac{s}{\beta_2}\exp\left(-\frac{E}{kT_2}\right) \tag{7}$$

From these two equations, the respective heating rates can be obtained as

$$\beta_1 = A_1 \exp\left(-\frac{E}{kT_1}\right) \tag{8}$$

$$\beta_2 = A_2 \exp\left(-\frac{E}{kT_2}\right) \tag{9}$$

where

$$A_1 = \frac{skT_1^2}{E - \alpha kT_1^2}$$

$$A_2 = \frac{skT_2^2}{E - \alpha kT_2^2}$$

Eqs. (8) and (9) can be arranged as follows

$$\frac{\beta_1}{\beta_2} = \frac{A_1 \exp\left(-\frac{E}{kT_1}\right)}{A_2 \exp\left(-\frac{E}{kT_2}\right)}$$

and then

$$\ln\left(\frac{\beta_1}{\beta_2}\right)=\ln\left(\frac{A_1}{A_2}\right)-\frac{E}{kT_1}+\frac{E}{kT_2}$$

from which

$$\frac{1}{T_2}=\frac{1}{T_1}+\frac{k}{E}\ln\left(\frac{\beta_1}{\beta_2}\right)-\frac{k}{E}\ln\left(\frac{A_1}{A_2}\right) \tag{10}$$

Equation (10) holds true for every temperature point of the peak at the same fraction of its maximum intensity and, of course, at I_M. Equation (10) can now be transformed as follows:

$$T_2=T_1-(T_1T_2)\frac{k}{E}\ln\left(\frac{\beta_1}{\beta_2}\right)+(T_1T_2)\frac{k}{E}\ln\left(\frac{A_1}{A_2}\right) \tag{11}$$

The shift of the peak from T_1 to T_2 as the heating rate increases, is given by the sum of the last two terms on the right of Eq. (11). Any way, the contribution of the second term is less than 5% of the total shift and so this term can be omitted and Eq. (11) simplify to

$$T_2=T_1-T_1T_2\frac{k}{E}\ln\left(\frac{\beta_1}{\beta_2}\right) \tag{12}$$

Taking β_1=1°C/s and β_2=50°C/s, the extreme value of $\ln\left(\frac{\beta_1}{\beta_2}\right)$ is 4 and the term T_1T_2 increases only a few per cent in the range (1 – 50) °C/s; therefore, the term $T_1T_2\frac{k}{E}$ can be considered as a constant and Eq. (12) assumes the final form of

$$T_2=T_1-c\ln\left(\frac{\beta_1}{\beta_2}\right) \tag{13}$$

The next step is to calculate the effective hating rate, β_{eff}, i.e. the rate of heating of the sample.

Let us indicate with T_g the peak maximum temperature of a peak received with temperature lag, with T_M the real value if there is no temperature lag, and with $\Delta T=T_g-T_M$ the difference. Both T_g and T_M are given by

$$T_g = T_0 + \beta \cdot T$$
$$T_M = T_g - \Delta T = T_0 + \beta_{eff} \cdot t \qquad (14)$$

where β is the heating rate of the heating element and T_0 is of the order of the room temperature.

From Eq. (14) we obtain

$$\beta_{eff} = \frac{T_g - T_0 - \Delta T}{T_g - T_0} \cdot \beta \qquad (15)$$

General-order kinetics

The intensity for a glow peak following a general-order kinetics is given by

$$I(T) = n_0 s \exp\left(-\frac{E}{kT}\right)\left[1 + \frac{s(b-1)}{\beta}\int_{T_0}^{T}\exp\left(-\frac{E}{kT'}\right)dT'\right]^{-\frac{b}{b-1}} \qquad (16)$$

which becomes

$$\ln I(T) = \ln(n_0 s) - \frac{E}{kT} - \frac{b}{b-1}\ln\left[1 + \frac{s(b-1)}{\beta}\int_{T_0}^{T}\exp\left(-\frac{E}{kT'}\right)dT'\right] \qquad (17)$$

As already done for the first-order case, two heating rates $\beta_1 < \beta_2$ are considered. Hence

$$\frac{d\ln(I_1)}{dT} = \frac{d\ln(I_2)}{dT} = \alpha$$

From Eq. (17) we get

$$\frac{d\ln(I_1)}{dT} = \frac{E}{kT_1^2} - \frac{b}{b-1} \cdot \frac{\frac{s(b-1)}{\beta_1}\exp\left(-\frac{E}{kT_1}\right)}{\left[1 + \frac{s(b-1)}{\beta_1}\int_{T_0}^{T_1}\exp\left(-\frac{E}{kT'}\right)dT'\right]} = \alpha$$

$$\frac{d\ln(I_2)}{dT}=\frac{E}{kT_2^2}-\frac{b}{b-1}\cdot\frac{\frac{s(b-1)}{\beta_2}\exp\left(-\frac{E}{kT_2}\right)}{\left[1+\frac{s(b-1)}{\beta_2}\int_{T_0}^{T_2}\exp\left(-\frac{E}{kT'}\right)dT'\right]}=\alpha$$

and then

$$\begin{aligned}\frac{E}{kT_1^2}-\frac{bs\exp\left(-\frac{E}{kT_1}\right)}{\beta_1\varepsilon_1}=\alpha\\ \frac{E}{kT_2^2}-\frac{bs\exp\left(-\frac{E}{kT_2}\right)}{\beta_2\varepsilon_2}=\alpha\end{aligned}\tag{19}$$

where ε_1 and ε_2 are the expressions with integral in the denominators.

Solving Eqs. (19) with respect to β_1 and β_2, we obtain

$$\begin{aligned}\beta_1=A_1\exp\left(-\frac{E}{kT_1}\right)\\ \beta_2=A_2\exp\left(-\frac{E}{kT_2}\right)\end{aligned}\tag{20}$$

with

$$A_1=\frac{bskT_1^2}{\varepsilon_1E\left(1-kT_1^2\alpha\right)}$$

$$A_2=\frac{bskT_2^2}{\varepsilon_2E\left(1-kT_2^2\alpha\right)}$$

Making the ratio and then its logarithm of Eq. (20), we finally obtain

$$\frac{1}{T_2}=\frac{1}{T_1}+\frac{k}{E}\ln\left(\frac{\beta_1}{\beta_2}\right)-\frac{k}{E}\ln\left(\frac{A_1}{A_2}\right)\tag{21}$$

which is similar to Eq. (10) obtained for the first-order kinetics. Therefore, Eq. (13) is also valid for the general-order kinetics.

Reference

Kitis G. and Tuyn J.W.N., J. Phys. D: Appl. Phys. **31** (1998) 2065

Temperature ratios T_1/T_M and T_2/T_M as a function of the kinetics order b

The following data have been obtained by synthetic peaks [1].

b	T_{MAX}	T_1	T_2	T_1/T_M	T_2/T_M
1.0001	384.575	367.216	397.059	0.9549	1.0325
1.0501	384.539	366.987	397.539	0.9544	1.0338
1.1001	384.503	366.764	398.013	0.9539	1.0351
1.1501	384.467	366.549	398.48	0.9534	1.0364
1.2001	384.431	366.339	398.941	0.9529	1.0377
1.2501	384.396	366.135	399.397	0.9525	1.039
1.3001	384.36	365.937	399.847	0.9521	1.0403
1.3501	384.325	365.743	400.292	0.9517	1.0415
1.4001	384.289	365.555	400.732	0.9513	1.0428
1.4501	384.254	365.37	401.168	0.9509	1.044
1.5001	384.219	365.191	401.599	0.9505	1.0452
1.5501	384.184	365.015	402.025	0.9501	1.0464
1.6001	384.15	364.843	402.448	0.9497	1.0476
1.6501	384.115	364.675	402.866	0.9494	1.0488
1.7001	384.081	364.51	403.28	0.949	1.05
1.7501	384.046	364.348	403.691	0.9487	1.0512
1.8001	384.012	364.19	404.098	0.9484	1.0523
1.8501	383.978	364.035	404.502	0.9481	1.0535
1.9001	383.943	363.883	404.902	0.9478	1.0546
1.9501	383.909	363.734	405.299	0.9474	1.0557
2.0001	383.876	363.587	405.693	0.9471	1.0568

T_1 and T_2 are the first and second half intensity temperatures and T_M is the temperature of the glow peak maximum.

As a convenient approximation, the average values of the ratios can be used, independently by the kinetics order:

$$\left(\frac{T_1}{T_M}\right) = 0.9507 \pm 2.39\times10^{-3}, \quad \left(\frac{T_2}{T_M}\right) = 1.0450 \pm 7.55\times10^{-3}$$

The following figures give the plots of T_M and of the ratios as a function of the kinetics order b.

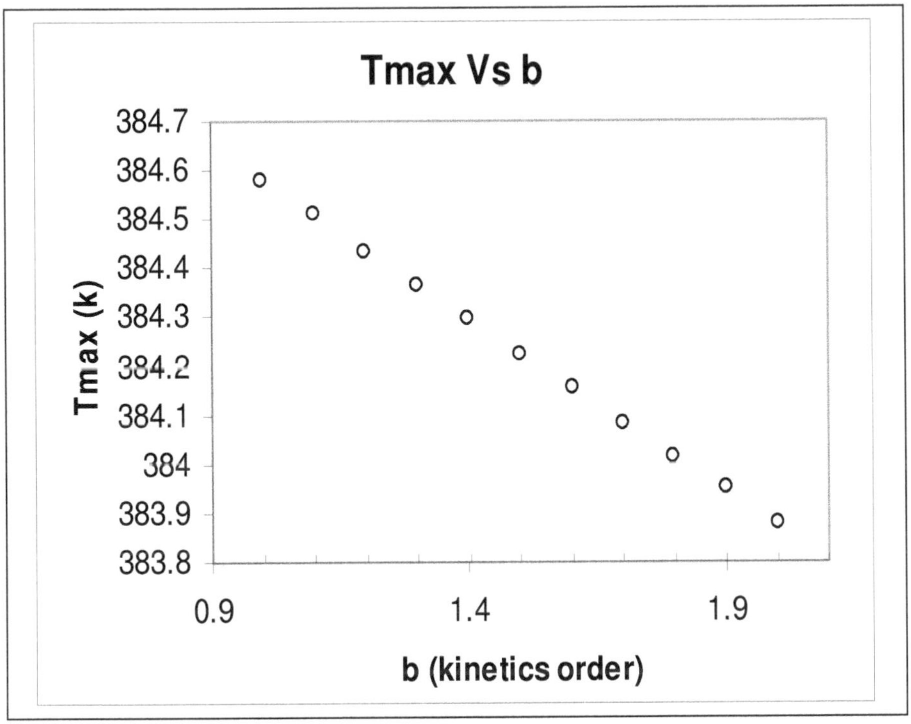

Fig. 1. Plot of the temperature at the maximum as a function of the kinetics order.

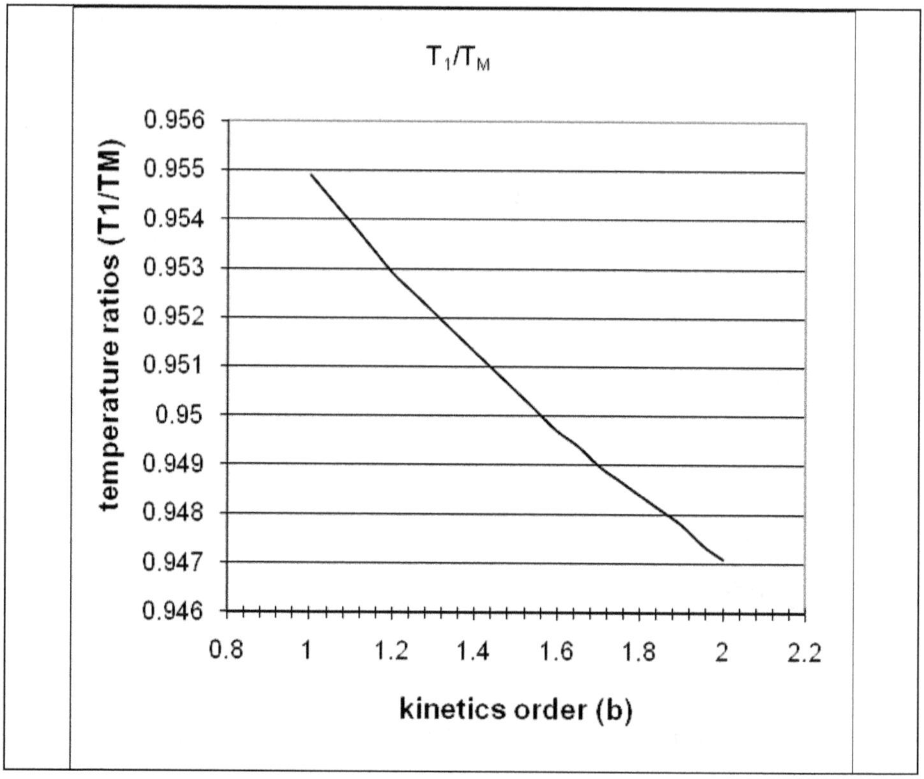

Fig. 2. Plot of the ratio T_1/T_M as a function of b.

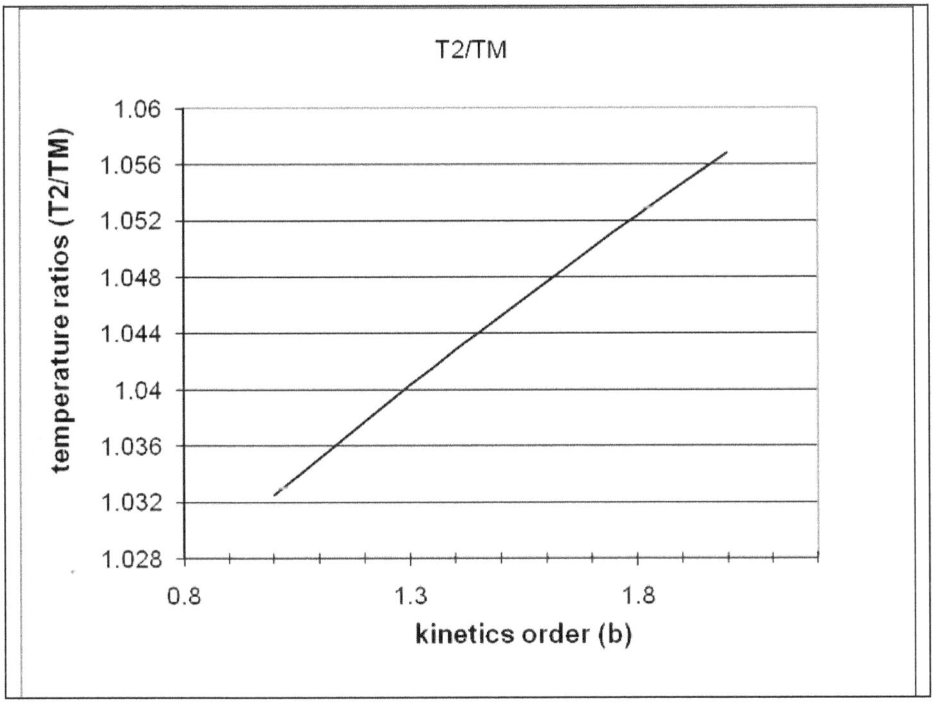

Fig. 3. Plot of the ratio T_2/T_M as a function of b.

Reference

1. Kitis G., private communication

Theory of thermoluminescent dose measurement

The following variables have to define:

- D_0 = absorbed dose
- $L_u(\lambda)$ = light generated in the dosemeter per unit of absorbed dose. It depends on:

- the mean phosphor sensitivity
- the mean absorbed dose
- the thermal and radiation history of the dosemeter
- the time interval between irradiation and readout
- the heating rate during readout

- k_1 = the geometrical-optical factor which depends on:

- μ, the mass absorption coefficient of the dosemeter material, in cm^2/gr

- $A(\lambda)$, the light absorption coefficient of the dosemeter, which is assumed uniform throughout the dosemeter
- G, the geometrical efficiency of light collection
- r, the tray reflectivity
- the dosemeter orientation in the readout system
  - k_2 = the fraction of the total light measured; i.e. when a part of the whole glow curve is considered, as a well isolated peak in a complex glow curve

The quantity of light, $L(\lambda)$, emitted from the dosemeter during the heating cycle, is then given by

$$L(\lambda) = L_u(\lambda) D_0 k_1 k_2 \tag{1}$$

Really the last expression should be modified to take into account the dosemeter background, i.e. the light due to an unirradiated dosemeter. It is assumed, in the following, that the readings have been already corrected by the background contribution.

The light emitted from the dosemeter, $L(\lambda)$, is modified by the optical filters between the dosemeter and the PM tube: $F(\lambda)$ indicates the transmission characteristic of the optical filters. The photons reaching the PM tube photocathode are converted to electrons according to the quantum efficiency, $Q(\lambda)$, of the PM tube. The electrical signal due to the electrons is amplified within the PM tube by the gain G. Thus, the output charge signal R, in coulombs, from the PM tube is

$$R = \int_0^\infty L(\lambda) \cdot F(\lambda) \cdot Q(\lambda) \cdot G \cdot d\lambda \tag{2}$$

This signal is often further amplified, i.e. by a DC amplifier or a high sensitivity charge-to-pulse converter.

The previous expression, substituting $L(\lambda)$ from Eq. (1), becomes

$$R = D_0 k_1 k_2 \int_0^\infty L_u(\lambda) \cdot F(\lambda) \cdot Q(\lambda) \cdot G \cdot d\lambda \tag{3}$$

The geometrical-optical factor k_1 can be evaluated for dosemeters of uniform cross-sectional area irradiated by a collimated beam incident normally on one surface of the dosemeter.

Referring to Fig. 1, let D_0 be the peak dose deposited at surface 1, T be the dosemeter thickness, t (in gr/cm^2) the distance from surface 1 ($(0 \le t \le T)$ and D_t the dose deposited at a distance t from surface 1. The dose distribution through the dosemeter is given by

$$D_t = D_0 \exp(-\mu \cdot t) \tag{4}$$

The light generated within the dosemeter will be attenuated in its path to the PM tube by absorption and scattering effects within the dosemeter material. The attenuation is given by

$$I_t = I_0 \exp\left[-A(\lambda) \cdot t\right] \tag{5}$$

where, referring to Fig. 2, I_0 is the light intensity at the emitted point O, I_t is the intensity at a distance t from O and $A(\lambda)$ the light absorption coefficient.

Fig. 2 also shows the path of the light rays which are considered approximately parallel if the thickness of the dosemeter is very much less than the distance from the dosemeter to the PM tube. With a good approximation all the angles are equal.

From Fig. 2 it can be observed that the light rays, from the point O, which reach the PM tube are emitted into the directions OA and BC after specular reflection at the dosemeter-tray interface. In the first case the distance through the dosemeter is greater than t and in the second case is greater than $2T$-t.

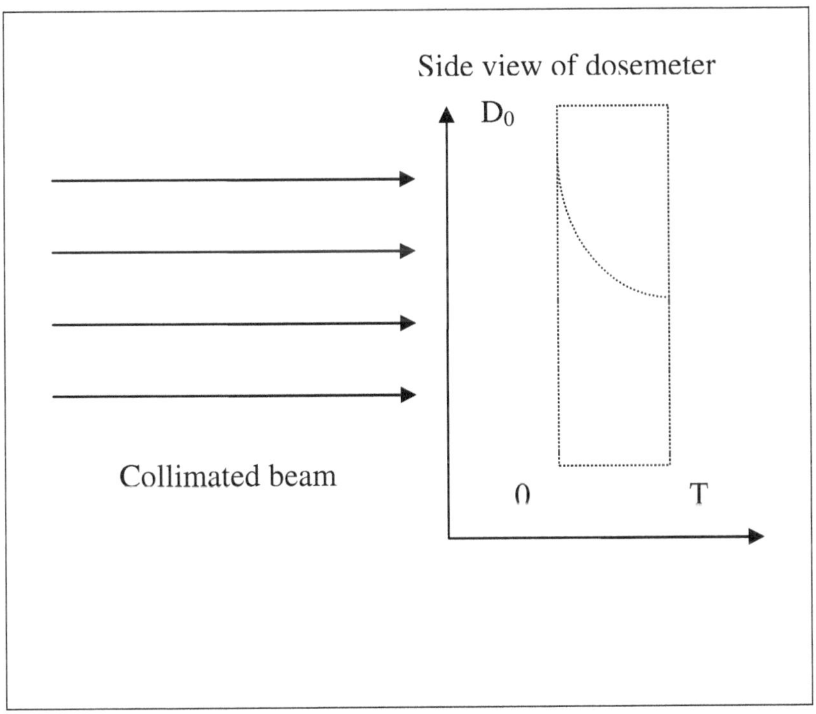

Fig. 1

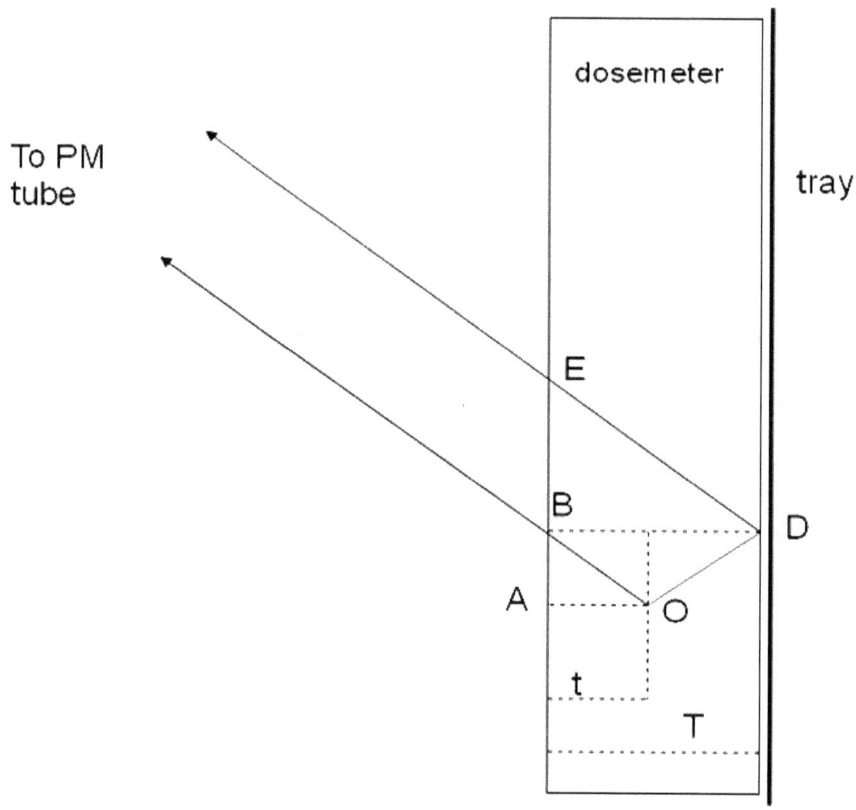

Fig. 2

Then

$$OA = \frac{t}{\cos\alpha} = c \cdot t \tag{6}$$

where c is a constant

$$OB + BC = \frac{T - t}{cso\alpha} + \frac{T}{\cos\alpha} = c(2T - t) \tag{7}$$

Similarly, the distance travelled through the dosemeter by rays from O which strike th PM tube at some other point, is given by similar expressions, just changing the constant *c*.

The previous quantities can be now used to give an explicit form to the light emitted by the dosemeter during readout, $L(\lambda)$. So, Eq. (1) can be rewritten as it follows

$$dL(\lambda) = \frac{L_u}{T} k_2 D_0 \exp(-\mu t)\{G_{eff} \exp[-A(\lambda)ct] + \\ + G_{eff} r \exp[-A(\lambda)c(2T - t)]\}dt$$

Putting $A(\lambda)c = a$, one obtains:

$$dL(\lambda) = \frac{L_u}{T} k_2 D_0 \exp(-\mu t)\{G_{eff} \exp[-at] + \\ + G_{eff} r \exp[-a(2T - t)]\}dt$$

and then

$$dL(\lambda) = \frac{L_u}{T} k_2 D_0 G_{eff} \{\exp[-(\mu + a)t] + \\ + r \exp(-2aT)\exp[-(\mu - a)t]\}dt$$

The total light output, $L(\lambda)$, is then given by

$$: L(\lambda) = \frac{L_u(\lambda)}{T} k_2 D_0 G_{eff} \int_0^T \{\exp[-(\mu + a) \cdot t] + \\ + r \cdot \exp(-2aT)\exp[-(\mu - a)t]\ \}dt \quad (8)$$

Comparing this equation with Eq. (1), one obtains the following expression for k_1:

$$k_1 = \frac{G_{eff}}{T}\left\{\frac{-\exp[-(\mu + a)T]}{(\mu + a)} - \frac{r \exp(-2aT)\exp[-(\mu - a)T]}{(\mu - a)}\right\}_0^T$$

and finally

$$k_1 = \frac{G_{eff}}{T}\left\{\frac{1 - \exp[-(\mu + a)T]}{(\mu + a)} - \frac{r \exp(-2aT)[1 - \exp[-(\mu - a)T]\ ]}{(\mu - a)}\right\} \quad (9)$$

Eq. (9) simplifies for different approximations:

1. $\mu \to 0$ (it is the case when the irradiation is performed using ^{60}Co)

$$k_1 = \frac{G_{eff}}{aT}\left[1 - \exp(-aT)\right] \cdot \left[1 + r\exp(-aT)\right] \tag{10}$$

2. $A(\lambda) \to 0 \quad (a \to 0)$

$$k_1 = \frac{G_{eff}}{\mu T}(1+r)\left[1 - \exp(-\mu T)\right]$$

3. $\mu \to 0, \quad A(\lambda) \to 0$

$$k_1 = G_{eff}(1+r) \tag{12}$$

Test for batch homogeneity

The batch homegeneity is concerned with the methods for the quality control on a new batch of dosimeters just received by users. Some quality tests can be carried out, each giving a different precision.

The simplest method is the following. The user screens all the samples by irradiating them with a known dose from a calibrated radiation source showing a good beam uniformity and making sure that all the samples have been inside the irradiation field. Any TL sample outside the specified tolerance limits should be rejected. The TL dosimeters can also be screened at periodic time intervals.

It must be noted that screening can only be used to determine acceptance or rejection of the samples. Indeed, there are two negative aspects of this procedure. Firstly, accepting a large range of responses (i.e. all responses which are within 20-30% of the mean response), large precision errors are introduced in the dose determination. This is very dangerous when the dosimeters are used in clinical applications. Secondly, the replacement of the rejected TLDs is difficult when the replacement dosimeters come from a different batch: a bias error can be introduced into the whole procedure for the dose assessment.

However, this test remains valid as a first step to know the characteristics of a new TLDs batch.

A quality control concerning the batch homogeneity for TLDs used in personnel dosimetry is suggested in the technical recommendations of the International Electrotechnical Commission (IEC) document. The procedure is given below with some examples.

Procedure for batch homogeneity.

All the **N** dosimeters of the same batch have to be annealed according to the annealing procedure used for the type of TL material under test.

At the end of the annealing procedure, all the dosimeters have to be irradiated using a calibrated gamma source under the appropriate electron equilibrium conditions. The given dose depends on the future use of the dosimeters; i.e., a dose of 5 mGy is suitable for personnel dosimetry, while 1 mGy is enough for environmental dosimetry.

Immediately after irradiation the TLDs are read to measure the TL emission (the readout cycle will be chosen as the best for the particular type of phosphor - see the section concerning the readout cycles) of each dosimeter. Let us indicate the values of the TL emission as

$$M_i \quad \text{with i} = 1, 2, 3, ..., \text{N}$$

The TLDs are now re-annealed and read again to measure the zero-reading (or the zero dose reading). This value should be the same as that already determined during the initialization procedure. In case the background levels are higher, the characteristics of the annealing oven must be checked (temperature uniformity inside the oven, correspondence between the temperature set and the actual temperature, etc.). Let us indicate these background values as

$$M_{0i} \quad \text{with i} = 1, 2, 3, ..., \text{N}$$

The net readout is then defined as

$$M_{i,net}=M_i - M_{0i} \quad \text{with i} = 1, 2, 3, ..., \text{N}$$

In such a series, the maximum and minimum values have to be identified and substituted into the equation

$$\Delta_{\max} = \frac{(M - M_0)_{\max} - (M - M_0)_{\min}}{(M - M_0)_{\min}} \cdot 100 \leq 30 \tag{1}$$

where Δ_{max} represents the uniformity index for the given batch. If such expression is not verified, namely the Δ_{max} of the batch is larger than 30, then some TLDs have to be rejected. Figure 1 shows, as an example, a histogram obtained from the readings of a batch of 1000 TL dosimeters.

The initial calculation of Δ_{max} gave:

n = 1000 $\quad \Delta_{max} = 48.5 > 30$

Since the uniformity index was larger than 30, some TLDs were progressively rejected. The results were:

Rejecting only 2 samples

n = 1000 - 2 $\Delta_{max} = 38.7 > 30$ not acceptable

Rejecting 4 samples

n = 1000 - 4 $\Delta_{max} = 33.7 > 30$ not acceptable

Rejecting 6 samples

n = 1000 - 6 $\Delta_{max} = 29.4 < 30$ acceptable.

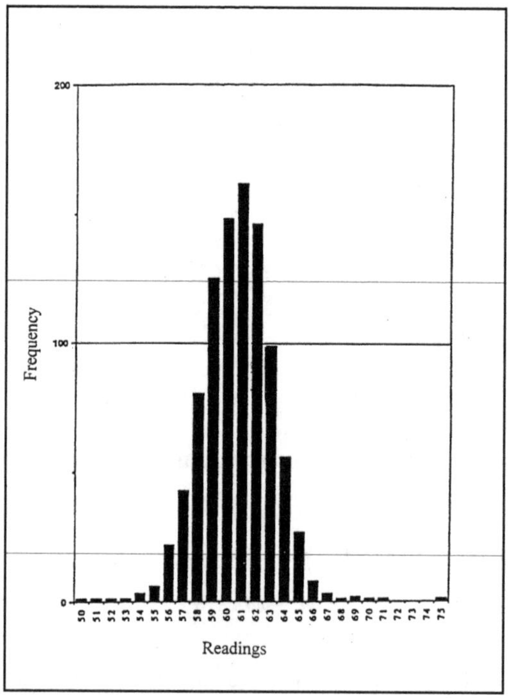

Fig. 1. Histogram of 1000 TLDs readings.

Another procedure can be used for this test (not included in the official recommendations). The average value of all readings is evaluated as

$$\overline{M} = \sum_{i=1}^{N} \frac{(M_i - M_{oi})}{N} = \sum_{i=1}^{N} \frac{M_{i_{net}}}{N} \quad (2)$$

and the following two quantities are evaluated

$$\overline{M} - \sigma_P \qquad \text{and} \qquad \overline{M} + \sigma_P$$

where σ_P is a predetermined value of the standard deviation. All dosimeters which exhibit a net TL readings outside the previous range are rejected.

Dos. N.	TL	Dos. N.	TL	Dos. N.	TL	Dos. N.	TL
1	8.468^{M}	21	7.601	41	6.765	61	144.9*
2	7.808	22	7.346	42	7.531	62	7.946
3	7.231	23	6.916	43	7.045	63	6.771
4	7.587	24	7.491	44	7.476	64	7.657
5	7.630	25	7.600	45	7.480	65	7.434
6	7.394	26	7.509	46	7.656	66	7.239
7	8.094	27	7.329	47	7.167	67	6.704
8	7.854	28	7.290	48	8.047	68	7.118
9	7.428	29	7.294	49	7.984	69	6.699^{m}
10	7.963	30	7.677	50	8.014	70	7.395
11	7.676	31	8.143	51	6.968	71	8.057
12	8.387	32	8.111	52	8.320	72	7.555
13	8.232	33	7.374	53	7.487	73	6.720
14	7.839	34	6.739	54	8.433	74	7.778
15	7.464	35	7.574	55	7.812	75	6.786
16	7.539	36	7.880	56	7.620	76	7.424
17	7.411	37	7.783	57	7.934	77	7.402
18	7.633	38	7.836	58	7.568	78	7.025
19	8.076	39	172.5*	59	7.872	79	7.580
20	7.085	40	7.912	60	7.167	80	7.334

Table 1. Example of data for the homogeneity test. The superscripts M and m indicate the maximum and minimum values, respectively. * indicates abnormal readings.

It can be noted here that it is not always possible or convenient to reject some dosimeters, i.e. when the batch is limited. In these cases all the samples are kept and their responses are corrected using the relative intrinsic sensitivity factor (also called individual correction factor). Any way, it has to be stressed that either some or more samples are rejected or all of the batch samples are considered, the correction factor must be calculated and used to achieve the best uniformity of the batch response.

Another example is reported here. The test has been carried out for a sample of 80 TLDs and the results show its usefulness in some particular cases. It must be noted that the background signal was obtained as an average value and subtracted from each reading. Table 1 lists the net values and the corresponding histogram is given in Fig. 2; among them, the responses of two TLDs are evidently abnormal and completely out of the range indicated by the test, so that their rejection is obvious.

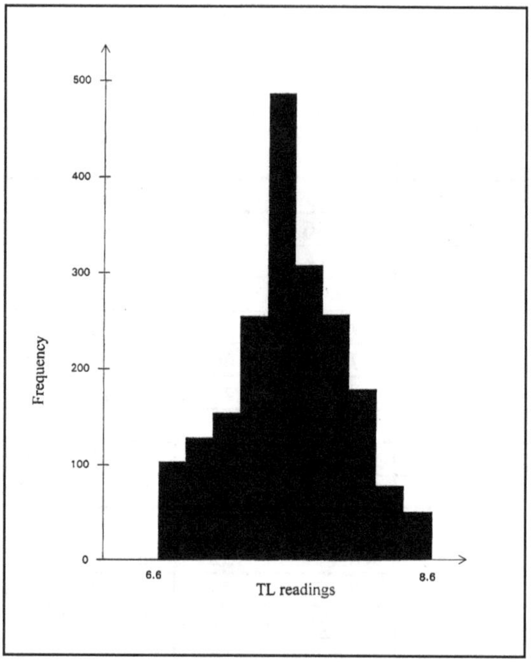

Fig. 2. Histogram of 80 readings.

Reference

Furetta C. and Weng P.S., Operational Thermoluminescence Dosimetry, World Scientific, 1998

Test for the reproducibility of a TL system (procedure)

- Select, randomly, a test group of 10 TLDs from a batch,
- Anneal the TLD samples according to the appropriate standard annealing,
- Irradiate the samples to a test dose of about 1 mGy: this dose is a compromise between high doses, which could give a residual TL in the successive cycles, and lower doses which could lower the reproducibility,
- Read out all the samples,

- Repeat point 4 for the background acquisition,
- Repeat the procedures 2 to 5 at least 10 times for statistical reasons,
- Complete the following Table:

TLD No.	readings Cycles No. 1	2	3-9	10	average (m_i)	S.D. (σ_i)	CV_i + μ_i
1	1938734	1943571	(omissis)	2022710	1968338	47149	2.4%
2	1633017	1654501	(omissis)	1657873	1648464	13483	0.8%
3	1668407	1584265	(omissis)	1683390	1645354	53432	3.2%
4	1848805	1861617	(omissis)	1802078	1837500	31338	1.7%
5	1810473	1888630	(omissis)	1963886	1887663	76711	4.1%
6	1474240	1436054	(omissis)	1503458	1471251	33801	2.3%
7	1245084	1278430	(omissis)	1262845	1262120	16685	1.3%
8	1274609	1313236	(omissis)	1300008	1295951	19630	1.5%
9	1284749	1234075	(omissis)	1281419	1266748	28344	2.2%
10	1835799	1926944	(omissis)	1809908	1857550	61475	3.3%
average	1601392	1612132	(omissis)	1628758			
rel. val.	1.000	1.007	(omissis)	1.017			

The coefficient of variation, for the i-th TLD, is given by

$$CV_i = \frac{\sigma_i}{m_i} \tag{1}$$

where σ_i and m_i are the standard deviation and the average values of the 10 repeated readings of the i-th dosimeter.

The half-width of the confidence interval, μ_i, is given by

$$\mu_i = CV_i \frac{t}{\sqrt{2(n-1)}} \tag{2}$$

where n is the number of repeated cycles and t is the value of the student test.

In the present case $n = 10$ and $t = 2.26$ at a confidence level of 95%. Then

$$\mu_i = CV_i \frac{2.26}{4.24} = 0.53CV_i \tag{3}$$

The reproducibility test, for each of the 10 doseneters, is then acceptable if

$$CV_i + \mu_i \leq 7.5\% \tag{4}$$

which transforms, considering Eq. (3), in the following acceptable level

$$CV_i \leq 5\%$$

So, to define a TL system as "reproducible" each dosimeter included in the test group should have a coefficient of variation no larger than 5%.

Reference

Scarpa G. in Corso sulla Termoluminescenza Applicata alla Dosimetria. Rome University "La Sapienza", Rome (Italy)

Thermal cleaning (peak separation)

The glow peaks in a glow curve are generally more or less overlapped. When the peaks are not too much overlapped, it is possible to use a thermal technique, called thermal cleaning, for getting a well defined and clearly separated peaks. This technique has been introduced and described by Nicholas and Woods (1964).

Let us imagine a phosphor showing a glow-curve with two, or more, overlapped peaks, each one having the maximum temperatures at $T_1 < T_2 < ... < T_n$. The sample is now heated to an appropriate intermediate temperature T just beyond the maximum temperature of the first peak and then rapidly cooled. This procedure erases the first peak, substantially emptying the traps responsible for it. The procedure is repeated for the subsequent peaks.

This technique is used when the **energy activation** E has to be determined by the **initial rise method**, where a clean initial rise of the peak is needed. It may also be used in any cases when a clean rise part of a peak (low temperature side of the peak) is necessary as in some methods based on **peak shape**.

The thermal procedure has to be applied very carefully. In fact, when a first-order process is supposed to be, the thermal cleaning allows to determine the temperature of the maximum, T_M, but not the intensity at the maximum because the cleaning thermal treatment could affect the number of the trapped charges. On the other hand, when a second order is assumed, even the value of T_M is not correct because the shift of the peak phosphor due to the reduction, as before, of the trapped charges.

Reference

Nicholas K.H. and Woods J., Br. J. Appl. Phys. **15** (1964) 783

Thermal fading (procedure)

Several equations can be obtained to take into account thermal fading effect as well as fading in competition with other effects, i.e., fading in competition with self-dose, fading and background irradiation, fading in accidental exposure and so on. The following first experiment is performed in the simplest way, just to determine the isothermal decay constant in a thermoluminescent material following a first order kinctics (Randall-Wilkins theory) and without any other effect in competition with fading.

- Determine the calibration factor of the reader as already explained in previous paragraph.
- Choose 30 similar TLDs that are expected to be used in the future dosimetric application.
- Identify each of them.
- Anneal the dosimeters according to the proper annealing procedure.
- At the end of the annealing the detectors have to be cooled in a reproducible way up to the room temperature.
- Read out the dosimeters with suitable heating cycle for the material chosen and determine the intrinsic background (zero dose reading).
- Note the values of the individual background, M_{i0} and calculate the average value $\overline{M}_0$ and its standard deviation σ_0. If σ_0 is less than 2%, use the average background value instead of the individual values.
- Irradiate at a test dose D_t for all 30 annealed dosimeters.
- Read immediately after irradiation in one session only.
- Determine the individual sensitivity factors for each of the 30 dosimeters

$$S_i = \frac{\overline{M}}{M_i - M_{i0}} \tag{1}$$

where

$$\overline{M} = \sum_{i=1}^{30} \frac{M_i - M_{i0}}{30} \tag{2}$$

(remember that Mi_0 can be substituted by the average background value if $\sigma_0 \leq 2\%$)

- Repeat five times the same procedure to determine S_i (anneal, irradiation, read) and calculate $\overline{S}_i$ and σ_i.
- Divide the group of 30 TLDs into three sub-groups of 10 dosimeters each, called group A, group B and group C.
- Anneal all the 30 TLDs.
- Irradiate group A only at a test dose D_t (about 0.1 Gy).
- Store all the three groups in a lead container in which the inside dose rate has to be very low.
- Temperature and relative humidity inside the container have to be monitored during the whole period of storage.
- The storage period, t_a, has to be chosen according to the specific needs for future dosimetric applications of the dosimeters.
- At the end of the storage period, group B is taken out of the container and irradiated at the test dose D_t.
- All the dosimeters are now read in one session only.
- Let us define the following quantities

$$\begin{aligned} \overline{M}_A &= \frac{1}{10}\sum_{i=1}^{10}\left(M_{Ai} - M_{Ai0}\right)\overline{S}_{Ai} \\ \overline{M}_B &= \frac{1}{10}\sum_{i=1}^{10}\left(M_{Bi} - M_{Bi0}\right)\overline{S}_{Bi} \\ \overline{M}_C &= \frac{1}{10}\sum_{i=1}^{10}\left(M_{Ci} - M_{Ci0}\right)\overline{S}_{Ci} \end{aligned} \tag{3}$$

  where

  $\overline{M}_A$ gives the measure of the TL at the end of the storage period

  $\overline{M}_B$ gives the reference prompt TL response at storage time t = 0

  $\overline{M}_C$ gives the measure of a possible increase of the TL emission, during the storage period, due to environmental and/or self irradiation.

- The isothermal decay constant is now calculated using the following expression

$$\lambda = \frac{1}{t_a} \ln \frac{\overline{\overline{M}_B - \overline{M}_C}}{\overline{\overline{M}_A - \overline{M}_C}} \qquad (h^{-1}) \tag{4}$$

It should be a good exercise to determine the value of λ as a function of different temperatures of storage.

Thermal quenching

Thermal quenching is the process such that the luminescence efficiency decreases with temperature, due to the increased probability of non-radiative transitions due to killer centers.

It has been shown that the luminescence efficiency can be expressed by the following equation

$$\eta = \frac{1}{1 + c\exp\left(-\frac{W}{kT}\right)} \tag{1}$$

where W can be evaluated from a plot of the luminescence intensity as a function of $1/T$.

Wintle discussed this problem extensively and she introduced the idea that the recombination process depends exponentially on the temperature: i.e., the recombination probability, A_m, is given by

$$A_m \propto \exp\left(\frac{W}{kT}\right) \tag{2}$$

where W is the energy depth of a non-radiative recombination level.

Equation (2) is a good approximation, in the high temperature range, of Eq. (1).

Thermally connected traps

Traps which are considered thermally connected belong to overlapping trap states having a very close energy difference and producing overlapped glow peaks. The thermally connected traps have been introduced by Sweet and Urquhart (**two trap model**) to explain their experimental results concerning ZnS single crystals.

Reference

Sweet M.A.S. and Urquhart D., Phys. Stat. Sol. (a) **59** (1980) 223

Thermally disconnected traps

A thermally disconnected trap is an extra energy level introduced by Dussel and Bube in 1967 for taking into account differences emerging in the results obtaining by simultaneous measurements of both thermoluminescence and thermally stimulated conductivity (TSC):

- the maxima of the two species of signals do not occur at the same temperature,
- important differences in the shape of the signals.

A thermally disconnected trap is a trap which can be filled by freed electrons produced by irradiation, but which has a trap depth which is much greater than the normal trapping levels. Thus, during heating the sample, only electrons in the shallower traps are freed, while the electrons trapped in the deeper levels (thermally disconnected) are not affected by heating. In other words, these trapping sites have a thermal stability of the trapped charges which is greater than that of the shallow traps related to the TL signal.

From an experimental point of view, it is quite difficult to prove the existence of the thermally disconnected traps because the limitation imposed by the black-body radiation background signal of the detection system, including TL sample, heating strip and surroundings.

Any way, for a theoretical interpretation of the experimental data obtained by both thermoluminescence and thermally stimulated conductivity measurements, it is realistic to include the thermally disconnected traps into the energy band scheme and then to modify the rate equations describing trap filling and trap emptying processes.

Reference

Dussel G.A. and Bube R.H., Phys. Rev. **155** (1967) 764

Thermoluminescence (thermodynamic definition)

Thermoluminescence requires the perturbation of a system from a state of thermodynamic equilibrium, via the absorption of external energy, into a metastable state. This is then followed by a thermally stimulated relaxation of the system back to its equilibrium condition.

Figure 3 shows an energy diagram for a crystal having a certain number of defects distributed between the conduction band (CB) and the valence band (VB). In thermal equilibrium condition, i.e. T = 0K, all the defect levels, up to the Fermi level F, are occupied by electrons. The other levels are empty (see Fig. 1a): electrons and crystal lattice are in thermal equilibrium.

The system can now be perturbed by an ionizing radiation. Under irradiation the electrons in the defect levels or in the valence band gain energy and rise into higher levels, beyond the Fermi level (Fig. 1b). After the irradiation a redistribution process takes place and the excited system goes back to equilibrium. The time required for going back to equilibrium may vary from milliseconds to years, depending on the material, its defects and the temperature.

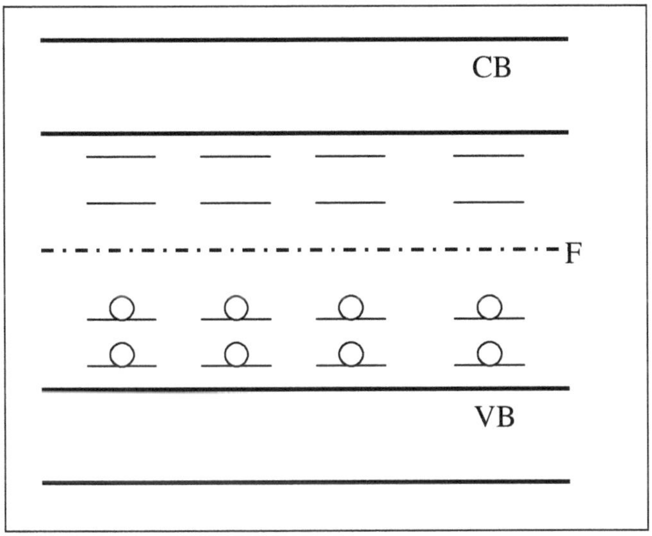

Fig. 3a. Energy diagram for a crystal having a certain number of defects distributed between the CB and the valence band VB. The open circles are the electrons.

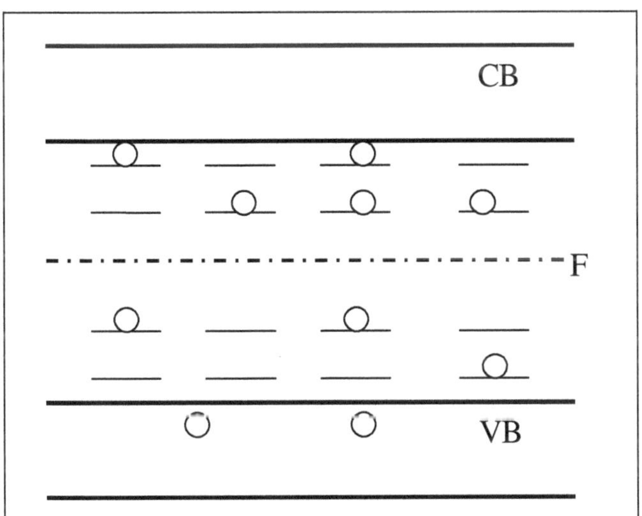

Fig. 3b. Redistribution of the trapped electrons due to the irradiation.

Thermoluminescence (TL)

From a microscopic point of view, thermoluminescence consists of a perturbation of the electronic system of insulating or semiconducting materials, from a state of thermodynamic equilibrium, via the absorption of external energy, i.e. produced by an ionizing radiation, into a metastable state. This is then followed by the thermally stimulated relaxation of the system back to its equilibrium condition.

Macroscopically, thermoluminescence is a temperature-stimulated light emission from a crystal, after removal of excitation (i.e. ionizing radiation); thermoluminescence is a case of phosphorescence observed under condition of steadily increasing temperature. A plot of the light intensity as a function of temperature is called **glow-curve**. A glow-curve may have one or more maxima, called **glow-peaks**, each corresponding to an energy level trap [1,2].

References

1. Mckeever S.W.S. and Chen R., "Luminescence Models", Rad. Measur. **27** (5/6) (1997) 625
2. Furetta C. and Weng P.S., "Operational Thermoluminescence Dosimetry" World Scientific, 1998

Thermoluminescent dosimetric system (definition)

A thermoluminescent dosimetric system consists of several parts as follows:

- the passive elements: the TL dosimeters (or detectors)
- a TL reader schematically consisting of a heating element, a PM tube, one or more electronic networks.
- an appropriate algorithm to convert the TL signal (response of the reader) to dose.
- ovens and/or furnaces to be used for thermal treatments of the dosimeters (annealing procedures).
- any other complementary instrumentation or facility which can be used for the right setting up and working for the system and/or for the implementation of the system (i.e. calibration sources; programme able to deconvolute the glow-curve, to make an automatic estimation of the background, to calculate the average TL values and so on).

Thermoluminescent materials: requirements

Several properties have to be examined for the choice of a TL material with respect to a specific application. In general, the more desirable properties of a TLD phosphor are listed as follows:

- a high concentration of traps and a high efficiency of light emission associated with the recombination process;
- a good storage stability of the trapped charges, as a function of storage time and temperature, so that a negligible fading affects the TL response. This should also be true for opposite extreme temperature values (i.e., tropical or artic climates);
- a very simple glow curve (i.e., a simple trap distribution) which allows the interpretation of the readings as simple as possible, without any thermal treatment after irradiation (post irradiation annealing). In case of more or less complex glow curve, the main peak (i.e., the dosimetric peak) should be well resolved among other possible peaks in the glow curve;
- a spectrum of the emitted TL light to which the detector system (photomultiplier and associated filters) responds well. A spectrum wavelengths between 300 and 500 nm seems the most desirable since it corresponds to the commercially available detector systems. Furthermore, the black body radiation does not interfer in this spectral range even at relatively high temperatures;
- the main peak should have a peak temperature at the maximum in the range 180°C ÷250°C. At higher temperatures the infrared emission from both TLD sample and TLD holder may interfere giving up to a source of errors in the reading interpretation;
- good resistance against disturbing environmental factors as light (optical fading), humidity, organic solvents, gases, moisture;
- the TL material should not suffer by radiation damage in the dose range of applications;
- the TL material should have a low photon energy dependence of response. For personnel and medical applications, tissue equivalent phosphors (effective atomic number of the tissue Z_{eff} =7.4), or approximated tissue equivalent, should be used to avoid energy corrections;
- a linear TL response over a wide range of doses is a desirable feature for most applications;

- the TL material should be non-toxic: this is very important for in-vivo medical applications;
- the TL response should be independent of dose rate and of the angle of radiation incidence;
- the lower limit of detection should be as low as possible for environmental monitoring;
- low self-irradiation due to natural radionuclides in the TLD materials for all kind of applications;
- the TLD phosphor should have a high/low thermal neutron sensitivity according to the specific use (i.e., monitoring around power plants, accelerators and so on);
- a good LET sensitivity may also be useful in some cases;
- high precision and high accuracy are required characteristics for any kind of applications;
- in case of need, the TL detectors should be suitable for postal service.

The above list cannot be fulfilled by only one type of TL phosphor. As a result, there is a serious limitation in the choice and the materials which can be used for dosimetry have properties which are a compromise among the various requirements. Any way, a material having very good performances for one or more specific applications can be easily found.

Three points method

M.S. Rasheedy has described a procedure for separating a complex glow curve into its glow peak components [1,2].

The empirical equation to describe the general-order kinetics is

$$I = -\frac{dn}{dt} = \frac{n^b}{N^{b-1}} s \exp\left(-\frac{E}{kT}\right) \tag{1}$$

where b is the kinetics order between 1 and 2.

The previous eqution assumes a different form considering that the concentration of populated traps at a temperature T_i is proportional to the area A_i under the glow peak between T_i and T_f, where T_f is the final temperature of the glow peak.

Considering the maximum intensity of the glow peak, I_M, Eq. (1) becomes:

$$I_M = \frac{A_M^b}{N^{b-1}} s \exp\left(-\frac{E}{kT_M}\right) \qquad (2)$$

where T_M is the temperature at the maximum intensity of the peak and A_M is the peak area between T_M and T_f.

From Fig. 1 three positions on the peak can be defined:

$$I_X \ \ at \ \ T_X$$

$$I_Y = \frac{I_X}{y} \quad at \quad T_Y$$

$$I_Z = \frac{I_X}{z} \quad at \quad T_Z$$

where:

I_Y and I_Z are portions of I_X

A_X, A_Y and A_Z are the areas under the peak from T_X, T_Y and T_Z to T_f.

Then, according to the previous quantities, three equations similar to Eq. (2) can be written:

$$I_X = \frac{A_X^b}{N^{b-1}} s \exp\left(-\frac{E}{kT_X}\right) \qquad (3)$$

$$I_Y = \frac{A_Y^b}{N^{b-1}} s \exp\left(-\frac{E}{kT_Y}\right) \qquad (4)$$

$$I_Z = \frac{A_Z^b}{N^{b-1}} s \exp\left(-\frac{E}{kT_Z}\right) \qquad (5)$$

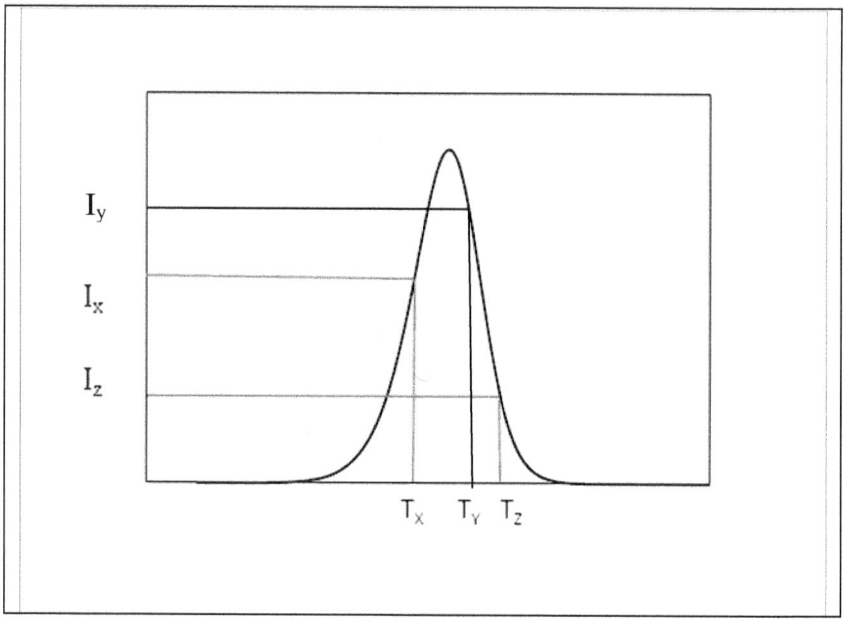

Fig. 1

From Eqs. (3) and (4) we get

$$E=\left[\ln Y-b\ln\left(\frac{A_X}{A_Y}\right)\right]\cdot\left(\frac{kT_XT_Y}{T_X-T_Y}\right) \tag{6}$$

From Eqs. (3) and (5) we get

$$E=\left[\ln Y-b\ln\left(\frac{A_X}{A_Z}\right)\right]\cdot\left(\frac{kT_XT_Z}{T_X-T_Z}\right) \tag{7}$$

Equating now Eqs. (6) and (7), the kinetics order b is obtained:

$$b=\frac{T_Y\left[T_X-T_Z\right]\ln(Y)-T_Z\left[T_X-T_Y\right]\ln(Z)}{T_Y\left[T_X-T_Z\right]\ln\left(\frac{A_X}{A_Y}\right)-T_Z\left[T_X-T_Y\right]\ln\left(\frac{A_X}{A_Z}\right)} \tag{8}$$

After that, the frequency factor S for the first order kinetics and the pre-exponential factor S" for the general order case are obtained from the following equations:

$$S = \left(\frac{\beta E}{kT_M^2}\right)\exp\left(\frac{E}{kT_M}\right) \quad (\sec^{-1}) \tag{9}$$

for the first order kinetics and

$$S'' = \frac{\beta E \exp\left(\frac{E}{kT_M}\right)}{\left(bkT_M^2\right) - (b-1)E\phi \exp\left(\frac{E}{kT_M}\right)} \tag{10}$$

where

$$\phi = \int_{T_0}^{T_M} \exp\left(-\frac{E}{kT'}\right) dT' \tag{11}$$

for the general order.

The last parameter which can be calculated is the value of n_0 (cm^{-3}). Its expression is obtained using the maximum intensity expression:

$$I_M = \frac{n_0 S'' \exp\left(-\frac{E}{kT_M}\right)}{\left[\frac{bkT_M^2 S''}{\beta E}\exp\left(-\frac{E}{kT_M}\right)\right]^{\frac{b}{b-1}}} \tag{12}$$

from which the value of n_0 is obtained as:

$$n_0 = \frac{I_M \exp\left(\frac{E}{kT_M}\right)}{S''}\left[\frac{bkT_M^2 S''}{\beta E \exp\left(\frac{E}{kT_M}\right)}\right]^{\frac{b}{b-1}} \tag{13}$$

It has to be noted that the values of I_M and T_M are obtained from the glow peak itself.

References

1. Rasheedy M.S., Thermochimica Acta **429** (2005) 143
2. Rasheedy M.S., El-Sherif M.A., Hefni M.A., NIM-B, **258** (2007) 440

Tissue equivalent phosphors

TL materials having an effective atomic number, Z_{eff}, similar to the one of the soft tissue (Z=7.4), are known as tissue-equivalent materials.

The tissue equivalence is a desirable feature for greater accuracy in biomedical, clinical and personal monitoring.

Tissue equivalence for photons requires that the mass energy absorption coefficients, $\frac{\mu_{en}}{\rho}$, for the dosimetry material match those for the tissue in which the dose is to be measured.

The cross-sections for photon interactions are directly proportional to the atomic number, Z, raised to some numerical power for each element in the dosemeter material, i.e. elemental cross-section $\propto Z^x$, where x depends on the type of interaction occurring and varies between 1 and 5. It has a value closed to 4 for the photoelectric effect [1,2].

A compound, as a thermoluminescent material is, may be regarded as a single element with an effective atomic number , Z_{eff}, given by

$$Z_{eff} = \left(\sum_i \alpha_i Z_i^x \right)^{1/x}$$

where α_i is the fractional electron content of element i-th in the compound.

For the photoelectric effect in muscle $Z = 7.4$, therefore materials with similar Z will have good tissue equivalence for low energy photons: their response will vary with photon energy in the same way as $\frac{\mu_{en}}{\rho}$ for tissue.

The requirements for tissue equivalence when a dosemeter is irradiated with neutrons or high LET particle is quite different. Neutrons entering tissue interact with H, C, N and O releasing secondary charged particles like protons, alpha particles and heavy recoil nuclei. For fast neutrons, the (n,p) reaction with H predominates contributing over 70% to the kerma, but few thermoluminescent materials contain H.

References

1. Mayneord W.V., The Significance of the Roentgen in Acta Int. Union Against Cancer **2** (1937) 271
2. Driscoll C.M.H. in Practical Aspects of Thermoluminescence Dosimetry, Proceedings of the Hospital Physicists' Meeting, University of Manchester (1984) Edited by A.P. Hufton

Trap characteristics obtained by fading experiments

A quick way for obtaining qualitative informations about the trap characteristics of a thermoluminescent material, consists in the comparison between glow-curves recorded at different time intervals after irradiation.

The following Fig. 4 shows the glow-curves of CVD diamond sample, irradiated with UV light. The first glow-curve, labelled 1 min UV, has been recorded immediately after UV irradiation; the second is the glow-curve recorded 23 hrs after the end of the UV irradiation. The third plot is the difference between the two previous glow-curves: this difference gives the indication of the TL lost during fading as a function of the glow-curve temperature.

Supposing the fading is an isothermal decay at room temperature, it is given by

$$I = I_0 \exp(-\lambda \cdot t) \tag{1}$$

where I_0 is the TL emission recorded immediately after irradiation, I is the TL after the fading time t, and λ is the decay constant of the process.

Hence

$$\lambda = -\frac{1}{t} \ln\left(\frac{I}{I_0}\right) \tag{2}$$

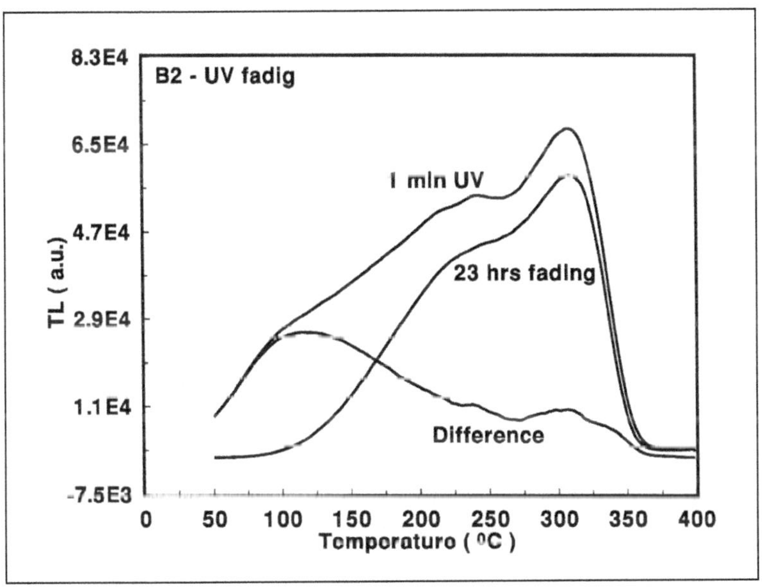

Fig. 4. Glow-curves of CVD diamond and their difference.

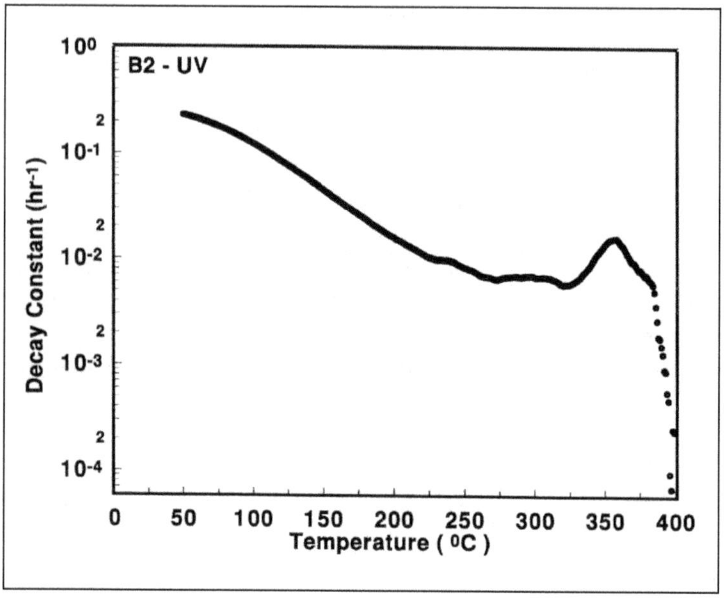

Fig. 5. Decay constant as a function of the glow-curve temperatures.

Figure 5 shows the plot of the resulting values of λ as a function of the glow-curve temperatures. In this figure it is possible to identify three different regions. In the first region λ decreases as the glowe-curve temperature increases; in the second region a clear plateau is observed and, finally, a third region, above 300°C, where the results are highly scattered because I and I_0 are more or less quite similar.

The behavior of λ in the first region is a clear indication of a continuous distribution of trapping levels, whereas the plateau region indicates a single trapping level.

Reference

Kitis G. private communication

Trap creation model

The supralinearity is explained via trap creation during irradiation.

Cameron used this model to explain the supralinearity in LiF. The model requires that the new traps created by irradiation are the same as those originally presented in the crystal srtucture. Furthermore, they suggested that luminescence centers are also created by irradiation and that the new centers are the same type of the original ones because the emission spectra are unchanged.

Reference

Cameron J.R., Suntharalingam N. and Kenney G.N., Thermoluminescence Dosimetry, University of Wisconsin Press, Madison (1968)

Trapping state

A trapping state is that for which the probability of thermal excitation from a localized state into the respective delocalized band is greater than the probability of recombination of the trapped charge with a free charge carrier of opposite sign.

Tunnelling

An electron trapped in a level A of an atom (Fig. 6) may recombine directly with a hole in a level B of another atom without involving the delocalized bands. Mikhailov gave in 1971 a model for tunnelling process [1-3].

The defects responsible for levels A and B must be closed to each other. This can occur when traps and recombination centers are in a very high concentration and when the two centers belong to the same defect site. This transition occurs through the potential barrier (tunnelling) which separates the electron in A from the hole in B. The recombination results in the emission of luminescence. The effect is athermal.

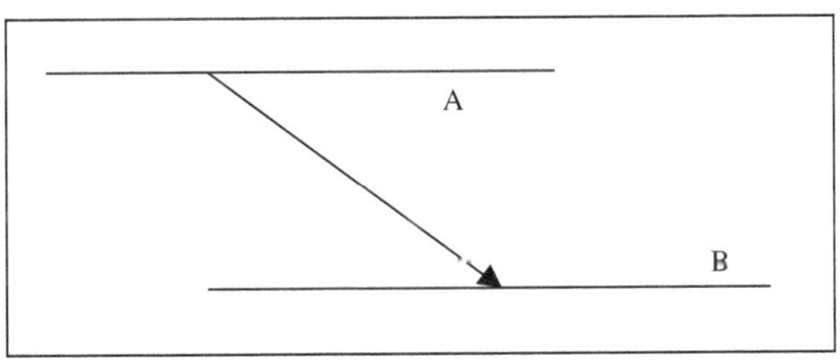

Fig. 6. Tunnelling between an electron in A and a hole in B.

The tunnelling process can explain the low temperature afterglow and the relationship between this and the TL lost in anomalous fading.

Visocekas et al (1976) [4] also considered the possibility that the electron can be first excited to a higher energy state and then, not having still enough energy to escape from the trap, recombines via tunnelling. This type of process is called thermally assisted tunnelling.

References

1. McKeever S.W.S., Thermoluminescence in Solids, Cambridge University Press (1985)
2. Chen R. and Kirsh Y., Analysis of Thermally Stimulated Processes, Pergamon Press (1981)
3. Chen R. and McKeever S.W.S., Theory of Thermoluminescence and Related Phenomena, World Scientific (1997)
4. Visocekas R., Leva T., Marti C., Lefaucheux F. and Robert M.C., Phys. Stat. Sol. (a) **35** (1976) 315

Two-trap model (Sweet and Urquhart)

This model has been proposed by Sweet and Urquhart to explain a situation where two peaks are so close that they appear as only one peak.

Let us define the following symbols:

E_1, E_2 $(\succ E\ )$ depth of two very closed traps (eV),

N_1, N_2 concentration of trapping centers (m^{-3}),

n_1, n_2 concentration of trapped electrons (m^{-3}),

M concentration of recombination centres (m^{-3}),

m concentration of trapped holes in recombination centres (m^{-3}),

n_c, m_v concentration of free electrons in CB and free holes in VB (m^{-3}); both are assumed negligible,

A_{n1}, A_{n2} trapping rate constants ($m^3\ s^{-1}$),

A_m recombination rate constant ($m^3\ s^{-1}$),

$\gamma_1, \gamma_2\left(=s_i \exp\left(-\frac{E_i}{kT}\right)\right)$ probability for electrons from trap to CB.

The following set of equations can be written:

$$\frac{dn_1}{dt} = -\gamma_1 n_1 + n_c A_{n1}\left(N_1 - n_1\right) \tag{1}$$

$$\frac{dn_2}{dt} = -\gamma_2 n_2 + n_c A_{n2}\left(N_2 - n_2\right) \tag{2}$$

$$\frac{dn_c}{dt} + \frac{dn_1}{dt} + \frac{dn_2}{dt} = -A_m n_c m \tag{3}$$

The condition of charge neutrality is now:

$$m = n_c + n_1 + n_2 \tag{4}$$

having considered $m_v = 0$.

Assuminig a lincar heating rate $T = T_i + \beta \cdot t$, where T_i is the initial temperature and $\beta = {dT}/{dt}$.

In order to solve numerically the previous equations, the following approximations are assumed:

$$n_c \prec\prec n_1, \frac{dn_c}{dt} \prec\prec \frac{dn_1}{dt} \tag{5}$$

$$n_c \prec\prec n_2, \frac{dn_c}{dt} \prec\prec \frac{dn_2}{dt} \tag{6}$$

Inserting (5) and (6) in Eq. (4) one gets

$$m = n_1 + n_2 \tag{7}$$

Then, using Eqs. (3) and (4), we obtain

$$-\frac{dm}{dt} = -\frac{dn_c}{dt} - \frac{dn_1}{dt} - \frac{dn_2}{dt} = A_m n_c m \tag{8}$$

Eq. (7) can be rewritten as

$$\frac{dm}{dt} \approx \frac{dn_1}{dt} + \frac{dn_2}{dt} \tag{9}$$

and then, inserting (9) into Eq. (8). We obtain

$$\frac{dn_c}{dt} \approx 0 \tag{10}$$

Inserting now (10) in Eq. (3) we get

$$\frac{dn_1}{dt} + \frac{dn_2}{dt} + A_m n_c m = 0 \tag{11}$$

Eqs. (1) and (2) can now be written as

$$\frac{dn_1}{dt}+\frac{dn_2}{dt}=-\gamma_1 n_1-\gamma_2 n_2+n_c A_{n1}\left(N_1-n_1\right)+n_c A_{n2}\left(N_2-n_2\right)$$

which can be rearranged, using (7) and (11), as

$$A_m n_c\left(n_1+n_2\right)=\gamma_1 n_1+\gamma_2 n_2-n_c A_{n1}\left(N_1-n_1\right)-n_c A_{n2}\left(N_2-n_2\right) \tag{12}$$

from which

$$n_c=\frac{\gamma_1 n_1+\gamma_2 n_2}{A_m\left(n_1+n_2\right)+A_{n1}\left(N_1-n_1\right)+A_{n2}\left(N_2-n_2\right)} \tag{13}$$

Because the glow-curve intensity is given by the decrease of the trapped holes during recombination, the TL intensity is

$$I(T)=-c\frac{dm}{dt} \tag{14}$$

Using Eqs. (7) and (8) and taking c = 1, Eq. (14) becomes

$$\begin{aligned} I(T) &= A_m n_c m = A_m n_c\left(n_1+n_2\right) \\ &= \frac{A_m\left(n_1+n_2\right)(\gamma_1 n_1+\gamma_2 n_2)}{A_m\left(n_1+n_2\right)+A_{n1}\left(N_1-n_1\right)+A_{n2}\left(N_2-n_2\right)} \end{aligned} \tag{16}$$

It has to be stressed that Eq. (16) becomes the equation of the first-order model by setting equal to zero the parameters with subscript 2 and considering that recombination dominates, i.e. $A_m n_1 \succ\succ A_{n1}\left(N_1-n_1\right)$. Equation (16) gives also the second order model considering the retrapping assumption and assuming to be far from the trap saturation, i.e. $N_1 \succ\succ n_1$.

Equation (16) has been computer-calculated for fitting the experimental results obtained from the study of ZnS. The parameters used for the best fit are

$$s_1=s_2=3\cdot 10^{15}s^{-1}, \frac{N_1}{n_1}=64, \frac{N_2}{n_2}=100, E_1=18meV, E_2=22.5meV,$$

$$\frac{A_{n1}}{A_m}=\frac{A_{n2}}{A_m}=90$$

Reference

Sweet M.A.S. and Urquhart D., Phys. Stat. Sol. (a), **59** (1980) 223

V

Various heating rates method (Bohum, Porfianovitch, Booth: first order)

-Bohum [1], Porfianovitch [2] and Booth [3], working independently, proposed a method based on two different heating rates for a first-order peak. Taking into account the **condition at the maximum** and using two different heating rates one obtains:

$$\frac{\beta_1 E}{kT_{M1}^2} = s\exp(-\frac{E}{kT_{M1}}) \quad (1)$$

$$\frac{\beta_2 E}{kT_{M2}^2} = s\exp(-\frac{E}{kT_{M2}}) \quad (2)$$

from which, by eliminating s, E is obtained according to the following expression

$$E = k\frac{T_{M1}T_{M2}}{T_{M1}-T_{M2}}\ln\left(\frac{\beta_1}{\beta_2}\right)\cdot\left(\frac{T_{M2}}{T_{M1}}\right)^2 \quad (3)$$

Therefore, in the assumption $b = 1$, E is easily evaluated by measuring the two peak temperatures corresponding to the maximum TL intensity for the two heating rates. If T_M can be measured within an accuracy of 1°C, the method yields E within 5%. The value of s can be then calculated by substituting the numerical value of E in one of the two equations (1) or (2).

References

1. Bohum A., Czech. J. Phys. **4** (1954) 91
2. Porfianovitch I.A., J. Exp. Theor. Phys. SSSR, **26** (1954) 696
3. Booth A.H., Canad. J. Chem. **32** (1954) 214

Various heating rates method (Chen-Winer: first order)

Chen and Winer reported a method using an approximation for the integral which appears in the first-order expression of $I(T)$. In fact, the integral can be approximated as following

$$\int_{T_0}^{T}\exp(-\frac{E}{kT})dT = (\frac{kT^2}{E})\exp(-\frac{E}{kT})(1-\Delta) \quad (1)$$

where

$$\Delta = \frac{2kT}{E} \approx 0.1$$

The insertion of (1) into the first-order expression for $I(T)$ yields

$$I = sn_0 \exp(-\frac{E}{kT}) \exp\left[-\frac{skT^2}{E\beta} \exp(-\frac{E}{kT})(1-\Delta) \right] \qquad (2)$$

Inserting now the **condition at the maximum** in Eq. (2), one obtains

$$I_M = n_0 s \exp\left(-\frac{E}{kT_M} \right) \exp -(1-\Delta_M)$$

or

$$I_M = \frac{n_0 s}{e} \exp\left(-\frac{E}{kT_M} \right) \exp(\Delta_M)$$

an then

$$I_M = \left(\frac{sn_0}{e} \right) (1+\Delta_M) \exp\left(-\frac{E}{kT_M} \right) \qquad (3)$$

Even with large variations of β, T_M changes only a few percent and therefore so Δ_M and $1+\Delta_M$ do consequently. Then, one can assume that the intensity is directly proportional to the exponential, considering as a constant the other quantities.

The plot of $\ln(I_M)$ against $1/T_M$ for various heating rates should get a straight line with a slope $-E/k$ from which E can be found. Owing to the approximated integration, which is true only for a linear heating rate, this method is valid only in this case.

Reference

Chen R. and Winer S.A.A., J. Appl. Phys. **41** (1976) 5227

Various heating rates method (Chen-Winer: second and general orders)

Chen and Winer showed how to apply the various heating rates method if the kinetics of any order is present, including the second-order kinetics. The expression of $I(T)$, in the general case, is given by

$$I(T) = sn_0 \exp\left(-\frac{E}{kT}\right)\left[1+\frac{s(b-1)}{\beta}\int_{T_0}^{T}\exp\left(-\frac{E}{kT'}\right)dT'\right]^{-\frac{b}{b-1}} \tag{1}$$

and the **condition of maximum emission**

$$\frac{kT_M^2 bs}{\beta E}\exp\left(-\frac{E}{kT_M}\right) = 1+\frac{s(b-1)}{\beta}\int_{T_0}^{T}\exp\left(-\frac{E}{kT'}\right)dT' \tag{2}$$

where $s = s'' n_0^{b-1}$.

The maximum value of $I(T)$ is obtained by inserting Eq. (2) in Eq. (1):

$$I_M - sn_0 \exp(-\frac{E}{kT})\left[\frac{kT_M^2 bs}{\beta E}\exp(-\frac{E}{kT_M})\right]^{-\frac{b}{b-1}} \tag{3}$$

which can be rewritten as

$$I_M^{b-1}\frac{T_M^{2b}}{\beta^b} = s^{-1} n_0^{b-1}\left(\frac{E}{kb}\right)^b \exp(\frac{E}{kT_M}) \tag{4}$$

From the ln of this expression one obtains

$$\ln\left[I_M^{b-1}\left(\frac{T_M^2}{\beta}\right)^b\right] = \frac{E}{kT_M}+c \tag{5}$$

with c constant.

By means of this equation it is possible to evaluate the quantity on the left side for different values of b and to obtain a set of experimental points. These points are then plotted as a function of $1/T_M$ on a semilog paper and fitted by a straight line whose slope is E/k. Of course, one must find the value of b for which the plot best approximates the linearity. In another way β^b may be included in the constant.

For the second order case, Eq. (5) gets

$$\ln\left[I_M\left(\frac{T_M^2}{\beta}\right)^2\right] = \frac{E}{kT_M}+c \tag{6}$$

This method is useful only when b is appreciably different from unity since for $b = 1$, T_M is independent of the initial concentration n_o of trapped electrons. Also in this case β^2 may be included in the constant.

As also suggested by Chen and Winer, one can consider the **condition of maximum emission** and the **integral approximation**. Remembering the condition of maximum emission for a general-order kinetics given by Eq. (2), one obtains:

$$\left(\frac{kT_M^2 bs}{\beta E}\right)\exp\left(-\frac{E}{kT_M}\right) \cong 1+\frac{s(b-1)kT_M^2}{\beta E}\exp\left(-\frac{E}{kT_M}\right)(1-\Delta_M) \tag{7}$$

which gives the final form

$$\left(\frac{\beta}{T_M^2}\right) \cong \exp\left(-\frac{E}{kT_M}\right)\left(\frac{ks}{E}\right)\left[1+(b-1)\Delta_M\right] \tag{8}$$

The quantity $[1+(b\text{-}1)\Delta_M]$ is close to unity and it can be considered as a constant, so that the plot of different values of left side expression versus $1/T_M$ should be well fitted by a straight line of slope $-E/k$. The new value of E can be compared with the former given by Eq. (6) to have an assessment of the error introduced by the integral approximation. The same procedure can be used in case of a second-order kinetics.

Reference

Chen R. and Winer S.A.A., J. Appl. Phys. **41** (1970) 5227

Various heating rates method (Gartia et al.: general order)

A new method using two heating rates has been proposed by Gartia et al. [1]. It is analogous to the Booth method, which is strictly valid for a first order peak, but it is applied to a non-first order TL peak and it is based on the variation of I_M with b, which variation is much more faster than the variation of T_M with b. Using the **general-order expression** [2], one can write:

$$\ln I_M = -\frac{E}{kT_M} + \ln sn_0 - \frac{b}{b-1}\ln\left[1+(b-1)\frac{sEE_2(u_m)}{ku_m\beta}\right] \tag{1}$$

where $u_m=E/kT_M$ and $E_2(u_m)$ is the second exponential integral [3]. The factor in square brackets is very close to unity; hence, using two linear heating rates β_1 and β_2 one obtains

$$\begin{aligned} \ln I_{m1} &\cong -\frac{E}{kT_{m1}} + \ln sn_0 \\ \ln I_{m2} &\cong -\frac{E}{kT_{m2}} + \ln sn_0 \end{aligned} \tag{2}$$

which give

$$E = \frac{kT_{m1}T_{m2}}{T_{m1} - T_{m2}} \ln \frac{I_{m1}}{I_{m2}} \tag{3}$$

The Authors pointed out that the maximum systematic error involved is less than 1% for any order of kinetics ($1.1 \le b \le 2.5$).

References

1. Gartia R.K., Ingotombi S., Singh T.S.G. and Mazumdar P.S., J. Phys. D: Appl. Phys. **24** (1991) 65
2. Singh T.C.S., Mazumdar P.S. and Gartia R.K., J. Phys. D: Appl. Phys. **23** (1990) 562
3. Gaustchi W. and Cahill W.F., Handbook of Mathematical Functions (Dover, N.Y., 1972)

Various heating rates method (Hoogenstraaten: first order)

Hoogenstraaten, starting form the **condition at the maximum** for a first order kinetics, suggested the of use several heating rates to obtain a linear relation as the following

$$\ln(\frac{T_M^2}{\beta}) = (\frac{E}{k})\frac{1}{T_M} + \ln\left(\frac{E}{sk}\right) \tag{1}$$

The resultant plot should yield a straight line with slope *E/k* and an intercept ln(*E/sk*).

Reference

Hoogenstraaten W., Philips Res. Repts. **13** (1958) 515

Various heating rates method (Rasheedy)

M.S. Rasheedy [1], proposed a method, based on three different heating rates, for obtaining the activation energy of any glow peak, independent on the kinetics order of the process. The method requires only the quantities at the maximum of the peak, i.e. I_M and T_M, then additional satellites on either sides of the peak would not influence the results. Furthermore, the E value is not affected by thermal quenching as in the initial rise method.

Starting from the general order equations proposed by May and Partridge [2]

$$I = -\frac{dn}{dt} = \frac{n^b}{N^{b-1}} s \exp\left(-\frac{E}{kT}\right) \tag{1}$$

where b is the kinetics order, its solution is given by

$$I = \frac{n_0^b s N^{1-b} \exp\left(-\frac{E}{kT}\right)}{\left\{1 + \left[\frac{(b-1)s\left(\frac{n_0}{N}\right)^{b-1}}{\beta}\right] \cdot \int_{T_0}^{T} \exp\left(-\frac{E}{kT'}\right) dT'\right\}^{\frac{b}{b-1}}} \tag{2}$$

The position of the maximum is then from Eq. (2) as:

$$1 + (b-1)\left(\frac{n_0}{N}\right)^{b-1} \left(\frac{s}{\beta}\right) \varphi = s\left(\frac{bkT_M^2}{\beta E}\right)\left(\frac{n_0}{N}\right)^{b-1} \exp\left(-\frac{E}{kT_M}\right) \tag{3}$$

where

$$\varphi = \int_{T_0}^{T_M} \exp\left(-\frac{E}{kT'}\right) dT' \tag{4}$$

Inserting Eq. (3) in Eq. (2), we obtain the maximum intensity of the peak:

$$I_M = \frac{n_0^b s N^{1-b} \exp\left(-\frac{E}{kT_M}\right)}{\left[s\left(\frac{bkT_M^2}{\beta E}\right)\left(\frac{n_0}{N}\right)^{b-1} \exp\left(-\frac{E}{kT_M}\right)\right]^{\frac{b}{b-1}}} \tag{5}$$

which can be re-written as

$$I_M^b \left(\frac{T_M^2}{\beta}\right)^b = I_M \left(\frac{N^{b-1}}{s}\right)\left(\frac{E}{bk}\right)^b \exp\left(\frac{E}{kT_M}\right) \tag{6}$$

Using now three different heating rates, Eq. (6) is re-written as:

$$I_{M1}^{b}\left(\frac{T_{M1}^{2}}{\beta_{1}}\right)^{b}=I_{M1}\left(\frac{N^{b-1}}{s}\right)\left(\frac{E}{bk}\right)^{b}\exp\left(\frac{E}{kT_{M1}}\right) \tag{7}$$

$$I_{M2}^{b}\left(\frac{T_{M2}^{2}}{\beta_{2}}\right)^{b}=I_{M2}\left(\frac{N^{b-1}}{s}\right)\left(\frac{E}{bk}\right)^{b}\exp\left(\frac{E}{kT_{M2}}\right) \tag{8}$$

$$I_{M3}^{b}\left(\frac{T_{M3}^{2}}{\beta_{3}}\right)^{b}=I_{M3}\left(\frac{N^{b-1}}{s}\right)\left(\frac{E}{bk}\right)^{b}\exp\left(\frac{E}{kT_{M3}}\right) \tag{9}$$

Dividing now Eq. (7) by Eq. (8) and Eq. (7) by Eq. (9) one gets

$$\left(\frac{I_{M1}T_{M1}^{2}\beta_{2}}{I_{M2}T_{M2}^{2}\beta_{1}}\right)^{b}=\left(\frac{I_{M1}}{I_{M2}}\right)\exp\left[\frac{E}{k}\left(\frac{T_{M2}-T_{M1}}{T_{M1}T_{M2}}\right)\right] \tag{10}$$

$$\left(\frac{I_{M1}T_{M1}^{2}\beta_{3}}{I_{M3}T_{M3}^{2}\beta_{1}}\right)^{b}=\left(\frac{I_{M1}}{I_{M3}}\right)\exp\left[\frac{E}{k}\left(\frac{T_{M3}\quad T_{M1}}{T_{M1}T_{M3}}\right)\right] \tag{11}$$

Let then

$$\theta_{1}=\frac{I_{M1}T_{M1}^{2}\beta_{2}}{I_{M2}T_{M2}^{2}\beta_{1}},\qquad \lambda_{1}=\frac{I_{M1}}{I_{M2}},\qquad \Psi_{1}=\frac{T_{M2}-T_{M1}}{T_{M1}T_{M2}}$$

$$\theta_{2}=\frac{I_{M1}T_{M1}^{2}\beta_{3}}{I_{M3}T_{M3}^{2}\beta_{1}},\qquad \lambda_{2}=\frac{I_{M1}}{I_{M3}},\qquad \Psi_{2}=\frac{T_{M3}-T_{M1}}{T_{M1}T_{M3}}$$

We get so, respectively for Eqs. (10) and (11):

$$b\ln\theta_{1}=\ln\lambda_{1}+\left(\frac{E}{k}\right)\cdot\Psi_{1} \tag{12}$$

$$b\ln\theta_{2}=\ln\lambda_{2}+\left(\frac{E}{k}\right)\cdot\Psi_{2} \tag{13}$$

From Eqs. (12) and (13) we get

$$\frac{\ln\theta_1}{\ln\theta_2}=\frac{\ln\lambda_1+\left(\frac{E}{k}\right)\cdot\Psi_1}{\ln\lambda_2+\left(\frac{E}{k}\right)\cdot\Psi_2} \tag{14}$$

From which the activation energy can be obtained as:

$$E=\frac{k\left[\left(\ln\theta_2\cdot\ln\lambda_1\right)-\left(\ln\theta_1\cdot\ln\lambda_2\right)\right]}{\left(\Psi_2\cdot\ln\theta_1\right)-\left(\Psi_1\cdot\ln\theta_2\right)} \tag{15}$$

Or, in explicit form:

$$E=\frac{k\left(T_{M1}T_{M2}T_{M3}\right)\cdot\left[\left(\ln\frac{I_{M1}T_{M1}^2\beta_3}{I_{M3}T_{M3}^2\beta_1}\cdot\ln\frac{I_{M1}}{I_{M2}}\right)-\left(\ln\frac{I_{M1}T_{M1}^2\beta_2}{I_{M2}T_{M2}^2\beta_1}\cdot\ln\frac{I_{M1}}{I_{M3}}\right)\right]}{\left[T_{M2}\left(T_{M3}-T_{M1}\right)\cdot\ln\frac{I_{M1}T_{M1}^2\beta_2}{I_{M2}T_{M2}^2\beta_1}\right]-\left[T_{M3}\left(T_{M2}-T_{M1}\right)\cdot\ln\frac{I_{M1}T_{M1}^2\beta_3}{I_{M3}T_{M3}^2\beta_1}\right]} \tag{16}$$

References

1. Rasheedy M.S., Int. J. Mod. Phys. B, **18**, N.s 20-21 (2004)
2. May C.E. and Partridge J.A., J. Chem. Phys. **40** (1964) 1401

Various heating rates method (Sweet-Urquhart: two trap model)

Sweet and Urquhart propose a variation of the **heating rate method**, very useful when two or more peaks are so closely overlapped that they appear as only one peak. The method has derived by the **two trap model** proposed by the same Authors.

The procedure is based on the measurements of various glow curves recorded with different heating rates.

According to the two trap model, the TL intensity is given by

$$I(T)=\frac{A_m\left(n_1+n_2\right)\left(\gamma_1 n_1+\gamma_2 n_2\right)}{A_m\left(n_1+n_2\right)+A_{n1}\left(N_1-n_1\right)+A_{n2}\left(N_2-n_2\right)}$$

where

N_1, N_2 concentration of trapping centres (m^{-3}),

n_1, n_2 concentration of trapped electrons (m^{-3}),

A_{n1}, A_{n2} trapping rate constants (m^3 s^{-1}),

A_m recombination rate constant (m^3 s^{-1}),

$\gamma_1, \gamma_2 \left(= s_i \exp\left(-\frac{E_i}{kT} \right) \right)$ probability for electrons from trap to CB. \bye

Considering the temperature dependence of $A_{n1}, A_{n2}, \frac{A_{n1}}{A_m}, \frac{A_{n2}}{A_m}, s_1, s_2$ to be small, the previous equation for different heating rates β_i can be rewritten as

$$\frac{I(T_{xi})}{\exp\left(-\frac{E}{kT_{xi}} \right)} = \text{constant for all indeces } i$$

where T_{xi} is the temperature at which the area under the glow curve on the low temperature side of T_{xi} is x% of the total area, i.e. $x = \frac{A_{xi}}{A_{TOT}}$. For different glow curves i, the temperature T_{xi} is chosen in such a way the x value is always the same. The corresponding values of $I(T_{xi})$ are also found. Changing the heating rate, T_x is seen to vary and a plot of $\ln I(T_{xi})$ against $1/T_{xi}$ is a straight line whose slope is E/k from which the activation energy can be found.

According to the results obtained by Sweet and Urqhuart, this method allows to determine the activation energies of two very close trapping states: when x is taken enough small (≈ 20%) the plot gives the shallow trap energy; on the contrary, when x is large (≈ 80%), the slope gives the energy of the deeper trap.

Reference

Sweet M.A.S. and Urquhart D., J. Phys. C: Solid State Phys., **14** (1981) 773

Various heating rate method when $s = s(T)$ (Chen and Winer: first- and general-ordcr)

Chen and Winer developed the method of the **various heating rate** in the case a temperature dependence of the **frequency factor** is suspected. The temperature dependence of the frequency factor is assumed as following

$$s = s_0 T^{\alpha} \tag{1}$$

- First order kinetics

In this case of first order kinetics and for $\alpha = 3/2$, the **maximum condition** when the frequency factor is temperature dependent can be written as following for two different heating rates

$$\frac{\beta_1}{T_{M1}^{7/2}} = \frac{ks_0}{E}\left(\frac{1}{1+\frac{3}{4}\Delta_{M1}}\right)\exp\left(-\frac{E}{kT_{M1}}\right) \tag{2}$$

$$\frac{\beta_2}{T_{M2}^{7/2}} = \frac{ks_0}{E}\left(\frac{1}{1+\frac{3}{4}\Delta_{M2}}\right)\exp\left(-\frac{E}{kT_{M2}}\right) \tag{3}$$

from which, by eleiminating s_0, one obtains

$$\left(\frac{\beta_1}{\beta_2}\right)\cdot\left(\frac{T_{M2}}{T_{M1}}\right)^{7/2} = \frac{1+\frac{3}{4}\Delta_{M2}}{1+\frac{3}{4}\Delta_{M1}}\exp\left[\left(-\frac{E}{k}\right)\cdot\left(\frac{1}{T_{M1}}-\frac{1}{T_{M2}}\right)\right] \tag{4}$$

In the previous equation, the coefficient of the exponential term can be assumed equal to unity, without introducing any consistent error; Eq. (4) is then reduced to the following form

$$E = k\frac{T_{M1}T_{M2}}{T_{M1}-T_{M2}}\ln\left[\left(\frac{\beta_1}{\beta_2}\right)\cdot\left(\frac{T_{M2}}{T_{M1}}\right)^{7/2}\right] \tag{5}$$

In the general case of any value of α, the relation becomes

$$E = k\frac{T_{M1}T_{M2}}{T_{M1}-T_{M2}}\ln\left[\left(\frac{\beta_1}{\beta_2}\right)\cdot\left(\frac{T_{M2}}{T_{M1}}\right)^{\alpha+2}\right] \tag{6}$$

From Eq. (6), when α is known by some independent measurements, it directly gives the value of *E*. It must be noted that, if *E* is already known by other methods, Eq. (6) can be used to evaluate α and hence s_0 from either Eqs. (2) or (3). In conclusion, a complete knowledge of the trap level under investigation is possible.

Using several different heating rates one has to plot

$$\ln\left(\frac{\beta}{T_M^{\alpha+2}}\right) \quad \text{vs} \quad \frac{1}{T_M}$$

which should yield a straight line, from whose slope the activation energy E is found.

In the case of a general order kinetics, supposing s'' depending on the temperature according to the relation

$$s'' = s_0'' T^{\alpha}$$

by substituting the **condition at the maximum** when the frequency factor is temperature dependent:

$$\frac{s_0'' b T_M^{\alpha+2} k \exp\left(-\frac{E}{kT_M}\right)}{\beta E\left(1+\frac{\alpha}{2}\Delta_M\right)} = \left[1+\frac{s_0''(b-1)}{\beta}\int_{T_0}^{T_M} T^{\alpha}\exp(-\frac{E}{kT})dT\right]$$

in

$$I(T) = s_0'' n_0 T^{\alpha}\exp(-\frac{E}{kT})\left[1+\frac{s_0''(b-1)}{\beta}\int_{T_0}^{T} T^{\alpha}\exp\left(-\frac{E}{kT}\right)dT\right]^{\frac{-b}{b-1}}$$

we obtain

$$I_M = s_0'' n_0 T^{\alpha}\exp\left(-\frac{E}{kT_M}\right)\left[\frac{s_0'' k b T_M^{\alpha+2}\exp\left(-\frac{E}{kT_M}\right)}{\beta E\left(1+\frac{\alpha}{2}\Delta_M\right)}\right]^{-\frac{b}{b-1}} \tag{7}$$

and rearranging

$$\frac{I_M^{b-1}\cdot T_M^{2b+\alpha}}{\beta^b} = \left(\frac{1}{s_0'' n_0}\right)\cdot\left[\frac{E\left(1+\frac{\alpha}{2}\Delta_M\right)}{k\cdot b}\right]^b \exp\left(\frac{E}{kT_M}\right) \tag{8}$$

The logarithm of Eq. (8) is then

$$\ln\left[I_M^{b-1}\cdot\frac{T_M^{2b+\alpha}}{\beta^b}\right] = \frac{E}{kT_M} + \ln\left(\alpha kT_M + E\right) + const \tag{9}$$

from which the activation energy can be experimentally evaluated with the usual procedure. In fact, since $\alpha \leq 2$ and $kT_M \prec\prec E$, the left-hand side of expression (9) is a linear function of $1/T_M$, with slope equal to E/k.

Reference

Chen R. and Winer S.A., J. Appl. Phys. **41** (1970) 5227

Z

Zirconium Oxide (ZrO_2)

Zirconia has very much attracted the attention of technologists and scientists owingto its combined electrical, chemical, optical and mechanical characteristics. All these properties make this material suitable for a large variety of applications, On the other hand, a little research has been done on its luminescent properties [1-14].

In order to obtain material suitable for thermoluminescence dosimetry, Zirconium TL phosphors have been synthesised by blending zirconium oxichloride ($ZrOCl{:}8H_2O$ by Merck) and ethhyl alcoohl. This solution is stirred for 15 minutes, then heated at 250°C for 30 minutes until full evaporation of the solvent. The amorphous powder is then submitted to different thermal treatments in an oxiding atmosphere (air) in order to stabilise the trap structure. Then the powder is crushed and sieved to select grains having a size between 100 and 300 μ. Sinterd ZrO_2+PTFE pellets of 5 mm in diameter and 0.8 mm in thickeness can also obtained.

The most attractive feature of ZrO_2 phosphor is its very high intrinsic sensitivity to UV radiation. The typical glow curve of ZrO_2, after UV irradiation, exhibits one single peak at 180°C.

After beta irradiation from $^{90}Sr/^{90}Y$ source, the glow curve presents two resolved peaks at 200°C and 250°C respectively. The TL response is linear from 2 to 60 Gy; the reprocucibility, over several repeated cycles of annealing, irradiation and readout, is better than 1.8% and fading at room temperature is 3.8% in one month.

The TL emission after X-ray irradiation of low energy, typically from 15 to 60 KV, shows two peaks at about 200°C and 280°C. At 60 KV X-rays, the TL response is linear from 0.04 Gy to 1.12 Gy.

References

1. Peters T.E., Pappalardo R.G and Hunt R.B., in *Solid State Luminescence*, edited by A.H. Kitai (Chapman & Hall, London, 1993).
2. Shionoya, in *Luminescence of Solids*, edited by D.R. Vij (Plenum Press, New York, 1998)
3. Bettinali C., Ferraresso G. and Manconi J.W., J. Chem. Phys. **50** (1969) 3957
4. Dhar A., Dewerd L.A. and Stoebe T.G., Med. Phys. **3** (1976) 415
5. Iacconi P., Keller P. and Caruba R., Phys. Status Solid (a) **50** (1978) 275
6. Shan-Chou Chang and Ching-Shen Su, Nucl. Trakcs. Radiat. Meas. **20** (3) (1992) 511
7. Azorín J., Rubio J., Gutiérrez A., González P. and Rivera T., J. Thermal. Anal. **39** (1993) 1107

8. Rivera T., Azorin J., Martínez E. and García Hipólito M. Desarrollo de nuevos materiales Termoluminiscentes para Dosimetría Personal y Ambiental de la Luz Ultravioleta. IV Congreso Regional Seguridad Radiológica y Nuclear IRPA, CUBA, 1998
9. Azorín J., Rivera T., Martínez E. and García M., Radiat. Meas. **29** (1998) 315
10. Azorin J. Rivera T., Falcony C., Martínez E. and García M., Rad. Prot. Dos. **85** (1999) 315
11. Azorín, J. Rivera T., Falcony C., García M and Martínez E., 10th Inter. Cong. Inter. Rad. Prot. Ass.. Hiroshima Japan (2000)
12. Rivera T., Azorin J., Falcony C., Martinez E. and García M., Radiat. Phys. Chem. **61** (2001) 421
13. Rivera T. Estudio de las propiedades termoluminiscentes y fotoluminiscentes del ZrO_2:TR y su aplicación a la dosimetría de la radiación ionizante. Tesis de Doctorado Universidad Autonóma Metropolitana. México D.F. (2002)
14. Rivera T., Azorin J., C. Falcony, M. and Martínez E., Rad. Prot. Dos. (In press).

APPENDIX 1

Appendix b=1

b=1	
K	TL
273	1.38E-02
274	1.61E-02
275	1.88E-02
276	2.19E-02
277	2.55E-02
278	2.97E-02
279	3.45E-02
280	4.00E-02
281	4.64E-02
282	5.37E-02
283	6.21E-02
284	7.17E-02
285	8.28E-02
286	9.54E-02
287	1.10E-01
288	1.27E-01
289	1.45E-01
290	1.67E-01
291	1.92E-01
292	2.20E-01
293	2.52E-01
294	2.88E-01
295	3.29E-01
296	3.76E-01
297	4.29E-01
298	4.89E-01
299	5.57E-01
300	6.34E-01
301	7.21E-01
302	8.19E-01
303	9.30E-01
304	1.05E+00
305	1.20E+00
306	1.35E+00
307	1.53E+00
308	1.73E+00
309	1.96E+00
310	2.21E+00
311	2.49E+00
312	2.81E+00
313	3.16E+00
314	3.56E+00
315	4.00E+00
316	4.49E+00
317	5.04E+00
318	5.66E+00
319	6.34E+00
320	7.10E+00
321	7.95E+00
322	8.90E+00
323	9.94E+00
324	1.11E+01
325	1.24E+01
326	1.38E+01
327	1.54E+01
328	1.72E+01
329	1.91E+01
330	2.12E+01
331	2.36E+01
332	2.62E+01
333	2.91E+01
334	3.23E+01
335	3.58E+01
336	3.96E+01
337	4.39E+01
338	4.85E+01
339	5.36E+01
340	5.92E+01
341	6.54E+01
342	7.21E+01
343	7.94E+01
344	8.74E+01
345	9.62E+01
346	1.06E+02
347	1.16E+02
348	1.27E+02
349	1.40E+02
350	1.53E+02
351	1.68E+02
352	1.83E+02
353	2.00E+02
354	2.18E+02
355	2.38E+02
356	2.59E+02
357	2.82E+02
358	3.06E+02
359	3.32E+02
360	3.60E+02
361	3.90E+02
362	4.21E+02
363	4.54E+02
364	4.89E+02
365	5.26E+02
366	5.64E+02
367	6.04E+02
368	6.46E+02
369	6.89E+02
370	7.33E+02
371	7.78E+02
372	8.24E+02
373	8.70E+02
374	9.15E+02
375	9.60E+02
376	1.00E+03
377	1.05E+03
378	1.08E+03
379	1.12E+03
380	1.15E+03
381	1.18E+03
382	1.20E+03
383	1.22E+03
384	1.23E+03
385	1.23E+03
386	1.22E+03
387	1.20E+03
388	1.18E+03
389	1.14E+03
390	1.10E+03
391	1.05E+03
392	9.88E+02
393	9.23E+02
394	8.51E+02
395	7.75E+02
396	6.97E+02
397	6.18E+02
398	5.39E+02
399	4.63E+02
400	3.91E+02
401	3.24E+02
402	2.63E+02
403	2.09E+02
404	1.63E+02
405	1.23E+02
406	9.13E+01
407	6.57E+01
408	4.59E+01
409	3.11E+01
410	2.04E+01
411	1.29E+01
412	7.83E+00
413	4.58E+00
414	2.57E+00
415	1.37E+00
416	7.00E-01
417	3.39E-01
418	1.55E-01
419	6.72E-02
420	2.73E-02

421	1.04E-02
422	3.68E-03
423	1.21E-03
424	3.68E-04
425	1.03E-04
426	2.64E-05
427	6.15E-06
428	1.30E-06
429	2.46E-07
430	4.16E-08
431	6.26E-09
432	8.30E-10
433	9.63E-11
434	9.70E-12

435	8.42E-13
436	6.24E-14
437	3.91E-15
438	2.05E-16
439	8.94E-18
440	3.20E-19
441	9.29E-21
442	2.16E-22
443	4.00E-24
444	5.78E-26
445	6.45E-28
446	5.47E-30
447	3.49E-32
448	1.64E-34

449	5.60E-37
450	1.37E-39
451	2.34E-42
452	2.76E-45
453	2.20E-48
454	1.16E-51
455	3.95E-55
456	8.53E-59
457	1.14E-62
458	9.20E-67
459	4.37E-71
460	1.19E-75
461	1.81E-80
462	1.49E-85

463	6.47E-91
464	1.43E-96
465	1.56E-102
466	8.13E-109
467	1.95E-115
468	2.08E-122
469	9.51E-130
470	1.79E-137
471	1.33E-145
472	3.74E-154
473	3.82E-163
474	1.35E-172
475	1.59E-182
476	5.89E-193

477	6.56E-204
478	2.08E-215
479	1.79E-227
480	3.95E-240
481	2.11E-253
482	2.59E-267
483	6.85E-282
484	3.68E-297

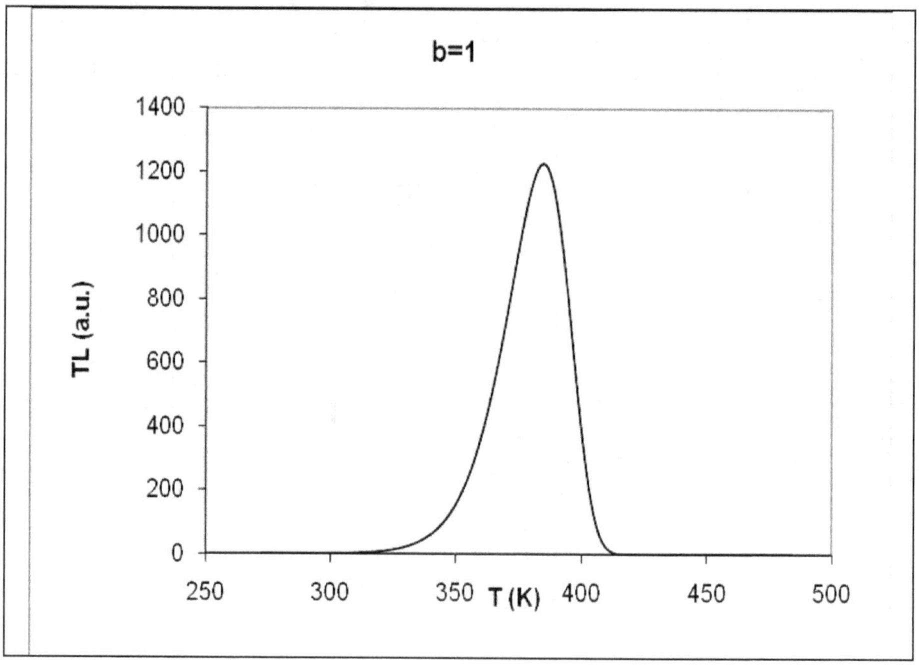

Appendix b=1.1

b=1.1	
K	TL
273	1.38E-02
274	1.61E-02
275	1.88E-02
276	2.19E-02
277	2.55E-02
278	2.97E-02
279	3.45E-02
280	4.00E-02
281	4.64E-02
282	5.37E-02
283	6.21E-02
284	7.17E-02
285	8.28E-02
286	9.54E-02
287	1.10E-01
288	1.27E-01
289	1.45E-01
290	1.67E-01
291	1.92E-01
292	2.20E-01
293	2.52E-01
294	2.88E-01
295	3.29E-01
296	3.76E-01
297	4.29E-01
298	4.89E-01
299	5.57E-01
300	6.34E-01
301	7.21E-01
302	8.19E-01
303	9.30E-01
304	1.05E+00
305	1.20E+00
306	1.35E+00
307	1.53E+00
308	1.73E+00
309	1.96E+00
310	2.21E+00
311	2.49E+00
312	2.81E+00
313	3.16E+00
314	3.56E+00
315	4.00E+00
316	4.49E+00
317	5.04E+00
318	5.66E+00
319	6.34E+00
320	7.10E+00
321	7.95E+00
322	8.89E+00
323	9.94E+00
324	1.11E+01
325	1.24E+01
326	1.38E+01
327	1.54E+01
328	1.72E+01
329	1.91E+01
330	2.12E+01
331	2.36E+01
332	2.62E+01
333	2.91E+01
334	3.23E+01
335	3.58E+01
336	3.96E+01
337	4.38E+01
338	4.85E+01
339	5.36E+01
340	5.91E+01
341	6.52E+01
342	7.19E+01
343	7.93E+01
344	8.72E+01
345	9.60E+01
346	1.05E+02
347	1.16E+02
348	1.27E+02
349	1.39E+02
350	1.52E+02
351	1.67E+02
352	1.82E+02
353	1.99E+02
354	2.17E+02
355	2.37E+02
356	2.57E+02
357	2.80E+02
358	3.04E+02
359	3.29E+02
360	3.57E+02
361	3.85E+02
362	4.16E+02
363	4.48E+02
364	4.82E+02
365	5.18E+02
366	5.55E+02
367	5.94E+02
368	6.34E+02
369	6.75E+02
370	7.17E+02
371	7.59E+02
372	8.02E+02
373	8.45E+02
374	8.88E+02
375	9.30E+02
376	9.70E+02
377	1.01E+03
378	1.04E+03
379	1.08E+03
380	1.10E+03
381	1.13E+03
382	1.15E+03
383	1.16E+03
384	1.17E+03
385	1.17E+03
386	1.16E+03
387	1.15E+03
388	1.12E+03
389	1.09E+03
390	1.06E+03
391	1.01E+03
392	9.64E+02
393	9.08E+02
394	8.48E+02
395	7.85E+02
396	7.19E+02
397	6.52E+02
398	5.85E+02
399	5.19E+02
400	4.56E+02
401	3.96E+02
402	3.39E+02
403	2.87E+02
404	2.40E+02
405	1.99E+02
406	1.62E+02
407	1.31E+02
408	1.04E+02
409	8.14E+01
410	6.30E+01
411	4.82E+01
412	3.63E+01
413	2.70E+01
414	1.98E+01
415	1.44E+01
416	1.03E+01
417	7.28E+00
418	5.08E+00
419	3.51E+00
420	2.39E+00

421	1.61E+00
422	1.08E+00
423	7.09E-01
424	4.63E-01
425	2.99E-01
426	1.92E-01
427	1.21E-01
428	7.64E-02
429	4.76E-02
430	2.95E-02
431	1.81E-02
432	1.10E-02
433	6.69E-03
434	4.03E-03
435	2.41E-03
436	1.44E-03
437	8.51E-04

438	5.02E-04
439	2.95E-04
440	1.72E-04
441	1.00E-04
442	5.83E-05
443	3.38E-05
444	1.95E-05
445	1.12E-05
446	6.47E-06
447	3.71E-06
448	2.13E-06
449	1.22E-06
450	6.95E-07
451	3.97E-07
452	2.26E-07
453	1.29E-07
454	7.36E-08

455	4.19E-08
456	2.39E-08
457	1.36E-08
458	7.75E-09
459	4.41E-09
460	2.52E-09
461	1.44E-09
462	8.19E-10
463	4.68E-10
464	2.68E-10
465	1.53E-10
466	8.76E-11
467	5.02E-11
468	2.88E-11
469	1.66E-11
470	9.53E-12
471	5.49E-12

472	3.16E-12
473	1.83E-12
474	1.06E-12
475	6.12E-13
476	3.55E-13
477	2.06E-13
478	1.20E-13
479	6.99E-14
480	4.08E-14
481	2.38E-14
482	1.40E-14
483	8.18E-15
484	4.81E-15
485	2.83E-15
486	1.67E-15
487	9.84E-16
488	5.82E-16

489	3.45E-16
490	2.05E-16
491	1.22E-16
492	7.25E-17
493	4.33E-17
494	2.58E-17
495	1.55E-17
496	9.28E-18
497	5.58E-18
498	3.36E-18
499	2.02E-18
500	1.22E-18

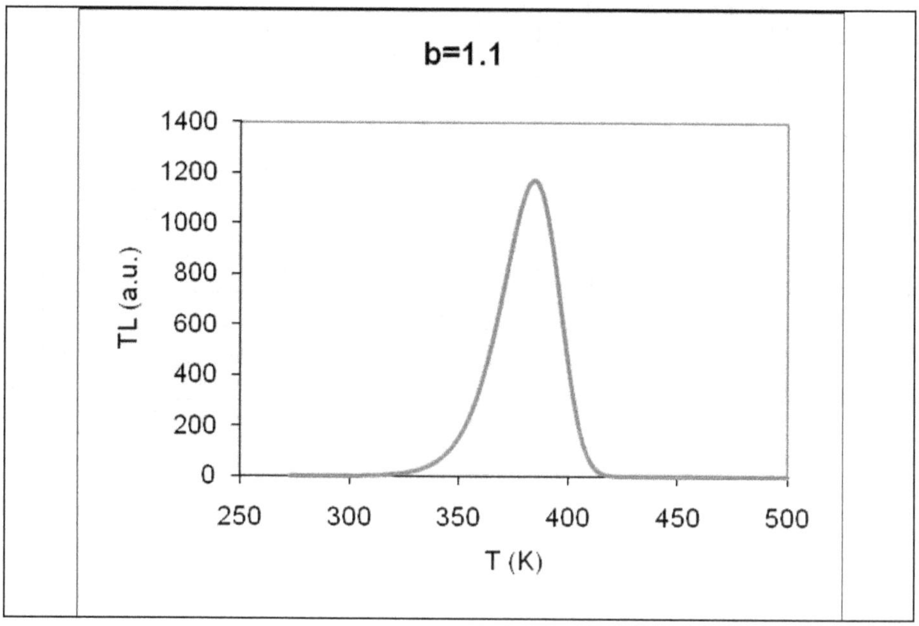

Appendix b=1.2

b=1.2									
K	TL	K	TL	K	TL	K	TL	K	TL
273	1.38E-02	301	7.21E-01	331	2.36E+01	361	3.81E+02	391	9.80E+02
274	1.61E-02	302	8.19E-01	332	2.62E+01	362	4.11E+02	392	9.37E+02
275	1.88E-02	303	9.30E-01	333	2.91E+01	363	4.43E+02	393	8.90E+02
276	2.19E-02	304	1.05E+00	334	3.22E+01	364	4.76E+02	394	8.38E+02
277	2.55E-02	305	1.20E+00	335	3.57E+01	365	5.10E+02	395	7.84E+02
278	2.97E-02	306	1.35E+00	336	3.96E+01	366	5.46E+02	396	7.28E+02
279	3.45E-02	307	1.53E+00	337	4.38E+01	367	5.83E+02	397	6.70E+02
280	4.00E-02	308	1.73E+00	338	4.84E+01	368	6.22E+02	398	6.12E+02
281	4.64E-02	309	1.96E+00	339	5.35E+01	369	6.61E+02	399	5.55E+02
282	5.37E-02	310	2.21E+00	340	5.91E+01	370	7.01E+02	400	4.99E+02
283	6.21E-02	311	2.49E+00	341	6.51E+01	371	7.41E+02	401	4.45E+02
284	7.17E-02	312	2.81E+00	342	7.18E+01	372	7.82E+02	402	3.94E+02
285	8.28E-02	313	3.16E+00	343	7.91E+01	373	8.23E+02	403	3.46E+02
286	9.54E-02	314	3.55E+00	344	8.71E+01	374	8.62E+02	404	3.01E+02
287	1.10E-01	315	4.00E+00	345	9.57E+01	375	9.01E+02	405	2.60E+02
288	1.27E-01	316	4.49E+00	346	1.05E+02	376	9.38E+02	406	2.23E+02
289	1.45E-01	317	5.04E+00	347	1.15E+02	377	9.74E+02	407	1.90E+02
290	1.67E-01	318	5.66E+00	348	1.27E+02	378	1.01E+03	408	1.60E+02
291	1.92E-01	319	6.34E+00	349	1.39E+02	379	1.04E+03	409	1.34E+02
292	2.20E-01	320	7.10E+00	350	1.52E+02	380	1.06E+03	410	1.12E+02
293	2.52E-01	321	7.95E+00	351	1.66E+02	381	1.08E+03	411	9.21E+01
294	2.88E-01	322	8.89E+00	352	1.81E+02	382	1.10E+03	412	7.56E+01
295	3.29E-01	323	9.94E+00	353	1.98E+02	383	1.11E+03	413	6.16E+01
296	3.76E-01	324	1.11E+01	354	2.16E+02	384	1.12E+03	414	4.99E+01
297	4.29E-01	325	1.24E+01	355	2.35E+02	385	1.12E+03	415	4.01E+01
298	4.89E-01	326	1.38E+01	356	2.56E+02	386	1.11E+03	416	3.21E+01
299	5.57E-01	327	1.54E+01	357	2.78E+02	387	1.10E+03	417	2.55E+01
300	6.34E-01	328	1.72E+01	358	3.01E+02	388	1.08E+03	418	2.02E+01
		329	1.91E+01	359	3.26E+02	389	1.05E+03	419	1.59E+01
		330	2.12E+01	360	3.53E+02	390	1.02E+03	420	1.25E+01

421	9.71E+00
422	7.54E+00
423	5.83E+00
424	4.49E+00
425	3.44E+00
426	2.63E+00
427	2.01E+00
428	1.53E+00
429	1.16E+00
430	8.76E-01
431	6.61E-01
432	4.98E-01
433	3.75E-01
434	2.81E-01
435	2.11E-01
436	1.58E-01
437	1.18E-01

438	8.81E-02
439	6.58E-02
440	4.91E-02
441	3.66E-02
442	2.72E-02
443	2.03E-02
444	1.51E-02
445	1.12E-02
446	8.36E-03
447	6.22E-03
448	4.63E-03
449	3.45E-03
450	2.56E-03
451	1.91E-03
452	1.42E-03
453	1.06E-03
454	7.90E-04

455	5.89E-04
456	4.39E-04
457	3.28E-04
458	2.45E-04
459	1.83E-04
460	1.37E-04
461	1.02E-04
462	7.67E-05
463	5.75E-05
464	4.31E-05
465	3.24E-05
466	2.43E-05
467	1.83E-05
468	1.38E-05
469	1.04E-05
470	7.81E-06
471	5.89E-06

472	4.45E-06
473	3.36E-06
474	2.54E-06
475	1.93E-06
476	1.46E-06
477	1.11E-06
478	8.40E-07
479	6.38E-07
480	4.86E-07
481	3.70E-07
482	2.82E-07
483	2.15E-07
484	1.64E-07
485	1.25E-07
486	9.59E-08
487	7.34E-08
488	5.63E-08

489	4.32E-08
490	3.31E-08
491	2.55E-08
492	1.96E-08
493	1.51E-08
494	1.16E-08
495	8.97E-09
496	6.93E-09
497	5.36E-09
498	4.14E-09
499	3.21E-09
500	2.49E-09

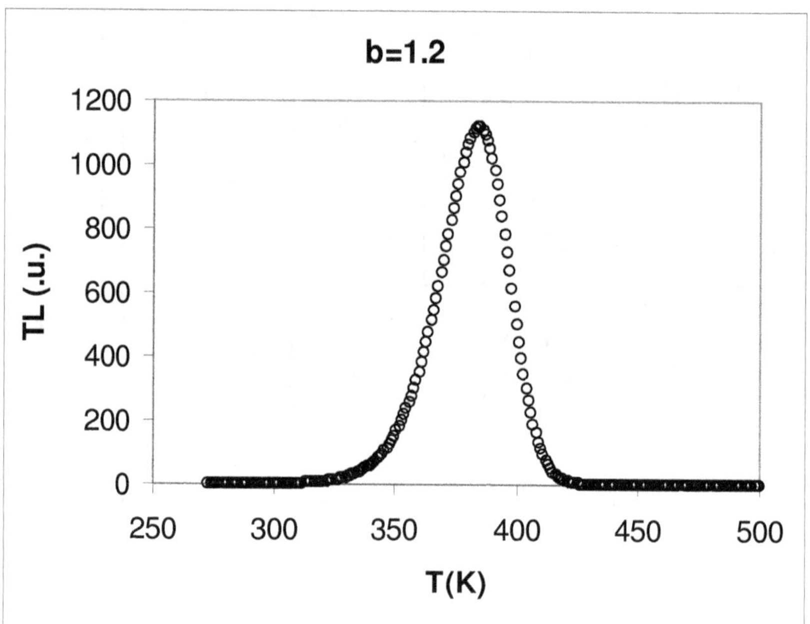

Appendix b=1.3

b=1.3	
K	TL
273	1.38E-02
274	1.61E-02
275	1.88E-02
276	2.19E-02
277	2.55E-02
278	2.97E-02
279	3.45E-02
280	4.00E-02
281	4.64E-02
282	5.37E-02
283	6.21E-02
284	7.17E-02
285	8.28E-02
286	9.54E-02
287	1.10E-01
288	1.27E-01
289	1.45E-01
290	1.67E-01
291	1.92E-01
292	2.20E-01
293	2.52E-01
294	2.88E-01
295	3.29E-01
296	3.76E-01
297	4.29E-01
298	4.89E-01
299	5.57E-01
300	6.34E-01
301	7.21E-01
302	8.19E-01
303	9.30E-01
304	1.05E+00
305	1.20E+00
306	1.35E+00
307	1.53E+00
308	1.73E+00
309	1.96E+00
310	2.21E+00
311	2.49E+00
312	2.81E+00
313	3.16E+00
314	3.55E+00
315	4.00E+00
316	4.49E+00
317	5.04E+00
318	5.65E+00
319	6.34E+00
320	7.10E+00
321	7.95E+00
322	8.89E+00
323	9.94E+00
324	1.11E+01
325	1.24E+01
326	1.38E+01
327	1.54E+01
328	1.71E+01
329	1.91E+01
330	2.12E+01
331	2.36E+01
332	2.62E+01
333	2.91E+01
334	3.22E+01
335	3.57E+01
336	3.95E+01
337	4.37E+01
338	4.84E+01
339	5.34E+01
340	5.90E+01
341	6.50E+01
342	7.17E+01
343	7.89E+01
344	8.69E+01
345	9.55E+01
346	1.05E+02
347	1.15E+02
348	1.26E+02
349	1.38E+02
350	1.51E+02
351	1.65E+02
352	1.81E+02
353	1.97E+02
354	2.15E+02
355	2.34E+02
356	2.54E+02
357	2.76E+02
358	2.99E+02
359	3.23E+02
360	3.50E+02
361	3.77E+02
362	4.06E+02
363	4.37E+02
364	4.69E+02
365	5.03E+02
366	5.37E+02
367	5.73E+02
368	6.10E+02
369	6.48E+02
370	6.86E+02
371	7.24E+02
372	7.63E+02
373	8.01E+02
374	8.38E+02
375	8.75E+02
376	9.09E+02
377	9.42E+02
378	9.71E+02
379	9.98E+02
380	1.02E+03
381	1.04E+03
382	1.06E+03
383	1.07E+03
384	1.07E+03
385	1.07E+03
386	1.06E+03
387	1.05E+03
388	1.03E+03
389	1.01E+03
390	9.81E+02
391	9.48E+02
392	9.11E+02
393	8.69E+02
394	8.25E+02
395	7.77E+02
396	7.28E+02
397	6.78E+02
398	6.28E+02
399	5.77E+02
400	5.28E+02
401	4.79E+02
402	4.33E+02
403	3.89E+02
404	3.47E+02
405	3.08E+02
406	2.72E+02
407	2.39E+02
408	2.09E+02
409	1.82E+02
410	1.57E+02
411	1.35E+02
412	1.16E+02
413	9.93E+01
414	8.46E+01
415	7.17E+01
416	6.06E+01
417	5.11E+01
418	4.29E+01
419	3.59E+01
420	3.00E+01

421	2.50E+01
422	2.08E+01
423	1.72E+01
424	1.43E+01
425	1.18E+01
426	9.73E+00
427	8.01E+00
428	6.59E+00
429	5.42E+00
430	4.45E+00
431	3.65E+00
432	2.99E+00
433	2.45E+00
434	2.00E+00
435	1.64E+00
436	1.34E+00
437	1.09E+00
438	8.93E-01
439	7.29E-01
440	5.95E-01
441	4.86E-01
442	3.97E-01
443	3.24E-01
444	2.64E-01
445	2.16E-01
446	1.76E-01
447	1.44E-01
448	1.18E-01
449	9.62E-02
450	7.86E-02
451	6.43E-02
452	5.26E-02
453	4.31E-02
454	3.53E-02
455	2.89E-02
456	2.37E-02
457	1.94E-02
458	1.59E-02
459	1.31E-02
460	1.07E-02
461	8.83E-03
462	7.26E-03
463	5.97E-03
464	4.92E-03
465	4.05E-03
466	3.34E-03
467	2.75E-03
468	2.27E-03
469	1.88E-03
470	1.55E-03
471	1.28E-03
472	1.06E-03
473	8.78E-04
474	7.28E-04
475	6.03E-04
476	5.00E-04
477	4.15E-04
478	3.45E-04
479	2.87E-04
480	2.38E-04
481	1.98E-04
482	1.65E-04
483	1.38E-04
484	1.15E-04
485	9.58E-05
486	8.00E-05
487	6.69E-05
488	5.59E-05
489	4.68E-05
490	3.92E-05
491	3.28E-05
492	2.75E-05
493	2.31E-05
494	1.94E-05
495	1.63E-05
496	1.37E-05
497	1.15E-05
498	9.67E-06
499	8.14E-06
500	6.86E-06

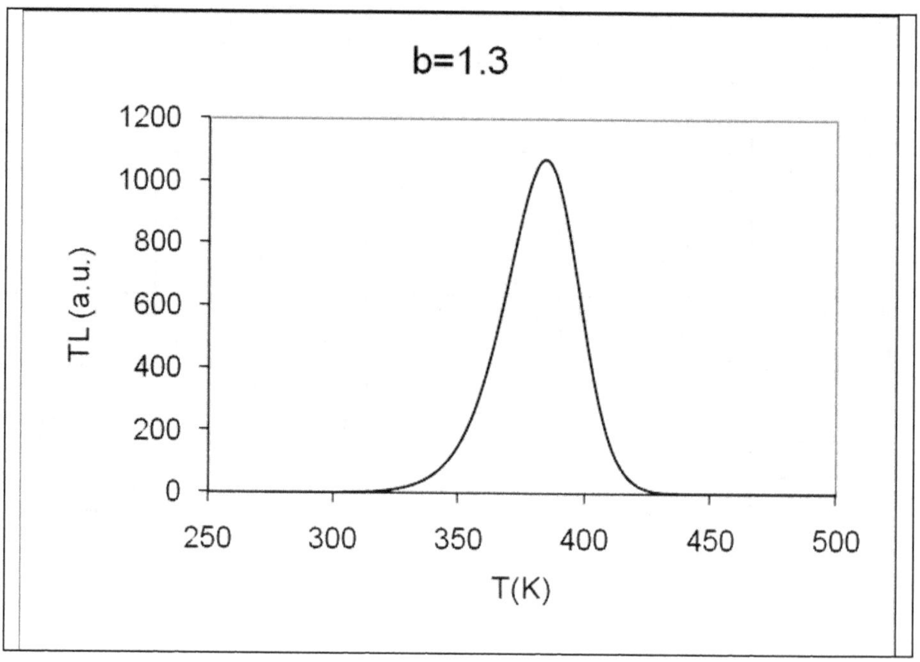

Appendix b=1.4

b=1.4	
K	TL
273	1.38E-02
274	1.61E-02
275	1.88E-02
276	2.19E-02
277	2.55E-02
278	2.97E-02
279	3.45E-02
280	4.00E-02
281	4.64E-02
282	5.37E-02
283	6.21E-02
284	7.17E-02
285	8.28E-02
286	9.54E-02
287	1.10E-01
288	1.27E-01
289	1.45E-01
290	1.67E-01
291	1.92E-01
292	2.20E-01
293	2.52E-01
294	2.88E-01
295	3.29E-01
296	3.76E-01
297	4.29E-01
298	4.89E-01
299	5.57E-01
300	6.34E-01
301	7.21E-01
302	8.19E-01
303	9.30E-01
304	1.05E+00
305	1.20E+00
306	1.35E+00
307	1.53E+00
308	1.73E+00
309	1.96E+00
310	2.21E+00
311	2.49E+00
312	2.81E+00
313	3.16E+00
314	3.55E+00
315	4.00E+00
316	4.49E+00
317	5.04E+00
318	5.65E+00
319	6.34E+00
320	7.10E+00
321	7.95E+00
322	8.89E+00
323	9.93E+00
324	1.11E+01
325	1.24E+01
326	1.38E+01
327	1.54E+01
328	1.71E+01
329	1.91E+01
330	2.12E+01
331	2.36E+01
332	2.62E+01
333	2.90E+01
334	3.22E+01
335	3.57E+01
336	3.95E+01
337	4.37E+01
338	4.83E+01
339	5.34E+01
340	5.89E+01
341	6.49E+01
342	7.16E+01
343	7.88E+01
344	8.67E+01
345	9.53E+01
346	1.05E+02
347	1.15E+02
348	1.26E+02
349	1.38E+02
350	1.51E+02
351	1.65E+02
352	1.80E+02
353	1.96E+02
354	2.13E+02
355	2.32E+02
356	2.52E+02
357	2.74E+02
358	2.96E+02
359	3.21E+02
360	3.46E+02
361	3.73E+02
362	4.02E+02
363	4.32E+02
364	4.63E+02
365	4.95E+02
366	5.29E+02
367	5.64E+02
368	5.99E+02
369	6.35E+02
370	6.72E+02
371	7.08E+02
372	7.45E+02
373	7.81E+02
374	8.16E+02
375	8.50E+02
376	8.82E+02
377	9.12E+02
378	9.39E+02
379	9.64E+02
380	9.85E+02
381	1.00E+03
382	1.02E+03
383	1.02E+03
384	1.03E+03
385	1.03E+03
386	1.02E+03
387	1.01E+03
388	9.93E+02
389	9.72E+02
390	9.47E+02
391	9.18E+02
392	8.84E+02
393	8.48E+02
394	8.09E+02
395	7.67E+02
396	7.24E+02
397	6.80E+02
398	6.35E+02
399	5.90E+02
400	5.46E+02
401	5.02E+02
402	4.60E+02
403	4.20E+02
404	3.81E+02
405	3.45E+02
406	3.11E+02
407	2.79E+02
408	2.49E+02
409	2.22E+02
410	1.97E+02
411	1.74E+02
412	1.54E+02
413	1.35E+02
414	1.19E+02
415	1.04E+02
416	9.09E+01
417	7.92E+01
418	6.89E+01
419	5.98E+01
420	5.19E+01

421	4.49E+01
422	3.88E+01
423	3.35E+01
424	2.89E+01
425	2.49E+01
426	2.14E+01
427	1.84E+01
428	1.58E+01
429	1.36E+01
430	1.16E+01
431	9.97E+00
432	8.54E+00
433	7.32E+00
434	6.27E+00
435	5.37E+00
436	4.60E+00
437	3.94E+00
438	3.37E+00
439	2.88E+00
440	2.47E+00
441	2.11E+00
442	1.81E+00
443	1.55E+00
444	1.33E+00
445	1.14E+00
446	9.75E-01
447	8.35E-01
448	7.16E-01
449	6.14E-01
450	5.27E-01
451	4.52E-01
452	3.88E-01
453	3.33E-01
454	2.86E-01
455	2.46E-01
456	2.12E-01
457	1.82E-01
458	1.57E-01
459	1.35E-01
460	1.16E-01
461	1.00E-01
462	8.63E-02
463	7.44E-02
464	6.42E-02
465	5.54E-02
466	4.79E-02
467	4.14E-02
468	3.58E-02
469	3.09E-02
470	2.68E-02
471	2.32E-02
472	2.01E-02
473	1.74E-02
474	1.51E-02
475	1.31E-02
476	1.14E-02
477	9.88E-03
478	8.58E-03
479	7.46E-03
480	6.49E-03
481	5.65E-03
482	4.92E-03
483	4.29E-03
484	3.74E-03
485	3.26E-03
486	2.84E-03
487	2.48E-03
488	2.17E-03
489	1.89E-03
490	1.66E-03
491	1.45E-03
492	1.27E-03
493	1.11E-03
494	9.72E-04
495	8.52E-04
496	7.47E-04
497	6.55E-04
498	5.75E-04
499	5.05E-04
500	4.44E-04

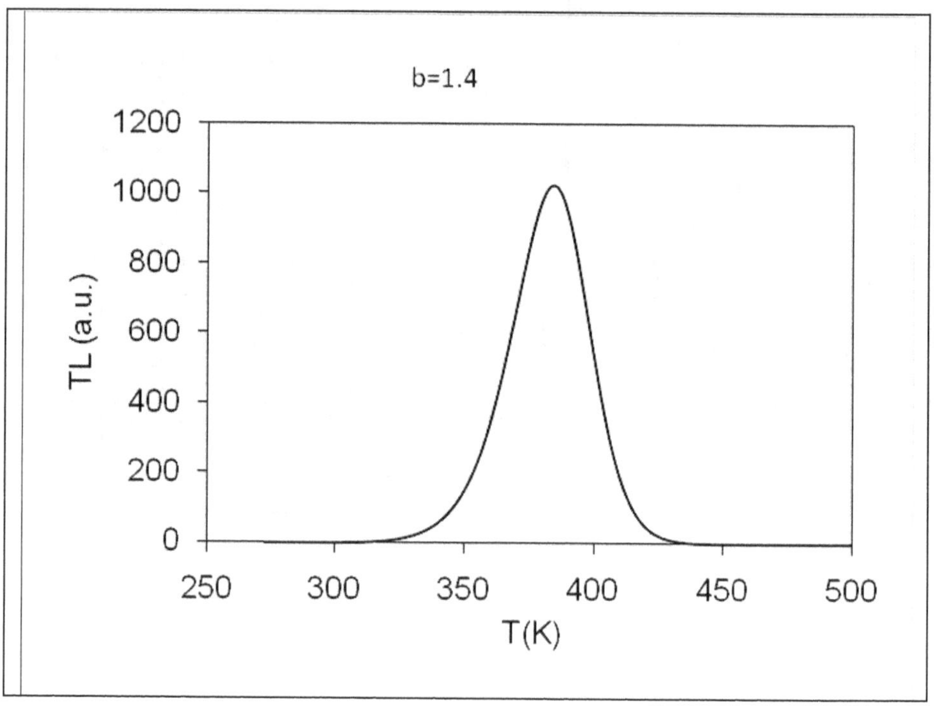

Appendix b=1.5

b=1.5

K	TL	K	TL	K	TL	K	TL	K	TL
273	1.38E-02	301	7.21E-01	331	2.36E+01	361	3.73E+02	391	9.18E+02
274	1.61E-02	302	8.19E-01	332	2.62E+01	362	4.02E+02	392	8.84E+02
275	1.88E-02	303	9.30E-01	333	2.90E+01	363	4.32E+02	393	8.48E+02
276	2.19E-02	304	1.05E+00	334	3.22E+01	364	4.63E+02	394	8.09E+02
277	2.55E-02	305	1.20E+00	335	3.57E+01	365	4.95E+02	395	7.67E+02
278	2.97E-02	306	1.35E+00	336	3.95E+01	366	5.29E+02	396	7.24E+02
279	3.45E-02	307	1.53E+00	337	4.37E+01	367	5.64E+02	397	6.80E+02
280	4.00E-02	308	1.73E+00	338	4.83E+01	368	5.99E+02	398	6.35E+02
281	4.64E-02	309	1.96E+00	339	5.34E+01	369	6.35E+02	399	5.90E+02
282	5.37E-02	310	2.21E+00	340	5.89E+01	370	6.72E+02	400	5.46E+02
283	6.21E-02	311	2.49E+00	341	6.49E+01	371	7.08E+02	401	5.02E+02
284	7.17E-02	312	2.81E+00	342	7.16E+01	372	7.45E+02	402	4.60E+02
285	8.28E-02	313	3.16E+00	343	7.88E+01	373	7.81E+02	403	4.20E+02
286	9.54E-02	314	3.55E+00	344	8.67E+01	374	8.16E+02	404	3.81E+02
287	1.10E-01	315	4.00E+00	345	9.53E+01	375	8.50E+02	405	3.45E+02
288	1.27E-01	316	4.49E+00	346	1.05E+02	376	8.82E+02	406	3.11E+02
289	1.45E-01	317	5.04E+00	347	1.15E+02	377	9.12E+02	407	2.79E+02
290	1.67E-01	318	5.65E+00	348	1.26E+02	378	9.39E+02	408	2.49E+02
291	1.92E-01	319	6.34E+00	349	1.38E+02	379	9.64E+02	409	2.22E+02
292	2.20E-01	320	7.10E+00	350	1.51E+02	380	9.85E+02	410	1.97E+02
293	2.52E-01	321	7.95E+00	351	1.65E+02	381	1.00E+03	411	1.74E+02
294	2.88E-01	322	8.89E+00	352	1.80E+02	382	1.02E+03	412	1.54E+02
295	3.29E-01	323	9.93E+00	353	1.96E+02	383	1.02E+03	413	1.35E+02
296	3.76E-01	324	1.11E+01	354	2.13E+02	384	1.03E+03	414	1.19E+02
297	4.29E-01	325	1.24E+01	355	2.32E+02	385	1.03E+03	415	1.04E+02
298	4.89E-01	326	1.38E+01	356	2.52E+02	386	1.02E+03	416	9.09E+01
299	5.57E-01	327	1.54E+01	357	2.74E+02	387	1.01E+03	417	7.92E+01
300	6.34E-01	328	1.71E+01	358	2.96E+02	388	9.93E+02	418	6.89E+01
		329	1.91E+01	359	3.21E+02	389	9.72E+02	419	5.98E+01
		330	2.12E+01	360	3.46E+02	390	9.47E+02	420	5.19E+01

T	TL	T	TL	T	TL	T	TL	T	TL
421	4.49E+01	438	3.37E+00	455	2.46E-01	472	2.01E-02	489	1.89E-03
422	3.88E+01	439	2.88E+00	456	2.12E-01	473	1.74E-02	490	1.66E-03
423	3.35E+01	440	2.47E+00	457	1.82E-01	474	1.51E-02	491	1.45E-03
424	2.89E+01	441	2.11E+00	458	1.57E-01	475	1.31E-02	492	1.27E-03
425	2.49E+01	442	1.81E+00	459	1.35E-01	476	1.14E-02	493	1.11E-03
426	2.14E+01	443	1.55E+00	460	1.16E-01	477	9.88E-03	494	9.72E-04
427	1.84E+01	444	1.33E+00	461	1.00E-01	478	8.58E-03	495	8.52E-04
428	1.58E+01	445	1.14E+00	462	8.63E-02	479	7.46E-03	496	7.47E-04
429	1.36E+01	446	9.75E-01	463	7.44E-02	480	6.49E-03	497	6.55E-04
430	1.16E+01	447	8.35E-01	464	6.42E-02	481	5.65E-03	498	5.75E-04
431	9.97E+00	448	7.16E-01	465	5.54E-02	482	4.92E-03	499	5.05E-04
432	8.54E+00	449	6.14E-01	466	4.79E-02	483	4.29E-03	500	4.44E-04
433	7.32E+00	450	5.27E-01	467	4.14E-02	484	3.74E-03		
434	6.27E+00	451	4.52E-01	468	3.58E-02	485	3.26E-03		
435	5.37E+00	452	3.88E-01	469	3.09E-02	486	2.84E-03		
436	4.60E+00	453	3.33E-01	470	2.68E-02	487	2.48E-03		
437	3.94E+00	454	2.86E-01	471	2.32E-02	488	2.17E-03		

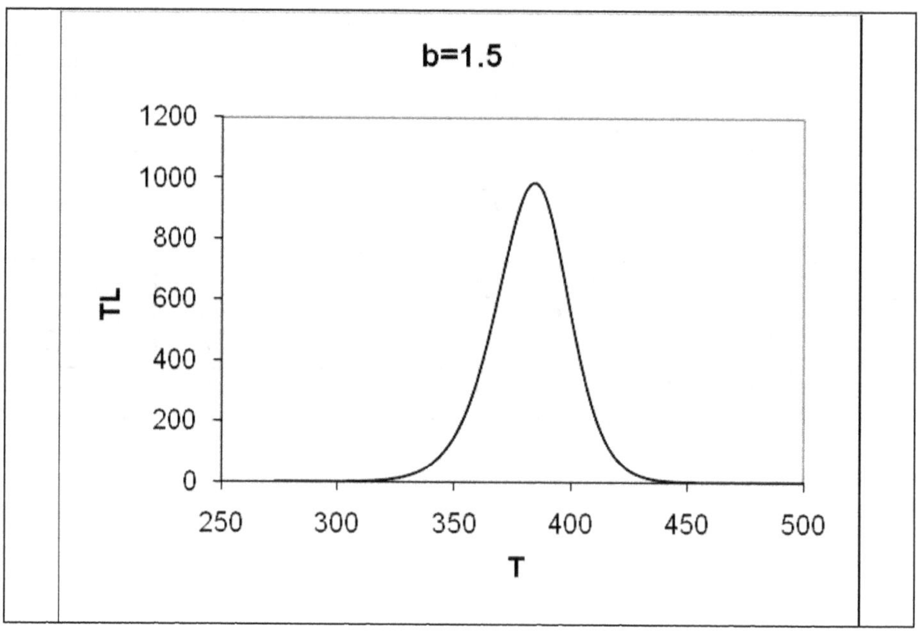

Appendix b=1.6

b=1.6

K	TL
273	1.38E-02
274	1.61E-02
275	1.88E-02
276	2.19E-02
277	2.55E-02
278	2.97E-02
279	3.45E-02
280	4.00E-02
281	4.64E-02
282	5.37E-02
283	6.21E-02
284	7.17E-02
285	8.28E-02
286	9.54E-02
287	1.10E-01
288	1.27E-01
289	1.45E-01
290	1.67E-01
291	1.92E-01
292	2.20E-01
293	2.52E-01
294	2.88E-01
295	3.29E-01
296	3.76E-01
297	4.29E-01
298	4.89E-01
299	5.57E-01
300	6.34E-01
301	7.21E-01
302	8.19E-01
303	9.30E-01
304	1.05E+00
305	1.20E+00
306	1.35E+00
307	1.53E+00
308	1.73E+00
309	1.96E+00
310	2.21E+00
311	2.49E+00
312	2.80E+00
313	3.16E+00
314	3.55E+00
315	4.00E+00
316	4.49E+00
317	5.04E+00
318	5.65E+00
319	6.34E+00
320	7.10E+00
321	7.94E+00
322	8.89E+00
323	9.93E+00
324	1.11E+01
325	1.24E+01
326	1.38E+01
327	1.54E+01
328	1.71E+01
329	1.91E+01
330	2.12E+01
331	2.35E+01
332	2.61E+01
333	2.90E+01
334	3.22E+01
335	3.56E+01
336	3.94E+01
337	4.36E+01
338	4.82E+01
339	5.32E+01
340	5.87E+01
341	6.47E+01
342	7.13E+01
343	7.85E+01
344	8.63E+01
345	9.48E+01
346	1.04E+02
347	1.14E+02
348	1.25E+02
349	1.37E+02
350	1.50E+02
351	1.63E+02
352	1.78E+02
353	1.94E+02
354	2.11E+02
355	2.29E+02
356	2.49E+02
357	2.70E+02
358	2.92E+02
359	3.15E+02
360	3.40E+02
361	3.66E+02
362	3.93E+02
363	4.21E+02
364	4.51E+02
365	4.81E+02
366	5.13E+02
367	5.45E+02
368	5.78E+02
369	6.11E+02
370	6.45E+02
371	6.78E+02
372	7.11E+02
373	7.43E+02
374	7.74E+02
375	8.04E+02
376	8.32E+02
377	8.58E+02
378	8.81E+02
379	9.02E+02
380	9.19E+02
381	9.33E+02
382	9.44E+02
383	9.50E+02
384	9.53E+02
385	9.52E+02
386	9.46E+02
387	9.37E+02
388	9.23E+02
389	9.06E+02
390	8.85E+02
391	8.61E+02
392	8.35E+02
393	8.05E+02
394	7.74E+02
395	7.41E+02
396	7.06E+02
397	6.71E+02
398	6.35E+02
399	5.98E+02
400	5.62E+02
401	5.27E+02
402	4.92E+02
403	4.58E+02
404	4.25E+02
405	3.93E+02
406	3.63E+02
407	3.35E+02
408	3.08E+02
409	2.82E+02
410	2.58E+02
411	2.36E+02
412	2.15E+02
413	1.96E+02
414	1.78E+02
415	1.62E+02
416	1.47E+02
417	1.33E+02
418	1.20E+02
419	1.09E+02
420	9.82E+01

421	8.86E+01
422	7.99E+01
423	7.20E+01
424	6.49E+01
425	5.84E+01
426	5.26E+01
427	4.73E+01
428	4.25E+01
429	3.82E+01
430	3.44E+01
431	3.09E+01
432	2.78E+01
433	2.49E+01
434	2.24E+01
435	2.01E+01
436	1.81E+01
437	1.63E+01
438	1.46E+01
439	1.31E+01
440	1.18E+01
441	1.06E+01
442	9.55E+00
443	8.58E+00
444	7.72E+00
445	6.95E+00
446	6.25E+00
447	5.63E+00
448	5.07E+00
449	4.56E+00
450	4.11E+00
451	3.70E+00
452	3.34E+00
453	3.01E+00
454	2.71E+00
455	2.45E+00
456	2.21E+00
457	1.99E+00
458	1.80E+00
459	1.63E+00
460	1.47E+00
461	1.33E+00
462	1.20E+00
463	1.09E+00
464	9.83E-01
465	8.90E-01
466	8.06E-01
467	7.30E-01
468	6.61E-01
469	5.99E-01
470	5.43E-01
471	4.93E-01
472	4.47E-01
473	4.06E-01
474	3.69E-01
475	3.35E-01
476	3.04E-01
477	2.77E-01
478	2.51E-01
479	2.29E-01
480	2.08E-01
481	1.89E-01
482	1.72E-01
483	1.57E-01
484	1.43E-01
485	1.30E-01
486	1.19E-01
487	1.08E-01
488	9.90E-02
489	9.04E-02
490	8.25E-02
491	7.53E-02
492	6.88E-02
493	6.29E-02
494	5.75E-02
495	5.26E-02
496	4.81E-02
497	4.40E-02
498	4.03E-02
499	3.69E-02
500	3.38E-02

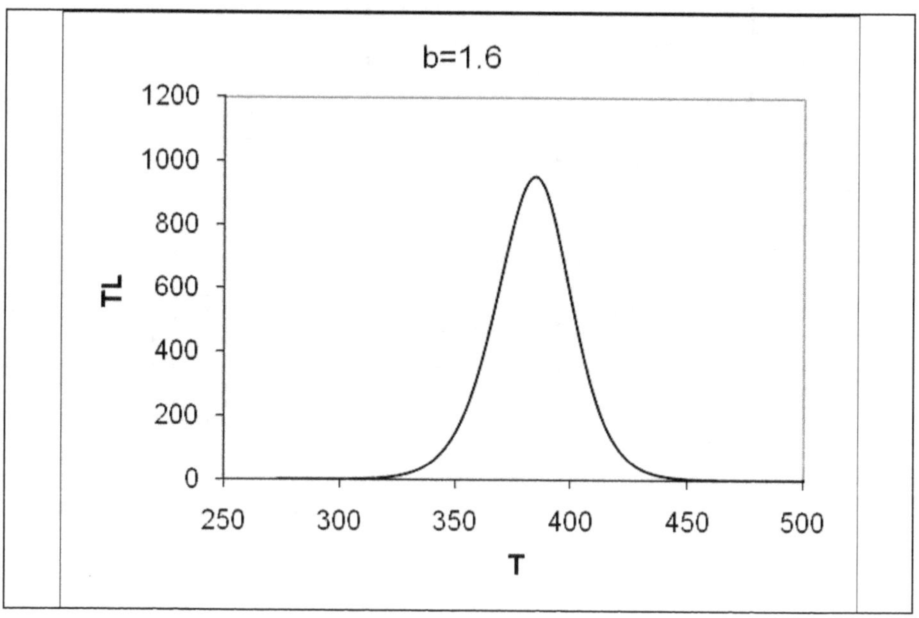

Appendix b=1.7

b=1.7	
K	TL
273	1.38E-02
274	1.61E-02
275	1.88E-02
276	2.19E-02
277	2.55E-02
278	2.97E-02
279	3.45E-02
280	4.00E-02
281	4.64E-02
282	5.37E-02
283	6.21E-02
284	7.17E-02
285	8.28E-02
286	9.54E-02
287	1.10E-01
288	1.27E-01
289	1.45E-01
290	1.67E-01
291	1.92E-01
292	2.20E-01
293	2.52E-01
294	2.88E-01
295	3.29E-01
296	3.76E-01
297	4.29E-01
298	4.89E-01
299	5.57E-01
300	6.34E-01
301	7.21E-01
302	8.19E-01
303	9.30E-01
304	1.05E+00
305	1.20E+00
306	1.35E+00
307	1.53E+00
308	1.73E+00
309	1.96E+00
310	2.21E+00
311	2.49E+00
312	2.80E+00
313	3.16E+00
314	3.55E+00
315	4.00E+00
316	4.49E+00
317	5.04E+00
318	5.65E+00
319	6.34E+00
320	7.10E+00
321	7.94E+00
322	8.88E+00
323	9.93E+00
324	1.11E+01
325	1.24E+01
326	1.38E+01
327	1.54E+01
328	1.71E+01
329	1.90E+01
330	2.12E+01
331	2.35E+01
332	2.61E+01
333	2.90E+01
334	3.21E+01
335	3.56E+01
336	3.94E+01
337	4.36E+01
338	4.81E+01
339	5.32E+01
340	5.86E+01
341	6.46E+01
342	7.12E+01
343	7.83E+01
344	8.61E+01
345	9.46E+01
346	1.04E+02
347	1.14E+02
348	1.25E+02
349	1.36E+02
350	1.49E+02
351	1.63E+02
352	1.77E+02
353	1.93E+02
354	2.10E+02
355	2.28E+02
356	2.47E+02
357	2.68E+02
358	2.89E+02
359	3.12E+02
360	3.37E+02
361	3.62E+02
362	3.89E+02
363	4.16E+02
364	4.45E+02
365	4.75E+02
366	5.05E+02
367	5.36E+02
368	5.68E+02
369	6.00E+02
370	6.32E+02
371	6.64E+02
372	6.95E+02
373	7.26E+02
374	7.55E+02
375	7.83E+02
376	8.09E+02
377	8.33E+02
378	8.55E+02
379	8.74E+02
380	8.90E+02
381	9.03E+02
382	9.12E+02
383	9.18E+02
384	9.20E+02
385	9.19E+02
386	9.13E+02
387	9.04E+02
388	8.92E+02
389	8.76E+02
390	8.57E+02
391	8.36E+02
392	8.11E+02
393	7.85E+02
394	7.56E+02
395	7.26E+02
396	6.95E+02
397	6.63E+02
398	6.30E+02
399	5.97E+02
400	5.64E+02
401	5.32E+02
402	5.00E+02
403	4.69E+02
404	4.38E+02
405	4.09E+02
406	3.81E+02
407	3.54E+02
408	3.28E+02
409	3.04E+02
410	2.81E+02
411	2.59E+02
412	2.39E+02
413	2.20E+02
414	2.02E+02
415	1.86E+02
416	1.70E+02
417	1.56E+02
418	1.43E+02
419	1.31E+02
420	1.20E+02

421	1.09E+02
422	1.00E+02
423	9.13E+01
424	8.34E+01
425	7.61E+01
426	6.94E+01
427	6.33E+01
428	5.77E+01
429	5.26E+01
430	4.79E+01
431	4.37E+01
432	3.98E+01
433	3.63E+01
434	3.31E+01
435	3.01E+01
436	2.75E+01
437	2.50E+01

438	2.28E+01
439	2.08E+01
440	1.90E+01
441	1.73E+01
442	1.58E+01
443	1.44E+01
444	1.31E+01
445	1.20E+01
446	1.09E+01
447	9.98E+00
448	9.12E+00
449	8.33E+00
450	7.61E+00
451	6.95E+00
452	6.35E+00
453	5.81E+00
454	5.31E+00

455	4.86E+00
456	4.45E+00
457	4.07E+00
458	3.73E+00
459	3.41E+00
460	3.13E+00
461	2.87E+00
462	2.63E+00
463	2.41E+00
464	2.21E+00
465	2.03E+00
466	1.86E+00
467	1.71E+00
468	1.57E+00
469	1.44E+00
470	1.32E+00
471	1.22E+00

472	1.12E+00
473	1.03E+00
474	9.47E-01
475	8.71E-01
476	8.02E-01
477	7.39E-01
478	6.80E-01
479	6.27E-01
480	5.78E-01
481	5.33E-01
482	4.91E-01
483	4.53E-01
484	4.18E-01
485	3.86E-01
486	3.56E-01
487	3.29E-01
488	3.04E-01

489	2.81E-01
490	2.60E-01
491	2.40E-01
492	2.22E-01
493	2.06E-01
494	1.90E-01
495	1.76E-01
496	1.63E-01
497	1.51E-01
498	1.40E-01
499	1.30E-01
500	1.20E-01

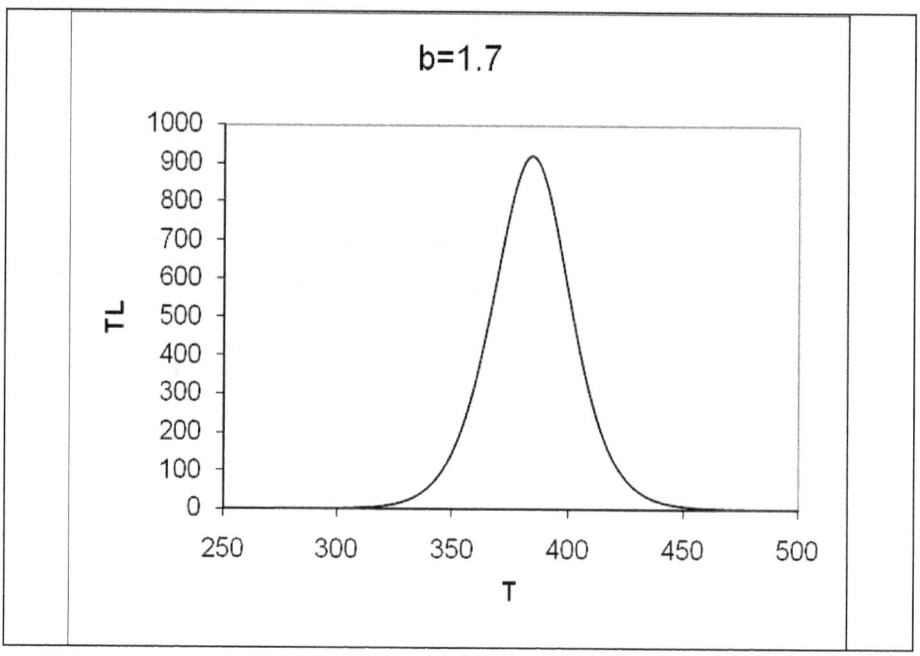

Appendix b=1.8

b=1.8	
K	TL
273	1.38E-02
274	1.61E-02
275	1.88E-02
276	2.19E-02
277	2.55E-02
278	2.97E-02
279	3.45E-02
280	4.00E-02
281	4.64E-02
282	5.37E-02
283	6.21E-02
284	7.17E-02
285	8.28E-02
286	9.54E-02
287	1.10E-01
288	1.27E-01
289	1.45E-01
290	1.67E-01
291	1.92E-01
292	2.20E-01
293	2.52E-01
294	2.88E-01
295	3.29E-01
296	3.76E-01
297	4.29E-01
298	4.89E-01
299	5.57E-01
300	6.34E-01
301	7.21E-01
302	8.19E-01
303	9.30E-01
304	1.05E+00
305	1.20E+00
306	1.35E+00
307	1.53E+00
308	1.73E+00
309	1.96E+00
310	2.21E+00
311	2.49E+00
312	2.80E+00
313	3.16E+00
314	3.55E+00
315	3.99E+00
316	4.49E+00
317	5.04E+00
318	5.65E+00
319	6.33E+00
320	7.10E+00
321	7.94E+00
322	8.88E+00
323	9.93E+00
324	1.11E+01
325	1.24E+01
326	1.38E+01
327	1.54E+01
328	1.71E+01
329	1.90E+01
330	2.12E+01
331	2.35E+01
332	2.61E+01
333	2.90E+01
334	3.21E+01
335	3.56E+01
336	3.94E+01
337	4.35E+01
338	4.81E+01
339	5.31E+01
340	5.86E+01
341	6.45E+01
342	7.11E+01
343	7.82E+01
344	8.60E+01
345	9.44E+01
346	1.04E+02
347	1.13E+02
348	1.24E+02
349	1.36E+02
350	1.48E+02
351	1.62E+02
352	1.77E+02
353	1.92E+02
354	2.09E+02
355	2.27E+02
356	2.46E+02
357	2.66E+02
358	2.87E+02
359	3.10E+02
360	3.33E+02
361	3.58E+02
362	3.84E+02
363	4.11E+02
364	4.39E+02
365	4.68E+02
366	4.98E+02
367	5.28E+02
368	5.59E+02
369	5.89E+02
370	6.20E+02
371	6.50E+02
372	6.80E+02
373	7.09E+02
374	7.37E+02
375	7.63E+02
376	7.88E+02
377	8.10E+02
378	8.30E+02
379	8.48E+02
380	8.63E+02
381	8.74E+02
382	8.83E+02
383	8.88E+02
384	8.90E+02
385	8.88E+02
386	8.83E+02
387	8.75E+02
388	8.63E+02
389	8.48E+02
390	8.31E+02
391	8.11E+02
392	7.89E+02
393	7.65E+02
394	7.39E+02
395	7.11E+02
396	6.83E+02
397	6.54E+02
398	6.24E+02
399	5.94E+02
400	5.64E+02
401	5.34E+02
402	5.05E+02
403	4.76E+02
404	4.48E+02
405	4.20E+02
406	3.94E+02
407	3.69E+02
408	3.44E+02
409	3.21E+02
410	2.99E+02
411	2.78E+02
412	2.59E+02
413	2.40E+02
414	2.23E+02
415	2.06E+02
416	1.91E+02
417	1.77E+02
418	1.63E+02
419	1.51E+02
420	1.40E+02

421	1.29E+02	438	3.21E+01	455	8.19E+00	472	2.24E+00	489	6.64E-01
422	1.19E+02	439	2.96E+01	456	7.57E+00	473	2.08E+00	490	6.19E-01
423	1.10E+02	440	2.72E+01	457	7.00E+00	474	1.94E+00	491	5.78E-01
424	1.01E+02	441	2.51E+01	458	6.48E+00	475	1.80E+00	492	5.40E-01
425	9.32E+01	442	2.31E+01	459	6.00E+00	476	1.67E+00	493	5.04E-01
426	8.59E+01	443	2.13E+01	460	5.55E+00	477	1.56E+00	494	4.71E-01
427	7.91E+01	444	1.97E+01	461	5.14E+00	478	1.45E+00	495	4.40E-01
428	7.29E+01	445	1.82E+01	462	4.76E+00	479	1.35E+00	496	4.11E-01
429	6.72E+01	446	1.67E+01	463	4.41E+00	480	1.25E+00	497	3.84E-01
430	6.19E+01	447	1.55E+01	464	4.09E+00	481	1.17E+00	498	3.59E-01
431	5.70E+01	448	1.43E+01	465	3.79E+00	482	1.09E+00	499	3.36E-01
432	5.25E+01	449	1.32E+01	466	3.51E+00	483	1.01E+00	500	3.14E-01
433	4.83E+01	450	1.22E+01	467	3.26E+00	484	9.42E-01		
434	4.45E+01	451	1.12E+01	468	3.02E+00	485	8.78E-01		
435	4.10E+01	452	1.04E+01	469	2.80E+00	486	8.18E-01		
436	3.78E+01	453	9.58E+00	470	2.60E+00	487	7.63E-01		
437	3.48E+01	454	8.86E+00	471	2.41E+00	488	7.12E-01		

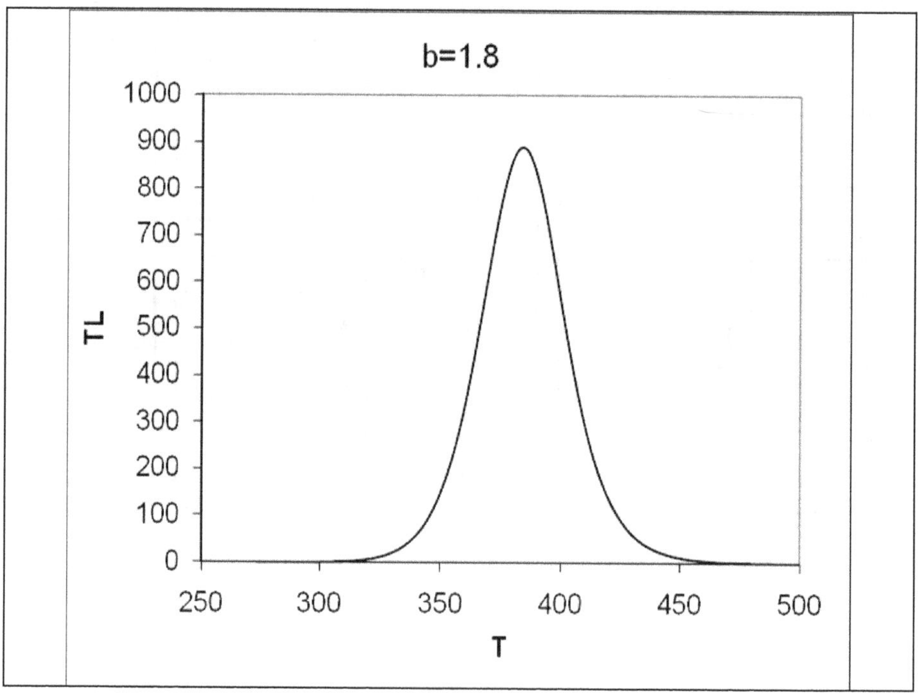

Appendix b=1.9

b=1.9	
K	**TL**
273	1.38E-02
274	1.61E-02
275	1.88E-02
276	2.19E-02
277	2.55E-02
278	2.97E-02
279	3.45E-02
280	4.00E-02
281	4.64E-02
282	5.37E-02
283	6.21E-02
284	7.17E-02
285	8.28E-02
286	9.54E-02
287	1.10E-01
288	1.27E-01
289	1.45E-01
290	1.67E-01
291	1.92E-01
292	2.20E-01
293	2.52E-01
294	2.88E-01
295	3.29E-01
296	3.76E-01
297	4.29E-01
298	4.89E-01
299	5.57E-01
300	6.34E-01
301	7.21E-01
302	8.19E-01
303	9.30E-01
304	1.05E+00
305	1.20E+00
306	1.35E+00
307	1.53E+00
308	1.73E+00
309	1.96E+00
310	2.21E+00
311	2.49E+00
312	2.80E+00
313	3.16E+00
314	3.55E+00
315	3.99E+00
316	4.49E+00
317	5.04E+00
318	5.65E+00
319	6.33E+00
320	7.09E+00
321	7.94E+00
322	8.88E+00
323	9.92E+00
324	1.11E+01
325	1.24E+01
326	1.38E+01
327	1.54E+01
328	1.71E+01
329	1.90E+01
330	2.12E+01
331	2.35E+01
332	2.61E+01
333	2.89E+01
334	3.21E+01
335	3.55E+01
336	3.93E+01
337	4.35E+01
338	4.80E+01
339	5.30E+01
340	5.85E+01
341	6.44E+01
342	7.10E+01
343	7.81E+01
344	8.58E+01
345	9.42E+01
346	1.03E+02
347	1.13E+02
348	1.24E+02
349	1.35E+02
350	1.48E+02
351	1.61E+02
352	1.76E+02
353	1.91E+02
354	2.08E+02
355	2.25E+02
356	2.44E+02
357	2.64E+02
358	2.85E+02
359	3.07E+02
360	3.30E+02
361	3.55E+02
362	3.80E+02
363	4.07E+02
364	4.34E+02
365	4.62E+02
366	4.91E+02
367	5.20E+02
368	5.49E+02
369	5.79E+02
370	6.08E+02
371	6.37E+02
372	6.66E+02
373	6.93E+02
374	7.19E+02
375	7.44E+02
376	7.67E+02
377	7.88E+02
378	8.07E+02
379	8.24E+02
380	8.37E+02
381	8.48E+02
382	8.55E+02
383	8.60E+02
384	8.61E+02
385	8.60E+02
386	8.55E+02
387	8.47E+02
388	8.36E+02
389	8.23E+02
390	8.07E+02
391	7.88E+02
392	7.68E+02
393	7.45E+02
394	7.22E+02
395	6.97E+02
396	6.71E+02
397	6.44E+02
398	6.16E+02
399	5.89E+02
400	5.61E+02
401	5.34E+02
402	5.07E+02
403	4.80E+02
404	4.54E+02
405	4.28E+02
406	4.04E+02
407	3.80E+02
408	3.57E+02
409	3.35E+02
410	3.14E+02
411	2.94E+02
412	2.75E+02
413	2.57E+02
414	2.40E+02
415	2.24E+02
416	2.09E+02
417	1.95E+02
418	1.81E+02
419	1.69E+02
420	1.57E+02

421	1.46E+02
422	1.36E+02
423	1.27E+02
424	1.18E+02
425	1.09E+02
426	1.02E+02
427	9.43E+01
428	8.76E+01
429	8.14E+01
430	7.56E+01
431	7.02E+01
432	6.52E+01
433	6.06E+01
434	5.63E+01
435	5.23E+01
436	4.85E+01
437	4.51E+01
438	4.19E+01
439	3.89E+01
440	3.62E+01
441	3.36E+01
442	3.13E+01
443	2.91E+01
444	2.70E+01
445	2.52E+01
446	2.34E+01
447	2.18E+01
448	2.03E+01
449	1.89E+01
450	1.76E+01
451	1.64E+01
452	1.52E+01
453	1.42E+01
454	1.32E+01
455	1.23E+01
456	1.15E+01
457	1.07E+01
458	9.99E+00
459	9.32E+00
460	8.70E+00
461	8.12E+00
462	7.58E+00
463	7.08E+00
464	6.61E+00
465	6.18E+00
466	5.77E+00
467	5.40E+00
468	5.05E+00
469	4.72E+00
470	4.41E+00
471	4.13E+00
472	3.87E+00
473	3.62E+00
474	3.39E+00
475	3.17E+00
476	2.97E+00
477	2.78E+00
478	2.61E+00
479	2.45E+00
480	2.29E+00
481	2.15E+00
482	2.02E+00
483	1.89E+00
484	1.78E+00
485	1.67E+00
486	1.57E+00
487	1.47E+00
488	1.38E+00
489	1.30E+00
490	1.22E+00
491	1.15E+00
492	1.08E+00
493	1.02E+00
494	9.55E-01
495	8.99E-01
496	8.46E-01
497	7.96E-01
498	7.50E-01
499	7.06E-01
500	6.65E-01

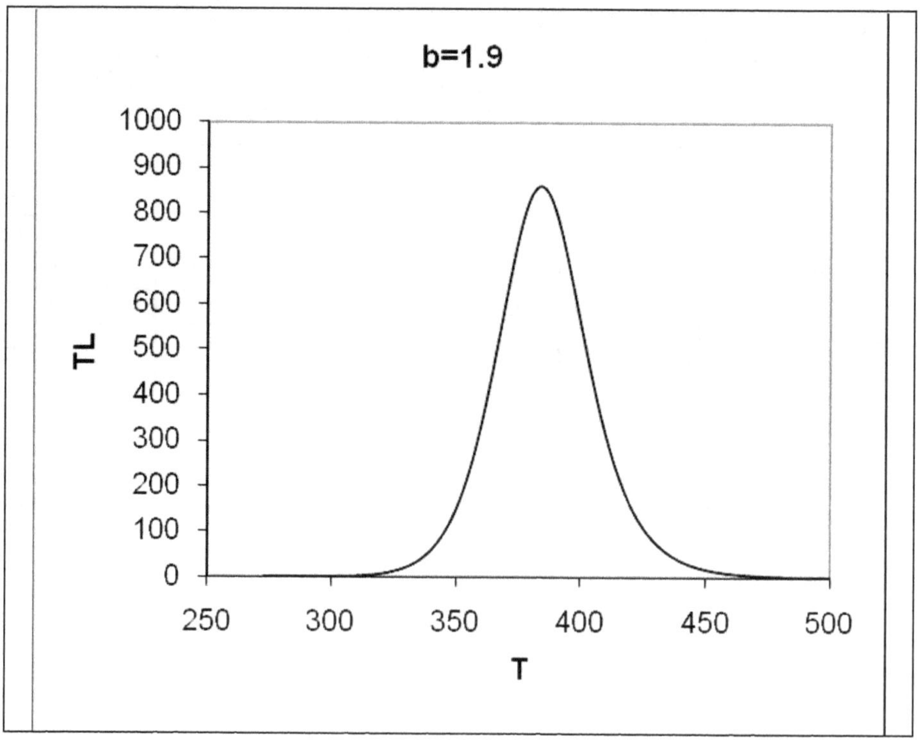

Appendix b=2

b=2	
K	TL
273	1.38E-02
274	1.61E-02
275	1.88E-02
276	2.19E-02
277	2.55E-02
278	2.97E-02
279	3.45E-02
280	4.00E-02
281	4.64E-02
282	5.37E-02
283	6.21E-02
284	7.17E-02
285	8.28E-02
286	9.54E-02
287	1.10E-01
288	1.27E-01
289	1.45E-01
290	1.67E-01
291	1.92E-01
292	2.20E-01
293	2.52E-01
294	2.88E-01
295	3.29E-01
296	3.76E-01
297	4.29E-01
298	4.89E-01
299	5.57E-01
300	6.34E-01
301	7.21E-01
302	8.19E-01
303	9.30E-01
304	1.05E+00
305	1.19E+00
306	1.35E+00
307	1.53E+00
308	1.73E+00
309	1.95E+00
310	2.21E+00
311	2.49E+00
312	2.80E+00
313	3.16E+00
314	3.55E+00
315	3.99E+00
316	4.49E+00
317	5.04E+00
318	5.65E+00
319	6.33E+00
320	7.09E+00
321	7.94E+00
322	8.88E+00
323	9.92E+00
324	1.11E+01
325	1.24E+01
326	1.38E+01
327	1.54E+01
328	1.71E+01
329	1.90E+01
330	2.11E+01
331	2.35E+01
332	2.61E+01
333	2.89E+01
334	3.21E+01
335	3.55E+01
336	3.93E+01
337	4.34E+01
338	4.80E+01
339	5.30E+01
340	5.84E+01
341	6.43E+01
342	7.08E+01
343	7.79E+01
344	8.56E+01
345	9.40E+01
346	1.03E+02
347	1.13E+02
348	1.23E+02
349	1.35E+02
350	1.47E+02
351	1.61E+02
352	1.75E+02
353	1.90E+02
354	2.07E+02
355	2.24E+02
356	2.42E+02
357	2.62E+02
358	2.83E+02
359	3.05E+02
360	3.27E+02
361	3.51E+02
362	3.76E+02
363	4.02E+02
364	4.29E+02
365	4.56E+02
366	4.84E+02
367	5.12E+02
368	5.40E+02
369	5.69E+02
370	5.97E+02
371	6.25E+02
372	6.52E+02
373	6.78E+02
374	7.03E+02
375	7.26E+02
376	7.48E+02
377	7.68E+02
378	7.85E+02
379	8.01E+02
380	8.13E+02
381	8.23E+02
382	8.30E+02
383	8.34E+02
384	8.35E+02
385	8.33E+02
386	8.29E+02
387	8.21E+02
388	8.11E+02
389	7.98E+02
390	7.83E+02
391	7.66E+02
392	7.48E+02
393	7.27E+02
394	7.05E+02
395	6.82E+02
396	6.58E+02
397	6.33E+02
398	6.08E+02
399	5.83E+02
400	5.57E+02
401	5.32E+02
402	5.07E+02
403	4.82E+02
404	4.57E+02
405	4.34E+02
406	4.10E+02
407	3.88E+02
408	3.67E+02
409	3.46E+02
410	3.26E+02
411	3.07E+02
412	2.89E+02
413	2.71E+02
414	2.55E+02
415	2.39E+02
416	2.24E+02
417	2.10E+02
418	1.97E+02
419	1.85E+02
420	1.73E+02

421	1.62E+02	438	5.19E+01	455	1.71E+01	472	5.98E+00	489	2.23E+00
422	1.52E+02	439	4.86E+01	456	1.61E+01	473	5.63E+00	490	2.11E+00
423	1.42E+02	440	4.55E+01	457	1.51E+01	474	5.31E+00	491	1.99E+00
424	1.33E+02	441	4.25E+01	458	1.42E+01	475	5.00E+00	492	1.88E+00
425	1.24E+02	442	3.98E+01	459	1.33E+01	476	4.71E+00	493	1.78E+00
426	1.16E+02	443	3.73E+01	460	1.25E+01	477	4.44E+00	494	1.68E+00
427	1.09E+02	444	3.49E+01	461	1.17E+01	478	4.19E+00	495	1.59E+00
428	1.02E+02	445	3.27E+01	462	1.10E+01	479	3.95E+00	496	1.51E+00
429	9.49E+01	446	3.06E+01	463	1.04E+01	480	3.73E+00	497	1.43E+00
430	8.88E+01	447	2.87E+01	464	9.73E+00	481	3.52E+00	498	1.35E+00
431	8.30E+01	448	2.69E+01	465	9.15E+00	482	3.32E+00	499	1.28E+00
432	7.76E+01	449	2.52E+01	466	8.61E+00	483	3.13E+00	500	1.21E+00
433	7.26E+01	450	2.36E+01	467	8.09E+00	484	2.96E+00		
434	6.79E+01	451	2.21E+01	468	7.62E+00	485	2.79E+00		
435	6.35E+01	452	2.07E+01	469	7.17E+00	486	2.64E+00		
436	5.94E+01	453	1.94E+01	470	6.75E+00	487	2.49E+00		
437	5.55E+01	454	1.82E+01	471	6.35E+00	488	2.36E+00		

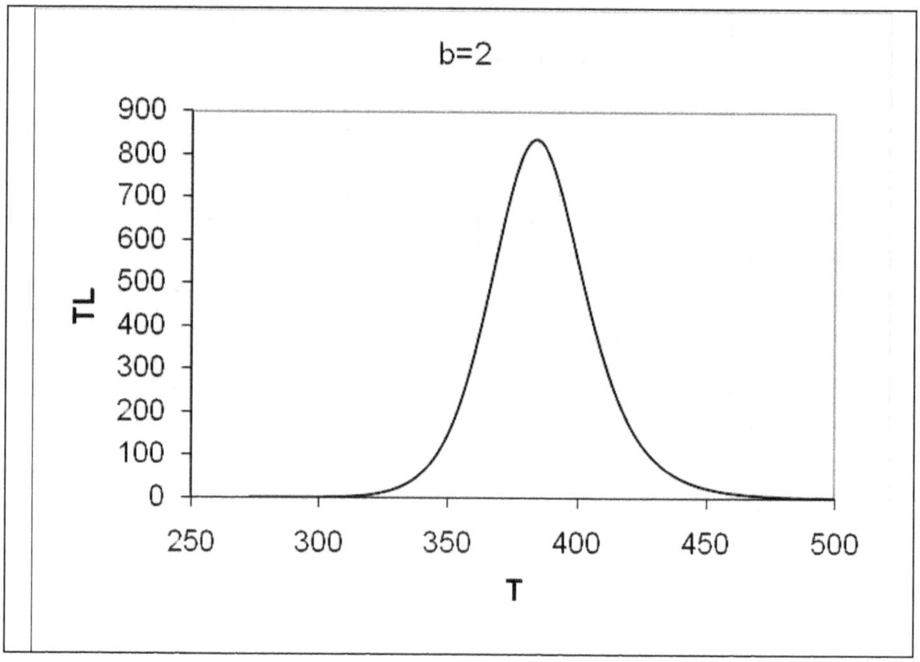

APPENDIX 2

Plots of the kinetics parameters as a function of the kinetics order

b	Tmax	ω	τ	δ	μ
1	384.58	29.825	17.335	12.49	0.418776
1.1	384.51	31.235	17.72	13.515	0.432688
1.2	384.435	32.595	18.07	14.525	0.44562
1.3	384.365	33.9	18.4	15.5	0.457227
1.4	384.295	35.17	18.71	16.46	0.468013
1.5	384.225	36.4	19.005	17.395	0.477885
1.6	384.155	37.605	19.285	18.32	0.487169
1.7	384.085	38.77	19.545	19.225	0.495873
1.8	384.015	39.91	19.795	20.115	0.504009
1.9	383.95	41.025	20.04	20.985	0.511517
2	383.88	42.11	20.265	21.845	0.51876

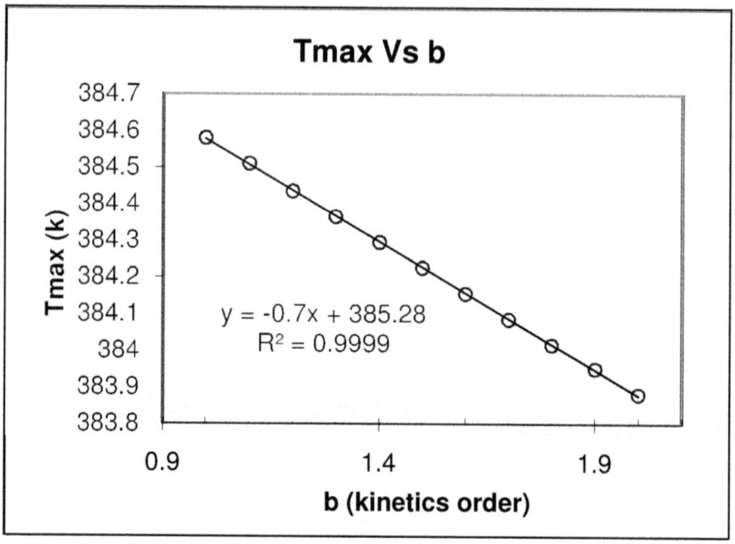

Tmax Vs b
384.7
384.6
384.5
384.4
384.3
384.2
384.1
384
383.9
383.8
Tmax (k)
y = -0.7x + 385.28
R² = 0.9999
0.9
1.4
1.9
b (kinetics order)

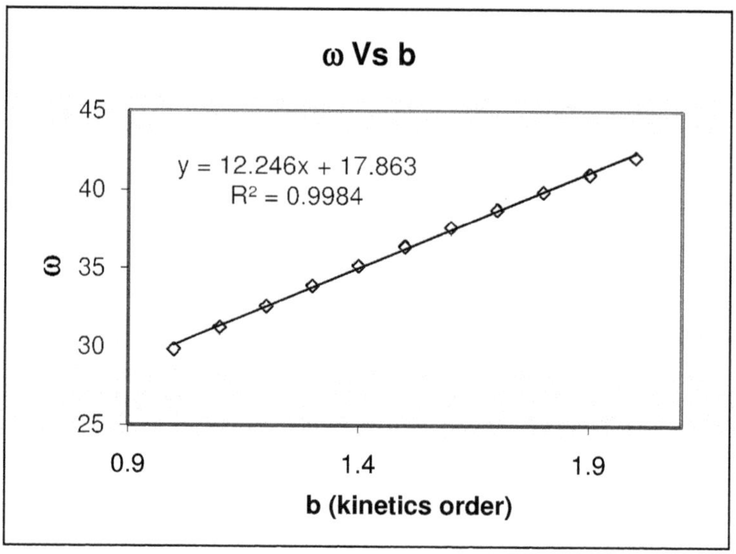

ω Vs b
45
40
35
30
25
ω
y = 12.246x + 17.863
R² = 0.9984
0.9
1.4
1.9
b (kinetics order)

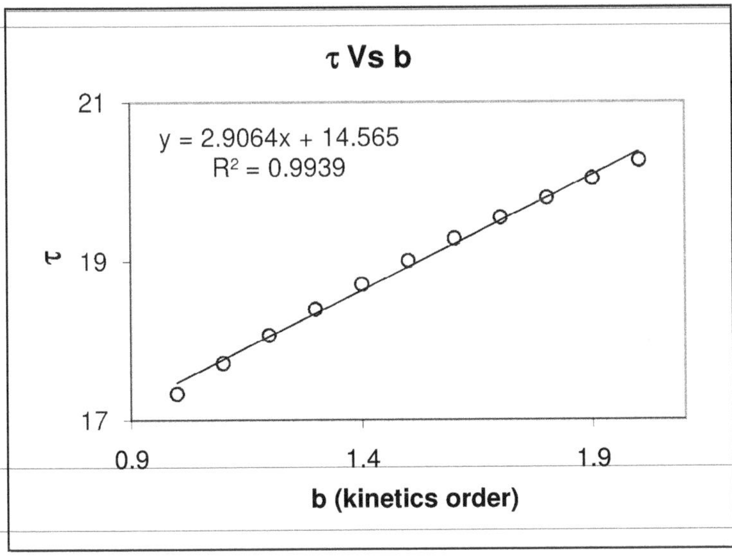
τ Vs b
y = 2.9064x + 14.565
R² = 0.9939
τ
21
19
17
0.9
1.4
1.9
b (kinetics order)

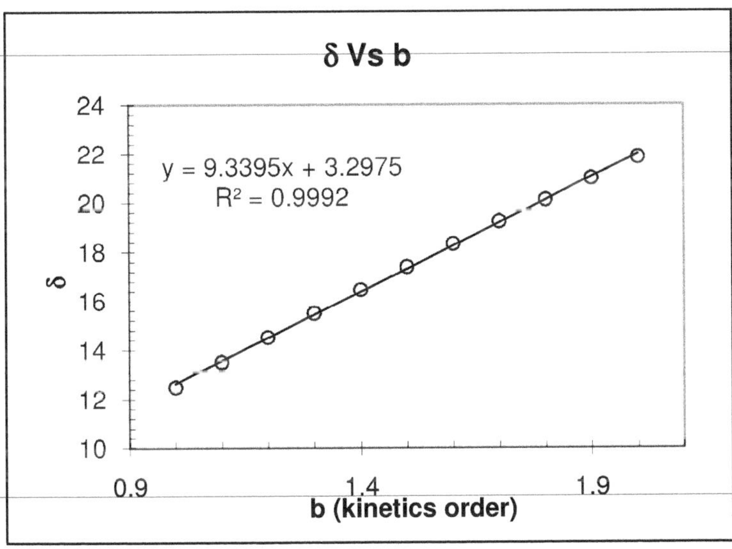
δ Vs b
y = 9.3395x + 3.2975
R² = 0.9992
δ
24
22
20
18
16
14
12
10
0.9
1.4
1.9
b (kinetics order)

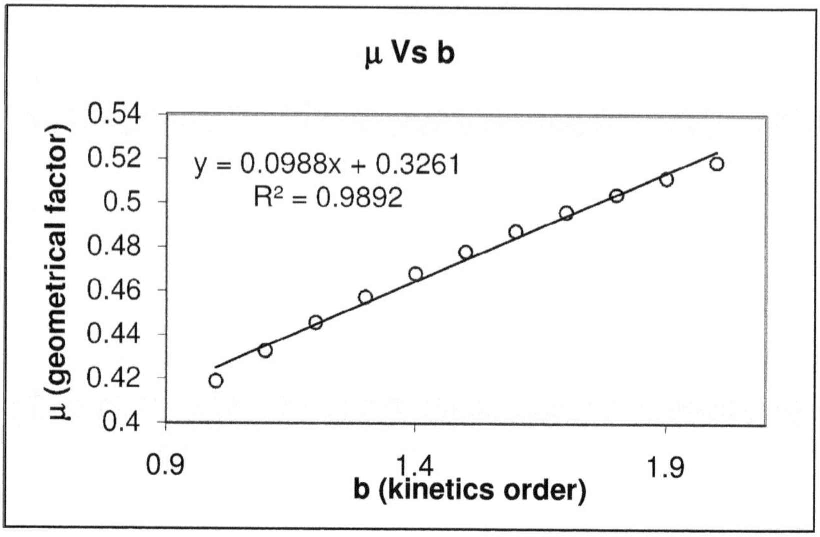
μ Vs b
y = 0.0988x + 0.3261
R² = 0.9892
μ (geometrical factor)
0.54
0.52
0.5
0.48
0.46
0.44
0.42
0.4
0.9
1.4
1.9
b (kinetics order)

APPENDIX 3

The OTOR Model

The simplest model in thermoluminescence consists of two energy levels: the electron traps and the recombination center (RC) shown in the Fig. 1. This is also known as the One-Trap-One-Recombination-Center (OTOR) model.

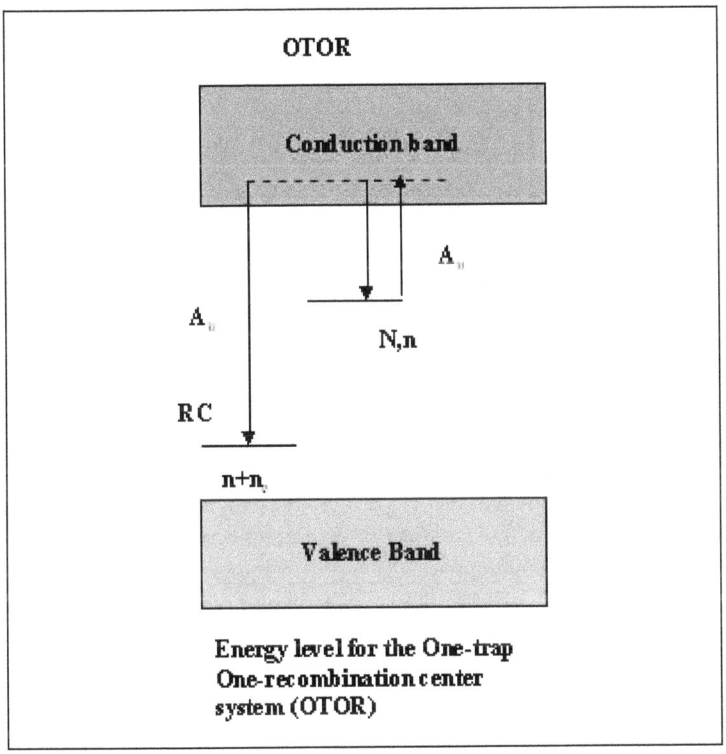

Fig. 1. The OTOR model.

A special case of the OTOR model leads to Thermoluminescence curves described by what is known as FIRST ORDER KINETICS, is discussed below.

The symbols used in the figures have the following meaning:

- N = total concentration of the electron traps in the crystal (cm^{-3}).
- n = concentration of the trapped electrons in the crystal traps (cm^{-3}).
- n_c = concentration of the free carriers in the conduction band CB (cm^{-3}).
- h = concentration of recombination centers (holes in centers) (cm^{-3}).
- E = activation energy of the electron traps (eV).

- s = frequency factor of the electron trap (s^{-1}). This quantity depends on the frequency of the number of hits of an electron in the trap. The expected maximum value of s should be similar to the lattice vibrational frequency (Debye frequency), i.e. 10^{12}–10^{14} s^{-1}. A possible range for s should be from 10^5 to 10^{13} s^{-1}.
- A_n = capture coefficient of the traps, also called retrapping probability ($cm^3 s^{-1}$),
- A_h = capture coefficient of the recombination center RC, also called recombination probability (cm^3 s^{-1}),
- I = TL emission.

The differential equations governing the traffic of electrons between the trap level, the recombination center and the conduction band are the following:

$$\frac{dn}{dt} = -ns\exp\left(-\frac{E}{kT}\right) + n_C(N-n)A_n \tag{1}$$

$$\frac{dn_C}{dt} = ns\exp\left(-\frac{E}{kT}\right) - n_C(N-n)A_n - n_C hA_h \tag{2}$$

$$I = -\frac{dh}{dt} = n_C hA_h \tag{3}$$

The first equation is related to the traffic of electrons in and out of the electron trap T; the electrons can leave the traps via thermal excitation, which is described mathematically by the term [$ns\exp(-E/kT)$]. The electrons can also be retrapped in the trap: this event is described mathematically by the retrapping term [$n_c(N-n)A_n$].

Eq. (2) describes the traffic of electrons in and out of the conduction band (CB); the electrons in the conduction band can be trapped in the recombination center RC, an event described mathematically by the term [$n_c hA_h$], or retrapped in N as given by the retrapping term [$n_c(N-n)A_n$]. The quantity h represents the total concentration of filled traps in the system at any moment. Eq. (2) is obtained by the charge neutrality condition:

$$\frac{dn_C}{dt} = \frac{dh}{dt} - \frac{dn}{dt} \tag{4}$$

The third equation gives finally the observed TL emission, related to a single peak in the glow curve, which is proportional to the amount of the emitted light.

Unfortunately the previous differential equations can not be solved in easily way and some simplified assumptions have to be made.

It is possible to arrive at an exact solution of the equations describing the OTOR model, by imposing the simplifying approximation called the quasiequilibrium condition:

$$\left|\frac{dn_C}{dt}\right| < \left|\frac{dn}{dt}\right|, \left|\frac{dh}{dt}\right| \tag{5}$$

which means that the number of free electrons in the conduction band is quasistationary.

Furthermore, if the initial concentration of the free electrons is assumed to be very small, i.e. $n_C << n$, then $\left|\frac{dn_C}{dt}\right| << \left|\frac{dn}{dt}\right|$ which means that the free charges do not accumulate in the conduction band, that is to say $dn_C = 0$.

In this case Eq. (2) transforms in

$$n_C = \frac{sn\exp(-\frac{E}{kT})}{nA_h + (N-n)A_n} \tag{6}$$

and the TL intensity is given by

$$I = -\frac{dh}{dt} = \frac{hA_h sn\exp\left(-\frac{E}{kT}\right)}{hA_h + (N-n)A_n} \tag{7}$$

It is now possible to modify the previous equation introducing the retrapping-recombination cross section ratio, given by

$$\sigma = \frac{A_n}{A_h} \tag{8}$$

Then, Equation (7) becomes

$$I = -\frac{dh}{dt} = ns\exp\left(-\frac{E}{kT}\right)\left[1 - \frac{\sigma(N-n)}{\sigma(N-n)+h}\right] \tag{9}$$

Equation (9) gives the general OTOR equation for the TL intensity. The term preceding the square brackets gives the number of electrons thermally released to the conduction band per unit of time. The term in the square brackets is the fraction of the electrons in the conduction band which undergoing recombination.

Using further assumptions it is possible to obtain the so called first and second order kinetics equations as it is shown in the following.

First Order Kinetics

The first order kinetics is the case when the probability of retrapping is negligible compared to the probability of recombination. The condition of a negligible retrapping is given by

$$hA_h >> (N-n)A_n$$

or by

$$\sigma = 0$$

The TL intensity given by Eq. (9) becomes then

$$I = -\frac{dh}{dt} = ns\exp\left(-\frac{E}{kT}\right) \tag{10}$$

From the charge neutrality expression, Eq. (4), and the condition $dn_C = 0$, we obtain

$$\frac{dh}{dt} = \frac{dn}{dt}$$

and then

$$h = n + \text{constant}$$

so that Eq. (10) can be rewritten as

$$I = -\frac{dn}{dt} = ns\exp\left(-\frac{E}{kT}\right) \tag{11}$$

which is the well-known Randall and Wilkins equation for the first kinetics order. Here *dn/dt* represents the rate of change of the electron concentration n(t) as the sample is heated during the thermoluminescence measurement. The electrons leave the traps via thermal excitation, which is described mathematically by the term [*nsexp*(-*E/kT*)]. The released electrons may be retrapped in the trap, but the probability for retrapping is assumed to be much smaller than the probability of recombination at the RC.

During a thermoluminescent measurement, the temperature T is changed linearly with the time, according to

$$T(t) = T_0 + \beta \cdot t \tag{12}$$

where T_0 is the starting temperature, usually the room temperature, and β is the heating rate expressed in °C/s.

Under this experimental condition, the above differential equation (11) can be solved to yield the following solution:

$$I(T) = n_0 s \exp\left(-\frac{E}{kT}\right) \exp\left[-\frac{s}{\beta}\int_{T_0}^{T} \exp\left(-\frac{E}{kT'}\right) dT'\right] \tag{13}$$

where n_0 is the initial concentration of filled traps at time t = 0.

The above expression can be evaluated by mean of numerical integration, and it yields a bell-shaped curve, as in Fig. 2, with a maximum intensity at a characteristic temperature T_M.

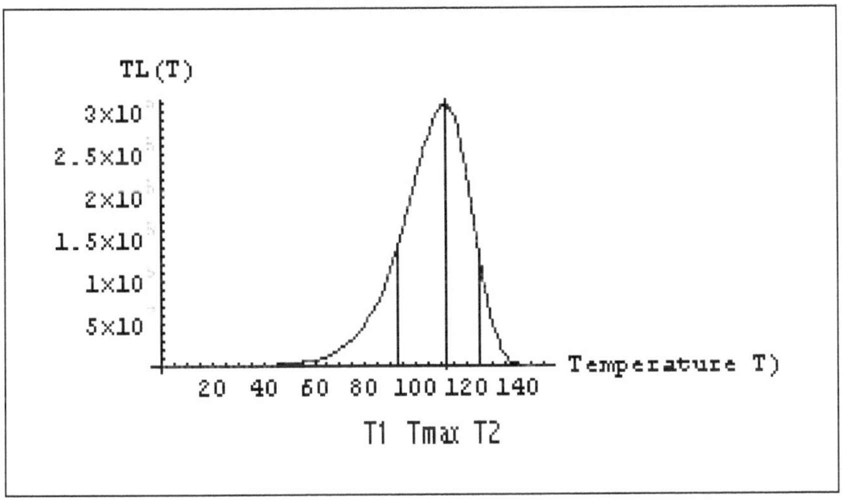

Fig. 2. A first order glow peak.

Some observation can be done on Eq. (13):

- *I(T)* depends on the three parameters *E, s* and *b*,
- *E* has values around 20*kT* in the range of occurrence of a TL peak,
- $\exp\left(-\frac{E}{kT}\right)$ is of the order of 10^{-7},
- when T is slightly greater than of T_0, the argument of the second exponential function is about equal to unity and decreases with increasing temperature. *I(T)* is then dominated by the first exponential and increases very fast as the temperature increases. At a certain temperature, T_M, the behavior of the two exponential functions cancel: at this temperature the maximum temperature occurs,

- above T_M, the decrease of the second exponential is much more rapid than the increase of the first exponential and $I(T)$ decreases until the traps are totally emptied.

We have to stress that a glow curve corresponding to a first order kinetics is not symmetric, as it is possible to observe in Fig. 2. Its asymmetry can be expressed by measuring T_M, T_1 and T_2, where T_1 and T_2 are respectively the temperatures on either side of the temperature at the maximum, corresponding at half of the maximum TL intensity. It is customary to define the following quantities:

$\tau = T_M - T_1$ is the half-width at the low temperature side of the peak (14)

$\delta = T_2 - T_M$ is the half-width towards the fall-off of the peak (15)

$\omega = T_2 - T_1$ is the total half-width (16)

$\mu = \dfrac{\delta}{\omega}$ is the geometrical factor (17)

The geometrical factor for a first order TL glow curve is equal to $\mu = 0.42$.

An important property of a first order glow curve is concerning the effect of the initial concentration of filled traps n_0 on it. Fig. 3 shows this: the initial concentration n_0 affects only the maximum height of the TL glow curve and leaves the shape and the temperature at the maximum unchanged.

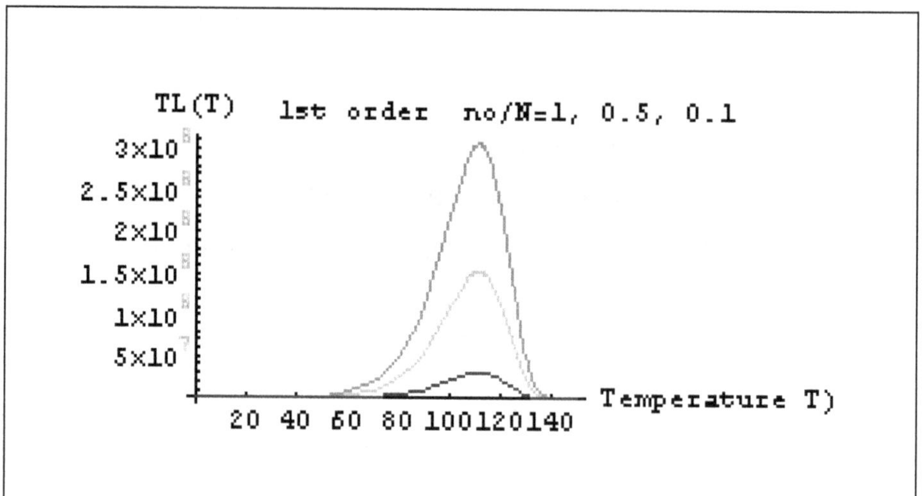

Fig. 3. Effect of n_0 on a first order glow curve.

Another property is concerning the effect of the activation energy. This is shown in Fig. 4: we can note that as the activation energy E is increased, the TL glow curve shifts towards higher temperature, but the curve maintains its overall shape.

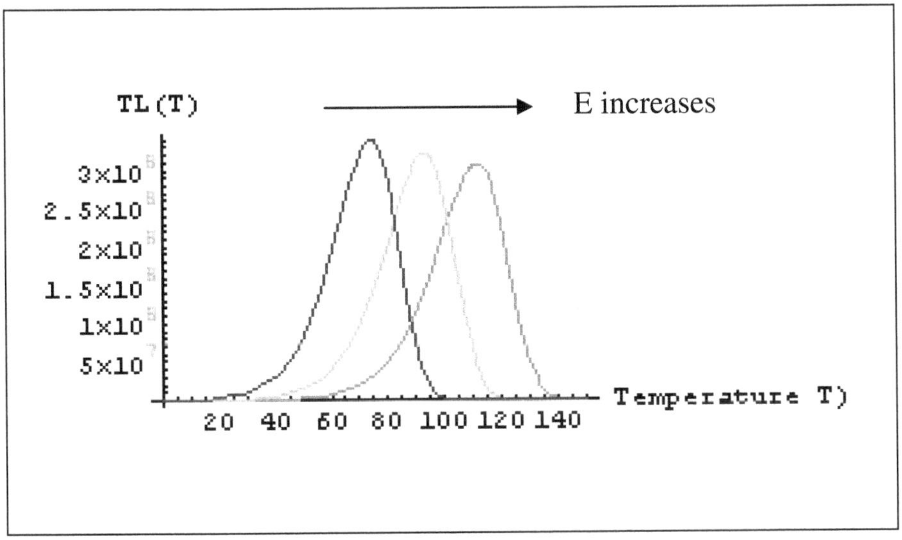

Fig. 4. The shift effect of the glow peak as a function of E.

Fig. 4 gives an example of three first order TL peaks, calculated using three different values of the activation energy E=0.90, 0.95 and 1.0 eV.

Second Order Kinetics

The second order expression can be derived from Eq. (7) considering now valid the retrapping assumption. This is expressed by

$$hA_h << (N-n)A_n \tag{18}$$

Then, from Eq. (7), the TL intensity is given by

$$I = -\frac{dh}{dt} = \frac{hA_h sn \exp\left(-\frac{E}{kT}\right)}{(N-n)A_n} \tag{19}$$

Two more assumptions have to be done now: the first is $h = n$, the neutrality condition, and the second is $N >> n$; the last one means to be far from saturation. Assuming also that the retrapping and the recombination probabilities are equal, Eq. (19) becomes the well know equation of Garlick and Gibson for the second order kinetics:

$$I = -\frac{dh}{dt} = n^2 s' \exp\left(-\frac{E}{kT}\right) \tag{20}$$

where s' = s/N.

As already done so far for the first order kinetics, the previous Eq. (20) can be solved assuming a linear rate of heating of the TL sample. The following equation is then obtained:

$$I(T) = -\frac{dh}{dt} = \frac{n_0^2 s' \exp\left(-\frac{E}{kT}\right)}{\left[1 + \frac{s' n_0}{\beta} \int_{T_0}^{T} \exp\left(-\frac{E}{kT'}\right) dT'\right]^2} \tag{21}$$

Fig. 5 is a graph of Eq. (21).

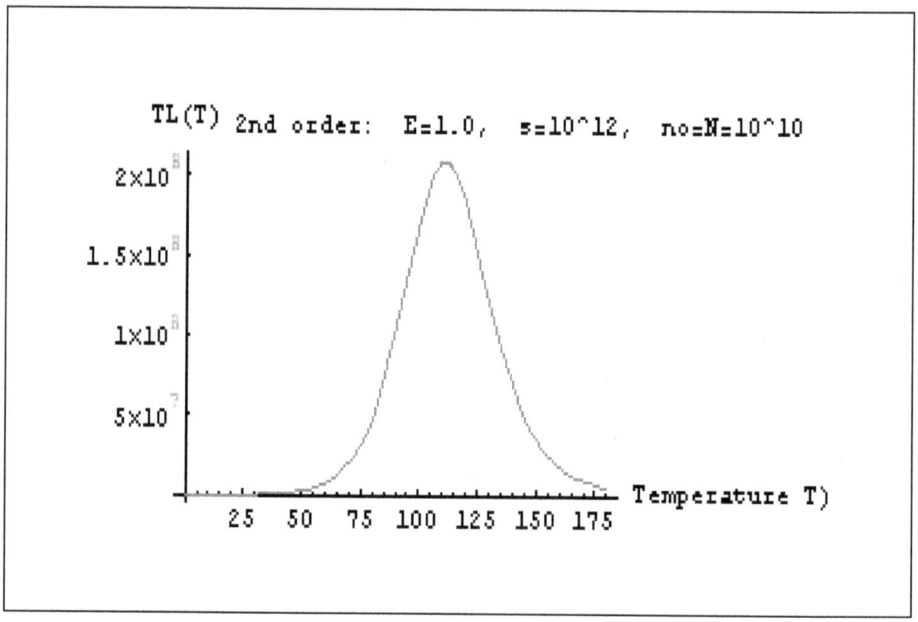

Fig. 5. A second order glow peak.

We have to note that a second order TL glow peak is much more symmetric than the first order peak. Just like in the case of the first order, we can introduce the peak parameters τ, δ and ω and the geometrical factor μ. The geometrical factor for a second order peak is equal to μ = 0.52.

We have also to note that the initial filled traps n_0 affects both the maximum height of the peak as well as its temperature at the maximum, but leaves the overall shape unchanged; in particular, as n_0 increases the peak temperature at the maximum tends to decrease (see Fig. 6).

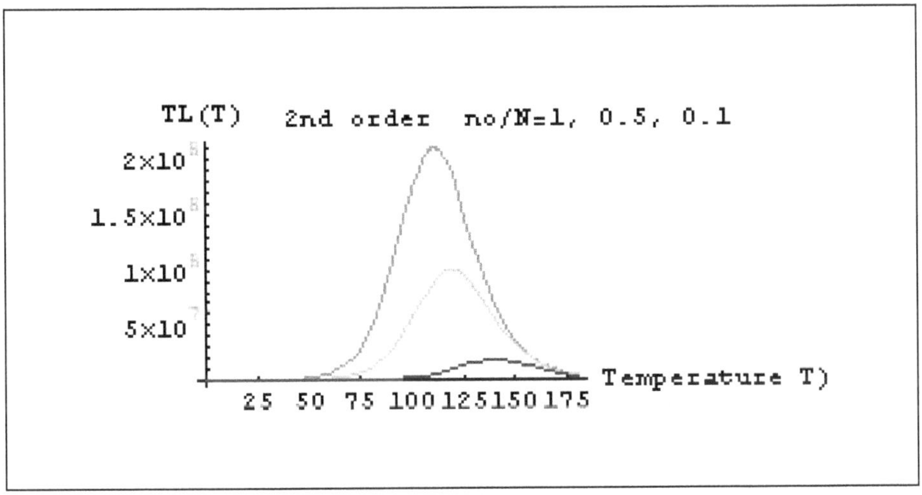

Fig. 6. Effect of n_0 on a second order peak.

The effect of the increase of the kinetics energy value on the second order curve is shown in the following Fig. 7

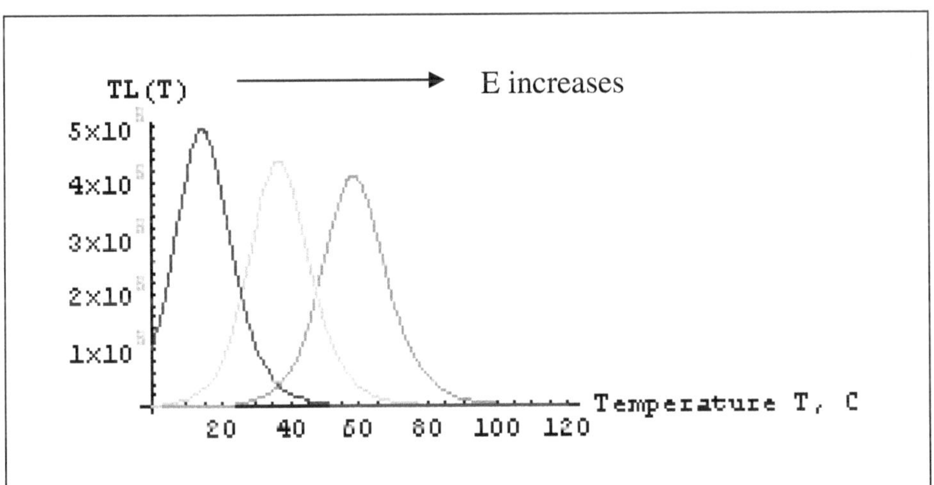

Fig. 7. The shift effect of the glow peak as a function of E.

A comparison between a first and a second order glow curves is shown in the following Fig.8. Both curves have been normalized by dividing each TL glow curve by its maximum TL value. (red curve =1st order, green =2nd order)

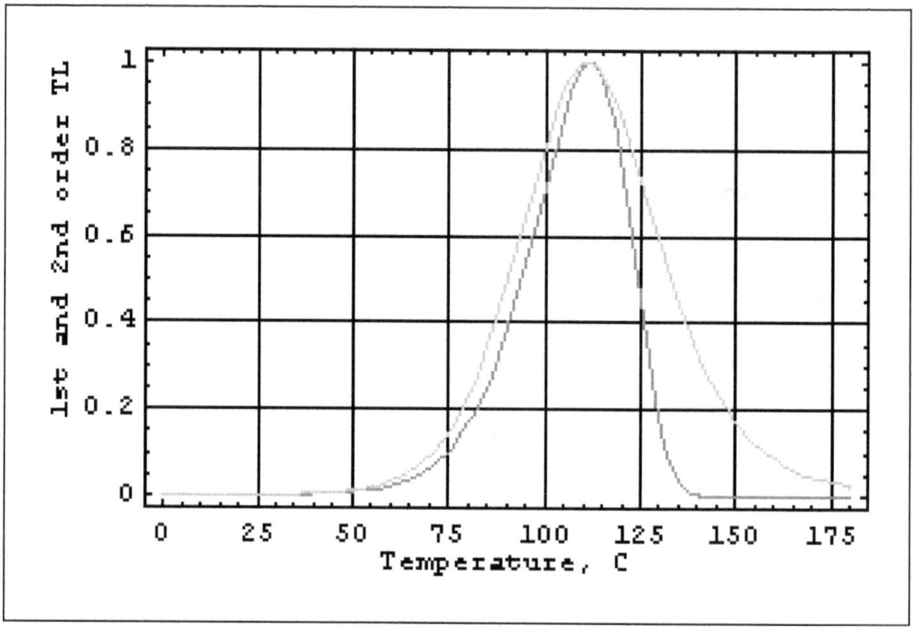

Fig. 8. Comparison between a first and a second order peaks.

<u>IMTS Model</u>

The IMTS model, Interactive Multi Trap System, is more complex and it consists of three energy levels: the electron trap, the recombination center RC and a thermally disconnected deeper trap, TDDT, which is not active at the considered temperature. The schema of this model is given in Fig. 9.

The traffic of electrons between the trap level, the recombination center, the conduction band and the thermally disconnected deeper trap is governing by the following set of differential equations:

$$\frac{dn}{dt} = -ns\exp\left(-\frac{E}{kT}\right) + n_C A_n (N-n) \tag{22}$$

$$\frac{dm}{dt} = n_C A_m (M-m) \tag{23}$$

$$\frac{dn_C}{dt} = ns\exp\left(-\frac{E}{kT}\right) - n_C\left[A_n(N-n) + A_m(M-m) + A_h(n+m+n_C)\right] \tag{24}$$

$$I = -\frac{dh}{dt} = n_C A_h (n+m+n_C) \tag{25}$$

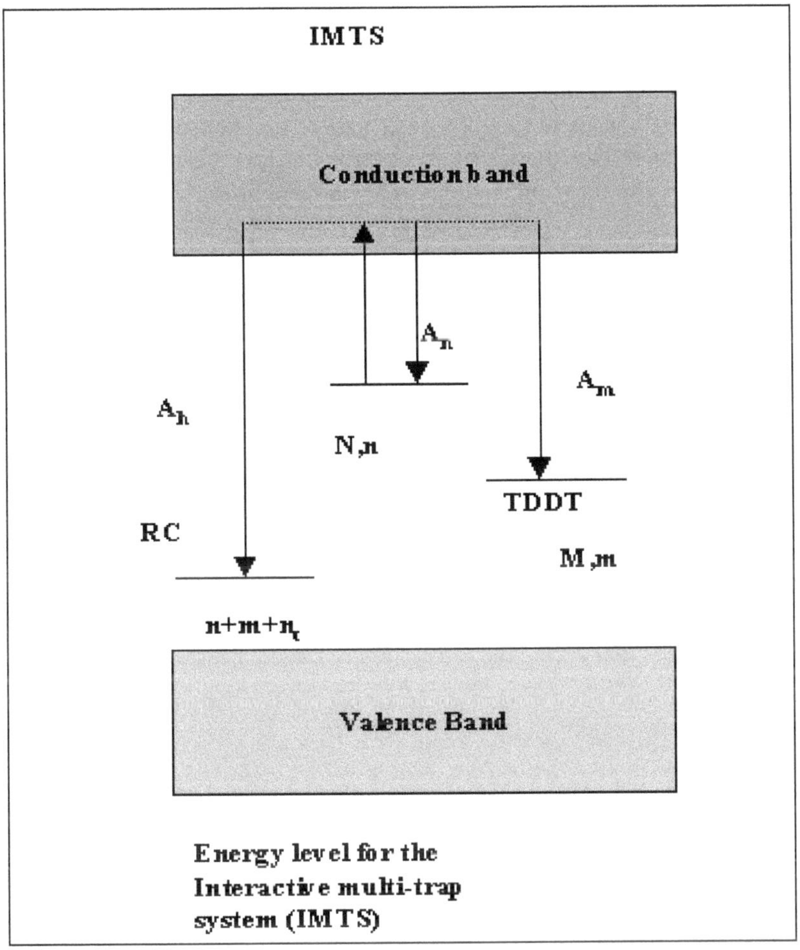

Fig. 9. Interactive Multi Trap System (IMTS).

The meaning of the variables in the IMTS model is:

- N is the total concentration of the electron traps in the crystal (cm^{-3})
- n is the concentration of the filled electron traps in the crystal (cm^{-3})

- n_C is the concentration of the free carriers in the conduction band (cm^{-3})
- M is the total concentration of the electron traps TDDT (cm^{-3})
- m is the concentration of the filled electron traps TDDT (cm^{-3})
- E is the activation energy of the electron traps (eV)
- s is the frequency factor of the electron trap (s^{-1})
- A_n is the capture coefficient of the traps ($cm^3\ s^{-1}$)
- A_h is the capture coefficient of the recombination center RC ($cm^3\ s^{-1}$)
- A_m is the capture coefficient of the TDDT ($cm^3\ s^{-1}$)
- The quantity $m+n+n_C$ represents the total concentration of filled traps in the system at any moment. Because the conservation of charge, this quantity is also equal to the total concentration of filled holes, h, in the recombination center.

Eq. (22) describes the traffic of electrons in and out of the electron trap. The electrons can leave the traps, active traps AT, via thermal excitation, which is described mathematically by the term $ns\exp\left(-\frac{E}{kT}\right)$. The electrons can also be retrapped in the trap: this fact is given by the retrapping factor $n_C A_n(N-n)$.

The second equation, Eq. (23), describes the traffic of electrons from the conduction band into the thermally disconnected deeper trap TDDT. The trapping of the electrons in the TDDT is given by the term $n_C A_m(M-m)$.

The third equation describes the traffic of electrons in and out of the conduction band: the electrons in the conduction band can be trapped in the recombination center, RC, an event described mathematically by the term $n_C A_h(n+m+n_C)$; they can be also retrapped in AT or trapped in TDDT.

The last equation gives finally the observed TL, which is proportional to the amount of light measured during the thermoluminescence measurement.

SUBJECT INDEX

ABOUT THE AUTHOR

Born October 1937, Dr. Claudio Furetta obtained the degree of Doctor of Physics from University of Rome "La Sapienza", Italy, where he has been research physicist since 1971.

He has held research assignments at BCMN of EURATOM in Belgium and at CERN of Geneva, Switzerland. He was visiting professor at the National Tsing Hua University, Hsinchu, Taiwan (ROC), at the Universidad Autonoma Metropolitana, Iztapalapa, Mexico D.F., at the Instituto de Ciencias Nucleares, Unidad de Irradiacion y Seguridad Radiologica, of the Universidad Nacional Autonoma de Mexico, Mexico D.F. and the Centro de Investigacion en Ciencia Aplicada y Tecnologia Avanzada, Mexico D.F

Dr. Furetta has done research in the field of nuclear detectors, radiation dosimetry and thermoluminescence and has published more than 200 research papers in international journals. In addition to this he has contributed to several books and has authored "Thermoluminescence Process: Theory and Methods", published by National Tsing Hua University, Taiwan, an "Elementary Course of Physiscs for Professional Nurses (in Italian), "Thermoluminescence and Dating" (in Italian), "Frequenty Asked Questions in TL and OSL" published by World Scientific, Singapore.

He has taught Thermoluminescent Dosimetry, Medical Physics and Thermoluminescent Dating at the Rome University and at the National Tsing Hua University.

www.ingramcontent.com/pod-product-compliance
Lightning Source LLC
LaVergne TN
LVHW080747300926
849162LV00008BA/1771

* 9 7 8 9 8 1 2 8 3 8 9 1 9 *